CW00376707

Lighting Engineering

Lighting Engineering

Applied calculations

R. H. Simons and A. R. Bean

OXFORD AUCKLAND BOSTON JOHANNESBURG MELBOURNE NEW DELHI

Architectural Press
An imprint of Butterworth-Heinemann
Linacre House, Jordan Hill, Oxford OX2 8DP
225 Wildwood Avenue, Woburn, MA 01801-2041
A division of Reed Educational and Professional Publishing Ltd

A member of the Reed Elsevier plc group

First published 2001

British Library Cataloguing in Publication Data
A catalogue record for this book is available from the British Library

Library of Congress Cataloguing in Publication Data
A catalogue record for this book is available on request

ISBN 0 7506 5051 6

Composition by Cambrian Typesetters, Frimley, Surrey
Printed and bound in Great Britain by MPG Books Ltd, Bodmin, Cornwall

Contents

Preface

The last decade has seen the universal application of personal computers to lighting engineering problems on a day-to-day basis. No longer is the power of the computer invoked just for that large prestige job, the computer is constantly in use in lighting engineering laboratories and lighting design offices. This means that many calculations that were previously impracticable are well within the reach of any engineer or, for that matter, any person who has access to an appropriate computer program. However, there can be dangers if the application engineer does not have a grasp of the underlying principles used in the calculation processes.

In this book we set out to give the reader the mathematical background to the calculational techniques used in illuminating engineering and link them to the applications in which they are used. In addition, we give details of photometric measurements, as a knowledge of these is necessary to understand the origin of the basic data used in the calculations and to check that the calculated performance requirements are met. We hope, also, the material will be of use to those who wish to write their own computer programs, and that it will be of didactic value.

To keep the treatment as concise as possible we have assumed that the reader has a basic knowledge of lighting engineering.

Some of the material in this book is an update of material used in a book we published in 1968 entitled *Lighting Fittings – Performance and Design* (Pergamon Press). That book was well received and it was the demand for a new edition from many of our colleagues that provided the primary spur for us to produce the present book. Motivation also came from our inner urge to write a book encompassing the modern calculation methods and from the encouragement of our publishers, in the persons of Eliane Wigzell and Kirsty Stroud. We sincerely hope that we have fulfilled our objectives.

There are many people we would like to thank. Jacques Lecocq of *Thorn Europhane* entered into long correspondence with Mr Simons on such topics as interpolation and the rotation of luminaire axes, and helped to put these on a sound mathematical basis. Kit Cuttle and Kevin Mansfield read the scripts and made many useful suggestions. Professor Peter Tregenza and Dr Paul Littlefair gave valuable advice on the chapter on daylight. Among the other people we consulted were Mr R. Lamb of the NPL and Mr J. A. Lynes, who also made valuable contributions.

Dr Bean is grateful to his wife for her extreme patience in decoding his manuscripts and so ably transferring them to the word processor. Both authors are grateful to their wives for providing moral support during the long time this book took to write.

A. R. Bean
R. H. Simons
13 December 1999

1

The Light Field of a Luminaire

1.1 Coordinate system

All light sources can be thought of as creating a light field, extending from the source in all, or some, directions (that is, a region in space filled with electromagnetic radiation within the visible range). In general, this characteristic field exists in space until interrupted or modified by a medium other than air, e.g. a wall.

If this light field distribution is to be studied and recorded in some way then a number of things must be done. First, a means of describing the position of any point in space relative to the light source must be established; second, a system of quantities and units must be devised to enable the lighting measurements to be taken in a meaningful way.

This first section deals with the first requirement, that is, a means of relating the point in space at which, say, a measurement is taken with respect to the position of the luminaire or lamp.

When a suitable spatial reference system is set up it would be convenient if the light source could be regarded as being small enough relative to the measurement apparatus as to be treated as a point lying at the origin of the reference system. The convenience is a mathematical one, since it greatly simplifies analysis.

Clearly, light sources may be of any size and the method adopted must be able to take this into account. The most common method is to ensure that the light path from the light source to the detector is sufficiently long for the light field to have taken on the form in that region of space of a theoretical point source of light of the same output and configuration.

It has been found that for many light sources this condition is fulfilled to an accuracy of 1% if the distance to the detector is five times the major dimension of the light source.

The implications are that, for large light sources, the measurement distance may need to be 12 m. For floodlights and luminaires with optical systems, these distances may need increasing up to 30 m (or more). This is because the light field distribution depends upon the flashing of the optical components more than the size of the luminaire (see Chapter 15).

Before we turn to the spacial reference systems adopted in practice, it is worth explaining how measurements taken at, say, 12 m can be used to calculate the performance of the luminaire at interior working distances which might well be 2 or 3 m.

The assumption is made that a large light source, say a luminaire, can be considered to consist of a number of small luminaires, each emitting the appropriate fraction of the total output and having the distribution characteristics measured at the correct (long) photometric distance for the whole luminaire (see Figure 1.1).

In this computer age the multiple calculations this entails present little difficulty.

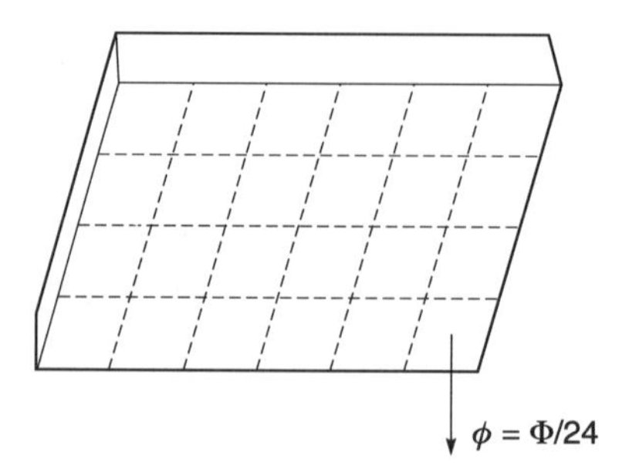

Fig. 1.1 The luminaire considered to consist of a number of small luminaires for calculational purposes (in this case 24)

1.2 Practical coordinate systems

Once the assumption can be made that the origin of the light distribution is a point, then that point can be considered to lie at the centre of a sphere, the size of which can be related to the measurement distance, i.e. the distance of the detector from the luminaire reference point.

Measurements taken in different directions could be marked on a diagram of a sphere with an appropriate set of angular coordinates (see Figure 1.2).

Three such angular reference systems are shown in Figure 1.3(a), (b), (c), (d), (e) and (f). All these systems are based on the familiar longitude and latitude system of spherical coordinates, but differ in orientation of the polar axis.

Figure 1.3(a) shows a system similar to that used on terrestrial maps.

Figure 1.3(b) is, in effect, the same system turned through 90°.

Figure 1.3(c) is Figure 1.3(b) rotated a further 90° so that it is, in effect, Figure 1.3(b) viewed 'end on'.

Systems based on these three ways of arranging angular coordinates have been described by the CIE (CIE 121–1996) and symbols have been agreed so that, once the symbols are quoted, the system referred to is defined.

The system shown in Figure 1.3(a) is designated the C, γ system.

The system shown in Figure 1.3(b) is designated the A, α system and the one shown in Figure 1.3(c) is designated the B, β system.

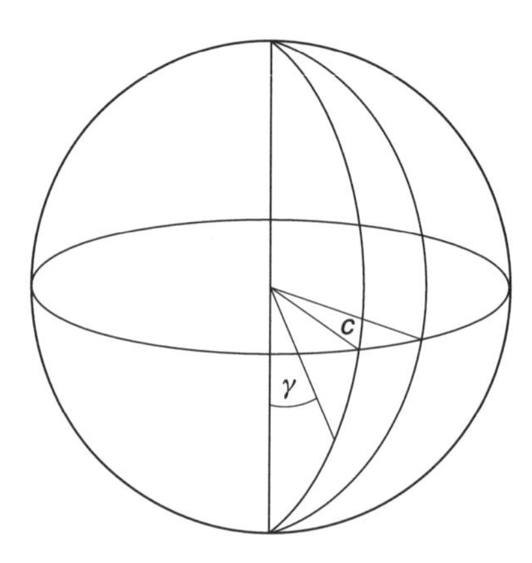

Fig. 1.2 A sphere with a possible system of angular coordinates

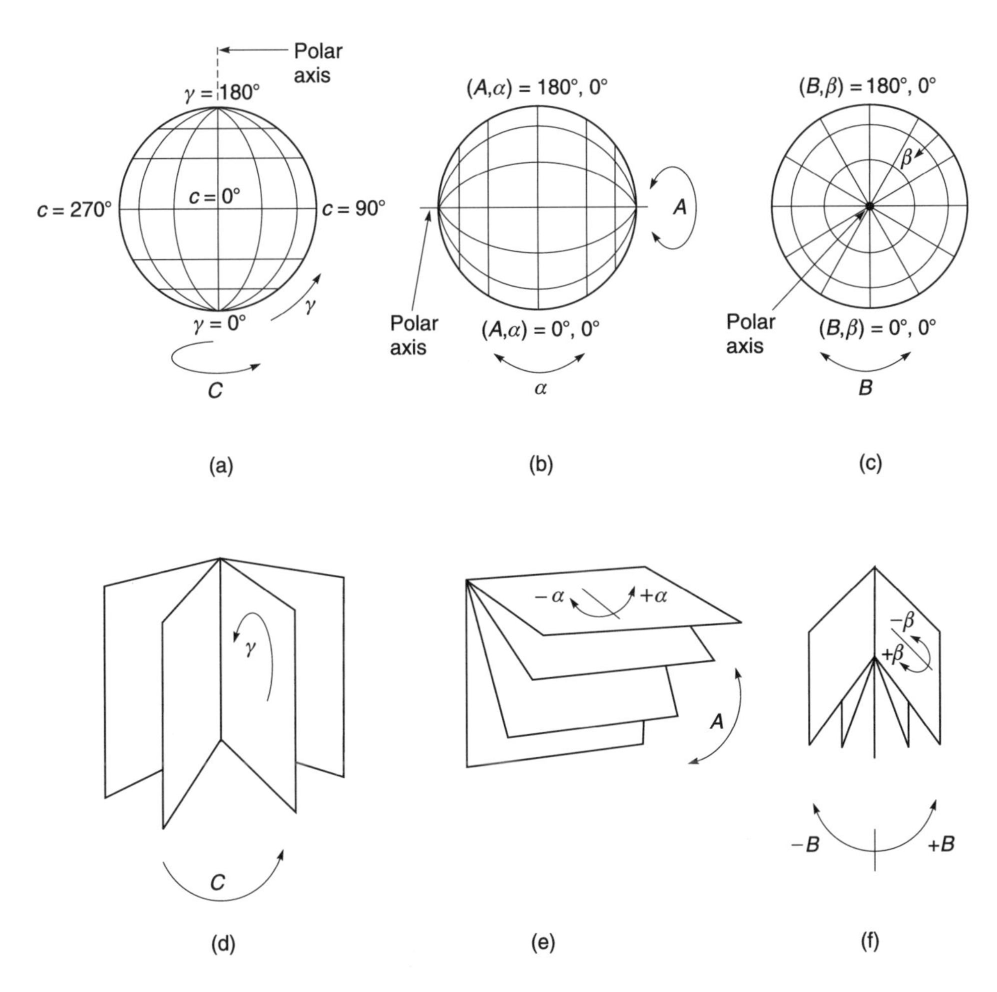

Fig. 1.3 (a) A coordinate system similar to that used on terrestrial globes; (b) the coordinate system turned through 90° (note the different origin for 'latitude' measurements); (c) the coordinate system rotated a further 90° to give an 'end on' view; (d) as (a) but using the 'book page' method of display; (e) as (b) but using the 'book page' method of display; (f) as (c) but using the 'book page' method of display

Since each of these systems relates to points on a sphere, measurements taken using a photometer following one system can be related, by appropriate angular conversions, to any of the others and expressed in terms of that system (i.e. a computer program that receives the data from the photometer could ensure that the data are printed out in the form required by the user, although this would introduce the possible inaccuracies of interpolation).

The first designating letter relates to the plane of measurement, e.g. *C* planes, *B* planes, *A* planes, while the Greek letters α, β and γ refer to the angles of measurement in the specified plane.

It is sometimes found helpful to think of the planes as though they were pages of a book and the CIE document uses this approach. Figure 1.3(d), (e) and (f) show this method of illustration.

The use of these three coordinate systems is restricted by the following conventions:

(1) The system of C planes is considered oriented rigidly in space and does not follow any tilt of the luminaire.
(2) The A and B planes systems are considered to be coupled rigidly to the luminaire and so would follow any tilt of the luminaire.

1.3 Transformation of coordinate systems

It was mentioned in the previous section that the three coordinate systems described so far are, of necessity, interrelated and so transformation from one system to another is straightforward. In this section, the details of these transformations are given together with the derivation of the necessary formulae.

1.3.1 TO TRANSLATE ANGLES MEASURED ON THE A, α SYSTEM TO THEIR EQUIVALENT IN THE C, γ SYSTEM AND VICE VERSA (see Figure 1.4)

To translate from A, α to C, γ from Figure 1.4:

$$\tan C = \mathrm{RP/RT}$$

$$= \frac{\mathrm{RP}}{\mathrm{QR}} \times \frac{\mathrm{QR}}{\mathrm{RT}}$$

So, $\tan C = \tan \alpha / \sin A$

Also,

$$\cos \gamma = \frac{\mathrm{QT}}{\mathrm{QP}}$$

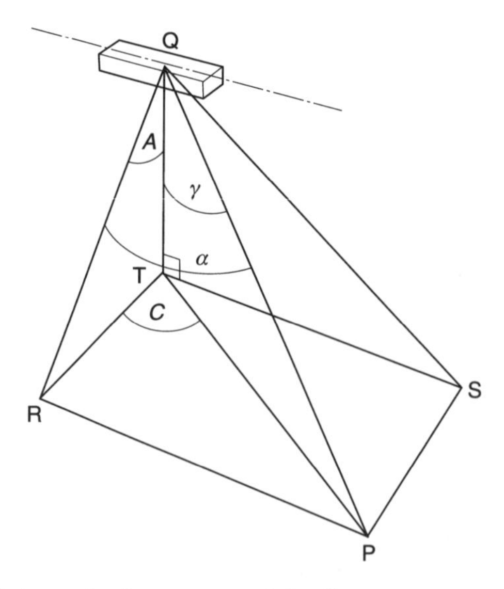

Fig. 1.4 The relationship between the A, α system and the C, γ system

$$= \frac{QT}{RQ} \times \frac{RQ}{QP}$$

So, $\cos\gamma = \cos A \times \cos\alpha$

To translate from C, γ to A, α from Figure 1.4:

$$\tan A = \frac{RT}{QT}$$

$$= \frac{RT}{TP} \times \frac{TP}{QT}$$

So, $\tan A = \cos C \times \tan\gamma$

Also,

$$\sin\alpha = \frac{RP}{QP}$$

$$= \frac{RP}{TP} \times \frac{TP}{QP}$$

So, $\sin\alpha = \sin C \times \sin\gamma$

1.3.2 TO TRANSLATE ANGLES MEASURED ON THE A, α SYSTEM TO THEIR EQUIVALENT ON THE B, β SYSTEM AND VICE VERSA (see Figure 1.5)

To translate from A, α to B, β from Figure 1.5:

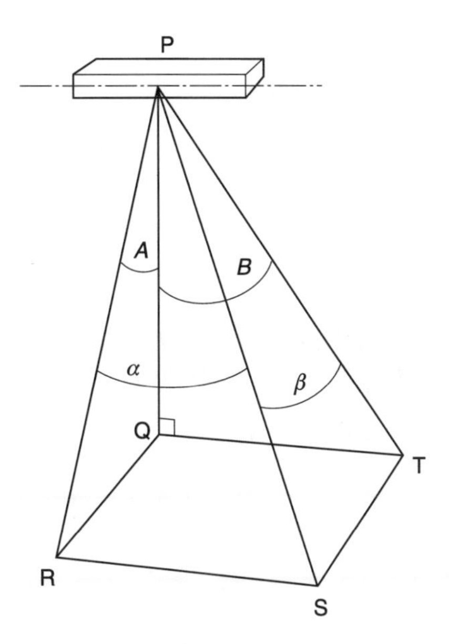

Fig. 1.5 The relationship between the A, α system and the B, β system

$$\tan B = \frac{QT}{QP}$$

$$= \frac{QT}{PR} \times \frac{PR}{QP}$$

So, $\tan B = \tan \alpha / \cos A$

Also,

$$\sin \beta = \frac{ST}{SP}$$

$$= \frac{ST}{PR} \times \frac{PR}{SP}$$

So, $\sin \beta = \sin A \times \cos \alpha$

To translate from B, β to A, α from Figure 1.5:

$$\tan A = \frac{QR}{PQ}$$

$$= \frac{QR}{PT} \times \frac{PT}{PQ}$$

So, $\tan A = \tan \beta / \cos B$

Also,

$$\sin \alpha = \frac{RS}{PS}$$

$$= \frac{RS}{PT} \times \frac{PT}{PS}$$

So, $\sin \alpha = \sin B \times \cos \beta$

1.3.3 TO TRANSLATE ANGLES MEASURED ON THE B, β SYSTEM TO THEIR EQUIVALENT ON THE C, γ SYSTEM AND VICE VERSA (see Figure 1.6)

To translate from B, β to C, γ from Figure 1.6:

$$\tan C = \frac{RS}{QR}$$

$$= \frac{QT}{ST} = \frac{QT}{PT} \times \frac{PT}{ST}$$

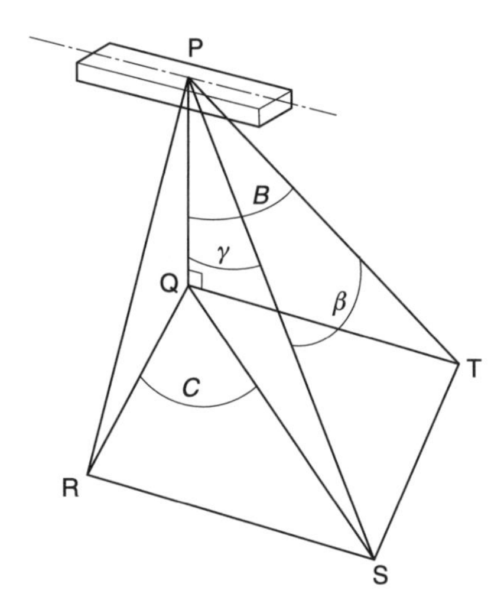

Fig. 1.6 The relationship between the B, β system and the C, γ system

So, $\tan C = \sin B / \tan \beta$

Also,

$$\cos \gamma = \frac{PQ}{PS} = \frac{PQ}{PT} \times \frac{PT}{PS}$$

So, $\cos \gamma = \cos B \times \cos \beta$

To translate from C, γ to B, β.

Using the method employed above and by reference to Figure 1.6 it can be shown that:

$$\tan B = \sin C \times \tan \gamma$$

and $\sin \beta = \cos C \times \sin \gamma$

1.3.4 APPLICATION OF THE TRANSLATION FORMULAE

Whilst the derivation of the formulae has been straightforward, their application presents three major problems. These are:

(1) the tangent of 90° is infinite, and therefore cannot be processed by a computer;
(2) division by zero, which occurs in certain instances, is inadmissible for computer calculations;
(3) the conditions for finding the quadrant in which the result lies are difficult to determine, especially for computer programming.

For manual calculations, both these problems can be solved by using the webs for the different systems, as in Figure 1.3(a), (b) and (c). The point to be transformed is plotted on the web for the initial coordinate system and then transferred to the web for the required coordinate system. This need not be done accurately as its purpose is simply to determine the correct quadrant for the result. For computer calculations, the conditions for determining the correct quadrant for translation of C, γ coordinates to B, β coordinates, and vice versa are given in Section 2.4.3.

The A, α coordinate system is rarely used and is not given further consideration.

1.4 Solid angle

In the previous sections the means of specifying the direction in which the light is travelling relative to the light source producing the light rays was considered and the coordinate systems in common use described.

For a practical transfer of energy to take place, a bundle of rays constituting light flux (Φ) emanating from the light source must also be considered and this requires a means of defining a multidimensional angle (called a solid angle) such as that shown in Figure 1.7.

When this solid angle is infinitesimally small then a particular direction can be uniquely defined. When the solid angle is large then an angular zone is defined.

Just as a plane angle is defined in terms of a circle and its magnitude taken as a ratio of the arc subtended on the circumference of the circle to the radius of the circle (radians), so, in an analogous way, the solid angle is defined as the area subtended on the surface of the enclosing sphere divided by the radius squared. That is:

$$\text{solid angle } \Omega = \frac{A}{r^2} \text{ steradians (see Figure 1.8)}$$

(Thus, the total solid angle subtended by the surface of a sphere at the centre of the sphere is 4π steradians.)

As can be seen from Figure 1.8, this definition relates simply to the solid angle subtended by the area at the centre of the sphere, so that equal areas of quite different shape can subtend the same value of solid angle.

As the range of directions contained within the solid angle becomes more restricted, the notation is generally modified so that

$$\Omega \rightarrow \delta\Omega \rightarrow d\Omega$$

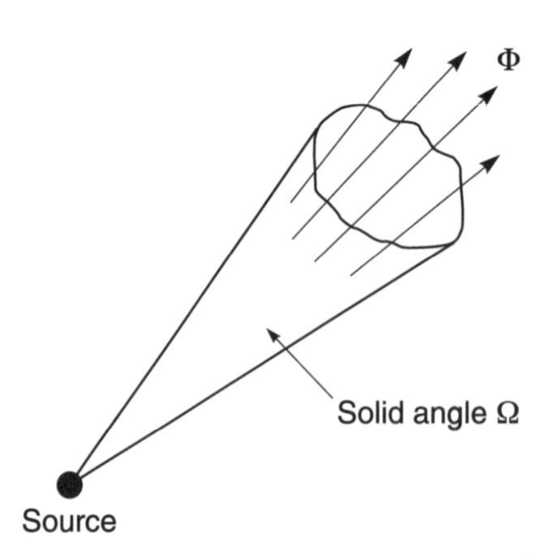

Fig. 1.7 The meaning of solid angle

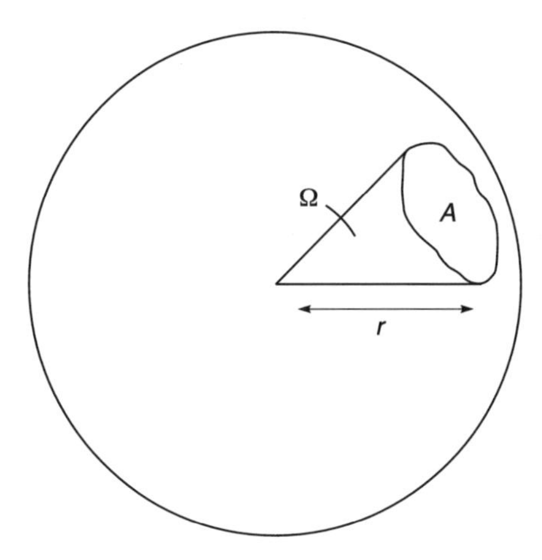

Fig. 1.8 Solid angle in relation to a sphere

where dΩ refers to an infinitesimally small solid angle and so it can be related to a specific direction.

1.4.1 THE SOLID ANGLE SUBTENDED BY A DISC (AT A POINT ON ITS AXIS)

The solid angle subtended by a disc of radius R is the same as that subtended by the cap of a sphere that has the same radius at its mouth and subtends the same angle from the centre of the sphere (see Figure 1.9).

The area of the spherical cap is given by:

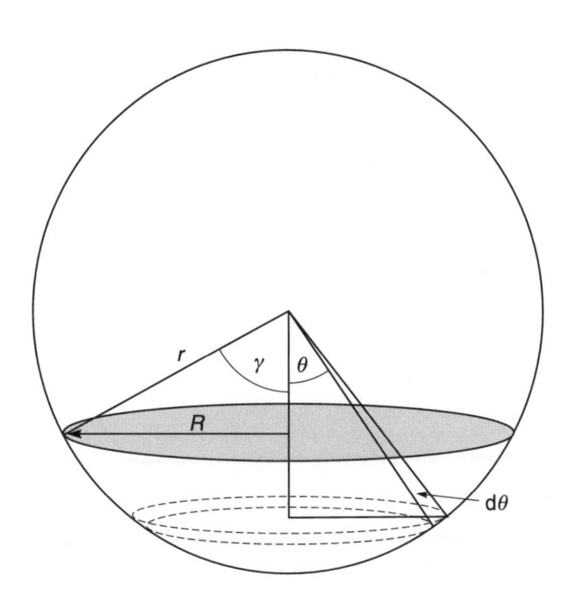

Fig. 1.9 The solid angle subtended by a disc

$$\int_0^{\gamma} 2\pi r \sin\theta \, r \, \mathrm{d}\theta$$

$$= 2\pi r^2 \int_0^{\gamma} \sin\theta \, \mathrm{d}\theta$$

$$= 2\pi r^2 [-\cos\theta]_0^{\gamma}$$

solid angle for the disc

$$\Omega = \frac{2\pi r^2}{r^2}[1 - \cos\gamma]$$

$$= 2\pi[1 - \cos\gamma]$$

This solid angle depends on the radius of the disc and the distance to the point from which the solid angle emanates, measured normal to the centre of the disc.

1.4.2 THE SOLID ANGLE SUBTENDED BY A RECTANGLE (AT A POINT)

The integration required to calculate the solid angle subtended by a rectangle is more complex than that carried out for a disc, so the result will be obtained by a method suggested by Gershun in his book *The Light Field* published in 1936.[1]

The method depends upon the fact that it is a simple matter to calculate the solid angle subtended at the centre of a sphere by a lune of the sphere (see Figure 1.10).

The solid angle of a lune formed by the intersection of two great circles on the surface of a sphere, as shown, is given by the ratio of the angle subtended by the lune at the centre of the sphere to the angle of the complete circle multiplied by the total solid angle of the sphere.

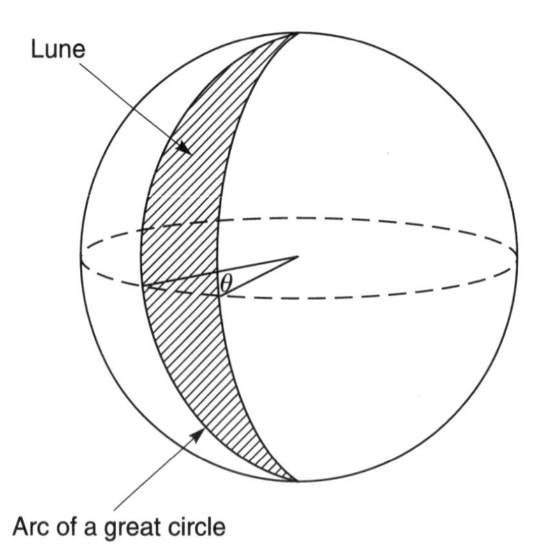

Fig. 1.10 The solid angle subtended by the lune of a sphere

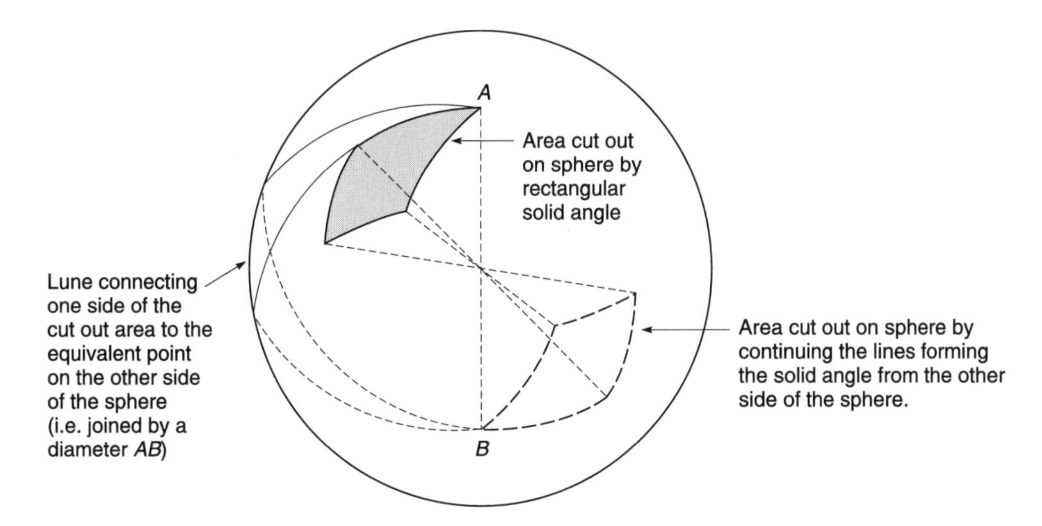

Fig. 1.11 A rectangular solid angle showing one of the enclosing lunes

So,

$$\text{solid angle of lune} = \frac{\theta}{2\pi} \times 4\pi = 2\theta \text{ steradians}$$

Now consider Figure 1.11.

This shows the area cut out by a rectangular solid angle when it intersects the surface of the sphere, and also the other area cut out on the other side of the sphere by an equivalent solid angle formed by continuing the lines of the first solid angle through to the other side of the sphere.

Also shown is a lune drawn to connect point A on the front cut-out to point B on the rear cut-out area located at the other end of the diameter connecting the two areas.

Similar lunes could be drawn connecting the four corners of each of the cut-out areas together. These lunes could then be expanded to fill all the area of the sphere not covered by the areas cut out by the rectangular solid angles. An end view of the sphere, showing one of the rectangular solid angles and one end of each of the four expanded lunes is given in Figure 1.12.

The solid angle for the two rectangular solid angles cut out from the sphere

$$= 4\pi - \sum 2\theta$$

or, since we only require one rectangular solid angle

$$\Omega = 2\pi - \sum \theta$$

For the four lunes of equal angle this becomes

$$\Omega = 2\pi - 4\theta$$

To obtain the solid angle it is necessary to determine θ. To do this we note that θ is the angle between the two planes that form the great circles and define the adjacent sides of the rectangular solid angle.

To calculate θ the rectangular solid angle is divided into four equal solid angles, as shown in Figure 1.13(a).

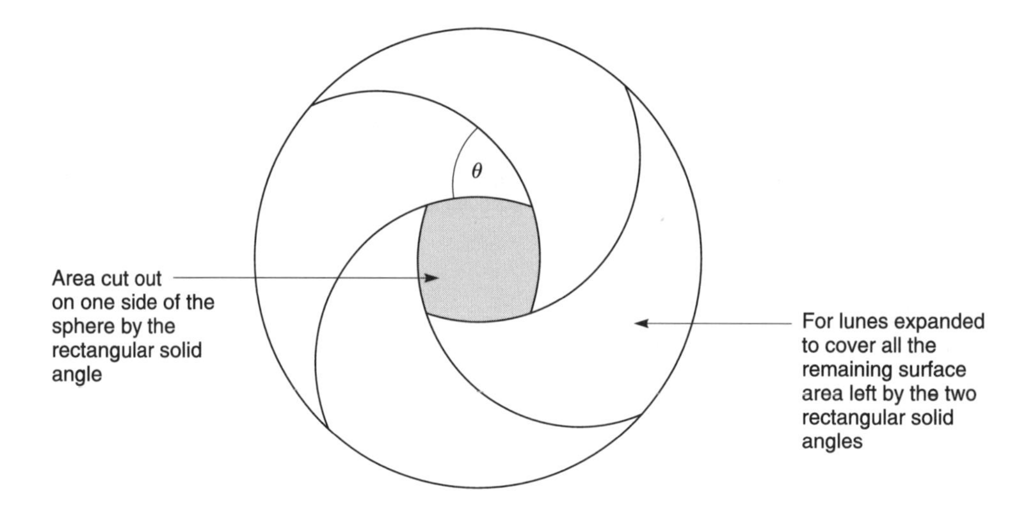

Fig. 1.12 A rectangular solid angle showing an end view of the expanded lunes

Since the same angle is repeated at the outer corner of each of the four solid angles only the outer angle of one of the solid angles needs to be considered. In Figure 1.13(a) the normals to the planes forming the external angle θ have been drawn so that a plane containing the interior angle to the two planes is included. This angle is $\pi - \theta$. Since two of the other angles of the quadrilateral are each equal to $\pi/2$ the fourth angle must be the required angle θ.

To determine θ, apply the cosine rule to the triangle AFC shown in Figure 1.13(b).

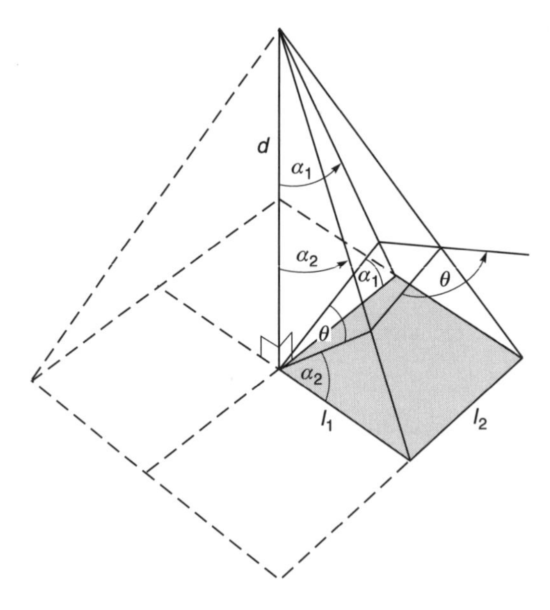

Fig. 1.13(a) The construction for the calculation of the angle between the two planes of the great circles (θ)

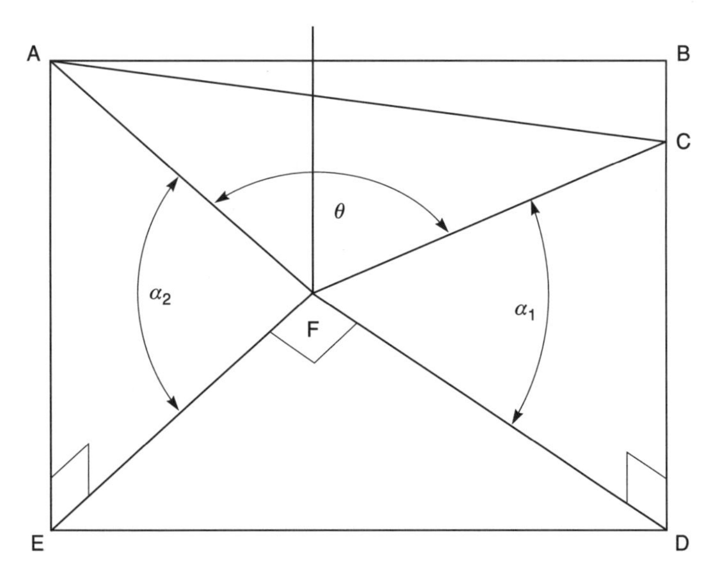

Fig. 1.13(b) The simplified construction for the application of the cosine rule

Let

$$\begin{aligned} AF &= FC = 1.0 \\ AC^2 &= AF^2 + FC^2 - 2AF{\cdot}FC\cos\theta \\ &= 2 - 2\cos\theta \end{aligned}$$

Also,

$$AC^2 = AB^2 + BC^2$$

And since

$$\begin{aligned} AB &= ED \\ AC^2 &= ED^2 + (BD - CD)^2 \\ &= \cos^2\alpha_2 + \cos^2\alpha_1 + (\sin\alpha_2 - \sin\alpha_1)^2, \text{ since } BD = AE \\ &= \cos^2\alpha_2 + \cos^2\alpha_1 + \sin^2\alpha_2 - 2\sin\alpha_2\sin\alpha_1 + \sin^2\alpha_1 \\ &= 1 - \sin^2\alpha_2 + 1 - \sin^2\alpha_1 + \sin^2\alpha_2 - 2\sin\alpha_2\sin\alpha_1 + \sin^2\alpha_1 \\ &= 2 - 2\sin\alpha_2\sin\alpha_1 \end{aligned}$$

Giving

$$2 - 2\cos\theta = 2 - 2\sin\alpha_2\sin\alpha_1$$

Hence,

$$\cos\theta = \sin\alpha_1\sin\alpha_2$$

$$\Omega = 2\pi - 4\theta \text{ (For the whole solid angle, i.e. the four equal solid angles added together)}$$

$$= 4\left(\frac{\pi}{2} - \theta\right)$$

$$= 4\sin^{-1}(\sin\alpha_1\sin\alpha_2) \text{ steradians}$$

or, when a rectangle is considered that has the apex of the pyramid over one corner,

$$\Omega = \sin^{-1}(\sin \alpha_1 \sin \alpha_2) \text{ steradians}$$

or, in terms of linear dimensions for the shaded rectangle in Figure 1.13(a),

$$\Omega = \tan^{-1} \frac{\frac{l_1}{d}\frac{l_2}{d}}{\sqrt{1+\left(\frac{l_1}{d}\right)^2+\left(\frac{l_2}{d}\right)^2}} \text{ steradians}$$

(Note $\sin^{-1}$ and $\tan^{-1}$ are evaluated in radians not degrees.)

1.5 Light flux, luminous intensity and illuminance

The light field created by a light source may be considered to consist of light flux emanating from the source at the origin of the coordinate system (see Figure 1.14).

The amount of light flux contained in a given solid angle, Ω, indicates the density of the light emanating from the source in the range of directions represented by the particular solid angle.

The quotient of flux contained in the solid angle by that solid angle is called the mean luminous intensity in that range of directions, i.e.

$$I = \frac{\Phi}{\Omega}$$

where Φ = luminous flux and Ω = solid angle.

Since, at any unobstructed distance from the light source, the light flux contained in the solid angle remains constant, the luminous intensity is independent of the distance of measurement from the source. Luminous intensity is therefore considered to be a property of the light source and can be used to indicate the performance of the light source.

To indicate the luminous intensity in a specific direction the solid angle must be made infinitesimally small, so

$$I = \frac{d\Phi}{d\Omega}$$

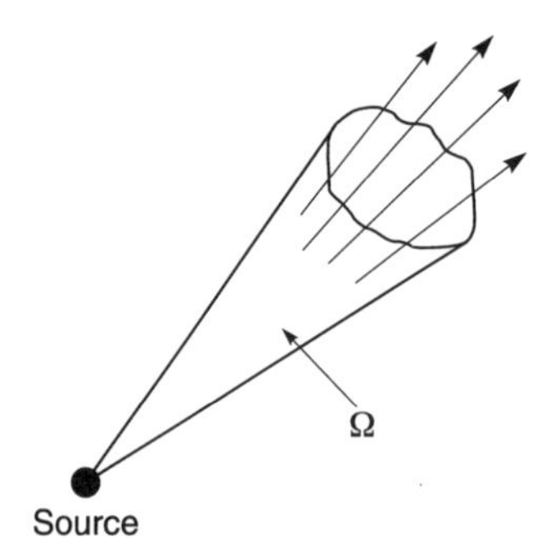

Fig. 1.14 The light flux contained in a given solid angle

The effect of this luminous intensity at different distances from the light source is indicated by the area density of the light flux at a particular point in space. This flux density is called the Illuminance (E):

$$E = \frac{\Phi}{A}$$

where Φ = luminous flux and A = area receiving the flux.

In the form given above, the illuminance indicated would be the average illuminance over area A. To indicate the illuminance at a point, the area is made infinitesimally small so that

$$E = \frac{\mathrm{d}\Phi}{\mathrm{d}A}$$

1.5.1 THE INVERSE-SQUARE LAW

The relationship between the luminous intensity and the illuminance it produces can be seen from Figure 1.15. In this diagram, two concentric spheres have been drawn with the source at the centre.

Area $\mathrm{d}A_1$ on the inner sphere receives the same amount of luminous flux as area $\mathrm{d}A_2$ on the outer sphere. As the solid angles are the same:

$$\mathrm{d}\Omega = \frac{\mathrm{d}A_1}{r_1^2} = \frac{\mathrm{d}A_2}{r_2^2}$$

and so the luminous intensity producing the illuminance on each area is the same.

The illuminance at distance $r_1 = \mathrm{d}\Phi/\mathrm{d}A_1 = E_1$.

The illuminance at distance $r_2 = \mathrm{d}\Phi/\mathrm{d}A_2 = E_2$.

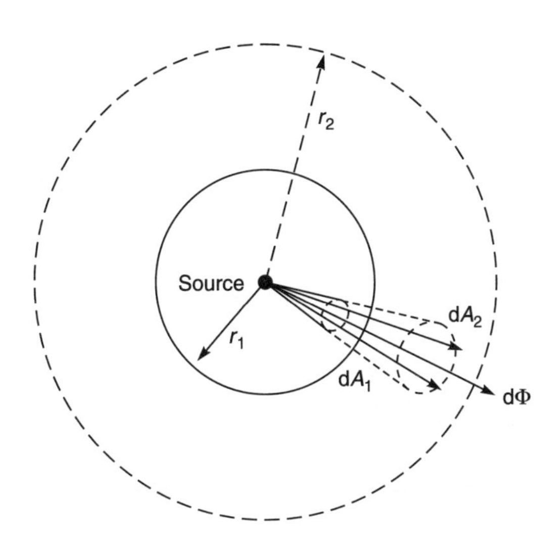

Fig. 1.15 The inverse-square law

Now

$$dA_2 = \frac{r_2^2}{r_1^2} dA_1$$

$$E_2 = \frac{d\Phi}{\left(\frac{r_2^2}{r_1^2} dA_1\right)} = \frac{r_1^2}{r_2^2} E_1$$

or

$$\frac{E_2}{E_1} = \frac{r_1^2}{r_2^2}$$

This indicates the well-known inverse-square law relationship for point sources of light. It is seen that the illuminance varies inversely as the square of the distance of the illuminated point from the source. Since large light sources can be considered to consist of a number of point sources, this relationship can be applied to all light sources, either directly or (as will be seen later) by developing formulae from the inverse-square law for the larger sources.

1.5.2 THE COSINE LAW

The surface density of the light flux received by an element of area varies not only with the distance from the light source, but also with the angle of the element of area with respect to the direction of the light flux.

The maximum illuminance occurs when the element of area receives the light flux normal to its surface (Figure 1.16).

When the element of area is tilted with respect to the direction of the light flux, the flux density on the element is reduced. This can be thought of in two ways.

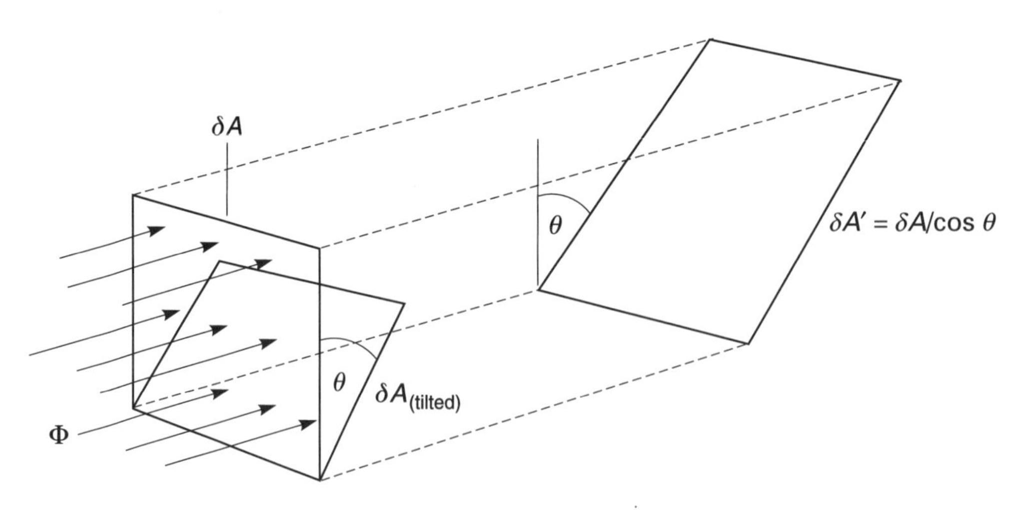

Fig. 1.16 The cosine law

(i) The tilted area no longer intercepts all the light flux it previously received and so the illuminance falls.
(ii) If the element is part of a larger surface, the original flux now falls on an element of area that is larger than before (δA_1). Accordingly, the illuminance falls.

For case (i) (see Figure 1.16), the intercepted flux is reduced in proportion to the orthogonal area receiving it. The amount of flux previously intercepted by δA was given by

$$\frac{\Phi}{\delta A} \times \delta A$$

When the element δA is tilted by angle θ the amount of flux intercepted by δA is given by

$$\Phi' = \frac{\Phi}{\delta A} \delta A \cos \theta = \Phi \cos \theta$$

and the flux received by δA is reduced by a factor $\cos \theta$.

For case (ii) (see Figure 1.16), if all the flux is intercepted it now falls on a larger element $\delta A'$:

$$\delta A' = \frac{\delta A}{\cos \theta}$$

Applying either of these approaches results in

$$E = E_{max} \cos \theta$$

In case (i) the reduced flux received by δA is now $\Phi \cos \theta$, so

$$E = \frac{\Phi \cos \theta}{\delta A}$$

In case (ii) the area receiving the flux Φ is now a larger area $\delta A'$:

$$E = \frac{\Phi}{\delta A'} = \frac{\Phi / \delta A}{\cos \theta} = \frac{\Phi \cos \theta}{\delta A}$$

which is the same result as before.

1.5.3 THE POINT SOURCE ILLUMINANCE EQUATION

The illuminance equation for a point source may be written as

$$E = \frac{I_\theta \cos \theta}{d^2}$$

where I_θ is the luminous intensity of the source in the direction of the illuminated point, θ is the angle between the normal to the plane containing the illuminated point and the line joining the source to the illuminated point, and d is the distance to the illuminated point (see Figure 1.17).

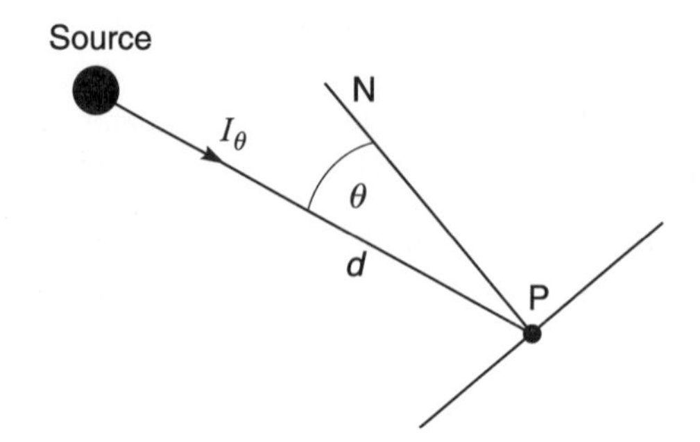

Fig. 1.17 The point source illuminance equation relationships

To give these concepts a means of practical application, it is necessary to define units in which they can be measured.

1.5.4 LUMINOUS INTENSITY

The unit of luminous intensity is the *candela* and the formal definition is:

Candela

The candela is the luminous intensity, in a given direction, of a source that emits monochromatic radiation of frequency 540×10^{12} Hz and that has a radiant intensity in that direction of 1/683 W per steradian: abbreviation cd.

This definition provides the link between radiant power (measured in watts) and the light that is one of the effects of that radiant power. The frequency has to be specified, since the response of the human eye to a particular amount of radiant power depends upon its frequency. This value is related to the other frequencies (or wavelengths) by an agreed table of values.

1.5.5 LUMINOUS FLUX

The unit of luminous flux (*lumen*) is defined very simply in terms of the candela

$$1 \text{ lumen} = 1 \text{ candela} \times 1 \text{ steradian}$$

So it follows that the luminous intensity in candelas is equal to the luminous flux in lumens contained in the solid angle under consideration divided by that solid angle in steradians; that is,

$$I = \frac{\Phi}{\Omega} \text{ candelas}$$

1.5.6 ILLUMINANCE

The illuminance is the luminous flux falling on a given area divided by that area,

$$E = \frac{\mathrm{d}\Phi}{\mathrm{d}A}$$

The unit of illuminance is the *lux*, which is equal to one lumen per square metre, abbreviation lx.

In the USA the *footcandle* is sometimes used and this is equal to one lumen per square foot.

1.6 Luminous intensity distribution diagrams

The performance of luminaires in terms of luminous intensity values related to a coordinate system is now frequently held on a computer database, often as part of a wider program for lighting calculations, and this is dealt with in Chapter 2. However, luminous intensity diagrams are still common and have a role to play in visually indicating the performance of a luminaire.

1.6.1 CARTESIAN COORDINATE DIAGRAMS

Perhaps the simplest form of intensity distribution diagram is one in which a distribution, which is symmetrical about the peak intensity, is plotted on Cartesian coordinates with angles from the beam axis plotted horizontally and intensity values plotted vertically. This type of diagram is particularly appropriate for highly directional projectors, where the rapid change of intensity with angle makes the other forms of diagram of less value (Figure 1.18). The limitation is that a separate diagram would be required for each plane unless the distribution is rotationally symmetrical.

1.6.2 POLAR COORDINATE DIAGRAMS

A long-established method of display is to plot the intensity values on polar coordinates. This has the advantage of giving a more easily appreciated visual impression of the intensity distribution from a luminaire.

This polar diagram (or polar curve) is produced for a specified plane passing through the centre of the luminaire. The intensity at any angle γ from the downward vertical is indicated by the distance from the origin to the curve, as shown.

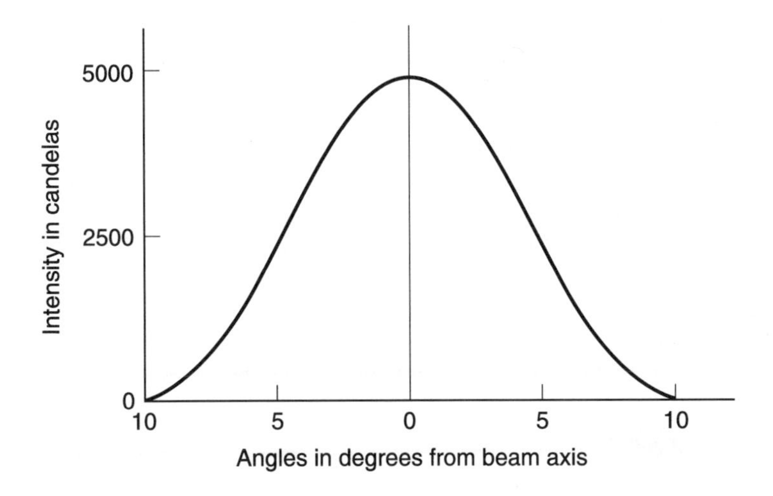

Fig. 1.18 The Cartesian coordinate luminous intensity diagram

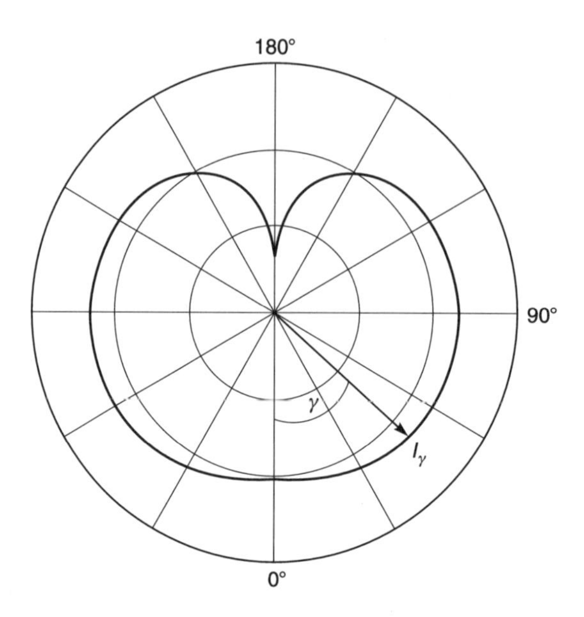

Fig. 1.19 Intensity distribution plotted on polar coordinates

Circles of equal intensity are usually included to simplify the computation of the intensity in particular directions.

The polar curve shown in Figure 1.19 has been assumed to be measured on one of the *C* planes of the $C\gamma$ coordinate system (see Section 1.2).

In some cases, the polar curve obtained has the same shape and magnitude on all the *C* planes passing through the luminaire.

Such a luminaire would be said to have a symmetrical distribution about the vertical axis.

If the curve shown in Figure 1.19 was for such a symmetrical distribution, it could be rotated about the vertical axis to produce a polar solid. Because of the shape of the polar curve chosen, the polar solid would be 'apple-shaped' and the polar curve could be likened to the cross-section revealed when an apple is cut in half down the centre of the core.

Polar curves are, in effect, cross-sections of a polar solid and, in many cases, these cross-sections have a different shape in different planes. A common example would be the polar curves for the axial and transverse planes of a linear fluorescent luminaire (Figure 1.20). Rotating either of the polar curves shown about the vertical axis would not produce the true polar solid for the luminaire.

Figure 1.20 shows the transverse polar curve and the axial polar curve for the vertical plane. Representation by transverse and axial polar curves relates to the A, α and B, β coordinate systems: references to Figure 1.3(b), (c) and (e), (f) illustrate these.

Normally, only one transverse curve is provided, which corresponds to the values for $\beta = 0°$ or $\alpha = 0°$ on each of the *B* or *A* planes. However, each *B* or *A* plane has its own axial polar curve.

It should be noted that in some road lighting literature, a form of polar curve is given for a cone through the peak intensity value (that is at constant γ angle).

1.6.3 ISOCANDELA DIAGRAMS

A more comprehensive means of displaying intensity values is to use an *isocandela diagram*. This form of diagram originates from considering the light source to be located at the centre of

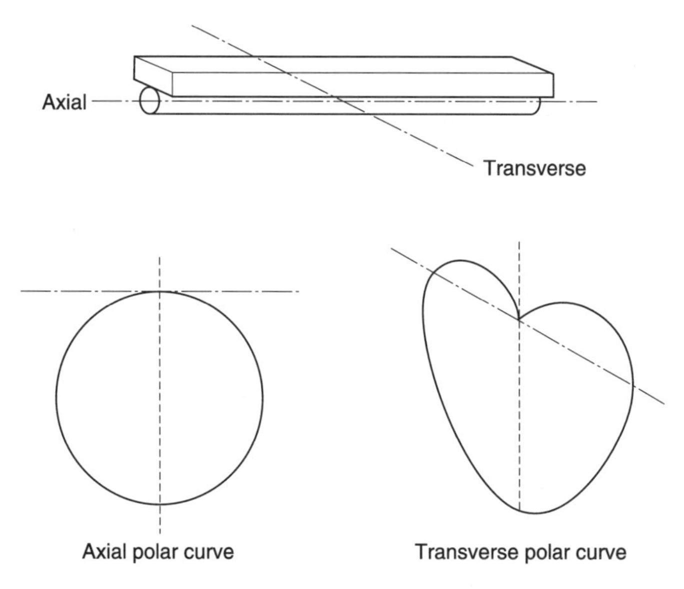

Fig. 1.20 Axial and transverse polar curves

a sphere and marking the values of the measured intensity on the surface of the sphere. The diagrams are called isocandela diagrams because points of equal intensity are joined to form lines similar to map contours. In Figure 1.21(a) and (b) two such plots are shown.

As long as we consider the sphere and not its projections, the direction of the angular coordinates has no effect upon the shape of the isocandela curves, but it does alter the way in which a particular point on the sphere is specified.

The advantage of this type of diagram is that it gives intensity information over a much wider

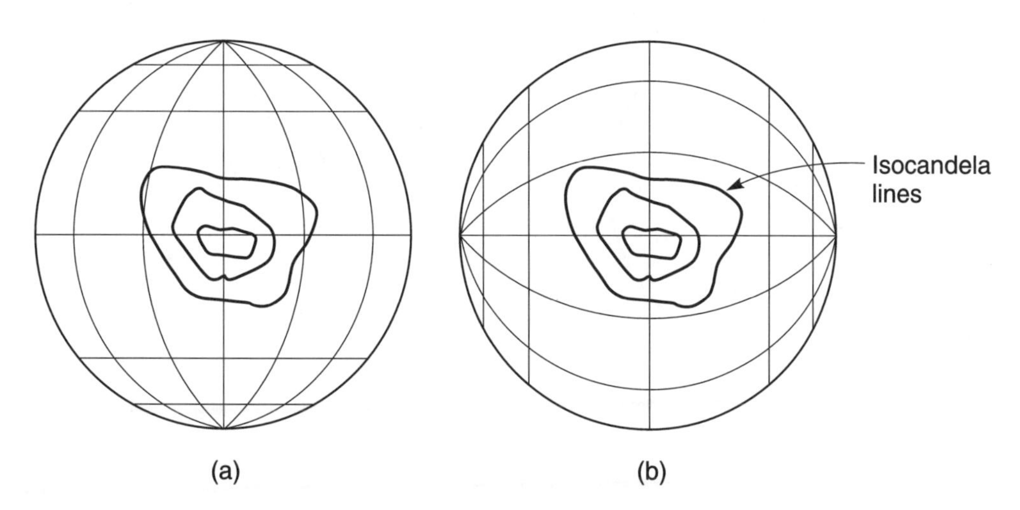

Fig. 1.21 (a) An isocandela diagram; (b) the same isocandela diagram but with the coordinate system rotated through 90°

range of directions than a polar or Cartesian plot of intensity values, which only relate to a single plane.

The disadvantage is the one also faced by map makers; that of adequately representing the curved surface of a sphere on a flat plane.

One of the most useful forms of projection is described below.

1.6.4 THE AZIMUTHAL PROJECTION

This form of projection has several advantages from the point of view of the lighting engineer; (1) it retains a circular boundary, (2) if the luminaire is tilted relative to the coordinate system the isocandela curves retain the same shape (neither of these features is present in the sinusoidal projection often used in the past),[2] and (3) areas on the projection are proportional to the solid angle.

The usual C, γ diagram is obtained in two stages:

(1) The sphere is viewed down the polar axis so that the parallels appear as circles and the meridians as straight lines. The circles are drawn such that the area of the circle is the same as that of the cap of the sphere existing on the real sphere between the parallel and the pole. Thus, in a simple way, an equal area diagram is produced related to the B, β system.

(2) The diagram is now rescaled and the B, β coordinates for each point are replaced by calculating the equivalent C, γ coordinates using the formulae developed at the beginning of this chapter (Section 1.3.1).

Stage 1

Consider Figure 1.22(a) and (b).

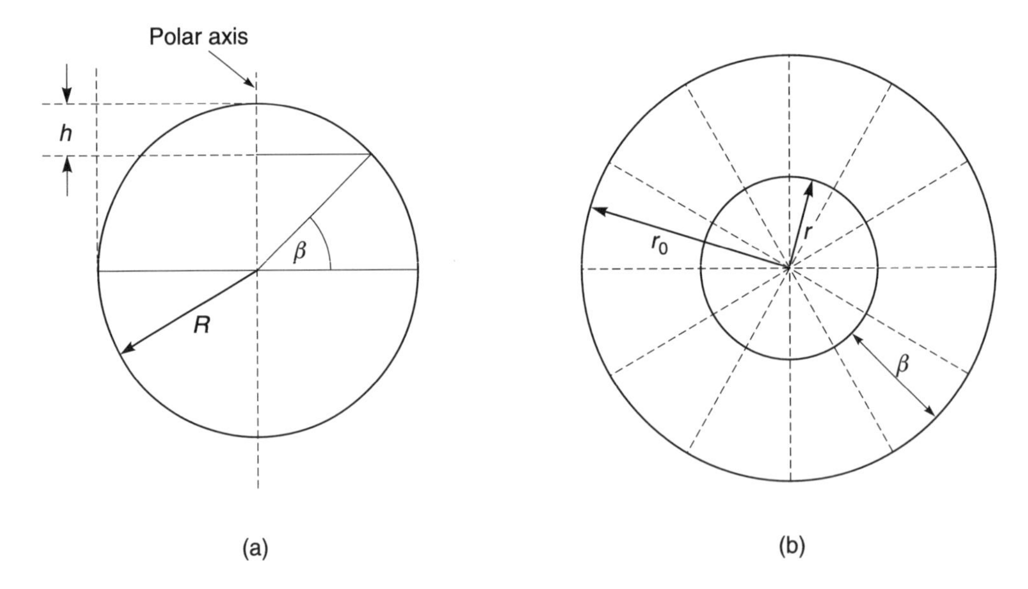

Fig. 1.22 (a) Side view of the sphere showing the spherical cap of height h and its relationship to angle β; (b) translation of the polar axis view into an equal area projection

The area of the spherical cap is equal to the solid angle subtended by the cap at the centre of the sphere multiplied by the radius squared (see Section 1.4.1), Figure 1.22(a):

$$\text{area of cap} = 2\pi R^2(1 - \sin\beta)$$

The polar projected view shown in Figure 1.22(b) must be equal to this area.

Therefore

$$\pi r^2 = 2\pi R^2(1 - \sin\beta)$$

$$r = \sqrt{2R^2(1 - \sin\beta)}$$

$$= \sqrt{2}R\sqrt{1 - \sin\beta}$$

When $\beta = 0°$

$$r_0 = \sqrt{2}R$$

Thus, the radius of the projected sphere is $\sqrt{2}$ times the radius of the sphere it represents. This is of little importance since the size of the projection is chosen to produce conveniently sized diagrams.

At any angle β the radius of the circle of latitude is given by

$$r = r_0\sqrt{1 - \sin\beta}$$

where r_0 is the radius of the perimeter of the projection.

Example:

Let $\beta = 30°$

$$r = r_0\sqrt{1 - 0.5}$$

$$= r_0\sqrt{0.5}$$

$$= 0.707r_0$$

Let $\beta = 60°$

$$r = r_0\sqrt{1 - 0.866}$$

$$= r_0\sqrt{0.134}$$

$$= 0.366r_0$$

Figure 1.23 shows the completed projection for 30° parallels and 30° meridians.

Stage 2

Translate C, γ values into B, β values (see Section 1.3.3).

For example:

Let $C = 30°$, $\gamma = 30°$

$$\tan B = \sin C \times \tan\gamma$$

$$= 0.5 \times 0.577$$

$$= 0.2885$$

$$B = 16.09°$$

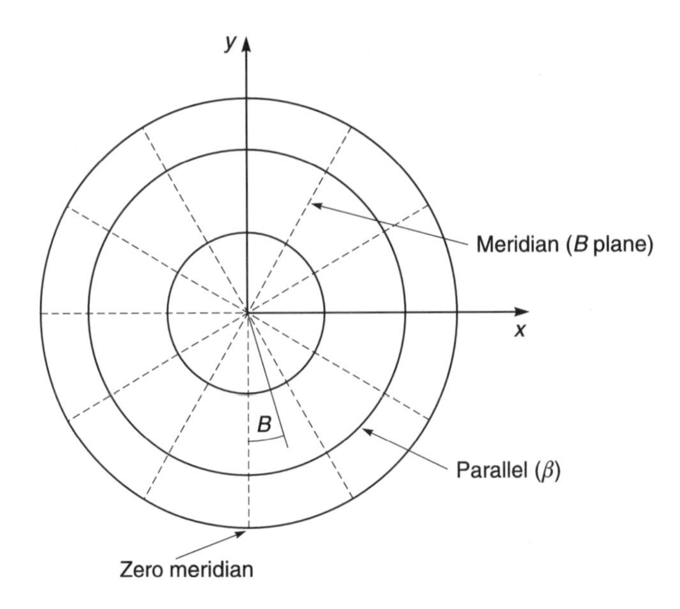

Fig. 1.23 The completed polar projection

$$\begin{aligned} \sin\beta &= \cos C \times \sin\gamma \\ &= 0.866 \times 0.5 \\ &= 0.433 \\ \beta &= 25.65° \end{aligned}$$

Since, for accurate plotting of the *C, γ* meridian and parallels, the relevant *B, β* meridians and circles would first need to be drawn, it is better to plot the diagram in terms of the *x* and *y* coordinates shown in Figure 1.23.

To plot the azimuthal projection on the *C, γ* system we note that on the *B, β* system

$$x = r \sin B$$

and

$$y = r \cos B$$

Also,

$$r = r_0 \sqrt{1 - \sin\beta}$$

giving

$$x = r_0 \sin B \sqrt{1 - \sin\beta}$$

and

$$-y = r_0 \cos B \sqrt{1 - \sin\beta}$$

These can be translated into the *C, γ* system by substituting $\sin\beta = \cos C.\sin\gamma$ (see Section 1.3.3) and $B = \tan^{-1}(\sin C.\tan\gamma)$.

A completed plot is shown in Figure 1.24.

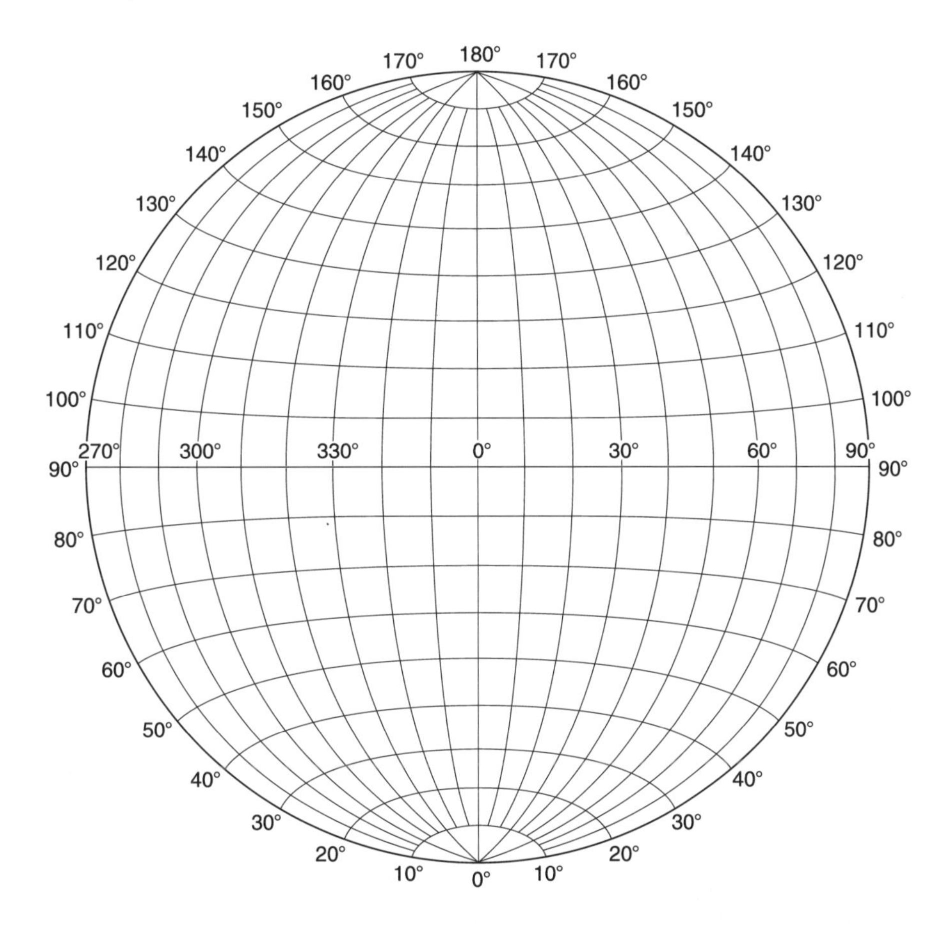

Fig. 1.24 The completed Azimuthal projection

1.7 Calculation of luminous flux

Since

$$I = \frac{d\Phi}{d\Omega}$$

the flux related to the intensity can be obtained from

$$\Phi = \int_0^{\Omega} I \, d\Omega$$

1.7.1 THE ZONE FACTOR METHOD

This method uses the mean vertical polar curve for the luminaire and calculates the solid angles for zones of the sphere through which intensity values read from the curve could be considered to be acting (see Figure 1.25).

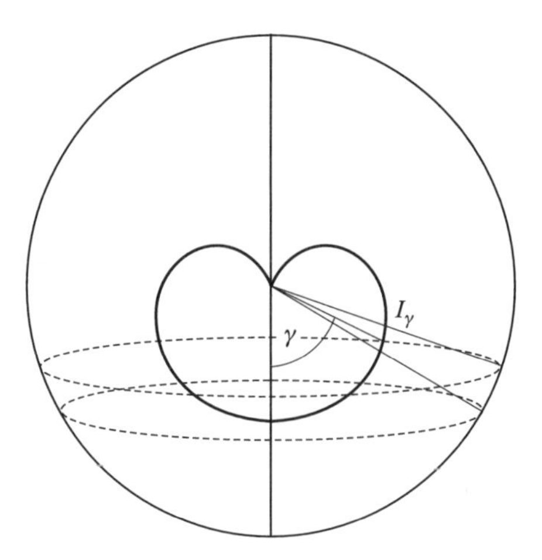

Fig. 1.25 Flux calculation from the mean vertical polar curve by the zone factor method

The zones are chosen to be small enough for the mid-zone intensity to be assumed to be equal to the average intensity acting across the zone. Once this is assumed, the equation may be rewritten as:

$$\Phi = I_\gamma \int d\Omega$$

A table of values of Ω (called *zone factors*) related to a series of angles in the vertical plane is then calculated and used to multiply the appropriate mid-zone intensities. The summation of the products gives the flux value required:

$$\Phi = \Sigma\, I_\gamma \Omega$$

In Section 1.4.1, the solid angle for the cap of a sphere was calculated as a means of determining the solid angle subtended by a disc. Consider Figure 1.26.

The zone defined by angles γ_1 and γ_2 can be considered to be the difference in the solid angle for γ_2 and that for γ_1:

The solid angle related to $\gamma_2 = 2\pi(1 - \cos\gamma_2)$
The solid angle related to $\gamma_1 = 2\pi(1 - \cos\gamma_1)$

The required solid angle is given by:

$$2\pi(1 - \cos\gamma_2) - 2\pi(1 - \cos\gamma_1)$$
$$\Omega_{\gamma_1,\gamma_2} = 2\pi(\cos\gamma_1 - \cos\gamma_2)$$

or

$$= 4\pi \sin\left(\frac{\gamma_2 - \gamma_1}{2}\right)\sin\left(\frac{\gamma_2 + \gamma_1}{2}\right)$$

The angular interval chosen for the zone factors depends upon the type of intensity distribution for which the calculation is to be made. If the distribution is changing rapidly in the vertical plane then 2° zones will be required, but with other distributions 5° or 10° zones might be used with similar accuracy. Zone factors for 2°, 5° and 10° intervals are given in Table 1.1.

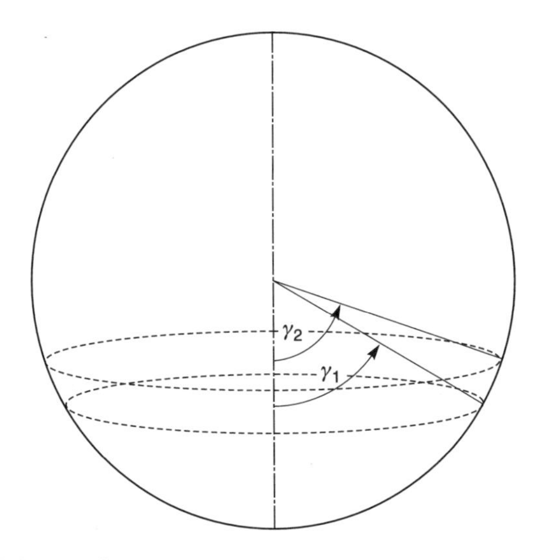

Fig. 1.26 Calculation of the zone factor

Table 1.1 Zonal factors for 2°, 5° and 10° zones

2° zones		5° zones			10° zones		
Zone limits (degrees)	Zonal factors	Zone limits (degrees)	Zone limits (degrees)	Zonal factors	Zone limits (degrees)	Zone limits (degrees)	Zonal factors
0–2	0.0038	0–5	175–180	0.0239	0–10	170–180	0.095
2–4	0.0115	5–10	170–175	0.0715	10–20	160–170	0.283
4–6	0.0191	10–15	165–170	0.1186	20–30	150–160	0.463
6–8	0.0267	15–20	160–165	0.1649	30–40	140–150	0.628
8–10	0.0343	20–25	155–160	0.2097	40–50	130–140	0.774
10–12	0.0418	25–30	150–155	0.2531	50–60	120–130	0.897
12–14	0.0493	30–35	145–150	0.2946	60–70	110–120	0.993
14–16	0.0568	35–40	140–145	0.3337	70–80	100–110	1.058
16–18	0.0641	40–45	135–140	0.3703	80–90	90–100	1.091
18–20	0.0714	45–50	130–135	0.4041			
		50–55	125–130	0.4349			
		55–60	120–125	0.4623			
		60–65	115–120	0.4862			
		65–70	110–115	0.5064			
		70–75	105–110	0.5228			
		75–80	100–105	0.5351			
		80–85	99–100	0.5434			
		85–90	90–95	0.5476			

If required, the solid angle associated with a particular interval of azimuth angle can be obtained by direct proportion, e.g.

$$\Omega_{\gamma_1,\gamma_2,C_1,C_2} = \frac{C_2 - C_1}{360^\circ} \times ZF_{\gamma_1,\gamma_2}$$

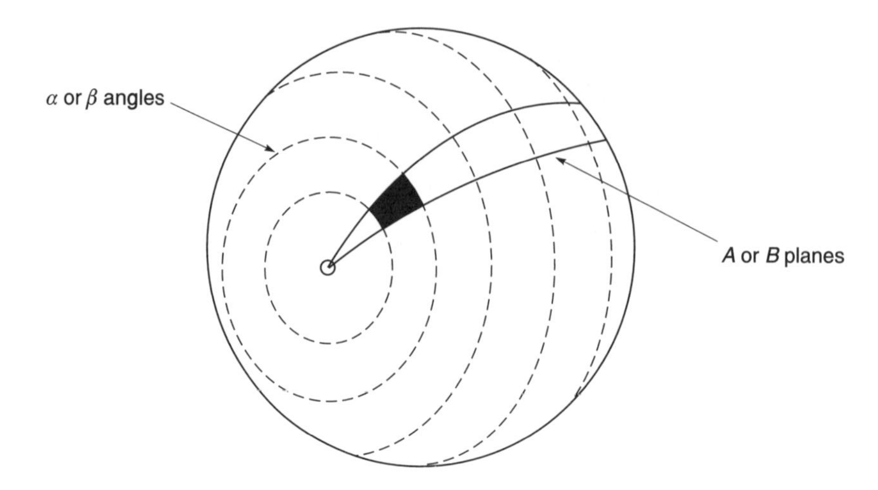

Fig. 1.27 Solid angles for A, α or B, β calculations

Thus, if $C_2 - C_1 = 10°$,

$$\Omega_{\gamma_1,\gamma_2,C_1,C_2} = \frac{1}{36} \times ZF_{\gamma_1,\gamma_2}$$

1.7.2 AN ALTERNATIVE APPROACH

If the intensity distribution has been measured on the B, β or A, α system, then the flux could be calculated as follows.

Consider the segment of the sphere outlined in Figure 1.27. It was pointed out in the previous section that the solid angle for a particular section of an annular zone (as shaded) can be calculated by direct proportion from the zone factor calculated for the whole zone. Thus, a set of new solid angle factors could be calculated for each segment of the sphere denoted by two A planes or B planes in terms of β or α increments. Each value would be used, for example, to multiply the intensity from the mid-zone A plane at the mid-zone α value (Figure 1.28). Hence the total flux in each segment of the whole spherical solid angle can be obtained. The formula is an adaptation of that given in Section 1.7.1 and could be in terms of α or β. β is used here.

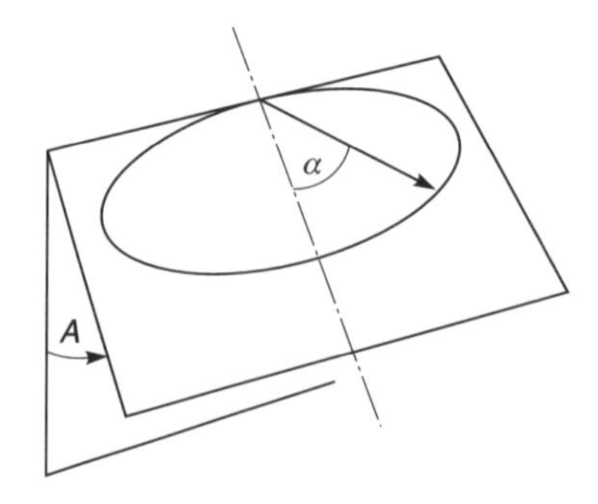

Fig. 1.28 Intensity curves on the A, α system

Table 1.2 Solid angles for 10° intervals

$\pm\,\alpha$ or β angular limits degrees	Solid angle steradians
0–10	30.31×10^{-3}
10–20	29.39×10^{-3}
20–30	27.58×10^{-3}
30–40	24.91×10^{-3}
40–50	21.50×10^{-3}
50–60	17.44×10^{-3}
60–70	12.86×10^{-3}
70–80	7.86×10^{-3}
80–90	2.64×10^{-3}

$$\Phi = 4\pi \sin\left(\frac{\beta_2 - \beta_1}{2}\right)\cos\left(\frac{\beta_2 + \beta_1}{2}\right) \times \left(\frac{\Delta B}{360}\right) \times I_{AV}$$

where I_{AV} is the average intensity (see Section 2.6).

Table 1.2 gives the solid angles for 10° intervals of both A and α or B and β.

1.7.3 RUSSELL ANGLES

If only the total flux from the light source is required then a simplification is obtained if the intensities are measured at the centre of equal solid angles.

The total flux is then calculated from

$$\Phi = \frac{4\pi}{n} \Sigma I_\gamma$$

where n is the number of equal solid angles into which the enclosing sphere is divided.

The following treatment assumes that the C, γ coordinate system is used.

The Russell angles are the γ values for the mid-points of a series of adjacent equal solid angles, where the mid-point divides the solid angle into two equal solid angles.

Beginning at 0°, the first Russell angle occurs after half the first solid angle has been traversed.

So,

$$\frac{4\pi}{2n} = 2\pi(1 - \cos\gamma_1)$$

$$\cos\gamma_1 = \frac{n - 1}{n}$$

This gives the first Russell angle.

To obtain the second Russell angle it is necessary to traverse a further complete interval of solid angle, i.e.

$$\frac{4\pi}{n}$$

So, the next angle is given by

$$\frac{3}{2}\frac{4\pi}{n} = 2\pi(1 - \cos\gamma_2)$$

The third angle is given by

$$\cos\gamma_3 = \frac{n-5}{n}$$

and so on, until the last angle in the lower hemisphere is obtained. At

$$\cos\gamma = \frac{n-(n-1)}{n} = \frac{1}{n}$$

Since the upper hemisphere is simply the reverse of the lower hemisphere, the Russell angles for the upper hemisphere are obtained by subtraction, e.g.

$$\gamma_{20} = 180^\circ - \gamma_1$$
$$\gamma_{19} = 180^\circ - \gamma_2$$

and so on.

For example,

Let $n = 20$

then the first Russell angle occurs at

$$\cos\gamma_1 = \frac{20-1}{20}$$
$$= 0.95$$
$$\gamma_1 = 18.2^\circ$$

So,

$$\gamma_{20} = 180^\circ - 18.2^\circ$$
$$= 161.8^\circ$$

$$\cos\gamma_2 = \frac{20-3}{20}$$
$$= 0.85$$
$$\gamma_2 = 31.8^\circ$$

So,

$$\gamma_{19} = 148.2^\circ$$

and so on until the ten angles for the lower hemisphere have been calculated and the ten angles for the upper hemisphere obtained by subtraction.

A complete set of 20 Russell angles is given in Table 1.3.

Table 1.3 20 Russell angles (degrees)

18.2	31.8	41.4	49.5	56.6	63.3	69.5
75.5	81.4	87.1	92.9	98.6	104.5	
110.5	116.7	123.4	130.5	138.6	148.2	161.8

1.7.4 WOHLAUER'S CONSTRUCTION

The various methods of computing the flux emitted by a light source already described are necessitated by the fact that the intensity distribution curve does not give a direct indication of the emitted flux.

A simple construction due to Wohlauer[3] enables the relative contributions of the different zones to be obtained directly from the polar curve.

Earlier, it was shown that the flux emitted in a zone can be calculated from

$$\Phi = \int I_\gamma \, d\Omega$$

$$= I_\gamma \times 4\pi \sin\left(\frac{\gamma_2 - \gamma_1}{2}\right) \times \sin\left(\frac{\gamma_1 - \gamma_2}{2}\right)$$

If zones of equal angular subtense are chosen then

$$4\pi \sin\left(\frac{\gamma_2 - \gamma_1}{2}\right)$$

is a constant and

$$I_\gamma \sin\left(\frac{\gamma_1 - \gamma_2}{2}\right)$$

can be obtained from a horizontal line drawn on a polar curve (Figure 1.29).

Using this construction makes it possible to see immediately the relative contributions to the

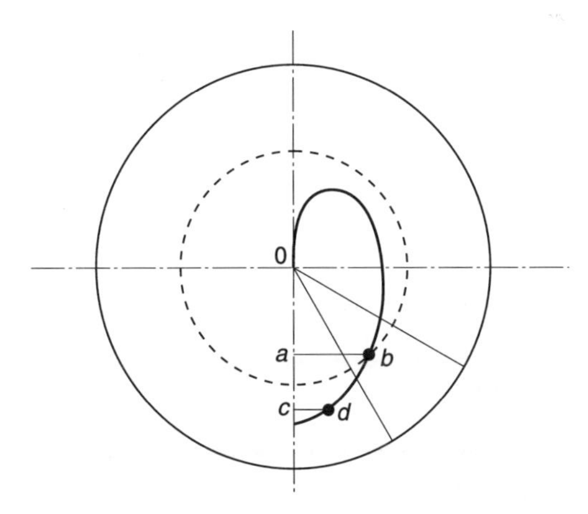

Fig. 1.29 Wohlauer's construction

total flux made by each zone. For example, on the diagram given in Figure 1.29 the relative flux contributions are represented by horizontal lines *ab* and *cd*.

Although the luminous intensity represented by distance *od* is greater than that represented by *ob*, the zonal flux represented by *ab* is greater than that represented by *cd*.

If required, the total flux could be calculated by adding together horizontal lines for each zone (measured in units of intensity) and multiplying the answer by $4\pi \sin[(\gamma_2 - \gamma_1)/2]$.

1.7.5 ILLUMINANCE GRID AND ISOLUX DIAGRAM

Two simple methods of determining the amount of light flux received by a surface are:

(1) the illuminance grid
(2) the isolux diagram

(1) The illuminance grid

This method is often used to compute the flux received by a surface of known area, e.g. a floor, wall or ceiling. The whole area is usually divided into a series of equal sized rectangular areas and the illuminances are calculated for a grid of points formed by the centre points of these rectangular areas.

The flux received is then

$$\Phi = E_{\text{average}} \times \text{Area of the surface}$$

With the developments in computing power this direct method has become more popular than the one which follows.

(2) The isolux diagram

This is a diagram on which contours of equal illuminance are drawn on a plane target surface.

The area between two contours is determined and multiplied by the average value of the illuminance between the contours to obtain the flux falling on this area. Repeating this procedure for each pair of contours, so that the entire area is covered, enables the total flux received to be determined by summation.

This method has sometimes proved useful for calculating the flux emitted by concentrating luminaires where all the flux can be intercepted by a surface of relatively small area.

Problems

1. Convert the following coordinates in the C, γ system to the B, β system.
   (a) (30°, 40°) [Answer (22.8°, 33.8°)]
   (b) (300°, 60°) [Answer (–56.3°, 25.7°)]
   (c) (50°, 120°) [Answer (127.0°, 33.8°)]
   (d) (270°, 150°) [Answer (–150.0°, 0°)]
2. Convert the following coordinates in the B, β system to the C, γ system.
   (a) (0°, 0°) [Answer (0°, 0°)]
   (b) (30°, 90°) [Answer (0°, 90°)]
   (c) (10°, 10°) [Answer (44.6°, 14.4°)]
   (d) (50°, 30°) [Answer (53.0°, 56.2°)]

3. A horizontal surface is illuminated by a source mounted 5 m above it. If the source emits 100 cd in all directions what is the illuminance on the surface
   (a) directly below the source? [Answer: 4 lx]
   (b) 5 m from the vertical through the source? [Answer: 1.41 lx]
   (c) 20 m from the vertical through the source? [Answer: 0.057 lx]

4. A point P is vertically above the centre of a horizontal disc of radius 3 m. Calculate the solid angle the disc subtends at P if the height of P above the disc is
   (a) 4 m [Answer: 1.26 sr]
   (b) 7 m [Answer: 0.508 sr]

5. A point P is vertically above one corner of a rectangle of dimensions 3 m by 4 m. Calculate the solid angle the rectangle subtends at P if the height P above the rectangle is
   (a) 5 m [Answer: 0.327 sr]
   (b) 10 m [Answer: 0.107 sr]

6. If, in Question 5, a source emitting 100 cd in all directions is at P, what is the luminous flux falling on the rectangle and the average illuminance?
   (a) [Answer: 32.7 lm, 2.73 lx]
   (b) [Answer: 10.7 lm, 0.891 lx]

7. A 5 m wide circular track of outside diameter 100 m is illuminated by a centrally placed luminaire emitting 10,000 cd in all directions towards the track. If the mounting height is 50 m, what is the average illuminance on the track?
   [Answer: 1.53 lx]

8. A point P is 5 m vertically above a horizontal rectangle of dimensions 6 m × 3 m. If the vertical from P is outside the boundary of the rectangle and 4 m from one narrow edge but in line with one wide edge, what solid angle does the rectangle subtend at P and what is its average illuminance if a source at P emits 1000 cd in all directions?
   [Answer: 0.1508 sr, 8.38 lx]

9. A cylindrical source has a luminous intensity I normal to its axis. By considering 10° zones, calculate the luminous flux it emits if its luminous intensity in any direction in any axial plane is proportional to the cosine of the angle between the direction and the normal to the cylindrical axis. The size of the cylinder can be regarded as infinitesimal.
   [Answer: $9.857I$ lm or $0.998\pi^2 I$] (See page 145, note 1)

10. A flat source has a luminous intensity I normal to its surface. The luminous intensity in any direction is proportional to the cosine of the angle that the direction makes with the normal (that is, the source is a uniform diffuser). By considering 10° zones, calculate the luminous flux emitted by the source, which can be regarded as infinitesimal.
   [Answer: $3.154I$ lm or $0.9987\pi I$] (See page 73)

References

1. Gershun, A. (1936) *The Light Field* (Moscow: Svetovoe Pole).
2. Keitz, H.A.E. (1971) *Light Calculations and Measurements* (Macmillan) p. 60.
3. Wohlauer, A.A. (1908) *Illum. Engng. (N.Y.)* **3**, 657.

2

The Luminous Intensity Table and Related Computer Applications

2.1 Introduction

In Chapter 1 we showed that the luminous intensity distribution of luminaires is required for carrying out lighting calculations. For computer calculations this information is presented in a standard layout, the *I*-table, so that it can be used with any of the computer programs available for lighting calculations. The purpose of this chapter is to describe the forms that *I*-tables take and the ancillary data that are included with them. A description is given of how interpolation may be used to find luminous intensities between the angular intervals given in the tables, together with methods of using the tables when the luminaire is turned about its photometric axes. Methods of calculating luminous flux in defined angular zones are given, followed by a description of file formats for the electronic transfer of luminaire data.

2.2 Layout of *I*-tables

Tables 2.1(a) and (b) show the layout of an *I*-table for a road lighting luminaire. The coordinate system is the (C, γ), which is typically used for this application. By implication it is assumed that the light distribution is symmetrical about the $C = 90° - 270°$ plane, otherwise the light distribution for $C = 90°$ through 180° to 270° would have been given (Figure 2.1, page 37).

In effect, therefore, the table gives the luminous intensity distribution for all directions around the luminaire and the boundary of the *I*-table is presentational. This usually applies to general purpose interior luminaires as well (Table 2.9, page 58, is an example). For certain classes of luminaire the luminous intensity distribution is only given over a restricted solid angle. This is usually the case for floodlights and spotlights, when the boundary of the *I*-table has a physical meaning (Table 2.11, page 60).

It will be noted that, in Tables 2.1(a) and (b), the angular spacings are uneven – they are smaller in the regions where the calculations demand the greatest accuracy. These spacings are recommended by CIE 30–2,[1] which was drafted at a time when computer memory had to be conserved and the number of luminous intensity readings had to be limited to allow a luminaire to be photometered in a reasonable time. Because of the increased power of computers and improvements in the automation of light distribution photometers, these restrictions no longer apply, and it is likely that future recommendations will ask for luminous intensity measurements to be made at regular and close angular intervals.

The luminous intensities for $\gamma = 0°$ and $\gamma = 180°$ are constant as C is varied. This is because these luminous intensities lie on the polar axis of the coordinate system so that changing the angle C does not change the direction of the luminous intensity.

Table 2.1(a) *I*-table for *C* from 270° to 355° for a road lighting luminaire. Luminous intensities in candelas per 1000 lumens

γ (deg)	Azimuth *C* (deg) 270.0	285.0	300.0	310.0	315.0	320.0	325.0	330.0	335.0	340.0	345.0	350.0	355.0
0.0	218	218	218	218	218	218	218	218	218	218	218	218	218
10.0	194	195	199	202	204	207	209	211	213	216	218	221	223
20.0	166	169	176	185	191	198	205	213	222	230	238	247	253
30.0	141	145	156	169	177	186	197	210	225	242	258	275	289
35.0	131	134	147	160	167	177	189	203	220	240	260	280	297
40.0	120	124	137	148	154	162	172	185	202	221	242	263	282
45.0	110	114	126	134	138	145	153	165	181	199	221	242	261
47.5	105	109	121	129	133	139	148	161	177	197	220	242	263
50.0	100	105	117	124	129	136	146	158	175	194	216	240	261
52.5	94	100	113	121	126	133	142	155	171	191	214	237	259
55.0	90	96	109	118	124	131	140	152	169	189	212	235	258
57.5	84	91	105	114	120	127	136	148	163	182	205	228	251
60.0	79	87	101	111	117	123	132	143	159	178	201	224	247
62.5	73	82	97	106	111	118	126	137	153	172	194	218	242
65.0	67	76	91	100	105	112	120	131	146	166	189	214	238
67.5	59	68	83	92	96	103	111	121	136	156	179	203	227
70.0	50	59	72	81	86	93	101	110	124	142	164	187	210
72.5	38	47	60	69	74	79	84	91	102	118	137	160	183
75.0	24	31	40	48	48	49	50	51	56	65	79	97	117
77.5	12	16	21	17	15	15	15	16	20	27	37	49	63
80.0	6	7	7	6	7	7	8	9	11	14	20	27	38
82.5	3	4	3	4	4	5	5	6	7	8	11	15	21
85.0	1	1	1	2	2	2	2	2	3	3	4	5	6
87.5	0	0	0	1	1	1	1	1	1	1	1	1	1
90.0	0	0	0	1	1	1	1	1	1	1	1	1	1
92.5	0	0	0	0	0	0	0	0	0	0	0	0	1
95.0	0	0	0	1	1	1	1	1	1	1	1	1	1
97.5	0	0	0	0	1	1	1	1	1	1	1	1	1
100.0	0	0	0	1	1	1	1	1	1	1	1	1	1
102.5	0	0	0	1	1	1	1	1	1	1	1	1	1
105.0	1	1	1	1	1	1	1	1	1	1	1	1	1
120.0	1	1	1	1	1	1	1	1	1	1	1	1	1
135.0	1	1	1	1	1	1	1	1	1	2	1	1	1
150.0	1	1	1	1	1	1	1	1	1	1	1	1	1
165.0	1	1	1	1	1	1	1	1	1	1	1	1	1
180.0	1	1	1	1	1	1	1	1	1	1	1	1	1

Table 2.1(b) Continuation of Table 2.1(a) for C from 0° to 90°

γ (deg)	Azimuth C (deg) 0.0	5.0	10.0	15.0	20.0	25.0	30.0	35.0	40.0	45.0	50.0	60.0	75.0	90.0
0.0	218	218	218	218	218	218	218	218	218	218	218	218	218	218
10.0	224	225	226	227	227	226	226	225	225	224	223	221	219	219
20.0	257	260	261	258	254	249	242	235	227	221	213	201	192	189
30.0	299	305	304	298	286	272	256	238	221	205	191	170	154	149
35.0	311	317	315	306	292	273	253	234	215	198	181	156	137	130
40.0	297	304	303	294	277	256	234	214	196	180	166	143	121	113
45.0	275	280	278	267	249	227	205	185	169	156	146	127	107	97
47.5	277	283	279	265	242	217	194	173	158	145	136	119	100	90
50.0	276	282	278	263	239	213	187	166	149	136	127	112	94	84
52.5	274	282	279	263	238	211	185	161	143	129	119	104	87	76
55.0	274	283	281	265	240	212	184	159	140	125	114	98	81	69
57.5	268	278	275	259	234	206	179	154	135	120	109	92	73	62
60.0	265	275	273	257	231	203	176	151	131	115	103	87	68	55
62.5	261	271	270	253	228	200	173	147	127	110	98	81	62	46
65.0	257	268	266	249	225	197	169	143	122	105	92	74	55	38
67.5	247	257	255	240	216	189	161	136	114	97	85	67	46	30
70.0	229	240	240	225	202	177	151	127	106	89	76	60	39	23
72.5	204	218	219	207	187	163	138	115	95	79	68	51	31	16
75.0	136	153	163	164	156	140	121	102	83	69	59	41	22	10
77.5	77	87	91	87	77	65	53	46	42	41	38	23	13	7
80.0	47	54	59	58	51	42	32	22	16	11	10	7	4	4
82.5	26	31	34	34	30	25	19	14	10	7	6	4	3	3
85.0	8	9	10	10	9	8	6	5	4	3	3	2	2	1
87.5	2	2	2	2	2	1	1	1	1	1	1	1	1	0
90.0	1	1	1	1	1	1	1	1	1	1	0	0	0	0
92.5	0	0	0	0	0	0	0	0	0	0	0	0	0	0
95.0	1	1	1	1	1	1	0	0	0	0	0	0	0	0
97.5	1	1	1	1	0	0	0	0	0	0	0	0	0	0
100.0	1	1	1	1	1	1	0	0	0	0	0	0	0	0
102.5	1	1	1	1	1	1	1	0	0	0	0	0	0	0
105.0	1	1	1	1	1	1	1	1	1	1	1	1	1	1
120.0	1	1	1	1	1	1	1	1	1	1	1	0	0	1
135.0	1	1	1	1	1	1	1	1	1	1	1	1	1	1
150.0	1	1	1	1	1	1	1	1	1	1	1	1	1	1
165.0	1	1	1	1	1	1	1	1	1	1	1	1	1	1
180.0	1	1	1	1	1	1	1	1	1	1	1	1	1	1

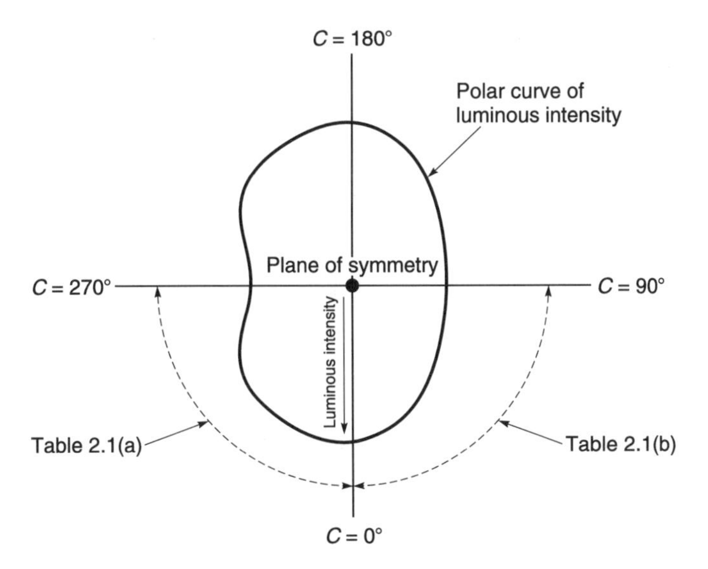

Fig. 2.1 Polar curve showing symmetry of the *I*-table

To apply the *I*-table, certain ancillary data are required by the user. These are:

- luminaire identification;
- identification of coordinate system. For instance, Table 2.11, page 60, is an example of an *I*-table in the *B*, β coordinate system
- attitude of the luminaire when photometered. For instance, for the above *I*-table, which is for a road lighting luminaire, the luminaire was photometrically tested with its spigot axis elevated by 5°;
- the units of the luminous intensity data. For most applications these are in candelas per kilolumen, although for some applications, such as airfield lighting, they may be stated in absolute units.

Table 2.13, page 62, gives a fuller list of the data needed for computer files of *I*-tables.

2.3 Interpolation in the *I*-table

Interpolation is used when a luminous intensity is required in a direction lying between directions given in the *I*-table. Extrapolation, the procedure by which a luminous intensity is found outside the directional limits of the *I*-table, is not recommended. In the overwhelming majority of calculations, interpolation is required in both directions of the chosen coordinate system. It is therefore important that the process chosen is as fast as possible so that the overall calculation time is not unduly lengthy.

Two interpolation procedures are used: linear and quadratic. In linear interpolation (Figure 2.2(a)), a straight line is fitted between two points lying on either side of the required direction and the luminous intensity is calculated from the equation of the line. In quadratic interpolation, a second degree curve is fitted between three points in directions straddling the required direction, and the required value is calculated from the equation of the line (Figure 2.2(b)).

The basic equation used for interpolation is due to Lagrange:

$$y(x) = \sum_{i=1}^{n} y_i \prod_{j=1,\, j\neq i}^{n} \left(\frac{x - x_j}{x_i - x_j} \right) \tag{2.1}$$

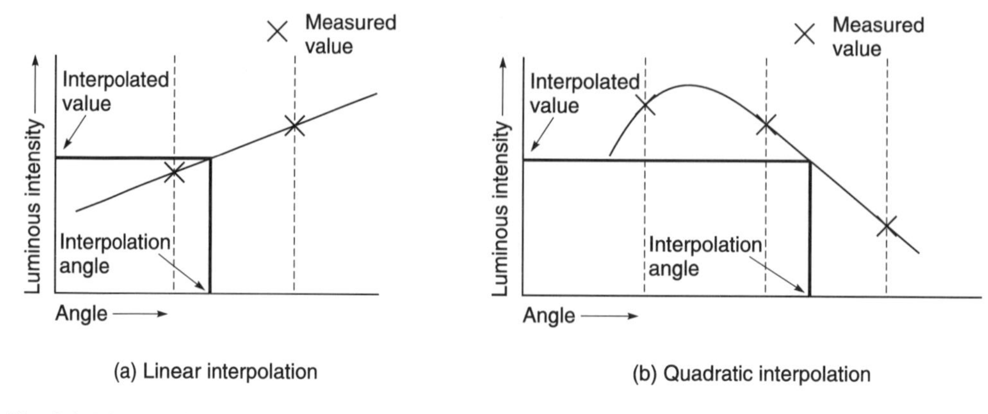

Fig. 2.2 Linear and quadratic interpolation: (a) linear interpolation; (b) quadratic interpolation

where

- $y(x)$ is the interpolated value of y at the point x;
- n is the number of points between which interpolation is required (2 for linear and 3 for quadratic);
- $(x_1, y_1), (x_2, y_2) \ldots (x_j, y_j) \ldots (x_n, y_n)$ are the points between which interpolation is required;
- $\sum$ signifies the summation of the terms;
- $\prod$ signifies the product of the terms.

In the next two sections we describe the application of this equation to linear and quadratic interpolation.

2.3.1 *LINEAR INTERPOLATION*

Figure 2.3 shows the angles required for linear interpolation in the (C, γ) system of coordinates. The m suffixes denote the column number of the I-table and the j suffixes the row number. For instance, C_{m+1} is the value of C in column m+1 of the I-table. The luminous intensity is required at (C, γ). For linear interpolation, we use (2.1) with n equal to 2. This then becomes:

$$y(x) = y_1\left(\frac{x - x_2}{x_1 - x_2}\right) + y_2\left(\frac{x - x_1}{x_2 - x_1}\right) \tag{2.2}$$

This equation can be applied to either C or γ first. When it is first applied to C, C is substituted for x:

$$\begin{aligned} x &= C \\ x_1 &= C_m \\ x_2 &= C_{m+1} \end{aligned} \tag{2.3}$$

From this substitution two constants (K_1 and K_2) can be defined:

$$K_1 = \frac{C - C_{m+1}}{C_m - C_{m+1}} \tag{2.4}$$

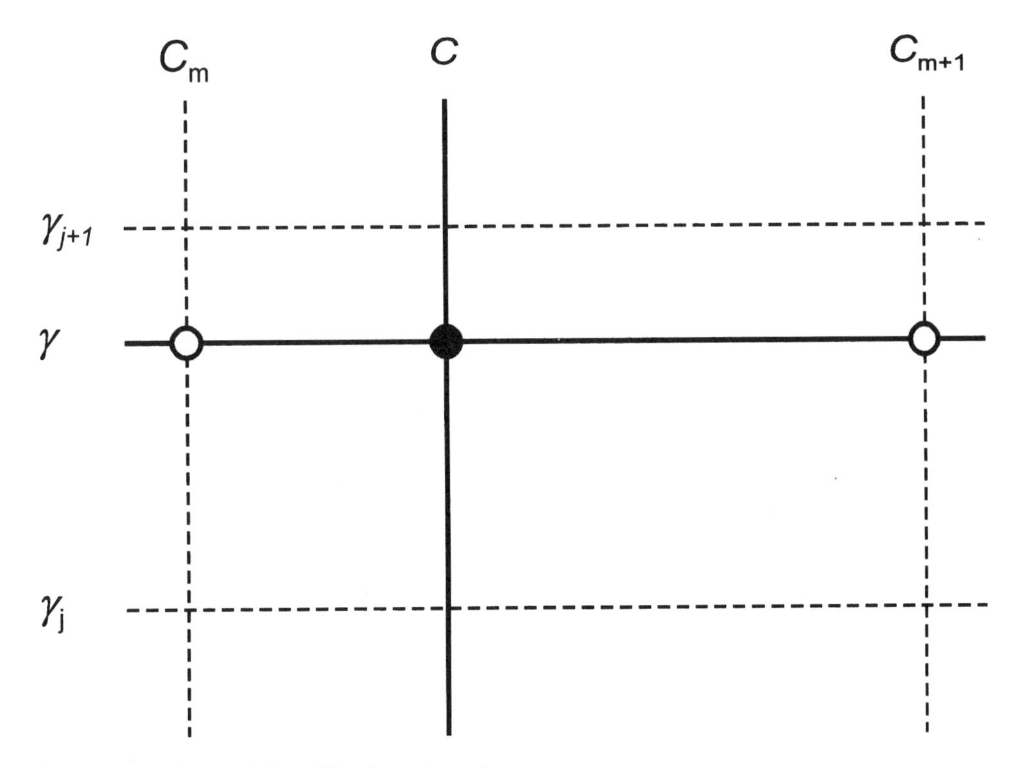

Fig. 2.3 Linear interpolation of luminous intensity

$$K_2 = \frac{C - C_m}{C_{m+1} - C_m}$$

It will be noticed that $K_1 + K_2 = 1$ so that the notation can be simplified by putting $K = K_1$, and $K_2 = 1 - K$.

It is now necessary to introduce additional notation to specify a luminous intensity in a direction. $I(C, \gamma)$ means the luminous intensity in the direction (C, γ), $I(C_m, \gamma_{j+1})$ means the luminous intensity in the direction of C in column m of the I-table, and of γ in row j+1.

We can substitute the following values in (2.2):

$$\begin{aligned} y(x) &= I(C, \gamma_j) \\ y_1 &= I(C_m, \gamma_j) \\ y_2 &= I(C_{m+1}, \gamma_j) \end{aligned} \tag{2.5}$$

After simplification this leads to

$$I(C, \gamma_j) = I(C_m, \gamma_j) + K[I(C_{m+1}, \gamma_j) - I(C_m, \gamma_j)] \tag{2.6}$$

Similarly,

$$I(C, \gamma_{j+1}) = I(C_m, \gamma_{j+1}) + K[I(C_{m+1}, \gamma_{j+1}) - I(C_m, \gamma_{j+1})] \tag{2.7}$$

For interpolation at a constant azimuth, C, a similar procedure produces the final result:

$$I(C, \gamma) = I(C, \gamma_j) + k[I(C, \gamma_{j+1}) - I(C, \gamma_j)] \tag{2.8}$$

where

$$k = \frac{\gamma - \gamma_j}{\gamma_{j+1} - \gamma_j} \tag{2.9}$$

If this procedure is reversed by first carrying out interpolation in the γ cones followed by interpolation in the C planes, the same result will be obtained. This fact can be used to check that a program for linear interpolation is correct.

For the procedure to be followed at the boundary of the I-table see Section 2.3.4, page 42.

2.3.2 QUADRATIC INTERPOLATION

Figure 2.4 shows the angles required for quadratic interpolation of luminous intensity in the direction (C, γ). As mentioned earlier, three angles are required for interpolation at each point. Two of the angles will lie on either side of the required point but the third point has to be chosen to achieve the best accuracy. This is done by following the empirically derived rules given next.

(1) The two tabular angles adjacent to the angle for interpolation are selected for insertion in the interpolation equations and the average calculated.
(2) If the angle for interpolation is smaller than this average then the third tabular angle is the next lower tabular angle (as shown for C in Figure 2.4); if the average is greater than this average then the third tabular angle is the next higher tabular angle (as shown for γ in Figure 2.4).

The formula for quadratic interpolation is obtained by putting n equal to 3 in (2.1), which gives:

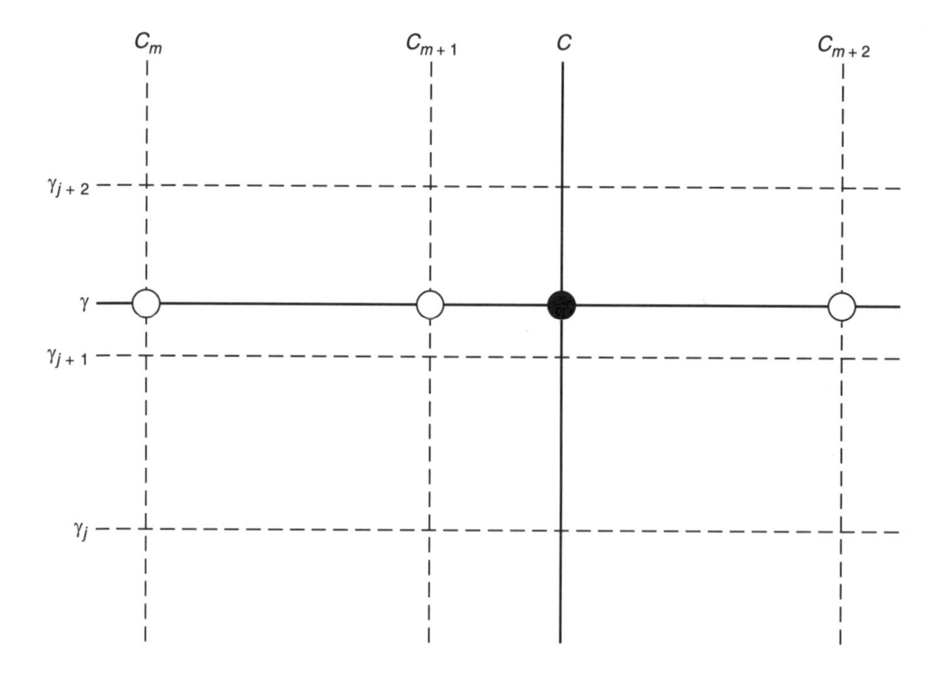

Fig. 2.4 Angles required for quadratic interpolation

$$y(x) = y_1\left(\frac{(x-x_2)(x-x_3)}{(x_1-x_2)(x_1-x_3)}\right) + y_2\left(\frac{(x-x_1)(x-x_3)}{(x_2-x_1)(x_2-x_3)}\right) + y_3\left(\frac{(x-x_1)(x-x_2)}{(x_3-x_1)(x_3-x_2)}\right) \quad (2.10)$$

where it will be noticed that there is circular permutation of the suffixes.

This interpolation can be applied to either C or γ. When it is first applied to C, this parameter is substituted for x in the above equation:

$$\begin{aligned} x &= C \\ x_1 &= C_m \\ x_2 &= C_{m+1} \\ x_3 &= C_{m+2} \end{aligned} \quad (2.11)$$

where

C is the angle at which I is to be found by interpolation;
m, $m+1$, $m+2$ are integers indicating the number of the columns in the I-table;
C_m, C_{m+1}, and C_{m+2} are values of C for the corresponding column numbers. These are chosen such that $C_{m+1}<C<C_{m+2}$ unless m is zero in which case $C_1<C<C_2$.

From this substitution three constants can be defined:

$$\begin{aligned} K_1 &= \frac{(C-C_{m+1})(C-C_{m+2})}{(C_m-C_{m+1})(C_m-C_{m+2})} \\ K_2 &= \frac{(C-C_m)(C-C_{m+2})}{(C_{m+1}-C_m)(C_{m+1}-C_{m+2})} \\ K_3 &= \frac{(C-C_m)(C-C_{m+1})}{(C_{m+2}-C_m)(C_{m+2}-C_{m+1})} \\ &= 1-K_1-K_2 \end{aligned} \quad (2.12)$$

Three equations can then be written allowing evaluation of I at the three values of γ:

$$\begin{aligned} I(C,\gamma_j) &= K_1 I(C_m,\gamma_j) + K_2 I(C_{m+1},\gamma_j) + K_3 I(C_{m+2},\gamma_j) \\ I(C,\gamma_{j+1}) &= K_1 I(C_m,\gamma_{j+1}) + K_2 I(C_{m+1},\gamma_{j+1}) + K_3 I(C_{m+2},\gamma_{j+1}) \\ I(C,\gamma_{j+2}) &= K_1 I(C_m,\gamma_{j+2}) + K_2 I(C_{m+1},\gamma_{j+2}) + K_3 I(C_{m+2},\gamma_{j+2}) \end{aligned} \quad (2.13)$$

For interpolation of the γ angles, further application of (2.1) gives three new constants:

$$\begin{aligned} k_1 &= \frac{(\gamma-\gamma_{j+1})(\gamma-\gamma_{j+2})}{(\gamma_j-\gamma_{j+1})(\gamma_j-\gamma_{j+2})} \\ k_2 &= \frac{(\gamma-\gamma_j)(\gamma-\gamma_{j+2})}{(\gamma_{j+1}-\gamma_j)(\gamma_{j+1}-\gamma_{j+2})} \\ k_3 &= \frac{(\gamma-\gamma_j)(\gamma-\gamma_{j+1})}{(\gamma_{j+2}-\gamma_j)(\gamma_{j+2}-\gamma_{j+1})} \\ &= 1-k_1-k_2 \end{aligned} \quad (2.14)$$

Finally,

$$I(C, \gamma) = k_1 I(C, \gamma_j) + k_2 I(C, \gamma_{j+1}) + k_3 I(C, \gamma_{j+2}) \tag{2.15}$$

which gives the required value of luminous intensity.

In the mathematical procedure described above, interpolation is first carried out for C and then for γ. As for linear interpolation, this procedure may be reversed after appropriate changes have been made to the equations, and the same result obtained. This fact can be used to check that a program is correct.

For the procedure to be followed at the boundary of the I-table see Section 2.3.4.

2.3.3 *CHOICE OF LINEAR OR QUADRATIC INTERPOLATION*

CIE lays down that linear interpolation may be used when the angular intervals are 2.5° or less, otherwise quadratic interpolation is to be used. However, this choice may be modified at the boundary of the I-table as described in the following section.

2.3.4 *INTERPOLATION AT THE BOUNDARY OF THE I-TABLE*

As pointed out on page 34, we may distinguish between two types of I-table boundary: presentational and physical. The type of interpolation differs for each of these.

Consider a case where the boundary is presentational. The method is best illustrated by means of examples. Suppose in Table 2.1(a), page 35, we require the luminous intensity at (C, γ) = (358°, 18°). Since the angular intervals are greater than 2.5°, quadratic interpolation has to be used, for which three values of luminous intensity are required for each interpolation. We need, therefore, to select the appropriate values from Table 2.1(a) and Table 2.1(b). A composite table can be drawn up as shown in Table 2.2.

The three values of C used for interpolation will be 355°, 0°, and 5° since 358° is nearer to 0° than it is to 355°. The same reasoning gives 10°, 20°, and 30° for γ.

The presentational boundary at the nadir (or zenith) of the I-table requires a different treatment. Suppose that the luminous intensity is required in the C meridian in Figure 2.5, which shows a view of the (C, γ) coordinate system as seen in the direction of the nadir. The C meridian illustrated continues along the C + 180° meridian on the 90° – 180° – 270° side of the hemisphere. Because of the bilateral symmetry of the luminaire, the luminous intensities on this meridian will be the same as those on the 360° – C meridian. This meridian can then be used for interpolation.

Worked example Find the luminous intensity at (C, γ) = (48°, 1°) from Tables 2.1(a) and 2.1(b), pages 35 and 36, by interpolation.

Answer Since the angular spacing is more than 2.5°, quadratic interpolation has to be used. Next, we have to decide the third angle to be selected when interpolation is carried out in C and

Table 2.2 Luminous intensities (cd/1000 lm) from Table 2.1

	From Table 2.1(a)	From Table 2.1(b)	
γ(°)	C = 355° ≡ –5°	C = 0°	C = 5°
10	223	224	225
20	253	257	260
30	289	299	305

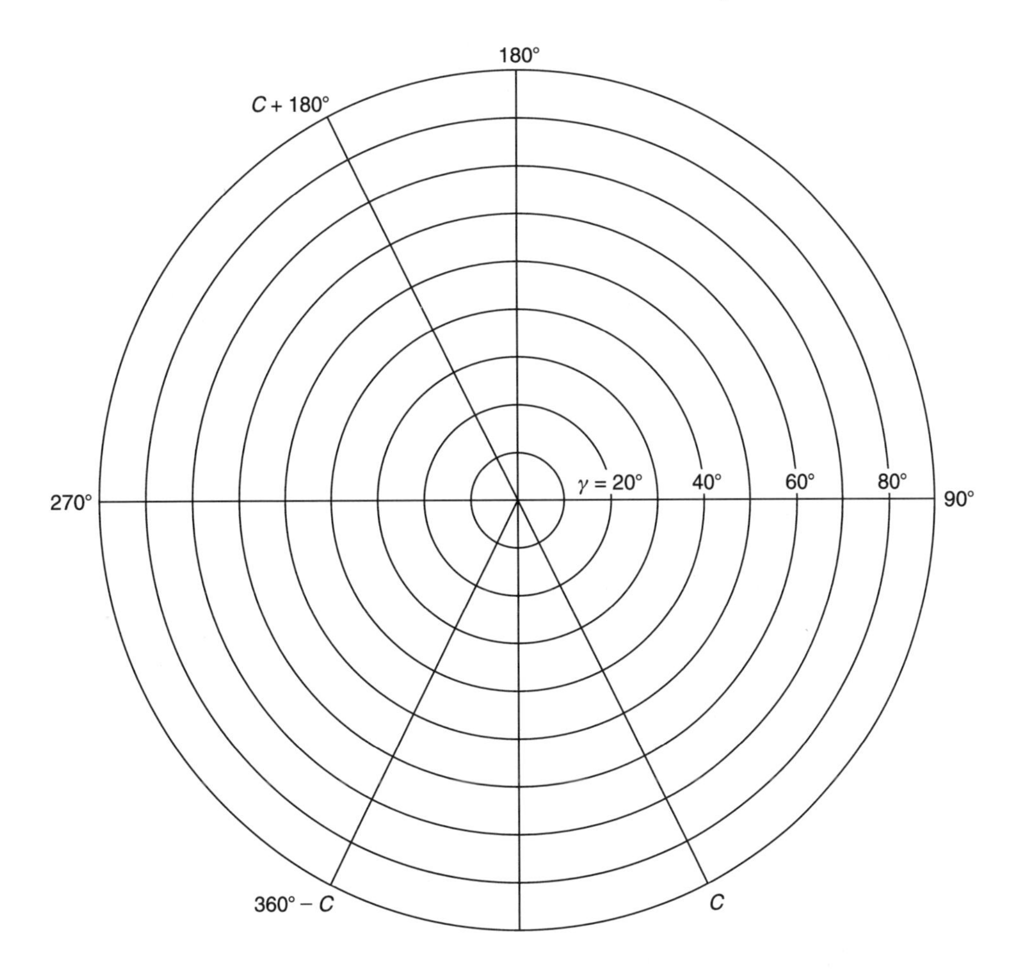

Fig. 2.5 Angles for interpolation viewed towards the nadir of the (C, γ) coordinate system

γ. For C: this angle lies between 45° and 50°, the average of which is 47.5°. Since C lies between this value and 50°, the third value is 60°. For γ: 1° lies closer to 0° than to 5°, so the third value of luminous intensity has to be chosen from the C values on the 90° through 180° to 270° side of the hemisphere. These are: 360° – 45° = 315°, 360° – 50° = 310°, and 360° – 60° = 300°. Table 2.3 is now drawn up. The value of γ_j has been made negative so that the differences between the values are of the same sign when taken in the same direction.

We can now substitute these in the interpolation equations:

$$K_1 = \frac{(C - C_{m+1})\,(C - C_{m+2})}{(C_m - C_{m+1})\,(C_m - C_{m+2})}$$

$$= \frac{(48 - 50)\,(48 - 60)}{(45 - 50)\,(45 - 60)} \tag{2.16}$$

$$= 0.32$$

Table 2.3 Intensity table for quadratic interpolation. Luminous intensities in cd/1000 lm

	$C_m = 45°$	$C_{m+1} = 50°$	$C_{m+2} = 60°$	Source
$\gamma_j = -10°$	204 (from C=315°)	202 (from C=310°)	199 (from C=300°)	Table 2.1(a)
$\gamma_{j+1} = 0°$	218	218	218	Table 2.1(a)
$\gamma_{j+2} = 10°$	224	223	221	Table 2.1(b)

$$K_2 = \frac{(C - C_m)(C - C_{m+2})}{(C_{m+1} - C_m)(C_{m+1} - C_{m+2})} \tag{2.16}$$

$$= \frac{(48 - 45)(48 - 60)}{(50 - 45)(50 - 60)}$$

$$= 0.72$$

$$K_3 = 1 - K_1 - K_2$$

$$= -0.04$$

Now,

$$I(C, \gamma_j) = K_1 I(C_m, \gamma_j) + K_2 I(C_{m+1}, \gamma_j) + K_3 I(C_{m+2}, \gamma_j)$$

So $I(48, 5) = (0.32 \times 0.204) + (0.72 \times 202) + (-0.04 \times 199)$
$= 202.76$ cd/1000 lm

$$I(C, \gamma_{j+1}) = K_1 I(C_m, \gamma_{j+1}) + K_2 I(C_{m+1}, \gamma_{j+1}) + K_3 I(C_{m+2}, \gamma_{j+1})$$

So $I(48, 0) = (0.32 \times 0.218) + (0.72 \times 218) + (-0.04 \times 218)$
$= 218$ cd/1000 lm

$$I(C, \gamma_{j+2}) = K_1 I(C_m, \gamma_{j+2}) + K_2 I(C_{m+1}, \gamma_{j+2}) + K_3 I(C_{m+2}, \gamma_{j+2})$$

So $I(48, -5) = (0.32 \times 0.224) + (0.72 \times 223) + (-0.04 \times 221)$
$= 223.4$ cd/1000 lm

Then,

$$k_1 = \frac{(\gamma - \gamma_{j+1})(\gamma - \gamma_{j+2})}{(\gamma_j - \gamma_{j+1})(\gamma_j - \gamma_{j+2})}$$

$$= \frac{(1 - 0)(1 - 10)}{(-10 - 0)(-10 - 10)}$$

$$= -0.045$$

$$K_2 = \frac{(\gamma - \gamma_j)(\gamma - \gamma_{j+2})}{(\gamma_{j+1} - \gamma_j)(\gamma_{j+1} - \gamma_{j+2})} \tag{2.17}$$

$$= \frac{(1 - (-10))(1 - 10)}{(0 - (-10))(0 - 10)}$$

$$= 0.99$$

$$k_3 = 1 - k_1 - k_2 \quad (2.17) \text{ continued}$$
$$= 1 - 0.045 - 0.99$$
$$= 0.055$$

Finally,

$$I(C, \gamma) = k_1 I(C, \gamma_j) + k_2 I(C, \gamma_{j+1}) + k_3 I(C, \gamma_{j+2})$$

So $$I(48°, 1°) = (-0.045 \times 202.76) + (0.99 \times 218) + (0.055 \times 223.4) \quad (2.18)$$
$$= 219.0 \text{ cd/1000 lm}$$

2.4 Turning the luminaire about the photometric axes in the (*C*, γ) coordinate system

In many applications, particularly road lighting, the luminaire is turned about its photometric axes when it is used in an application. In principle, there are two ways of allowing for this for the calculation of derived data from *I*-tables; by matrix multiplication or by use of the coordinate transformation formulae described in Section 1.3, page 4.

The matrix multiplication method is easier to develop and apply, especially for turning movements about the three axes of a coordinate system, and is for this reason preferred. However, the input is in terms of distances. Where the tilt of an *I*-table has to be changed, the coordinate transformation method is advantageous as the input is purely in terms of angles. For this reason it will also be described.

First, we need to establish sign conventions with regard to the sense of turning and with regard to directions in space.

The conventions are shown in Figure 2.6. The first axis of the luminaire coincides with the poles of the (C, γ) coordinate system, and the two other axes are perpendicular to this and to each other. Conventionally, the third axis of the luminaire is taken along the long axis for interior fluorescent luminaires and in the vertical plane of the spigot axis for road lighting luminaires.

The angle through which the luminaire is turned about the first photometric axis is called *orientation* (υ), about the second photometric axis *tilt* (δ), and about the third photometric axis *rotation* (ψ).[a] The sense in which the luminaire is turned about the axes is such that the system of axes is *right-handed*,[2] that is the turning of a right-handed screw in the positive direction of Ox corresponds to a translation in the sense $y \rightarrow z$. Similarly, the turning of a screw in the positive direction of Oy corresponds to a translation in the sense $z \rightarrow x$, and the turning of a screw in the positive direction of Oz corresponds to a translation in the sense $x \rightarrow y$. These turning directions are indicated by the arrows on the circles for υ, δ, and ψ.

Figure 2.7 shows the coordinate system as it is conventionally portrayed in mathematical texts. Figure 2.7 shows the situation before and after turning has taken place. γ_f is the angle that P subtends after turning has taken place, and the corresponding azimuth is C_f. The luminous intensity directed towards P is I, which after turning (Figure 2.7(b)) becomes I_f.

We will assume that an *I*-table is available with measurements taken with the third luminaire axis horizontal. The correction where there is a *tilt in measurement*, δ_m, will be made later.

[a] The fact that the word *rotation* is used in this specialized sense means that we will avoid using it in other senses; the word *turning* will be used instead.

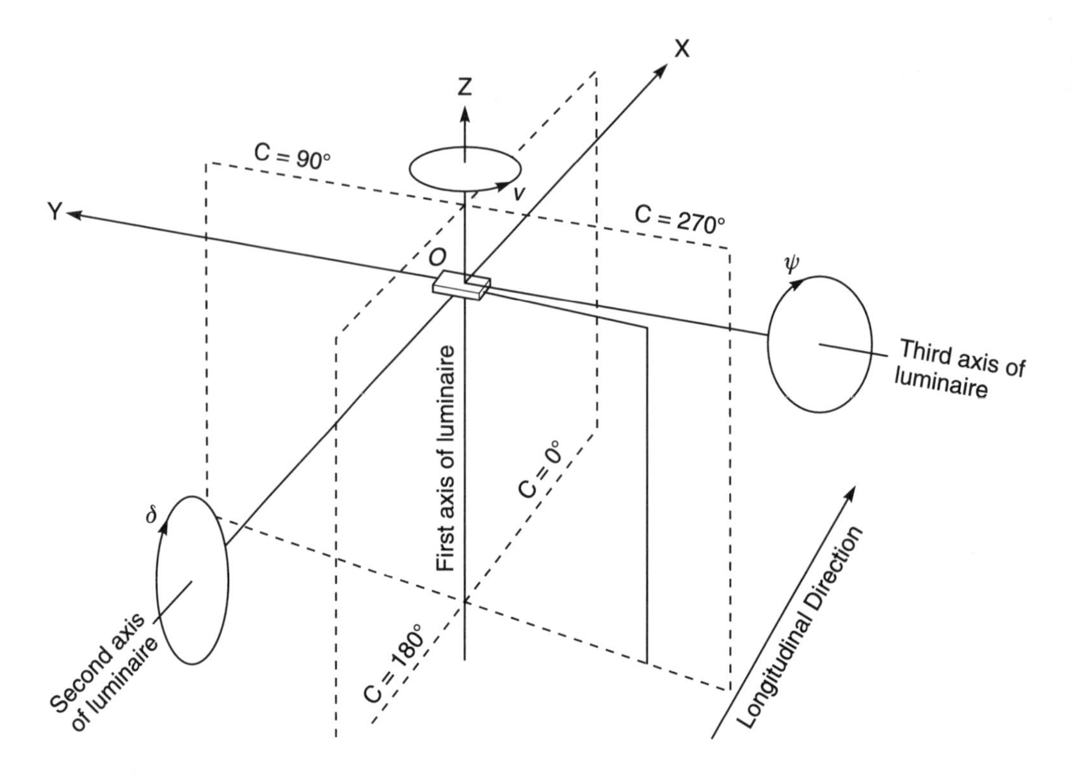

Fig. 2.6 Axes of rotation

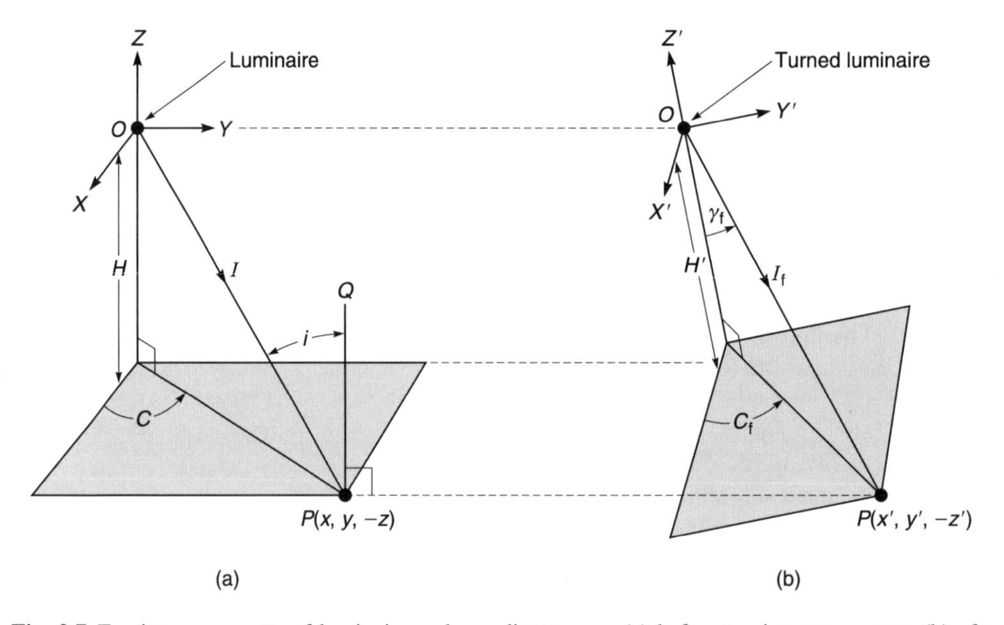

Fig. 2.7 Turning movements of luminaire and coordinate axes: (a) before turning movements; (b) after turning movements

2.4.1 MATRIX REPRESENTATION OF TURNING MOVEMENTS

In the generally available texts on using matrices for representing turning movements, the turning of a vector with respect to a fixed coordinate system is considered. We have the inverse problem; there is a fixed point P (Figure 2.8) and the coordinate system is turned through an angle θ so that the coordinates of the point P, which are (u, v) referred to Ou and Ov, become (u', v') referred to the turned axes, Ou' and Ov'. The problem is to express u' and v' in terms of u, v and θ. This can be done by means of the following matrix multiplication:

$$\begin{bmatrix} u' \\ v' \end{bmatrix} = \begin{bmatrix} \cos\theta & -\sin\theta \\ \sin\theta & \cos\theta \end{bmatrix} \begin{bmatrix} u \\ v \end{bmatrix} = \begin{bmatrix} u\cos\theta - v\sin\theta \\ u\sin + v\cos\theta \end{bmatrix} \tag{2.19}$$

which leads to

$$(u', v') = (u \cos \theta - v \sin \theta, u \sin \theta + v \cos \theta) \tag{2.20}$$

It should be emphasized that the turning movement of the coordinate system takes place in an clockwise direction. In fact, the matrix needed to turn a coordinate system is the transpose of the matrix needed to turn a point about the origin through the same angle.

In applying this to the turning of luminaires, we need three dimensions. If the third axis is Ow, (2.19) becomes:

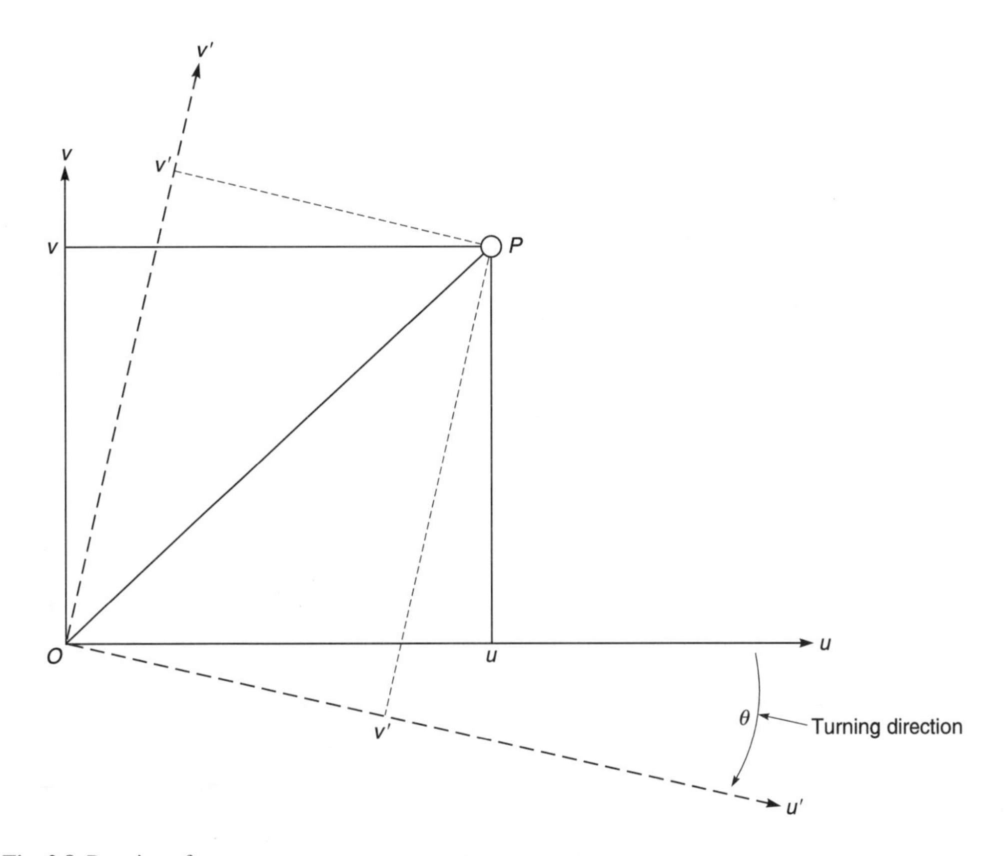

Fig. 2.8 Rotation of axes

$$\begin{bmatrix} u' \\ v' \\ w' \end{bmatrix} = \begin{bmatrix} \cos\theta & -\sin\theta & 0 \\ \sin & \cos & 0 \\ 0 & 0 & 1 \end{bmatrix} \begin{bmatrix} u \\ v \\ w \end{bmatrix} = \begin{bmatrix} u\cos\theta - v\sin\theta \\ u\sin\theta + v\cos\theta \\ w \end{bmatrix} \tag{2.21}$$

which gives $(u', v', w') = (u \cos \theta - v \sin \theta, u \sin \theta + v \cos \theta, w)$.

This establishes the algebra for turning the axes. We now can apply it to turning the (C, γ) coordinate system about the three axes. This is done in two stages; derivation of the matrices for turning the coordinate system about *Ox, Oy* and *Oz* in Figure 2.6, and multiplying them together or composing them in the correct order to give the final result. Particular care has to be taken to determine the correct signs for the direction of turning in each case.

In some texts[3] the order of multiplication of the matrices is reversed and the transpose of the transformation matrix is used. This does not change the result as is shown below:

$$\begin{bmatrix} u' \\ v' \\ w' \end{bmatrix} = \begin{bmatrix} u & v & w \end{bmatrix} \begin{bmatrix} \cos\theta & \sin\theta & 0 \\ -\sin\theta & \cos\theta & 0 \\ 0 & 0 & 1 \end{bmatrix} = \begin{bmatrix} u\cos\theta - v\sin\theta \\ u\sin\theta + \cos\theta \\ w \end{bmatrix} \tag{2.22}$$

Rotation of the luminaire ψ

As already stated, this turning movement takes place about *Ox* (Figure 2.6). Figure 2.9(a) views the luminaire in the direction of the *Ox* and Figure 2.9(b) shows how the polar axis moves in relation to the direction *OP*, which is stationary.

We can now use (2.21) as a pattern for writing the matrix for turning the axes of the coordinate system through the angle ψ. From Figures 2.8 and 2.9(c), we find $Ox \equiv Ou$, $Oz \equiv Ov$, similarly for the primed coordinates, and $\psi \equiv \theta$ so

$$\begin{bmatrix} x' \\ y' \\ z' \end{bmatrix} = \begin{bmatrix} \cos\psi & 0 & -\sin\psi \\ 0 & 1 & 0 \\ \sin\psi & 0 & \cos\psi \end{bmatrix} \begin{bmatrix} x \\ y \\ z \end{bmatrix} = \begin{bmatrix} x\cos\psi - z\sin\psi \\ y \\ x\sin\psi - z\cos\psi \end{bmatrix} \tag{2.23}$$

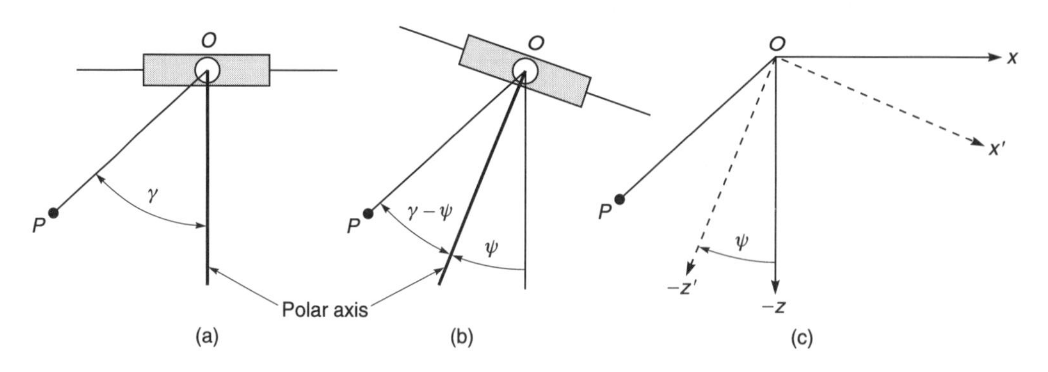

Fig. 2.9 Rotation of the luminaire, viewed along the spigot axis

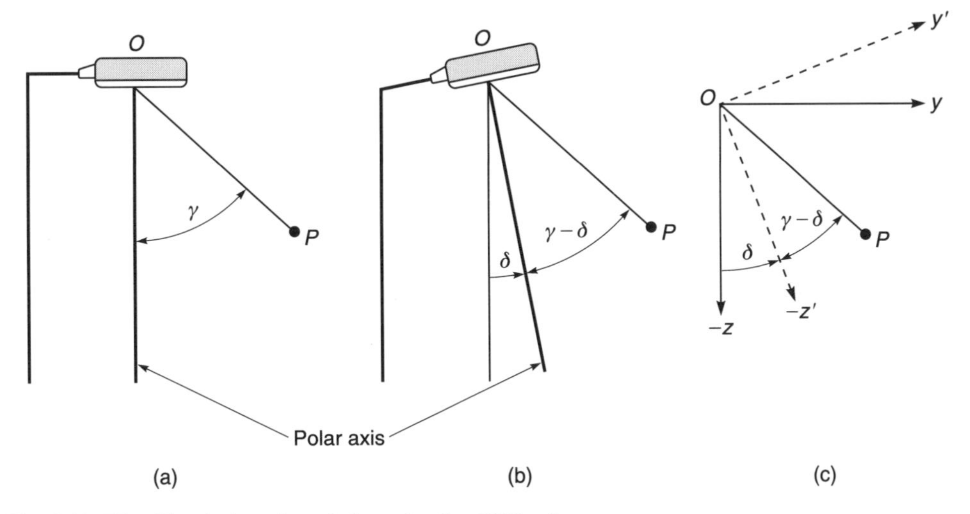

Fig. 2.10 Tilt of luminaire, viewed along the $C = 180°$ axis

Tilt in application δ

Tilt in application is the angular tilt given to a luminaire from its horizontal inclination. This turning movement takes place about Ox in Figure 2.6, page 46. Figure 2.10 shows the angles involved.

OP is the direction in which the luminous intensity is required. Mathematically, the luminaire is regarded as being turned about its photometric centre so that its height above the road is unaltered as is its distance from P. Figure 2.10(a) shows the initial position and it is clear from Figure 2.10(b) that the new angle that OP makes with the downward vertical, or ordinate, is $\gamma - \delta$. Figure 2.10(c) shows the turning movement of the axes.

We can now use (2.21) as a pattern for writing the matrix for turning Oy and Oz through an angle δ.

$$\begin{bmatrix} x' \\ y' \\ z' \end{bmatrix} = \begin{bmatrix} 1 & 0 & 0 \\ 0 & \cos\delta & \sin\delta \\ 0 & -\sin\delta & \cos\delta \end{bmatrix} \begin{bmatrix} x \\ y \\ z \end{bmatrix} = \begin{bmatrix} x \\ y\cos + z\sin\delta \\ -y\sin\delta + z\cos\delta \end{bmatrix} \tag{2.24}$$

Worked example A point has (x, y) coordinates (0, 10) in metres. If a luminaire is mounted at a height of 10 m and tilted up by 15°, what are the values of γ and C which have to be looked up in the I-table? What is the angle of incidence and azimuth of the light at the point?

Answer The data given are:

$$\begin{aligned} x &= 0 \\ y &= 10 \text{ m} \\ z &= -10 \text{ m} \\ \delta &= 15° \end{aligned}$$

Note that z is negative.

From (2.24)

$$
\begin{aligned}
x' &= x \\
&= 0 \text{ m} \\
y' &= y \cos \delta + z \sin \delta \\
&= 10 \cos 15 - 10 \sin 15 \\
&= 7.071 \text{ m} \\
z' &= -y \sin \delta + z \cos \delta \\
&= -10 \sin 15 - 10 \cos 15 \\
&= -12.247 \text{ m}
\end{aligned}
$$

From Figure 2.7(b), we find:

$$
\begin{aligned}
\gamma &= \tan^{-1} \frac{y'}{-z'} \\
&= \tan^{-1} \frac{7.071}{12.247} \\
&= 30°
\end{aligned}
$$

In Figure 2.7(a) we see that the angle of incidence $\angle OPQ$ is i, so

$$
\begin{aligned}
i &= \tan^{-1} \frac{-z}{\sqrt{x^2 + y^2}} \\
&= \tan^{-1} \frac{10}{\sqrt{10^2}} \\
&= 45°
\end{aligned}
$$

As a check on the previous result;

$$
\begin{aligned}
\gamma &= i - \delta \\
&= 45° - 15° \\
&= 30°
\end{aligned}
$$

which is the same as before.

From Figure 2.7(b):

$$
\begin{aligned}
C &= \tan^{-1} \frac{y'}{x'} \\
&= \tan^{-1} \frac{7.071}{0} \\
&= 90°
\end{aligned}
$$

Orientation of the luminaire v This turning movement takes place about Oz in Figure 2.6, page 46. Figure 2.11 shows the angles involved. Note that the luminaire is viewed from below so that the diagram is in conformity with the previous similar diagrams, where the line of sight is from the negative axis towards the positive axis.

OP is the direction in which the luminous intensity is required. Figure 2.11(a) shows the initial position and it is clear from Figure 2.11(b) that the new angle that *OP* makes with the $C = 0$ axis is $C - v$. Figure 2.11(c) shows the turning movement of the axes.

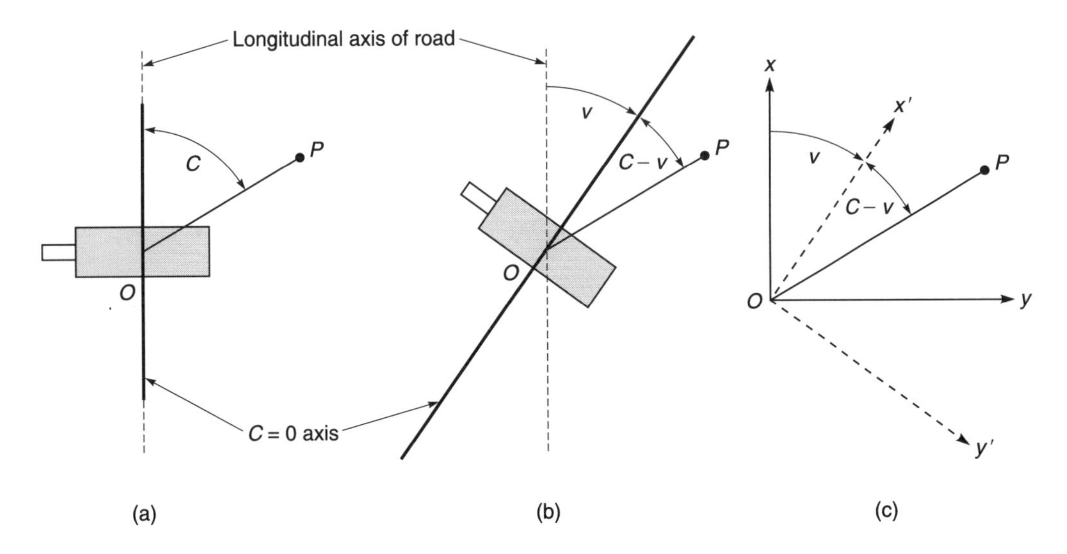

Fig. 2.11 Orientation of the luminaire, as seen from below

Using similar reasoning to that used for the previous turning movements we find:

$$\begin{bmatrix} x' \\ y' \\ z' \end{bmatrix} = \begin{bmatrix} \cos v & \sin v & 0 \\ -\sin v & \cos v & 0 \\ 0 & 0 & 1 \end{bmatrix} \begin{bmatrix} x \\ y \\ z \end{bmatrix} = \begin{bmatrix} x\cos v - y\sin v \\ x\sin v + z\cos v \\ z \end{bmatrix} \tag{2.25}$$

which gives:

$$\begin{aligned} x' &= x \cos v - y \sin v \\ y' &= x \sin v + y \cos v \\ z' &= z \end{aligned} \tag{2.26}$$

2.4.2 COMPOSITION OF TURNING MOVEMENTS

When turning movements about different axes are applied in succession to a luminaire, it is important that they are applied in the correct order. Rotation has to be applied first, then tilt, followed by orientation. If these are interchanged an incorrect result will be obtained. Rotation has to be applied first since, for this turning movement, the luminaire has to be turned about its own axis, which would be impossible if one of the other turning movements were applied first. Tilt has to be applied next as the movement is in the vertical plane through the luminaire, which would be impossible if orientation were applied first. So the final matrix is:

$$\begin{bmatrix} x' \\ y' \\ z' \end{bmatrix} = \begin{bmatrix} \cos\psi & 0 & -\sin\psi \\ 0 & 1 & 0 \\ \sin\psi & 0 & \cos\psi \end{bmatrix} \begin{bmatrix} 1 & 0 & 0 \\ 0 & \cos\delta & \sin\delta \\ 0 & -\sin\delta & \cos\delta \end{bmatrix} \begin{bmatrix} \cos v & \sin v & 0 \\ -\sin v & \cos v & 0 \\ 0 & 0 & 1 \end{bmatrix} \begin{bmatrix} x \\ y \\ z \end{bmatrix} \tag{2.27}$$

This leads to:

$$
\begin{aligned}
x' &= x(\cos v \cos \psi - \sin v \sin \delta \sin \psi) + y(\sin v \cos \psi + \cos v \sin \delta \sin \psi) \\
&\quad -z(\cos \delta \cos \psi) \\
y' &= -x \sin v \cos \delta + y \cos v \cos \delta + z \sin \delta \\
z' &= x(\cos v \sin \psi + \sin v \sin \delta \cos \psi) + y(\sin v \sin \psi - \cos v \sin \delta \cos \psi) \\
&\quad + z \cos \delta \cos \psi
\end{aligned}
\tag{2.28}
$$

z is measured upwards from the luminaire, which differs from the usual convention, where H is measured upwards from the plane of interest. We therefore substitute H for $-z$, and $H' = -z'$ (Figure 2.7), and the equations become:

$$
\begin{aligned}
x' &= x(\cos v \cos \psi - \sin v \sin \delta \sin \psi) + y(\sin v \cos \psi + \cos v \sin \delta \sin \psi) \\
&\quad + H(\cos \delta \cos \psi) \\
y' &= -x \sin v \cos \delta + y \cos v \cos \delta - H \sin \delta \\
H' &= -x(\cos v \sin \psi + \sin v \sin \delta \cos \psi) - y(\sin v \sin \psi - \cos v \sin \delta \cos \psi) \\
&\quad + H \cos \delta \cos \psi
\end{aligned}
\tag{2.29}
$$

The reader should note that when P is in a horizontal plane above the photometric centre of the luminaire, H is negative.

From Figure 2.7, page 46 we can determine C:

$$
C = \tan^{-1} \frac{y'}{x'} \tag{2.30}
$$

The correct quadrant in which to place C is determined from Figure 2.12.

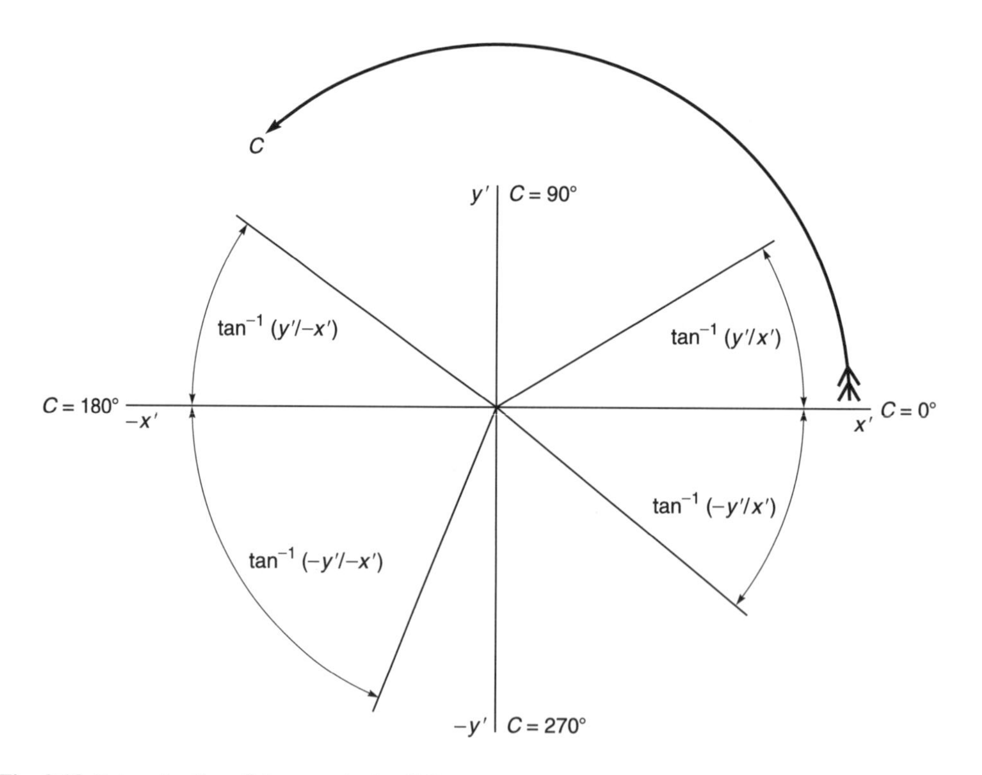

Fig. 2.12 Determination of the magnitude of C

$$\text{If } x' > 0 \text{ and } y' \geq 0 \text{ then } C' = \tan^{-1}\frac{y'}{x'}$$

$$\text{If } x' < 0 \text{ and } y' > 0 \text{ then } C' = \tan^{-1}\frac{y'}{x'} + 180°$$

$$\text{If } x' < 0 \text{ and } y' \leq 0 \text{ then } C' = \tan^{-1}\frac{y'}{x'} + 180° \tag{2.31}$$

$$\text{If } x' > 0 \text{ and } y' \leq 0 \text{ then } C' = \tan^{-1}\frac{y'}{x'} + 360°$$

However, in *I*-tables *C* has the limited range $270° \leq C < 360°$ and $0° \leq C \leq 90°$ so Table 2.4 is used to bring *C* within these ranges.

Also,

$$\gamma = \tan^{-1}\frac{\sqrt{(x')^2 + (y')^2}}{H'} \tag{2.32}$$

As calculated, this will be in the range $-90° \leq \gamma \leq 90°$. To find the correct quadrant we use Table 2.5.

2.4.3 CORRECTING THE I-TABLE FOR TILT

This method is useful when a luminaire has been measured at one tilt and an *I*-table is required at another tilt. Figure 2.13 is a schematic showing the principle involved. As in the treatment

Table 2.4 Values of *C* to be used in *I*-table

Calculated value of *C* (deg)	*C* to be used in *I*-table (deg)
$0 \leq C \leq 90$	C
$90 < C \leq 180$	$180 - C$
$180 < C \leq 270$	$540 - C$
$270 \leq C < 360$	C
360	0

Table 2.5 Calculation of γ

H'	γ (deg)	Range of γ (deg)
> 0	$\tan^{-1}\frac{\sqrt{(x')^2 + (y')^2}}{H'}$	$0 \geq \gamma < 90$
0	90	90
< 0	$\tan^{-1}\frac{\sqrt{(x')^2 + (y')^2}}{H'} + 180$	$90 > \gamma \leq 180$

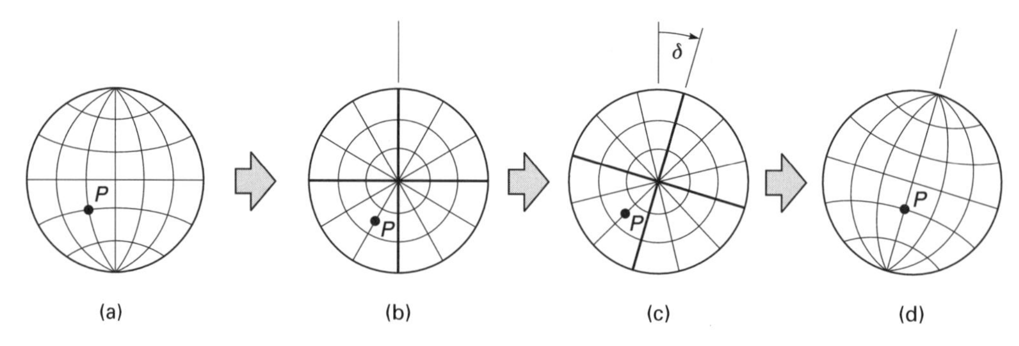

Fig. 2.13 Correction for tilt: (a) point in (C, γ) coordinates; (b) point transferred to (B, β) coordinates; (c) web tilted; (d) point transferred to (C, γ) coordinates

using matrices, the point P is stationary and we have to consider the turning of the coordinate web. Basically the method consists of expressing the position of the point P in (B, β) coordinates (Figure 2.13(b)), allowing for the tilt (Figure 2.13(c)), and transferring the point back to (C, γ) coordinates (Figure 2.13(d)).

Transferring to (B, β) coordinates

For this we use the formulae:

$$\tan B = \sin C_0 \times \tan \gamma_0$$
$$\sin \beta = \cos C_0 \times \sin \gamma_0 \tag{2.33}$$

In these equations the suffix zero is used to indicate that the value to which it is attached is the initial value, before the tilt has been applied.

A computer will not be able to evaluate the first equation when γ is equal to 90°. However, for this case B is equal to 90°. β as determined from these formulae will lie in the first quadrant, which is correct as we are considering only one half of the coordinate system since bilateral symmetry is assumed. $90° \leq B \geq 90°$ has to be assigned to the correct quadrant, which is given by B'' in Table 2.6.

Allowing for tilt

$$B_\delta = B'' + \delta \tag{2.34}$$

where

δ is the angle of tilt;
B_δ is the angle of the B plane after tilt has been applied.

Table 2.6 Conditions for determining the quadrants for B''

γ_0 (deg)	C_0 (deg)	B'' (deg)
≤90	≤90	B
>90	≤90	$B + 180$
<90	≥270	B
≥90	≥270	$B - 180$
90	0	Tilt has no effect

Table 2.7 Conditions for finding the quadrant of B', γ', and C'

	Conditions to be fulfilled	Action
A	$B_\delta = -180°$	$\gamma = 180°$. $C = C_0$
B	$B_\delta = 180°$	$\gamma = \gamma_0$. $C = C_0$
C	$B_\delta > 180°$ and $B'' \leq 180°$	$B'_\delta = -360° + B + \delta$. Go to (2.39)
D	$B_\delta < -180°$, $B'' \geq -180°$, and $\delta < 0$	$B'_\delta = 360 + B + \delta$. Go to (2.39)
E	All other conditions	$B'_\delta = B$. Go to (2.39)

Table 2.8 Conditions for finding the quadrant of C and γ

B (deg)	C (deg)	γ (deg)
< 0	$C'' + 180$	γ''
$\lvert B \rvert > 90$	C''	$180 - \gamma''$
All other conditions	C''	γ''

Transferring to (C, γ) coordinates

In Table 2.7, C and γ are the final values of these variables. Lines A and B cover the case when γ is equal to 180°. Clearly no further calculation is needed to find C or γ. Lines C and D cover the case of the B plane moving through the zenith when δ is added to or subtracted from B. Line E deals with all other cases, for which it is necessary to apply the transformation formulae:

$$C'' = \tan^{-1}\left(\frac{\sin B'_\delta}{\tan \beta}\right) \tag{2.35}$$

$$\gamma'' = \cos^{-1}(\cos B'_\delta \times \cos \beta)$$

It should be noted that when $\beta = 0$ then $C'' = 90°$, and when $\beta = 90°$ then $C'' = 0$.

Table 2.8 takes account of the value of B for finding C and γ from C'' and γ''.

Worked example A road lighting luminaire is mounted at a height of 10 m. The illuminance is required at a point P situated at (x, y) coordinates (3, 4) in metres. An I-table is available for the luminaire tilted at 5°. What are the (C, γ) coordinates of P if the luminaire is used at zero tilt?

Answer To solve this problem we may use either the matrix or the coordinate transformation methods. For illustration purposes we shall use both methods.

Matrix method

For this we need to use (2.29), page 52, in which:

$$H = 10 \text{ m}$$
$$x = 3 \text{ m}$$
$$y = 4 \text{ m}$$
$$\delta = -5°$$
$$\psi = 0°$$
$$\nu = 0°$$

Hence:

$$x' = x(\cos \nu \cos \psi - \cos \nu \sin \delta \sin \psi) + y(\sin \nu \cos \psi + \sin \nu \sin \delta \sin \psi) + H(\cos \nu \sin \psi)$$

$$= 3(1 + 0) + 4(0) + 10(0)$$
$$= 3 \text{ m}$$

$$y' = -x \sin v \cos \delta + y \cos v \cos \delta - H \sin \delta$$
$$= -3(0) + 4(\cos 0 \times \cos -5) - 10 \sin -5$$
$$= 4.856336 \text{ m}$$

$$H' = -x(\cos v \sin \psi + \sin v \sin \delta \cos \psi) - y(\sin v \sin \psi - \cos v \sin \delta \cos \psi) + H(\cos v \cos \psi)$$
$$= -3(0 + 0) - 4(0 - \cos 0 \times \sin -5 \times \cos 0) + 10(\cos -5 \times \cos 0)$$
$$= 9.613324 \text{ m}$$

As $x' > 0$ and $y' > 0$, $C = \tan^{-1}(y'/x')$ from (2.31), so $C = \tan^{-1}(3/4.856)$ or 46.0610°.

Also,

$$\gamma = \tan^{-1} \frac{\sqrt{(x')^2 + (\gamma')^2}}{H'}$$
$$= \tan^{-1} \frac{\sqrt{(3)^2 + (4.856)^2}}{10}$$
$$= 22.7493°$$

Coordinate transformation method

To use this method we need to find C and γ.

$$C = \tan^{-1} \frac{y}{x}$$
$$= \tan^{-1} \frac{4}{3}$$
$$= 53.1301°$$

$$\gamma = \tan^{-1} \frac{\sqrt{x^2 + y^2}}{H}$$
$$= \tan^{-1} \frac{\sqrt{3^2 + 4^2}}{10}$$
$$= 26.5631°$$

The quadrant for C is checked by using equations similar to those in (2.31). In this case it is the first quadrant.

$$\tan B = \sin C \times \tan \gamma$$
$$= \sin 53.1301 \times \tan 26.5651$$
$$= 0.2974$$
$$\sin \beta = \cos C \times \sin \gamma$$
$$= \cos 53.1301 \times \sin 26.5651$$
$$= 0.2683$$

From these results we find $B = 21.8014°$ and $\beta = 15.5648°$. Also from Table 2.6 it is evident that B'' lies in the first quadrant and is equal to 21.0814°.

From (2.34):

$$\begin{aligned} B_\delta &= B'' + \delta \\ &= 21.8014 + (-5) \\ &= 16.8014° \end{aligned}$$

We now consider Table 2.7 to find which condition applies for the next step in the calculation. Line E applies, so that values can now be entered in (2.35),

$$\begin{aligned} C'' &= \tan^{-1}\left(\frac{\sin B'_\delta}{\tan \beta}\right) \\ &= \tan^{-1}\left(\frac{\sin 16.8014}{\tan 15.5648}\right) \\ &= 46.061° \end{aligned}$$

$$\begin{aligned} \gamma'' &= \cos^{-1}(\cos B'_\delta \times \cos \beta) \\ &= \cos^{-1}(\cos 16.8014 \times \cos 15.5648) \\ &= 22.7493° \end{aligned}$$

From Table 2.8, where the last line applies, we find that $C = 46.061°$ and $\gamma = 22.7493°$, which agree with the results obtained by the matrix method.

2.5 Turning the luminaire about the photometric axes in the (*B*, *β*) coordinate system

The (*B*, *β*) coordinate system is used for intensity tables for floodlights. This is convenient for calculation purposes because, when the floodlight is tilted about the polar axis for aiming, only a change in *B* is required. In addition, orientation can easily be allowed for because this can be directly calculated in terms of *β*. More difficult is allowing for inclining the polar axis, which is hardly, if ever, required. It may be achieved by using (2.33, page 54), and then using the (x', y', z') coordinates to find the (*B*, *β*) coordinates to be looked up in the intensity table. This is dealt with in detail in Section 12.5, page 417.

2.6 Calculation of luminous flux from *I*-tables

In this section we will consider the calculation of luminous flux from *I*-tables in the *C*, *γ* and B, *β* coordinate systems. This will be done by means of examples worked on spreadsheets, with explanations.

2.6.1 *C*, *γ* COORDINATE SYSTEM

Spreadsheet	**Column**	**Explanation**
Table 2.9	1	For general purpose luminaires, readings are usually taken at 5° intervals in *γ*. The zones for the calculation of luminous flux are then taken as subtending 10°, with the luminous intensity at each mid-zone angle being taken as the average for that zone.

Table 2.9 Calculation of zonal luminous flux on the *C*, *γ* coordinate system

1	2	3	4	5	6	7	8
Angle	Azimuth (°)				Average	Zone	Luminous
γ°	0	30	60	90	intensity	factor	flux
0	150.00	150.00	150.00	150.00	150.00		
5	149.86	149.71	149.57	149.43	149.69	0.10	14.29
10	149.43	148.86	148.29	147.73	148.77		
15	148.72	147.44	146.18	144.93	147.24	0.28	41.74
20	147.72	145.48	143.27	141.09	145.12		
25	146.44	142.97	139.58	136.28	142.44	0.46	65.93
30	144.89	139.95	135.18	130.58	139.21		
35	143.06	136.44	130.12	124.10	135.48	0.63	85.11
40	140.95	132.45	124.47	116.96	131.29		
45	138.58	128.03	118.29	109.28	126.67	0.77	98.10
50	135.95	123.21	111.67	101.20	121.66		
55	133.05	118.02	104.68	92.86	116.33	0.90	104.37
60	129.90	112.50	97.43	84.38	110.71		
65	126.51	106.70	89.99	75.89	104.86	0.99	104.09
70	122.87	100.65	82.45	67.54	98.84		
75	119.00	94.41	74.90	59.42	92.68	1.06	98.05
80	114.91	88.02	67.43	51.65	86.44		
85	110.59	81.54	60.12	44.32	80.18	1.09	87.48
90	106.07	75.00	53.03	37.50	73.93	Sum 0°–90°	699.14
95	101.34	68.46	46.25	31.25	67.74	1.09	73.91
100	96.42	61.98	39.84	25.61	61.66		
105	91.31	55.59	33.84	20.60	55.71	1.06	58.94
110	86.04	49.35	28.31	16.24	49.94		
115	80.59	43.30	23.27	12.50	44.39	0.99	44.06
120	75.00	37.50	18.75	9.38	39.06		
125	69.26	31.98	14.77	6.82	34.00	0.90	30.50
130	63.39	26.79	11.32	4.79	29.22		
135	57.40	21.97	8.41	3.22	24.75	0.77	19.17
140	51.30	17.55	6.00	2.05	20.59		
145	45.11	13.56	4.08	1.23	16.76	0.63	10.53
150	38.82	10.05	2.60	0.67	13.28		
155	32.47	7.03	1.52	0.33	10.15	0.46	4.70
160	26.05	4.52	0.79	0.14	7.38		
165	19.58	2.56	0.33	0.04	4.97	0.28	1.41
170	13.07	1.14	0.10	0.01	2.94		
175	6.54	0.29	0.01	0.00	1.28	0.10	0.12
180	0.00	0.00	0.00	0.00	0.00	Sum 0°–180°	912.57

Spreadsheet	**Column**	**Explanation**
Table 2.9	2 to 5	For fluorescent luminaires, readings are conventionally taken in just one quadrant as being representative of all four quadrants. The luminous intensity readings are in unscaled units unless the photometer has been scaled before the measurement of the luminaire by means of the bare lamp. See Section 2.2, page 34 for an *I*-table for a road lighting luminaire.

Table 2.9 6 If readings are taken in all four quadrants an unweighted average can be taken, but in the illustration a weighted average is required. The luminous intensities at 0° and 90° would be duplicated at 180° and 270° of azimuth respectively, whereas at the remaining angles the luminous intensities would be replicated four times. The weighted average I_{AV} is therefore given by

$$I_{AV} = \frac{I_0 + I_{30} + I_{60} + I_{90} + I_{120} + I_{150} + \ldots + I_{330}}{12}$$

$$= \frac{I_0 + 2I_{30} + \ldots + 2I_{60} + I_{90}}{6}$$

Table 2.9 7 The zone factors (ZF) are calculated from one of the formulae in Section 1.7.1, page 25.

$$\text{zone factor} = 4\pi \sin\left(\frac{\gamma_2 - \gamma_1}{2}\right) \sin\left(\frac{\gamma_2 + \gamma_1}{2}\right)$$

$$= 4\pi \sin 5° \sin \gamma_{AV}$$

where γ_2 and γ_1 are the upper and lower bounds of a zone respectively, and γ_{AV} is the angle of the mid-zone luminous intensity, that is, 5°, 15°, 25°, etc.

Table 2.9 8 The average luminous intensity for each zone is multiplied by ZF to give the luminous flux in arbitrary units. These are summed to give the luminous flux for the upper and lower hemispheres.

Table 2.10 2 The luminous flux values for the upper and lower hemispheres are transferred from Table 2.9. Also entered is the bare lamp luminous flux that is obtained from a separate set of measurements made on the bare lamp without any changes being made to the sensitivity of the photometer.

Table 2.10 3 The total, up, and down light output ratios are found by dividing by the bare lamp luminous flux.

Table 2.10 4 The scaling factor to convert the luminous intensity readings into candelas per kilolumen (cd/klm) is found by dividing 1000 by the bare lamp luminous flux in unscaled units. This can be used to convert the unscaled luminous intensity readings to cd/klm by multiplication.

Table 2.10 Calculation of light output ratios and scaling factor

1	2	3	4
Angular zone (°)	Luminous flux unscaled units	Light output ratios	Scale factor
0–90	699.14	46.13	0.6598
90–180	213.43	14.08	
0–180	912.57	60.21	
Bare lamp	1515.66		

Table 2.11 *I*-table based on the *B*, *β* coordinate system

1	2	3	4	5	6	7	8	9	10	11
B (°) of mid-zone	*B* of mid-zone (deg)									
	22.5	17.5	12.5	7.5	2.5	–2.5	–7.5	–12.5	–17.5	–22.5
	Luminous intensity (cd/klm)									
20	18	25	30	35	39	35	34	25	24	17
15	20	188	225	336	445	440	330	227	181	21
10	25	307	806	1888	2566	2560	1893	818	313	23
5	30	406	1112	2784	3241	3243	2788	1120	412	31
0	45	665	1886	3960	4603	4607	3956	1895	666	46
–5	40	500	1605	3602	4300	4308	3609	1605	509	41
–10	38	445	996	2244	2866	2869	2244	990	444	39
–15	31	245	556	1554	1886	1880	1547	563	249	30
–20	26	151	188	218	288	285	220	180	150	25
–25	18	32	55	111	145	145	111	51	30	18

2.6.2 B, β COORDINATE SYSTEM

Spreadsheet	**Column**	**Explanation**
Table 2.11	1	Only the mid-zonal *B* angles are shown. This is to keep the presentation of the table small. Normally, luminous intensity readings would be taken at the boundary of the zones, that is at 0°, 5°, 10°, etc. to enable an accurate isocandela diagram to be drawn. *B* angles below the beam are taken as negative.
Table 2.11	2 to 11	Only the mid-zonal *β* angles are shown, once again to keep the presentation of the table small. Luminous intensities at the boundaries need to be recorded to enable an accurate isolux diagram to be drawn as for the *B* angles.
Table 2.12	2 to 6	The luminous intensities from Table 2.11 are averaged in these columns for corresponding values of *β* at any given *B*.
Table 2.12	7 to 11	The luminous flux in each of the 'box' zones is calculated. The formula from Section 1.7.2 page 28 is used.

flux = zone factor × average intensity

$$= 4\pi \sin\left(\frac{\beta_2 - \beta_1}{2}\right) \cos\left(\frac{\beta_2 - \beta_1}{2}\right) \times \left(\frac{\Delta B}{360}\right) I_{AV}$$

$$= 4\pi \sin\frac{\Delta\beta}{2} \cos\beta_{AV} \times \left(\frac{5°}{360°}\right) \times I_{AV}$$

where β_1 and β_2 are the angular bounds of the zone; ΔB, β_{AV} is the average of β_1, and β_2, in degrees of arc, is the angular spacing between the *B* planes; $\Delta\beta$ is the angular spacing between the *β* cones; and I_{AV} is the average luminous intensity in a zone. I_{AV} is taken to be equal to the luminous intensity in the centre of a zone.

Table 2.12 Calculation of zonal luminous flux for the B, β coordinate system

1	2	3	4	5	6	7	8	9	10	11	12
B (°)	B (deg) of mid-zone					B (deg) of mid-zone					
of mid-	2.5	7.5	12.5	17.5	22.5	2.5	7.5	12.5	17.5	22.5	
zone	Average luminous intensity (cd/klm)					Luminous flux (lm/klm lamp flux)					Sums
22.5	37	35	28	25	18	0.3	0.3	0.2	0.2	0.1	1.1
17.5	443	333	226	185	21	3.4	2.5	1.7	1.4	0.2	9.2
12.5	2563	1891	812	310	24	19.5	14.4	6.2	2.4	0.2	42.6
7.5	3242	2786	1116	409	31	24.7	21.2	8.5	3.1	0.2	57.7
2.5	4605	3958	1891	666	46	35.1	30.1	14.4	5.1	0.3	85.0
–2.5	4304	3606	1605	505	41	32.8	27.4	12.2	3.8	0.3	76.6
–7.5	2868	2244	993	445	39	21.8	17.1	7.6	3.4	0.3	50.1
–12.5	1883	1551	560	247	31	14.3	11.8	4.3	1.9	0.2	32.5
–17.5	287	219	184	151	26	2.2	1.7	1.4	1.1	0.2	6.6
–22.5	145	111	53	31	18	1.1	0.8	0.4	0.2	0.1	2.7
					Sums	155.2	127.3	56.8	22.6	2.2	364.2

Table 2.12 — The *negative B* angles, contrary to the usual mathematical convention, are shown on the right-hand side to be in conformity with Figure 1.6, page 7.

2.7 File formats for the electronic transfer of luminaire photometric data

This section deals with the way in which photometric data are stored in an electronic file for use in computer programs. The medium on which the data are stored may be any one of those which can be used by a computer, such as a disk. There is obviously an advantage to be gained if file formats from all sources have the same presentation, since this facilitates interchange of data.

A number of national and international bodies have published or are in the process of publishing file formats. These bodies include CIE, CIBSE, CEN and IESNA. Whilst their file formats have a strong family resemblance, no doubt as a result of their being based on the IESNA file format,[4] which was the first in the field, there are differences that have come about because of the different needs in different countries and organizations. The purpose of this section is to describe their main features.

The file format can conveniently be considered in three parts. The first part consists of the identification of the luminaire, together with details of its mechanical and electrical features. These may include the lamp type, ballast lumen factor, input voltage, as well as the luminous areas of the luminaire in directions required for the calculation of discomfort glare together with a shape code to indicate the shape of the luminaire, also used in the calculation of discomfort glare. The second part gives numerical details required for the interpretation of the luminous intensity data that follow and constitute the third part.

For illustration, the lines in the second and third parts of CIE 102[5] are given in Table 2.13. Asterisks indicate that a new line must be commenced. Two asterisks indicate a key line, which must be included even if no data are given. Letters such as ‘APOS’ and ‘TLME’ identify the line. The actual data are included between the signs ‘<’ and ‘>’. The explanation column is not part of the file format and has been put in to help the reader.

ASCII format is used with a maximum of 78 characters per line to which must be added a return line feed sequence. Each item of data may be separated by a comma or one or more space characters. However, some programs may only accept one or the other convention, in which case the file will have to be edited.

Table 2.14 gives an example of how the data in Table 2.1(a) and Table 2.1(b) are represented in the CIE File Format. The reader will notice that, in effect, the columns in Tables 2.1(a) and

Table 2.13 Presentation of luminous intensity data for CIE File Format

Status	Line name	Data to be included	Explanation
**	PTYP=	<photometric type>	Coordinate system: PTYP = $C(C, \gamma)$ PTYP = $B(B, \beta)$ PTYP = $A(A, \alpha)$
*	APOS=	<angle position code>	Orientation of the coordinate system with respect to the luminaire. The *C* series is used for the (C, γ) coordinate system. For example, *C1* is used for road lighting luminaires, *C2* for general purpose luminaires, and so on up to *C7*. The *B* series is used for the (B, β) coordinate system. For example *B1* is used for floodlights symmetrical about a central axis, *B2* for floodlights with one plane of symmetry, and so on up to *B4*. Similarly for the *A* series.
*	TLME=	<tilt during measurement>	Angle of tilt during measurement in degrees. The default value is zero.
*	LUBA=	<lumen basis of photometry>	Normally the luminous intensity data is based on candelas per 1000 lamp lumens, which is the default value. Other figures may be used for convenience in reducing the number of figures to be quoted for each luminous intensity.
*	MULT=	<multiplier>	A number by which the luminous intensity values must be multiplied to obtain the real values. A negative number indicates values in candelas. The default value is unity.
**	NCON=	<number of cone angles>	The cone angles are the α, β or γ angles, so the number of α, β or γ angles in each half-plane is required.
**	NPLA=	<number of half-plane angles>	Number of *A*, *B*, or *C* half-planes.
**	CONA=	<cone angles> <1st half plane angle followed by luminous intensities at all angles in half-plane> <2nd half plane angle followed by luminous intensities at all angles in half-plane> .. .. <last half plane angle followed by luminous intensities at all angles in half-plane>	The cone 'angles' are the α, β or γ angles. The number of lines ending in a carriage return-line feed sequence is one more than the number of half plane angles because of the inclusion of the first line giving the cone angles.

Table 2.14 Luminous intensity data in CIE File Format

PTYP = *C*
APOS = *C*1
TLME = 5
LUBA = 1000
MULT = 1
NCON = 36
NPLA = 27
CONA = 0.0 10.0 20.0 30.0 35.0 40.0 45.0 47.5 50.0 52.5 55.0
57.5 60.0 62.5 65.0 67.5 70.0 72.5 75.0 77.5 80.0 82.5 85.0
87.5 90.0 92.5 95.0 97.5 100.0 102.5 105.0 120.0 135.0 150.0 165.0
180.0

0.0	218	224	257	299	311	297	275	277	276	274	274	268	265
261	257	247	229	204	136	77	47	26	8	2	1	0	1
1	1	1	1	1	1	1	1	1					
5.0	218	225	260	305	317	304	280	283	282	282	283	278	275
271	268	257	240	218	153	87	54	31	9	2	1	0	1
1	1	1	1	1	1	1	1	1					
10.0	218	226	261	304	315	303	278	279	278	279	281	275	273
270	266	255	240	219	163	91	59	34	10	2	1	0	1
1	1	1	1	1	1	1	1	1					
15.0	218	227	258	298	306	294	267	265	263	263	265	259	257
253	249	240	225	207	164	87	58	34	10	2	1	0	1
1	1	1	1	1	1	1	1	1					
20.0	218	227	254	286	292	277	249	242	239	238	240	234	231
228	225	216	202	187	156	77	51	30	9	2	1	0	1
0	1	1	1	1	1	1	1	1					
25.0	218	226	249	272	273	256	227	217	213	211	212	206	203
200	197	189	177	163	140	65	42	25	8	1	1	0	1
0	1	1	1	1	1	1	1	1					
30.0	218	226	242	256	253	234	205	194	187	185	184	179	176
173	169	161	151	138	121	53	32	19	6	1	1	0	0
0	0	1	1	1	1	1	1	1					
35.0	218	225	235	238	234	214	185	173	166	161	159	154	151
147	143	136	127	115	102	46	22	14	5	1	1	0	0
0	0	0	1	1	1	1	1	1					
40.0	218	225	227	221	215	196	169	158	149	143	140	135	131
127	122	114	106	95	83	42	16	10	4	1	1	0	0
0	0	0	1	1	1	1	1	1					
45.0	218	224	221	205	198	180	156	145	136	129	125	120	115
110	105	97	89	79	69	41	11	7	3	1	1	0	0
0	0	0	1	1	1	1	1	1					
50.0	218	223	213	191	181	166	146	136	127	119	114	109	103
98	92	85	76	68	59	38	10	6	3	1	0	0	0
0	0	0	1	1	1	1	1	1					
60.0	218	221	201	170	156	143	127	119	112	104	98	92	87
81	74	67	60	51	41	23	7	4	2	1	0	0	0
0	0	0	1	0	1	1	1	1					
75.0	218	219	192	154	137	121	107	100	94	87	81	73	68
62	55	46	39	31	22	13	4	3	2	1	0	0	0
0	0	0	1	0	1	1	1	1					
90.0	218	219	189	149	130	113	97	90	84	76	69	62	55
46	38	30	23	16	10	7	4	3	1	0	0	0	0
0	0	0	1	1	1	1	1	1					
270.0	218	194	166	141	131	120	110	105	100	94	90	84	79
73	67	59	50	38	24	12	6	3	1	0	0	0	0
0	0	0	1	1	1	1	1	1					
285.0	218	195	169	145	134	124	114	109	105	100	96	91	87
82	76	68	59	47	31	16	7	4	1	0	0	0	0
0	0	0	1	1	1	1	1	1					
300.0	218	199	176	156	147	137	126	121	117	113	109	105	101
97	91	83	72	60	40	21	7	3	1	0	0	0	0
0	0	0	1	1	1	1	1	1					

Table 2.14 (*continued*)

310.0	218	202	185	169	160	148	134	129	124	121	118	114	111
106	100	92	81	69	48	17	6	4	2	1	1	0	1
0	1	1	1	1	1	1	1	1					
315.0	218	204	191	177	167	154	138	133	129	126	124	120	117
111	105	96	86	74	48	15	7	4	2	1	1	0	1
1	1	1	1	1	1	1	1	1					
320.0	218	207	198	186	177	162	145	139	136	133	131	127	123
118	112	103	93	79	49	15	7	5	2	1	1	0	1
1	1	1	1	1	1	1	1	1					
325.0	218	209	205	197	189	172	153	148	146	142	140	136	132
126	120	111	101	84	50	15	8	5	2	1	1	0	1
1	1	1	1	1	1	1	1	1					
330.0	218	211	213	210	203	185	165	161	158	155	152	148	143
137	131	121	110	91	51	16	9	6	2	1	1	0	1
1	1	1	1	1	1	1	1	1					
335.0	218	213	222	225	220	202	181	177	175	171	169	163	159
153	146	136	124	102	56	20	11	7	3	1	1	0	1
1	1	1	1	1	1	1	1	1					
340.0	218	216	230	242	240	221	199	197	194	191	189	182	178
172	166	156	142	118	65	27	14	8	3	1	1	0	1
1	1	1	1	1	2	1	1	1					
345.0	218	218	238	258	260	242	221	220	216	214	212	205	201
194	189	179	164	137	79	37	20	11	4	1	1	0	1
1	1	1	1	1	1	1	1	1					
350.0	218	221	247	275	280	263	242	242	240	237	235	228	224
218	214	203	187	160	97	49	27	15	5	1	1	0	1
1	1	1	1	1	1	1	1	1					
355.0	218	223	253	289	297	282	261	263	261	259	258	251	247
242	238	227	210	183	117	63	38	21	6	1	1	1	1
1	1	1	1	1	1	1	1	1					

(b) become rows, and each row is preceded by the γ angle. It should be noted that the full file format would include the name of the luminaire, its identification number, information on the ballast and circuit, and details about the lamp or lamps, as stated previously.

The first block of data listed under CONA= constitutes the cone or γ angles. Each successive block of data consists of the C angle followed by the luminous intensities at that angle. To ease reading the table, each C angle is printed in this book in bold type; this is not a requirement of the file format.

In the UK, the CIBSE format described in Technical Memorandum No 14 is used.[6] This differs from the CIE system in two important respects. Only two coordinate systems are allowed, the (C, γ) and the (B, β) systems, the latter being described as the (H, V) system (see page 418). The horizontal angles are presented on a separate line from the luminous intensity data, as indicated in Table 2.15.

The reader will notice that, unlike the CIE format, the CIBSE format does not name lines. This means that the stipulated information must appear on the specified line numbers. The first ten lines are concerned with information about the luminaire and then the following lines are as indicated in the table. CIBSE provides a second file format in which derived data such as utilization factors and glare indices may be included.

The draft documents which CEN have produced on file formats indicate that the CEN File Format is likely to resemble that of the CIE closely.[7]

If data are to be presented according to the requirements of a particular file format issued by a lighting, or other, body the relevant documents should be consulted to obtain full details.

Table 2.15 Presentation of luminous intensity data for CIBSE File Format

Status	Line number	Data to be included	Explanation
**	11	<design attitude>	The angle at which the luminaire is designed to operate, measured with respect to the horizontal plane. Typically 0° for indoor luminaires and 5° for road lighting luminaires.
**	12	<number of vertical angles><number of horizontal angles>	Number of γ or B angles followed by the number of C or β angles
**	13	<vertical angles>	The γ or B angles
**	14	<horizontal angles>	The C or β angles
**	15 on	<luminous intensities at 1st C or β angle> <luminous intensities at 2nd C or β angle> <————————> <————————> <luminous intensities at last C or β angle>	The number of lines, ending in a carriage return-line feed sequence is equal to the number of C or β angles

Problems

1. From the table below, which shows luminous intensities in candelas, use linear interpolation to find the luminous intensity at the following angles: (C, γ) = (a) (359.0°, 11.5°), (b) (3.1°, 11.2), (c) (1.0°, 12.5°).

   Answer: [(a) 596.5 cd, (b) 649.28 cd, (c) 633.75 cd]

γ (deg)	Azimuth C (deg) 358	0	2	4
10	560	600	621	630
11	570	615	641	655
12	581	620	645	657
13	587	623	647	650

2. Find the luminous intensity from Table 2.1 at the following angles by using quadratic interpolation: (C, γ) = (a) (8.5°, 38.0°), (b) (354.0°, 54.0°), (c) (321.0°, 2.0°).

   Answer: [(a) 310.506 cd/klm, (b) 255.146 cd/klm, (c) 216.171 cd/klm]
3. Verify that the calculated data in the table below are correct.

Given data						Calculated data				
x (m)	y (m)	H (m)	ν (°)	δ (°)	ψ (°)	x' (m)	y' (m)	H' (m)	C (°)	γ (°)
5	5	10	5	0	0	5.4168	4.5452	10	40	35.2644
5	5	10	0	5	0	5	4.1094	10.3977	39.4162	31.9001
5	5	10	0	0	5	5.8525	5	9.5262	40.5083	38.9397
5	5	10	5	5	5	6.2989	3.6563	9.8466	30.1340	36.4892
–5	5	10	5	5	5	–3.6185	4.5246	10.7905	128.6509	28.2320

4. A luminaire is photometrically tested with a tilt upwards of 10° but is used with a titlt of 5°. Find the coordinates that have to be looked up in an *I*-table to find the luminous intensity directed towards points with (C, γ) coordinates of (358°, 20°) and (340°, 91°).

   Answer: [(11.5750°, 20.4205) and (339.9881°, 89.2885°)]

References

1. CIE 30–2 (TC–4.6) (1982). *Calculation and Measurement of Luminance and Illuminance in Road Lighting*.
2. BS 5775 Part 11: 1993 (ISO 31–11: 1992) *Specification for Quantities, Units and Symbols.* p. 20.
3. CIE Draft (1999) *Road Lighting Calculations*.
4. LM–63–86. IES Recommended Standard File Format for the Electronic Transfer of Photometric Data. *New York, Illuminating Engineering Society of North America* (1986). Latest version is LM–63–95 issued in 1995.
5. CIE 102–1993. *Recommended File Format for Electronic Transfer of Luminaire Photometric Data*. Technical report.
6. CIBSE TM 14 (1988) *CIBSE Standard File Format for the Electronic Transfer of Luminaire Photometric data* (1988).
7. (1997) Draft European Standard prEN 13032–1.

3

Direct Illuminance from Point, Line and Area Sources

Illuminance can be treated as a vector quantity, since it has both magnitude and direction and obeys the laws of vector addition. However, care must be exercised when interpreting the results – see the comments at the end of Section 3.1.

3.1 Illuminance as a vector quantity

Consider a point source illuminating a surface at an angle θ to the normal (Figure 3.1).

$$E_p = \frac{I \cos \theta}{d^2} = E_{max}.\cos \theta$$

where E_{max} is the maximum illuminance that the source could produce at point P, i.e. when θ = 0°.

E_p can therefore be considered to be a component of the vector E_{max} acting along the normal to P *(N)*.

For a point source, the direction of this illumination vector E_{max} is from the point source to the illuminated point, but for an area source the direction of the vector at the illuminated point is not always clear. The illumination vector at a point, produced by a particular light source, is in the direction of the maximum illuminance that the source can produce at that point and is equal in magnitude to that maximum illuminance.

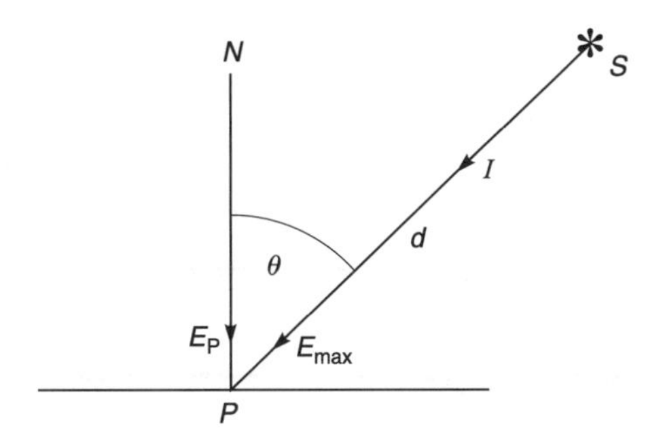

Fig. 3.1 A point source illuminating a surface at an angle θ to the normal

If an area source is considered to be an array of point sources, then the illumination vector for the area source is the resultant of the illumination vectors produced by this array of point sources at the illuminated point. In a similar way, any array of individual sources each producing its own illumination vector can produce a resultant vector.

The use of this vector concept can sometimes simplify a calculation. However, care must be taken in a complex situation to ensure that only sources that can illuminate the point are taken into account. For example, in the case of a large area source, the plane of the illuminated point may intersect the plane of the light source. If this happens the result of including all the source in the calculation would be a calculated value representing the difference of the illuminances on each side of the plane containing the point (see Chapter 8).

3.2 Illuminance on an oblique plane

The vector concept is useful in developing expressions for the illuminance on an oblique plane, tilted with respect to the horizontal plane.

When the direction and magnitude of E_{max} are known, then the illuminance at the point P on any plane can be obtained by multiplying E_{max} by the cosine of the angle between the direction of E_{max} and the normal to that plane at P, i.e.

$$E_p = E_{max} . \cos \theta_n$$

However, it is often more convenient to work in terms of components of E_{max} that are on the horizontal and vertical planes and are routinely calculated.

The right-handed coordinate system commonly adopted for specifying the x, y and z directions and senses of the vectors is not particularly convenient for the physical realities of the situation, but it is considered wise to conform to this and deal with any difficulties by using appropriate notation (Figures 3.2 and 3.3).

x, y corresponds to the horizontal plane and y, z and x, z to the vertical planes.

The point P is taken as the origin of the coordinate system, i.e. ($x = 0, 0$, $y = 0, 0$, $z = 0, 0$). The z axis is the axis of the horizontal component of illuminance, but since the illuminance vector acts downwards it is denoted as $E_{(-z)}$ and similarly the x and y components, which act onto vertical planes, are denoted by $E_{(-x)}$ and $E_{(-y)}$. Also shown in Figure 3.3 is the normal to the illuminated plane at P. This normal lies in the plane indicated by angle β and is at an angle ϕ to the horizontal illuminance vector $E_{(-z)}$.

If the components $E_{(-x)}$, $E_{(-y)}$ and $E_{(-z)}$ are themselves resolved into orthogonal components

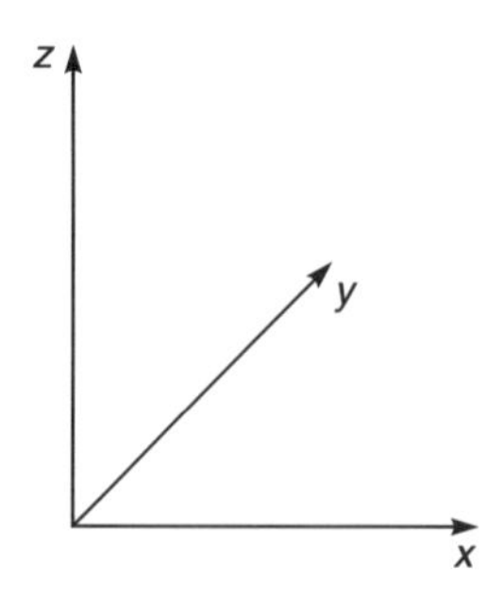

Fig. 3.2 The right-handed coordinate system

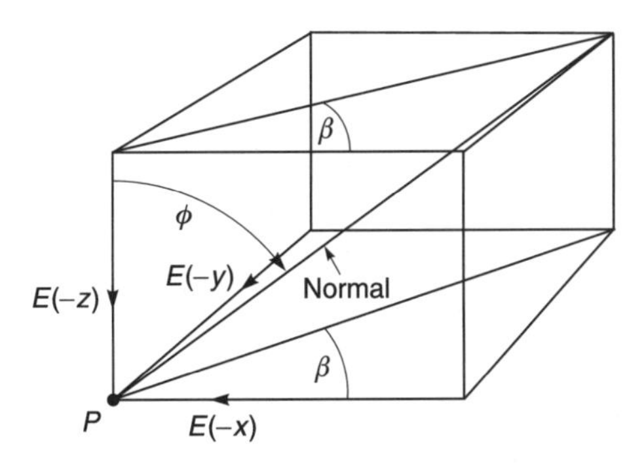

Fig. 3.3 Notation and relationships for illuminance on an oblique plane

with one of the components coincident with the normal to the plane of P, then the other components of $E_{(-x)}$, $E_{(-y)}$ and $E_{(-z)}$ will be parallel to the plane of P and so contribute zero illuminance. For this reason, only the components normal to the oblique plane need be considered.

Using the notation given in Figure 3.3 gives

$$E_p = E_{(-z)} \cos \phi + E_{(-x)} \cos \beta \sin \phi + E_{(-y)} \sin \beta \sin \phi$$

3.2.1 POINT SOURCES

Consider Figure 3.4, here for convenience the y axis has been set parallel to the plane containing point P as shown. In this case $\beta = 0°$ and so,

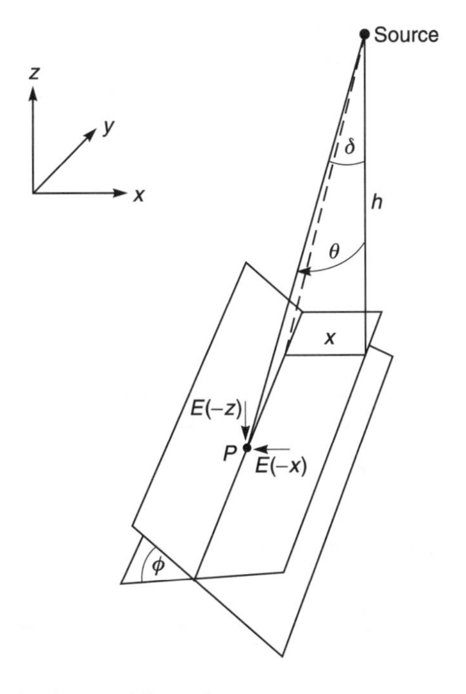

Fig. 3.4 A point source illuminating an oblique plane

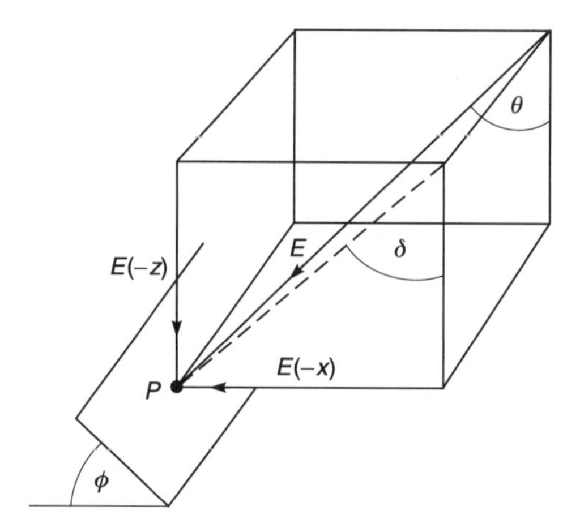

Fig. 3.5 The relationship between $E_{(-x)}$ and $E_{(-z)}$

$$E_p = E_{(-z)} \cos \phi + E_{(-x)} \sin \phi$$

Also, since $E_{(-x)} = E_{(-z)} \tan \delta$ (Figure 3.5),

$$E_p = E_{(-z)} \left(\cos \phi + \frac{x}{h} \sin \phi \right)$$

where $E_{(-z)}$ is the horizontal plane illuminance at P.

This expression has been developed for a plane tilted towards the light source (i.e. the normal is moved from the vertical towards the source). If the tilt is away from the light source then $\sin \phi$ becomes negative and so the vertical component is subtracted, not added.

3.2.2 LINE SOURCES

Consider Figure 3.6.

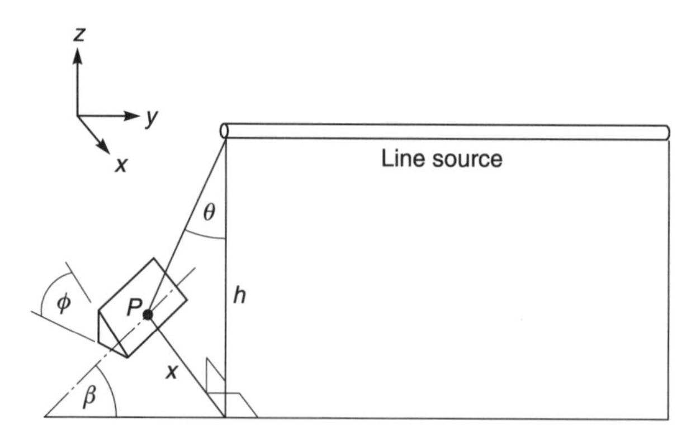

Fig. 3.6 A line source illuminating an oblique plane

It was shown earlier that

$$E_p = E_{(-z)} \cos \phi + E_{(-x)} \cos \beta \sin \phi + E_{(-y)} \sin \beta \sin \phi$$

If the y axis is arranged to coincide with the axis of the line source (Figure 3.6), then $E_{(-x)}$ becomes the vertical illuminance at P on a plane parallel to the source and $E_{(-y)}$ becomes the vertical illuminance at P on a plane perpendicular to the source axis.

Since the vertical illuminance on a plane parallel to a line source

$$E_{(-x)} = E_{(-z)} \tan \theta$$

$$= E_{(-z)} \frac{x}{h}$$

$$E_p = E_{(-z)} \left(\cos \phi + \frac{x}{h} \cos \beta \sin \phi \right) + E_{(-y)} \sin \beta \sin \phi$$

Note: the advantage of this expression is that $E_{(-z)}$ and $E_{(-y)}$ can often be calculated using the aspect factor method (see Section 3.7.3) and ϕ, β, x and h are easy to determine.

3.2.3 AREA SOURCES

The general expression applies to area sources, but since the inclined plane frequently cuts the plane of the source and changes not only its size, but also its shape with respect to point P (Figure 3.7), the calculation is more difficult and requires careful consideration to ensure a correct solution.

If the change of size and shape is not taken into account the value obtained will be the difference in the illuminance on the two sides of the plane at P.

The vector that represents the difference of the illuminance on opposite sides of a plane has its uses and this is developed in Chapter 8, on interior lighting.

Later in this chapter the vector method is applied to a particular case for a uniformly diffusing rectangular area source.

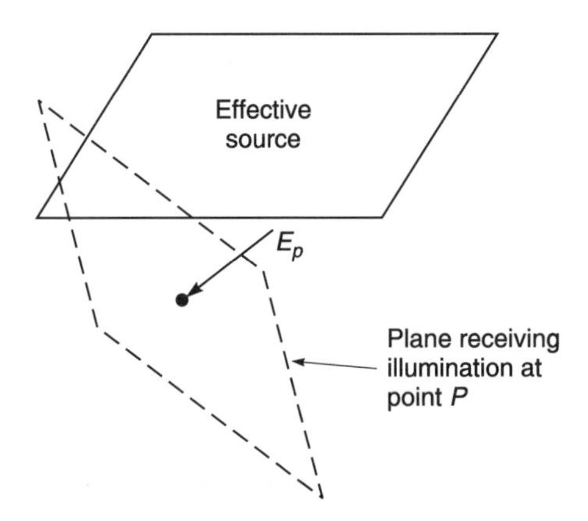

Fig. 3.7 When the plane of the illuminated point cuts through the area source the calculation is more difficult

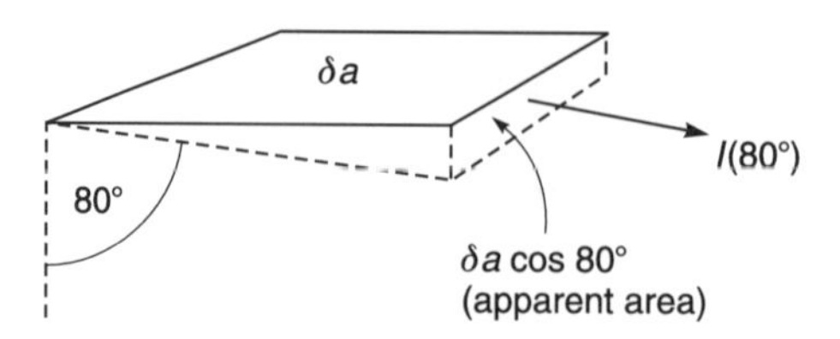

Fig. 3.8 A large area may appear small because of the angle of view

3.3 Luminance and luminous exitance

In Chapter 1, the concept of luminous intensity, which relates to the illuminating power of a light source, either primary or secondary (i.e. by reflection) was introduced. It was also shown how this concept gives rise to the inverse-square law for illuminance produced by point sources of light.

Luminance extends this concept to consider the illuminating power of an element of a light source or surface. It therefore relates the luminous intensity produced to the area responsible for it. As seen earlier, when a source has a luminous intensity that varies with direction (the most common case) then it is important for this direction to be specified. It is equally important when specifying luminance to specify this direction in relation to the normal to the surface element producing the luminous intensity, since the intensity will appear to come from the projection of the area in the direction concerned. This is important since the luminous intensity may come from a large element of area which, when viewed at an extreme angle to the normal (say, 80°), appears very small.

Luminance is therefore defined as the luminous intensity per unit projected area of the source (or surface) in the specified direction (see Figure 3.8).

$$L_\theta = \frac{I_\theta}{\delta a \cos \theta} \text{ cd m}^{-2}$$

3.4 A special case – uniform diffusion

When the intensity per unit projected area (L) is the same from all directions of view of the element, i.e. 0° to +90° in elevation and over 360° in azimuth, the surface or source is said to produce uniform diffusion. This is a very important case, since (a) many surfaces and sources approximate to this distribution, a matt painted surface, for example; (b) this assumption can greatly simplify many calculations.

A simple example of (b) is that once this assumption of uniform diffusion is made, the intensity distribution of the element is also fixed. This distribution can be determined in the following way:

$$L = \frac{I_\theta}{\delta a \cos \theta}$$

but since δa is fixed then for L to remain constant I_θ must equal $I_{max} \cos \theta$.

So it is possible to say that for the condition of uniform diffusion

$$I_\theta = I_{max} \cos \theta$$

where I_{max} is the intensity normal to the surface element δa (Figure 3.9).

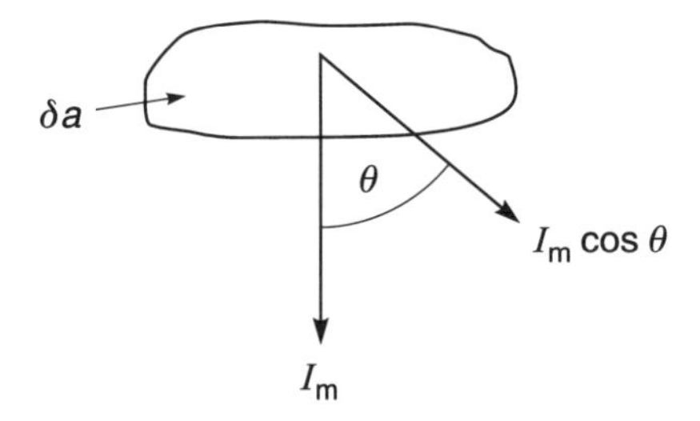

Fig. 3.9 The relationship between the maximum intensity and the intensity at angle θ for uniform diffusion

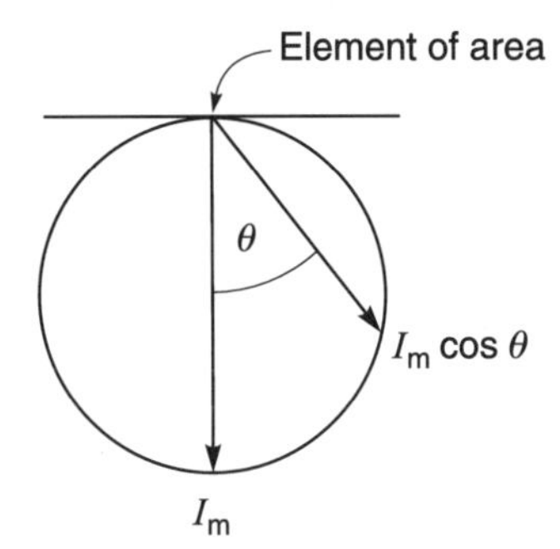

Fig. 3.10 The polar curve for uniform diffusion is a tangential circle

The polar curve for such a distribution is a circle tangential to the origin (Figure 3.10).

If we consider all directions this becomes a polar solid, which is spherical.

Once the intensity distribution has been determined it becomes a simple matter to determine the relationship between I_m and the total flux emitted by the element.

From Chapter 1 the solid angle for an elemental zone of a sphere is

$$\frac{2\pi r \sin\theta r \, d\theta}{r^2} = 2\pi \sin\theta \, d\theta$$

So that

$$\Phi = \int_{\theta=0}^{\pi/2} 2\pi I_{max} \cos\theta \sin\theta \, d\theta$$

$$= 2\pi I_{max} \left[\frac{1}{2} \sin^2\theta \right]_0^{\pi/2}$$

$$= \pi I_{max}$$

So, for a uniform diffuser, there is a simple relationship between the maximum intensity and the flux emitted, and hence between the luminance and the flux emitted per unit area.

For a uniform diffuser

$$L = \frac{I_{max}}{\delta a}$$

$$\pi L = \frac{\pi I_{max}}{\delta a} = \frac{\Phi}{\delta a} \text{ lm m}^{-2}$$

$\Phi/\delta a$ is given the symbol M and is called the luminous exitance of the surface element. It is the luminous flux emitted per unit area.

Because of the simple relationship between luminous exitance and luminance in the common case of the uniform diffuser, luminance is sometimes confused with luminous exitance. This is partly because the concept of luminous exitance is often omitted from courses (and books).

For example, when an illuminance of E lux is reflected from a surface, the luminous exitance M is given by

$$M = \rho E$$

where ρ is the reflectance of the surface.

If the surface gives uniformly diffuse reflectance then:

$$M = \pi L \quad \text{or} \quad L = \frac{M}{\pi}$$

However, if the surface is *not* a uniform diffuser, then the relationship becomes complex, and

$$L \neq \frac{M}{\pi}$$

since L is not constant with angle of view.

3.5 An important tool: the principle of equivalence

Because the assumption of uniform diffusion is often an acceptable approximation, the principle of equivalence may be applied to many lighting problems.

The principle may be demonstrated in the following way (Figure 3.11).

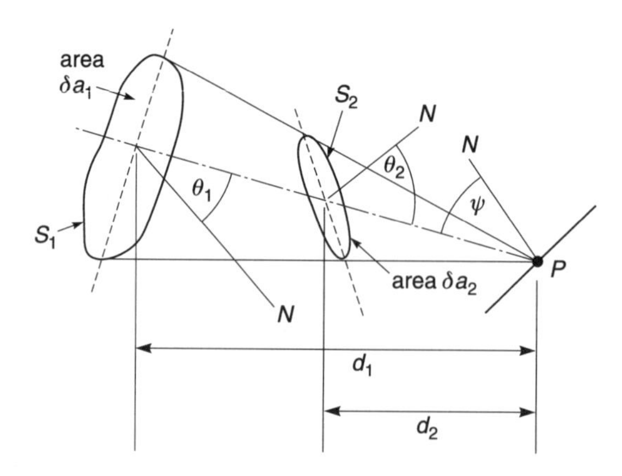

Fig. 3.11 Illustration of the principle of equivalence

When viewed from point P both light sources S_1 and S_2 appear to have the same boundary. If the inverse-square law is applied to each source, then:

$$E_{ps_1} = \frac{L\delta a_1 \cos\theta_1 \cos\phi}{d_1^2}$$

$$E_{ps_2} = \frac{L\delta a_2 \cos\theta_2 \cos\phi}{d_2^2}$$

Also,

$$\frac{\delta a_1 \cos\theta_1}{d_1^2} = \frac{\delta a_2 \cos\theta_2}{d_2^2} \quad \text{(the same solid angle)}$$

so that

$$E_{ps_1} = E_{ps_2}$$

3.6 Uniformly diffuse sources

Although many light sources do not give a uniformly diffuse distribution, before we consider non-uniformly diffuse sources the simplification introduced by assuming uniform diffusion will be illustrated (1) by introducing the unit hemisphere method, and (2) by applying that method to three cases; the disc source, the triangular source and the rectangular source.

The disc source approximation can be useful for a quick estimate of the illuminance produced by a source of similar area but of a more complex shape. The triangular source can be used as a component to build up illuminance equations for other shapes of source and, particularly, the rectangular source. The rectangular source is used to calculate the reflected illuminance from the rectangular room surfaces which can usually be assumed to have uniformly diffuse reflectance.

3.6.1 UNIT HEMISPHERE METHOD

Consider an element δs of a source S, having a uniformly diffuse distribution (Figure 3.12). Let the illuminance at P due to this element be δE_p. By the principle of equivalence the illuminance at P would be the same if it was illuminated by the element of the surface area of the hemisphere $\delta s'$, providing it has the same luminance as δs.

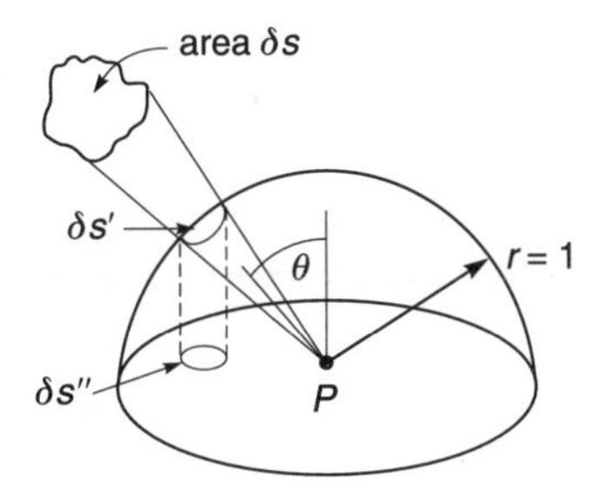

Fig. 3.12 The unit hemisphere method

So that

$$\delta E_p = \frac{L\delta s' \cos\theta}{1^2}$$

Also,

$$\delta s'' = \delta s' \cos\theta$$

Therefore,

$$\delta E_p = L\delta s''$$

The same argument may be extended to each element of area of source S and so

$$E_p = L \sum \delta s'' = LS''$$

3.6.2 THE DISC SOURCE

Calculation of the illuminance on the horizontal plane at a point P directly beneath the disc (see Figure 3.13).

$$\begin{aligned} E_p &= LS'' \\ &= L(\pi x^2) \end{aligned}$$

$$E_p = \pi L \sin^2 \frac{\alpha}{2}$$

Note: this result could also be expressed as:

$$E_p = \pi L \left(\frac{R^2}{R^2 + h^2} \right) \quad \text{or} \quad E_p = \frac{\pi L}{2} (1 - \cos\alpha)$$

In the latter form it can be extended to relate to any point on a plane parallel to a uniformly diffusing disc source. This is done in three stages.

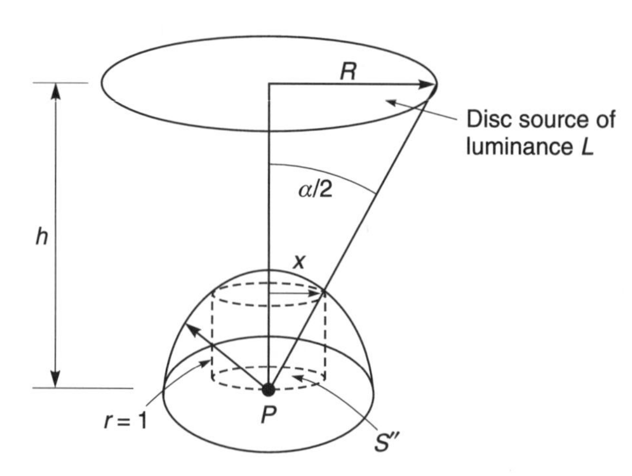

Fig. 3.13 The unit hemisphere method applied to a disc source

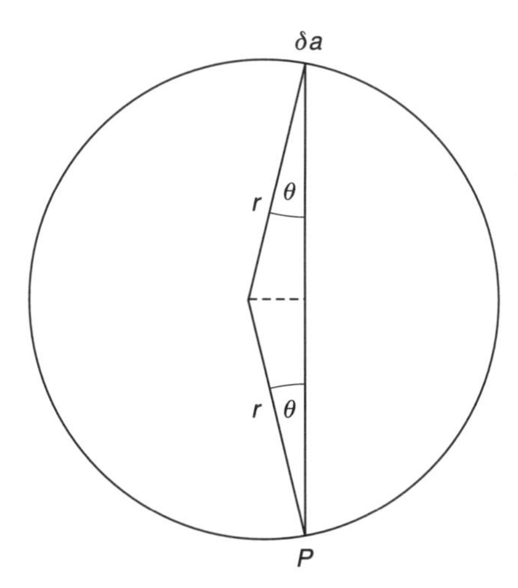

Fig. 3.14 The first stage in extending the use of the disc source formula

Stage 1 (Figure 3.14)
Consider a uniformly diffusing luminous element δa on the inside of a sphere.

The illuminance at point P produced by this element δa on the inside of the sphere is given by:

$$E_p = \frac{L\delta a \cos\theta . \cos\theta}{(2r\cos\theta)^2}$$

$$= \frac{L\delta a}{4r^2}$$

Since this expression does not contain θ it means that the element may be located at any point on the inside of the sphere and will produce the same value of illuminance at all points on the inside of the sphere.

Stage 2
Consider Figure 3.15. Let the diffusing disc be placed within the sphere so as to have a boundary coincident with that of the spherical cap *ABC*. Because of the principle of equivalence we can assume that if the spherical cap *ABC* had the same luminance as the disc and the disc was removed, the illuminance produced at point *P* would remain the same. Further, from Stage 1 we know that each element of the spherical cap would produce the same value of illuminance at each point on the inside of the sphere below the disc boundary. This means that the disc itself would have the same effect. That is, the illuminance on the inner surface of the enclosing sphere at any point *P* would be equal to that at *O*. Let this illuminance be E_q.

In passing, we note that since angles *APC* and *AOC* are subtended by the same chord within the circle, they must be equal.

So, the illuminance

$$E_q = \frac{\pi L}{2}(1 - \cos\alpha) \qquad \text{where } \alpha = \text{angle } APC$$

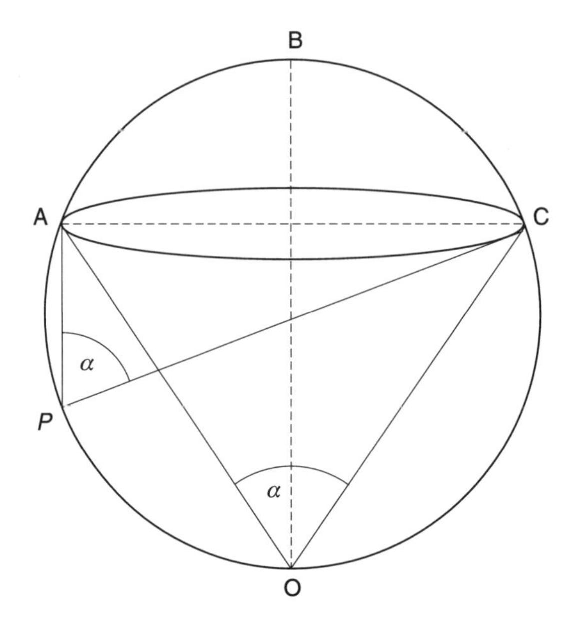

Fig. 3.15 The second stage in extending the use of the disc source formula

Stage 3

Consider Figure 3.16. Returning to the principle of equivalence we can now conclude that the illuminance at *P* could be considered to be produced not just by the disc, or the spherical cap, but by the base of an elliptical cone of major axis *DFG*. (This cone will have a different ellipse as its base according to the position of *P* on the circle, but it does not affect the argument given

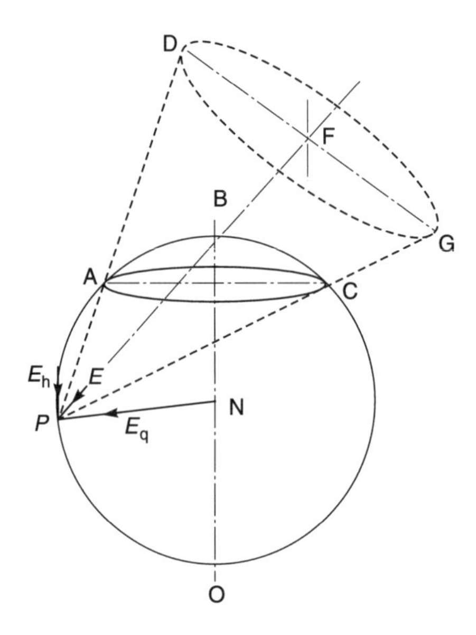

Fig. 3.16 The third stage in extending the use of the disc source formula

below). This generation of an elliptical cone from an oblique circular cone was first demonstrated by Appollonius who lived from 260–200 BC.[1]

In Section 3.1, we introduced the concept of the illumination vector. The direction of the illumination vector is that in which all illuminance components at right-angles to that direction are equal and so cancel. The symmetry of the elliptical source predicates that the illumination vector lies on the axis of the elliptical cone, i.e. on line *FP* and passes through point *B* (because angles *APB* and *BPC* are equal, once again subtended by equal chords). We note that *BN* and *NP* are radii.

Finally, since *BN* is parallel to the direction of the horizontal component of illuminance E_h and the angle between E_h and the illumination vector *E* must have the same value as the angle *BPN* (i.e. the angle between the component E_q, representing the sphere illuminance and the illuminance vector *E*), we can conclude that

$$E_h = E_q$$

So that

$$E_h = \frac{\pi L}{2}(1 - \cos\alpha)$$

The outcome of this rather extended section greatly increases the usefulness of the disc formula. In addition, it demonstrates the importance of the principle of equivalence.

3.6.3 THE TRIANGULAR SOURCE

The expression for the illuminance on the horizontal plane at a point *P* directly beneath the apex of a right-angled triangle is developed below (Figure 3.17).

The projection of sector *M* onto the base of the unit hemisphere gives the same value for *S″* as the projection of *S′*.

$$\text{Area of sector } M = \pi r^2 \times \frac{B_1}{2\pi}$$

but $r = 1$, so area of sector $M = B_1/2$, where B_1 is the sector angle in radians.

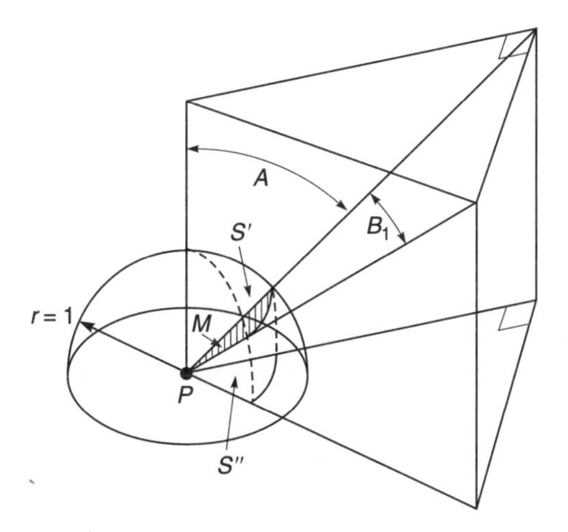

Fig. 3.17 The unit hemisphere method applied to a triangular source

Therefore,

$$S'' = \frac{B_1}{2} \sin A$$

where A is the angle between the normal at P and the plane of sector M and

$$E_p = \frac{L}{2} (B_1 \sin A)$$

3.6.4 THE RECTANGULAR SOURCE

This result can be extended to a rectangular source by considering it to consist of two triangular sources (Figure 3.18).

$$E_p = \frac{L}{2} (B_1 \sin A + A_1 \sin B)$$

Perhaps, surprisingly, the result for a triangular source parallel to the horizontal plane can be extended to the case of a rectangular source on a plane perpendicular to that of the illuminated point. Once again, we invoke the principle of equivalence (Figure 3.19).

The illuminated point is located opposite one corner of the perpendicular or vertical rectangular area light source (*ABCD*).

An observer at point *P* viewing diffusing panel *ABCD* would receive the same impression (and illuminance) as if they had been looking through a window at a triangular light source of infinite extent defined by the directions of lines *NA* and *NB*, except that the section of the triangular source on their side of the window would be missing, i.e. triangle *NAB*.

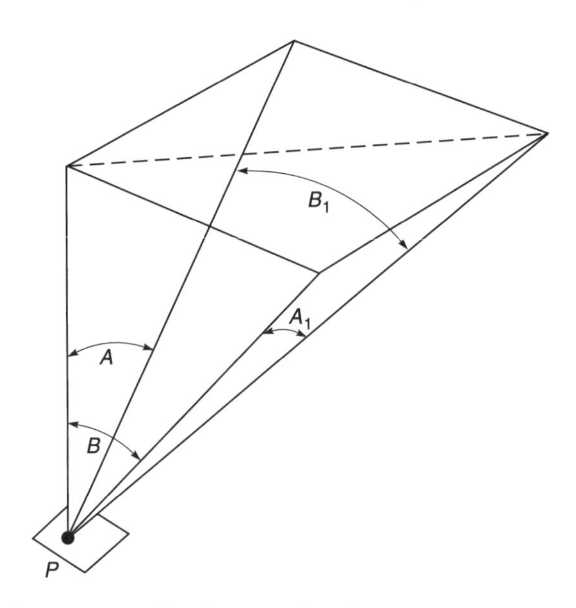

Fig. 3.18 The rectangular source considered as two triangular sources

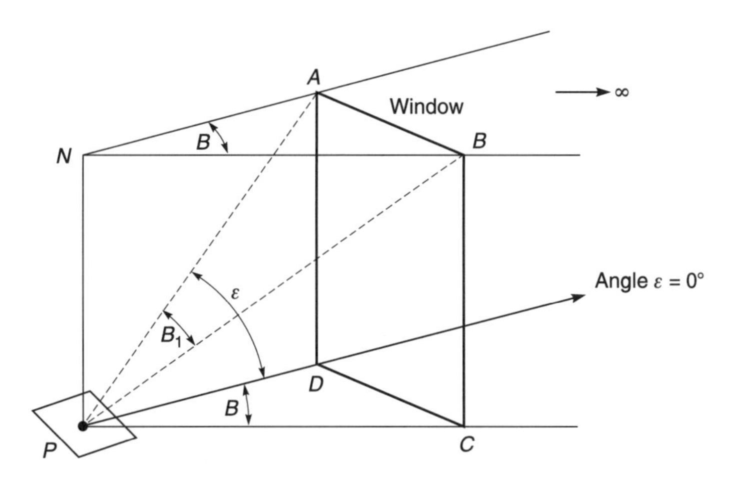

Fig. 3.19 The principle of equivalence used to develop the formula for a perpendicular area source from the triangular source formula for the parallel plane

So, the illuminance at P produced by area $ABCD$ would be given by the difference of the illuminances produced by these two triangular sources.

For the triangle NAB

$$E_1 = \frac{L}{2}(B_1 \cos \varepsilon) \qquad \text{where } \varepsilon = \left(\frac{\pi}{2} - A\right)$$

For the triangle of infinite extent $\varepsilon = 0°$, $\cos \varepsilon = 1.0$.

So,

$$E_2 = \frac{L}{2} B$$

and

$$E_p = \frac{L}{2}(B - B_1 \cos \varepsilon)$$

Important

A difficulty introduced by developing the formula for the perpendicular plane from that for a triangular source illuminating a parallel plane is that, as the angles are measured in the diagram, ε is the complimentary angle to A. For this reason the more usual form of the formula is, therefore,

$$E_p = \frac{L}{2}(B - B_1 \cos A)$$

and this is in the form quoted in the use of the formula in Section 8.5.1, and Figure 8.8(b) is labelled accordingly.

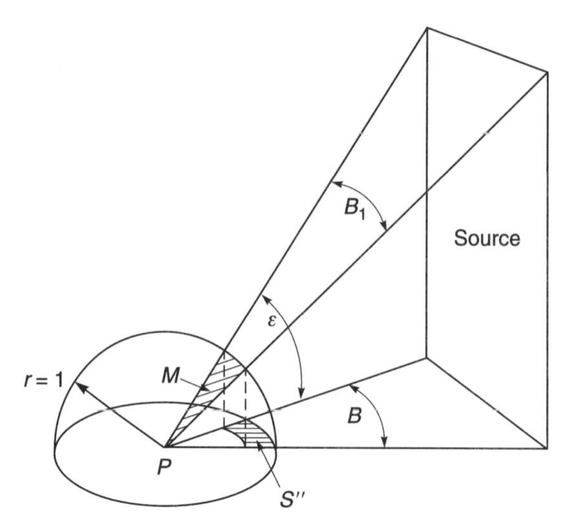

Fig. 3.20 The unit hemisphere method applied to the perpendicular area source

The real source in Figure 3.20 is, of course, in the vertical plane and so the angle ε would correspond to angle *A* if a horizontal rectangular source (Figure 3.18) had been rotated through 90° to take this position.

This result could have been obtained by the direct use of the unit hemisphere method (Figure 3.20).

The projection of the solid angle intersection with the unit hemisphere *S''*, is equal to the area of the sector subtending angle *B*, minus the projected area of the shaded sector *M*.

$$\text{Area of sector with angle } B = \frac{B}{2}$$

$$\text{Projected area of sector } M = \frac{B_1}{2} \cos \varepsilon$$

So that

$$E_{\text{p}} = \frac{L}{2}(B - B_1 \cos \varepsilon)$$

as before.

The results for uniformly diffuse rectangular sources are of particular value since they can be used to determine the distribution of reflected flux from walls and ceilings onto the working plane assuming uniform wall or ceiling illuminance and reflectance.

3.6.5 THE METHOD OF COMPONENT SOURCES

It has been convenient to calculate the illuminance under one corner of a right-angled triangular source and, in deriving the illuminance for a uniformly diffusing rectangle perpendicular to the plane of the illuminated point, we used the device of subtracting the illuminance that would be produced by one triangular source from that produced by a larger triangular source (Section 3.6.4). This method can be used in many cases and so we will describe it as the method of component sources. It has the great advantage of allowing the equations related to simple cases to be used for more complicated cases.

For example, the equations for a rectangular source where the illuminated point is under one corner of the source can be extended to any point on the parallel plane beneath the source by considering the source to consist of a number of rectangular sources, see Figures 3.21 and 3.22.

From Figure 3.21 it will be seen that the illuminance at *P* from rectangular source *ABCD* is given by summing the contributions from the assumed component sources *1, 2, 3, 4*.

So,

$$E_p = E_1 + E_2 + E_3 + E_4$$

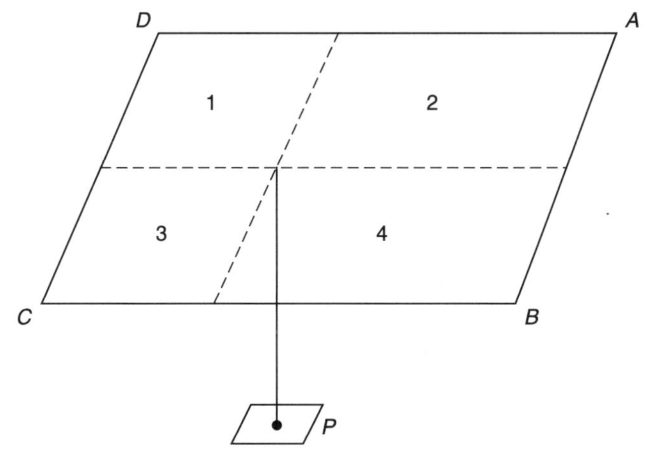

Fig. 3.21 The method of component sources

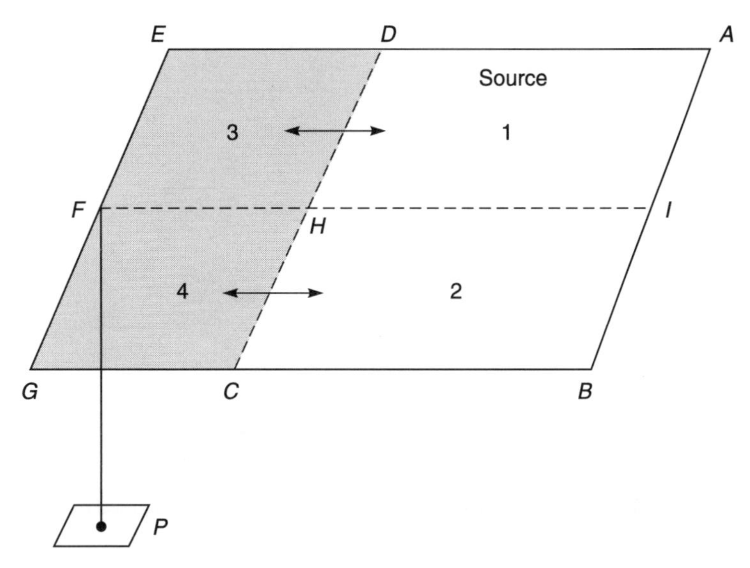

Fig. 3.22 The case where the point is not under the rectangular source

From Figure 3.22 it will be seen that the illuminance at *P* from source *ABCD* is given by the illuminance from assumed sources *AIFE* + *BIFG* minus the illuminance from the non-existant component sources *EDHF* + *GCHF*.

$$E'_p = E'_{(1+3)} + E'_{(2+4)} - (E_3 + E_4)$$

3.6.6 GENERAL ILLUMINANCE EQUATION FOR A UNIFORMLY DIFFUSING RECTANGULAR SOURCE TILTED AT ANY ANGLE FROM 0° TO 90° TO THE ILLUMINATED PLANE, WITH ONE EDGE PARALLEL TO THE ILLUMINATED PLANE

The two results obtained in Section 3.6.4 for the illuminances (1) at a point on a plane parallel to the source (E_1); (2) at a point on a plane perpendicular to the source (E_2) may be used to develop a general illuminance equation for a rectangular uniformly diffusing source.

Such a general equation might be used for a source mounted on a sloping ceiling or for a sloping roof light, shown shaded (Area $R \times T$) in Figure 3.23.

The two equations are:

$$E_1 = \frac{L}{2}(B_1 \sin A + A_1 \sin B) \quad (3.1)$$

$$E_2 = \frac{L}{2}(B - B_1 \cos A) \quad (3.2)$$

Applying the method of component sources (see Section 3.6.5) the illuminance at the point defined (with respect to one corner of the source) by x, y and z is given by

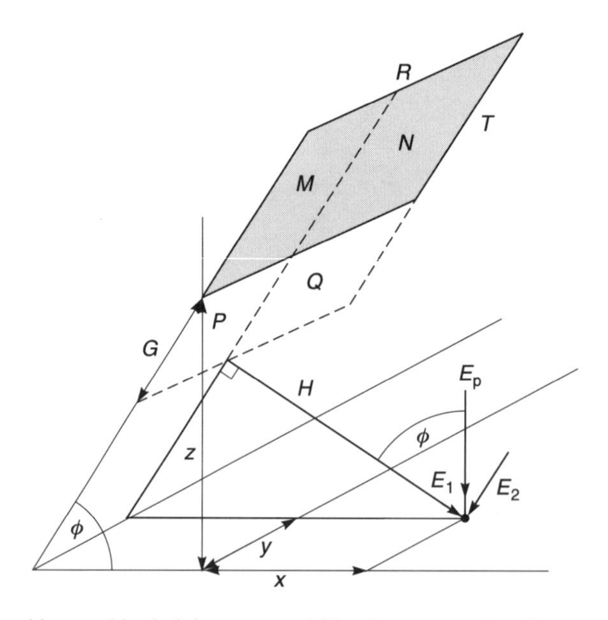

Fig. 3.23 The relationships used in deriving a general illuminance equation for a rectangular source

$$E_p = [E(MP) + E(NQ)] - (E(P) + E(Q)]$$

or, in terms of the vector components, *normal* to the illuminated plane:

$$E_p = [(E_1(MP) + E_1(NQ)) - (E_1(P) + E_1(Q))] \cos\phi + [(E_2(MP) + E_2(NQ)) - (E_2(P) + E_2(Q))] \sin\phi$$

For each E_1 term there is an expression in the form of equation (3.1) and for each E_2 term an expression in the form of equation (3.2).

When these expressions are written in terms of the dimensions given in Figure 3.23 and gathered together, the illuminance E_p is given by:

$$E_p = \frac{L}{2}\frac{y\cos\phi}{\sqrt{H^2+y^2}}\left(\tan^{-1}\frac{(T+G)}{\sqrt{H^2+y^2}} + \tan^{-1}\frac{G}{\sqrt{H^2+y^2}}\right)$$

$$+\frac{(T+G)\cos\phi = H\sin\phi}{\sqrt{H^2+(T+G)^2}}\left(\tan^{-1}\frac{y}{\sqrt{H^2+(T+G)^2}} + \tan^{-1}\frac{(R-y)}{\sqrt{H^2=(T+G)^2}}\right)$$

$$+\frac{(R-y)\cos\phi}{\sqrt{H^2+(R-y)^2}}\left(\tan^{-1}\frac{(T+G)}{\sqrt{H^2+(R-y)^2}} - \tan^{-1}\frac{G}{\sqrt{H^2+(T+G)^2}}\right)$$

$$+\frac{H\sin\phi - G\cos\phi}{\sqrt{H^2+G^2}}\left(\tan^{-1}\frac{y}{\sqrt{H^2+(R-y)^2}} + \tan^{-1}\frac{(R+y)}{\sqrt{H^2+G^2}}\right)$$

Combining the angles using

$$(\tan^{-1} A - \tan^{-1} B) = \tan^{-1}\left(\frac{A-B}{1+AB}\right) \quad \text{and} \quad \tan^{-1} A + \tan^{-1} B = \tan^{-1}\left(\frac{A+B}{1+AB}\right)$$

gives,

$$E_p = \frac{L}{2}\frac{y\cos\phi}{\sqrt{y^2+H^2}}\tan^{-1}\left(\frac{T\sqrt{y^2+H^2}}{y^2+H^2+G^2+TG}\right) + \frac{T\cos\phi - x}{\sqrt{x^2+z^2+T^2+2TG}}$$

$$\times\tan^{-1}\left(\frac{R\sqrt{x^2+z^2+T^2+2TG}}{x^2+y^2+z^2-yR+T^2+2TG}\right)$$

$$+\frac{(R-y)\cos\phi}{\sqrt{(R-y)^2+H^2}}\tan^{-1}\left(\frac{T\sqrt{(R-y)+H^2}}{(R-y)^2+H^2+G^2+TG}\right)$$

$$+\frac{x}{\sqrt{x^2+z^2}}\tan^{-1}\left(\frac{R\sqrt{x^2+z^2}}{x^2+y^2+z^2-Ry}\right)$$

Dimensions H and G shown in Figure 3.23 relate the position of the illuminated point to the position of light source on the sloping plane, and since

$$G = z \sin \phi - x \cos \phi \quad \text{and} \quad H = z \cos \phi + x \sin \phi$$

the illuminated point and the source can be related more easily to the relevant horizontal and vertical surfaces in a practical situation. This is particularly useful if this general expression is incorporated in a larger computer program.

Note: as already pointed out in Section 1.3.4, care has to be taken with the inclusion of this type of expression in a computer program, since incorrect results can occur when the $\tan^{-1}$ argument denominator is zero, or when both the numerator and denominator are zero.

The equation above is similar to that published by Pierpoint and Hopkins.[2] (Noting that $M/\pi = L$ for a uniformly diffuse source.)

However, they derived the expression by employing a double integration, while the above derivation has employed the principle of equivalence to obtain the rectangular source formulae, the method of component sources, and the concept of illuminance as a vector. This is a further illustration of the value of these techniques.

It should be noted that the x, y and z coordinates used in Figure 3.23, to relate the equation to the work of Pierpoint and Hopkins, are not the same as those used in the opening section of this chapter.

In Figure 3.23, the x, y and z coordinates have been related to the illuminated plane, as in the Pierpoint and Hopkins' work, in order to make comparison between the two sets of equations easier.

In the opening section of this chapter, the x, y and z coordinates relate simply to the horizontal and vertical planes and not the illuminated plane (see Section 3.2). The difference occurs because, in Section 3.2, the plane of illuminance is tilted with respect to the horizontal plane, whereas in the Pierpoint and Hopkins' work the rectangular source is tilted relative to the horizontal plane.

3.6.7 *UNIFORMLY DIFFUSING SPHERICAL SOURCE*

To complete the consideration of large uniformly diffusing sources we use the principle of equivalence to treat the uniformly diffusing sphere as an equivalent uniformly diffusing disc source (Figure 3.24).

The disc source equation gives

$$E_p = \pi L \frac{r^2}{r^2 + h^2}$$

but

$$\frac{h^2}{r^2} = \frac{d^2 - a^2}{a^2}$$

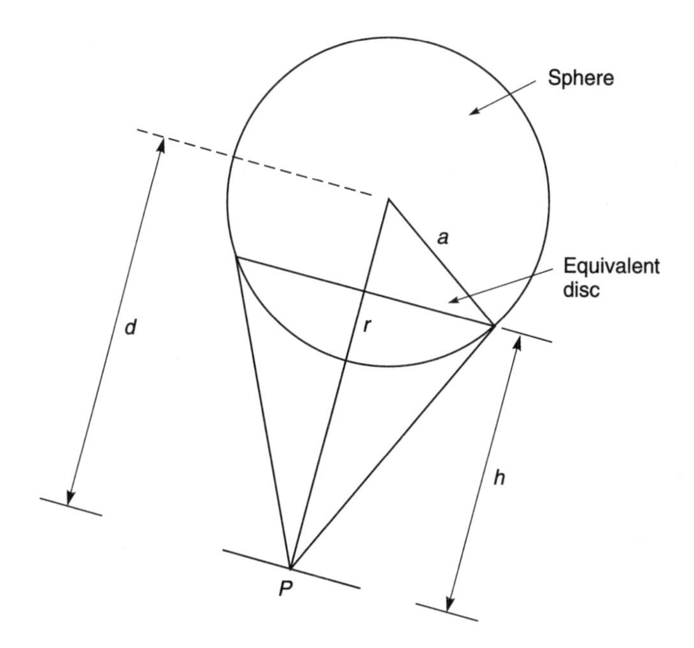

Fig. 3.24 The illuminance from a spherical source

So,

$$E_\mathrm{p} = \pi L \frac{a^2}{a^2 + d^2 - a^2}$$

$$= \frac{\pi L a^2}{d^2}$$

But, $\pi La^2 = I$ where I = luminous intensity of the sphere in any direction.

So,

$$E_\mathrm{p} = \frac{I}{d^2}$$

Thus the spherical uniformly diffuse source obeys the inverse-square law at all distances, provided that the distance is measured from the centre of the sphere (i.e. d).

The above result is a consequence of the fact that the illumination vector always acts from the centre of a diffusing sphere.

The illuminance on a tilted plane for any source is obtained by multiplying by the cosine of the angle between the normal to the illuminated plane and the direction of the illumination vector, giving

$$E_p = E_{\mathrm{max}} \cos\theta$$

$$= \frac{I}{d^2} \cos\theta$$

provided that the illuminated plane does not pass through the source when extended (if so, it gives the difference of the illuminance on each side of the plane).

An alternative and simple proof that a uniformly diffusing spherical source obeys the inverse-square law at all distances is indicated below.

The flux received by a sphere of radius r will be the same whatever the value of r. The illuminance on any point on the inside of an enclosing sphere is, therefore,

$$E = \frac{\phi}{4\pi r^2}$$

From this equation it is clear that the illuminance varies as the square of the distance of the point of measurement from the centre of the spherical source. It is interesting to note that this very simple proof does not depend upon the sphere being uniformly diffusing, but only upon uniform emission per unit area of source.

3.7 Non-uniformly diffuse area sources

Many light sources do not have uniformly diffusing properties and the luminance does vary with the angle of view. With the advent of powerful computing capacity, any area light source can be considered to consist of many elemental point sources and inverse-square law computations carried out and summed to give an accurate result, provided that the luminous intensity distribution for each element of area is known.

On the other hand, another approach with, or without, the aid of a computer is to consider the light source distribution to be approximated by a cosine power distribution, such as $\cos^2 \theta$ or $\cos^3 \theta$. Such approximations have been used to represent a louvred luminaire distribution ($\cos^2 \theta$) or luminaires with deep louvres ($\cos^3 \theta$). Also, those distributions that cannot be approximated in this simple way can usually be represented by a cosine series, such as:

$$I_\theta = A + B \cos \theta + C \cos^2 \theta + D \cos^3 \theta$$

It is, therefore, useful to develop formulae for area sources with these various cosine power distributions. However, point by point calculations are much more sensitive to errors due to such approximations and so they are more useful for flux calculations. An example of a flux calculation is given in Section 4.3.

3.7.1 THE DISC SOURCE

Consider the calculation of the illuminance on the horizontal plane at a point P directly beneath the disc (Figure 3.25).

The intensity of the element of the source $\mathrm{d}r \times \mathrm{d}l$ in the direction of P is given by

$$\mathrm{d}E_\mathrm{p} = \frac{L_\mathrm{m}\, \mathrm{d}l\, \mathrm{d}r \cos^n \theta \cos \theta}{D^2}$$

where L_m is the maximum luminance (when $\theta = 0°$).

Now $r = h \tan \theta$, so, $\mathrm{d}r = h \sec^2 \theta\, \mathrm{d}\theta$ and $D = h \sec \theta$ giving,

$$\mathrm{d}E_\mathrm{p} = \frac{L_\mathrm{m}\, \mathrm{d}l\, h \sec^2 \theta\, d\theta \cos^{(n+1)} \theta}{h^2 \sec^2 \theta}$$

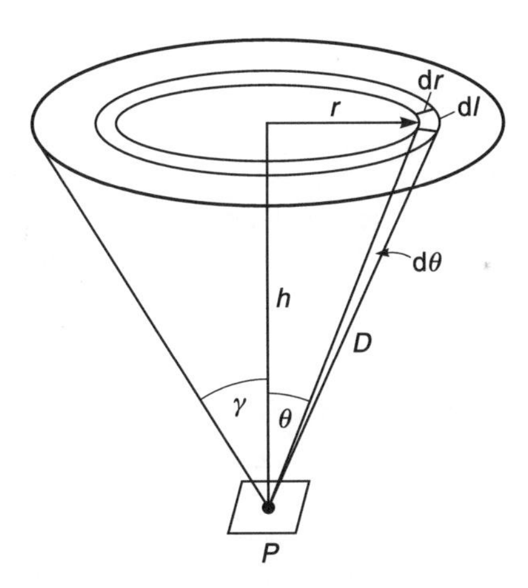

Fig. 3.25 A disc source with a cosine power intensity distribution

$$= \frac{L_m}{h} \, dl \cos^{(n+1)} \theta \, d\theta$$

The element of length dl produces the same illuminance at P as all the other elements of length dl making up the annular ring $2\pi r$ dr.

So,

$$dE_p = \frac{L_m}{h} 2\pi r \cos^{(n+1)} \theta \, d\theta$$

$$= 2\pi L_m \tan\theta \cos^{(n+1)} \theta \, d\theta$$

$$= 2\pi L_m \sin\theta \cos^n \theta \, d\theta$$

$$E_p = 2\pi L_m \int_{\theta=0}^{\gamma} \sin\theta \cos^n \theta \, d\theta$$

$$E_p = \frac{2\pi L_m}{(n+1)} [-\cos^{(n+1)} \theta]_{\theta=0}^{\gamma}$$

$$= \frac{2\pi L_m}{(n+1)} (1 - \cos^{(n+1)} \gamma)$$

when $n = 0$, $E_p = 2\pi L_m(1 - \cos\gamma)$.

Note: $2\pi(1 - \cos\gamma)$ is the solid angle.

When $n = 1$ (uniformly diffusing source),

$$E_p = \pi L \, (1 - \cos^2 \gamma)$$

$$= \pi L \sin^2 \gamma$$

where $\gamma = \alpha/2$ when this expression was deduced for a uniformly diffusing disc.

When $n = 2$,

$$E_p = \frac{\pi L_m}{3}(1 - \cos^3 \gamma)$$

These solutions only apply to a point directly under the centre of the disc and, except in the case of uniform diffusion (n=1), cannot be extended to relate to any point on the parallel plane.

3.7.2 THE TRIANGULAR SOURCE

The solution for the disc source (Figure 3.25) gives a convenient starting point for developing an illuminance equation for the triangular source (and hence the rectangular source) (Figures 3.26 and 3.27).

In Figure 3.27, the axis of the diagram has been rotated through 90° (compared with Figure 3.26) to enable the various angles and dimensions to be more clearly indicated.

Consider Figure 3.27:

$$dE_p = \frac{2\pi L}{(n+1)}(1 - \cos \gamma^{(n+1)}) \frac{d\theta}{2\pi}$$

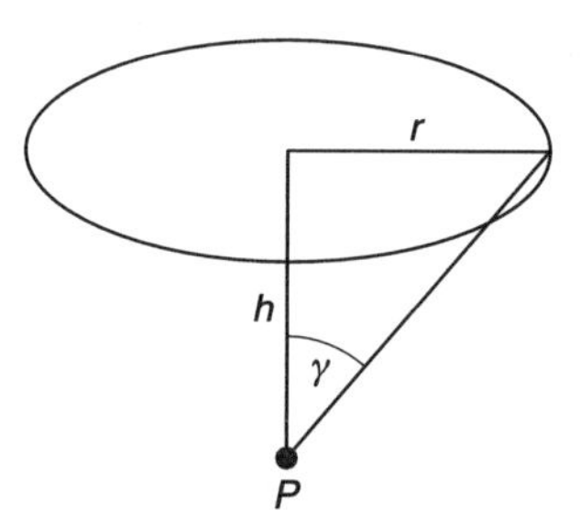

Fig. 3.26 The disc source relationships used in developing the triangular source formula

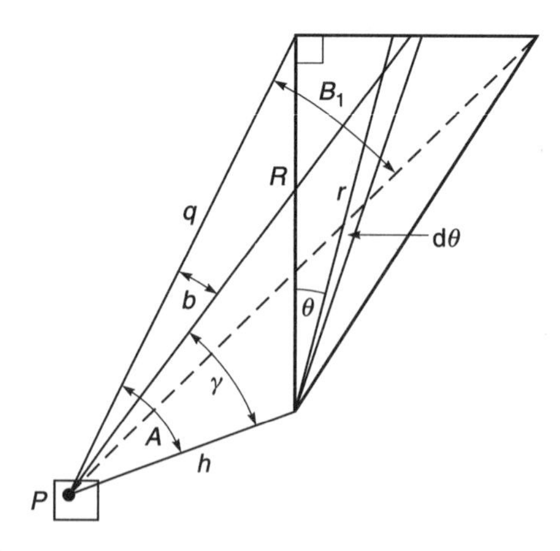

Fig. 3.27 Illuminance on a parallel plane from a cosine power distribution triangular source

Change the variable from θ to b:

$$\mathrm{d}\theta = (\sec^2 b \operatorname{cosec} A / \sec^2 \theta)\ \mathrm{d}b$$
$$= \left[\sec^2 b \operatorname{cosec} A \Big/ \frac{(R^2+h^2)\tan^2 b + R^2}{R^2}\right] \mathrm{d}b$$
$$= (\sec^2 b \operatorname{cosec} A / \operatorname{cosec}^2 A \tan^2 b + 1)\ \mathrm{d}b$$

Let cosec $A = C$:

$$\mathrm{d}\theta = (C\sec^2 b / (C^2 \tan^2 b + 1)\ \mathrm{d}b$$
$$\cos\gamma = \frac{h}{(h^2+r^2)^{\frac{1}{2}}} = \frac{h}{(h^2+(R^2+h^2)\tan^2 b + R^2)^{\frac{1}{2}}}$$
$$= \frac{1}{(1+((R/h)^2+1)\tan^2 b + (R/h)^2)^{\frac{1}{2}}}$$
$$= \frac{1}{(\sec^2 A \tan^2 b + \sec^2 A)^{\frac{1}{2}}}$$

Let sec $A = S$:

$$\cos\gamma = \frac{1}{(S^2 \tan^2 b + S^2)^{\frac{1}{2}}}$$
$$= \frac{1}{S\sec b}$$

So,

$$dE_{\mathrm{p}} = \frac{L_m}{(n+1)}\left(1 - \frac{1}{(S\sec b)^{(n+1)}}\right)\left(\frac{C\sec^2 b}{C^2\tan^2 b + 1}\right)\mathrm{d}b$$
$$E_{\mathrm{p}} = \frac{L_m C}{(n+1)}\int_0^{B_1}\left(1 - \frac{1}{(S\sec b)^{(n+1)}}\right)\left(\frac{\sec^2 b}{C^2\tan^2 b + 1}\right)\mathrm{d}b$$

Solving for $n = 0$:

$$E_{\mathrm{p}} = L_m C\int_0^{B_1}\left(1 - \frac{1}{S\sec b}\right)\left(\frac{\sec^2 b}{1 + C^2\tan^2 b}\right)\mathrm{d}b$$
$$= \frac{L_m S}{C}\int_0^{B_1}\left(\frac{1}{S + \cos b}\right)\mathrm{d}b$$

Let $t = \tan(b/2)$ so, $\mathrm{d}b = 2\mathrm{d}t/(1+t^2)$:

$$E_{\mathrm{p}} = \frac{L_{\mathrm{m}} S}{C} \int \frac{2\,\mathrm{d}t/(1+t^2)}{S+(1-t^2)/(1+t^2)}$$

Since,

$$\cos b = \frac{\cos^2 \frac{b}{2} - \sin^2 \frac{b}{2}}{\cos^2 \frac{b}{2} + \sin^2 \frac{b}{2}} = \frac{1-t^2}{1+t^2}$$

then

$$E_{\mathrm{p}} = \frac{2L_{\mathrm{m}} S}{C} \int \frac{\mathrm{d}t}{(S+1)+(S-1)t^2}$$

$$= \frac{2L_{\mathrm{m}} S}{C\sqrt{\frac{S+1}{S-1}}(S-1)} \int \frac{\sqrt{\frac{S+1}{S-1}}\,\mathrm{d}t}{\left(\frac{S+1}{S-1}\right)+t^2}$$

$$= 2L_{\mathrm{m}} \left[\tan^{-1}\left(\frac{\tan\frac{b}{2}}{\left(\frac{S+1}{S-1}\right)} \right) \right]_0^{B_1}$$

$$= 2L_{\mathrm{m}} \tan^{-1}\left(\tan\left(\frac{B_1}{2}\right) \tan\left(\frac{A}{2}\right) \right)$$

Solving for $n = 1$:

$$E_{\mathrm{p}} = \frac{L_{\mathrm{m}} C}{2} \int_0^{B_1} \left(1 - \frac{1}{(S\sec b)^2}\right)\left(\frac{\sec^2 b}{1+C^2\tan^2 b}\right) \mathrm{d}b$$

$$= \frac{L_{\mathrm{m}} C}{2} \int_0^{B_1} \left(\frac{S^2(1+\tan^2 b)-1}{S^2}\right)\left(\frac{1}{1+C^2\tan^2 b}\right) \mathrm{d}b$$

and since

$$S^2 = \left(\frac{C^2}{C^2-1}\right)$$

$$= \frac{L_m C}{2} \int_0^{B_1} \left(\frac{1 + C^2 \tan^2 b}{C^2} \right) \left(\frac{1}{1 + C^2 \tan^2 b} \right) db$$

$$= \frac{L_m}{2C} \int_0^{B_1} db$$

$$= \frac{L_m}{2} (B_1 \sin A)$$

Solving for $n = 2$:

$$E_p = \frac{L_m C}{3} \int_0^{B_1} \left(1 - \frac{1}{(S \sec b)^3} \right) \left(\frac{\sec^2 b}{1 + C^2 \tan^2 b} \right) db$$

$$= \frac{L_m}{3SC} \int_0^{B_1} \left(\frac{S^3 \sec^3 b - 1}{\sec b (S^2 \sec^2 b - 1)} \right) db$$

$$= \frac{L_m}{3SC} \int_0^{B_1} \left(\cos b + \frac{S^2}{S + \cos b} \right) db$$

$$= \frac{L_m}{3} \left(\sin A \cos A \sin B_1 + 2 \tan^{-1} \left(\tan \frac{A}{2} \tan \frac{B_1}{2} \right) \right)$$

Solving for $n = 3$:

$$E_p = \frac{L_m C}{4} \int_0^{B_1} \left(1 - \frac{1}{(S \sec b)^4} \right) \left(\frac{\sec^2 b}{1 + C^2 \tan^2 b} \right) db$$

$$= \frac{L_m C}{4S^4} \int_0^{B_1} \frac{\left(2 \frac{S^2}{C^2} (1 + C^2 \tan^2 b) + \frac{S^4}{C^4} (1 + C^2 \tan^2 b)^2 \right)}{\sec^2 b (1 + C^2 \tan^2 b)}$$

$$= \frac{L_m}{4CS^2} \left(S^2 B_1 + \int_0^{B_1} \cos^2 b \; db \right)$$

$$= \frac{L_m}{4} \sin A \cos^2 A \left(\frac{B_1}{\cos^2 A} + \frac{1}{2} B_1 + \frac{1}{2} \sin B_1 \cos B_1 \right)$$

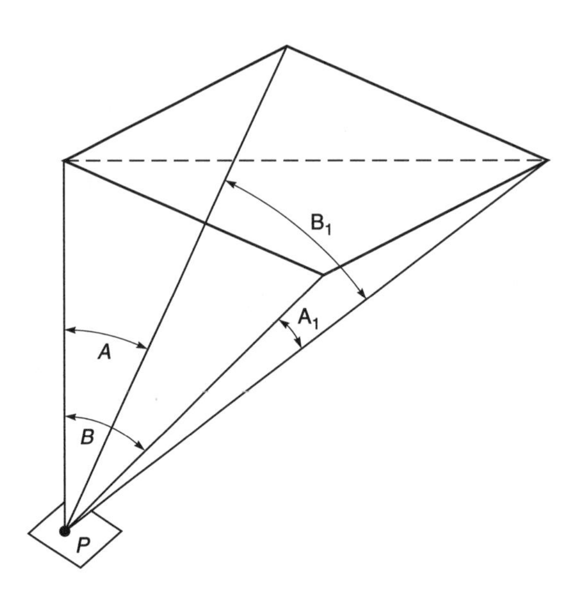

Fig. 3.28 The cosine power distribution rectangular source considered as two triangular sources

As in the section on diffuse sources, the above results can be extended to the case of a rectangle by considering the rectangle to be two right-angled triangles (Figure 3.28).

The illuminance expression then becomes:

$n = 0$

$$E_p = 2L_m\left(\tan^{-1}\left(\tan\left(\frac{A}{2}\right)\tan\left(\frac{B_1}{2}\right)\right) + \tan^{-1}\left(\tan\left(\frac{B}{2}\right)\tan\left(\frac{A_1}{2}\right)\right)\right)$$

$n = 1.0$ (uniformly diffuse source)

$$E_p = \frac{L_m}{2}(B_1 \sin A + A_1 \sin B)$$

$n = 2.0$

$$E_p = \frac{L_m}{3}\left[\sin A \cos A \sin B_1 + 2\tan^{-1}\left(\tan\left(\frac{A}{2}\right)\tan\left(\frac{B_1}{2}\right)\right) + \sin B \cos B \sin A_1 + 2\tan^{-1}\left(\tan\left(\frac{B}{2}\right)\tan\left(\frac{A_1}{2}\right)\right)\right]$$

$n = 3.0$

$$E_p = \frac{L_m}{4}\left[\sin A \cos^2 A\left(\frac{B_1}{\cos^2 A} + \frac{B_1}{2} + \frac{1}{2}\sin B_1 \cos B_1\right) + \sin B \cos^2 B\left(\frac{A_1}{\cos^2 B} + \frac{A_1}{2} + \frac{1}{2}\sin A_1 \cos A_1\right)\right]$$

An alternative derivation of these formulae is given by Bell.[3]

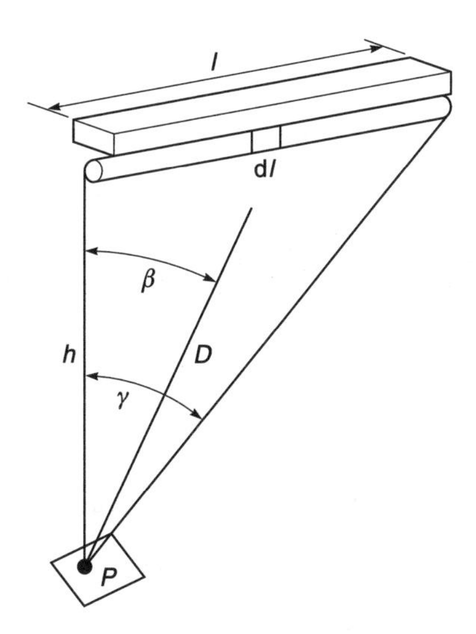

Fig. 3.29 Illuminance at a point opposite one end of a line source

3.7.3 LINE SOURCES

Consider the point P opposite one end of a line source (Figure 3.29). Many sources are long and narrow compared with the distance from the source to the illuminated point.

In this case, solutions may be obtained by the summation of the contributions from each element of length dl, to the illuminance at P. To do this we apply the inverse-square law to each element of the linear source and integrate.

Each element of length dl is assumed to have an intensity given by dividing the total intensity of the source, at the angle subtended from the element dl to the illuminated point, by the length of the source and multiplying this by dl (Figure 3.25).

This assumes that each element of the source contributes the same amount to the total intensity of the source. (This is usually an acceptable assumption, but is not always accurate enough if the point is close to the source, since the ends of a practical line source usually give a much reduced output within 0.15 m of the lamp cathodes.)

The intensity of element dl in direction β is given by

$$\mathrm{d}I_\beta = I_\beta \frac{\mathrm{d}l}{l}$$

Applying the inverse-square law, we obtain

$$\mathrm{d}E_\mathrm{p} = \frac{I_\beta \cos\beta \,\mathrm{d}l}{lD^2}$$

Now $S = h \tan\beta$ so, $\mathrm{d}l = h \sec^2\beta \,\mathrm{d}\beta$ which gives:

$$\mathrm{d}E_\mathrm{p} = \frac{I_\beta}{l} \times \frac{h \sec^2\beta \cos\beta \,\mathrm{d}\beta}{h^2 \sec^2\beta}$$

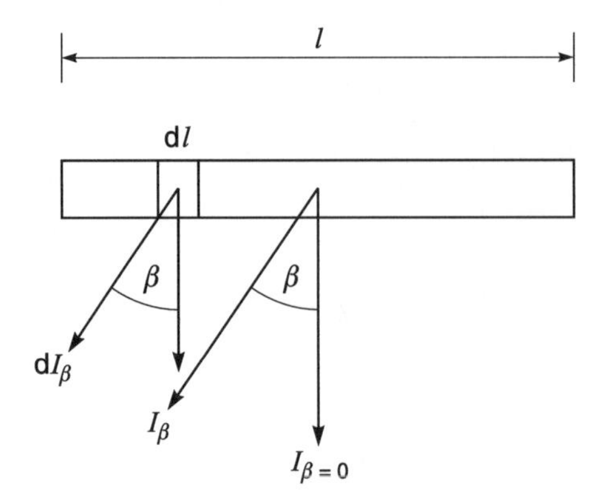

Fig. 3.30 The assumption that each equal length element contributes the same amount to the total intensity of the source

$$E_\text{p} = \int_{\beta=0}^{\gamma} \frac{I_\beta}{lh} \cos\beta \, \text{d}\beta$$

The solution depends upon the variation of I_β, given by the axial polar curve, over the range of the integral (Figures 3.31 and 3.32). The method fails if the axial intensity curve changes shape significantly with the transverse plane angle B (Figure 3.33).

(For further information about the B, β coordinate system see Chapter 1.)

If I_β is expressed in terms of I_B then

$$E_\text{p} = \frac{I_B}{1} \int_0^{\gamma} \left(\frac{I_\beta}{I_B}\right) \cos\beta \, \text{d}\beta$$

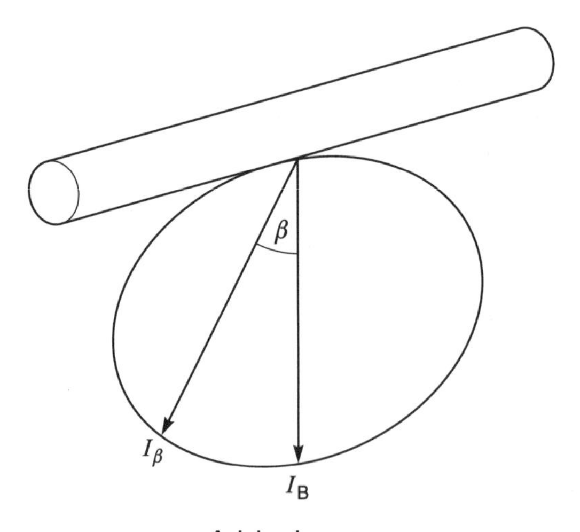

Fig. 3.31 The axial intensity curve

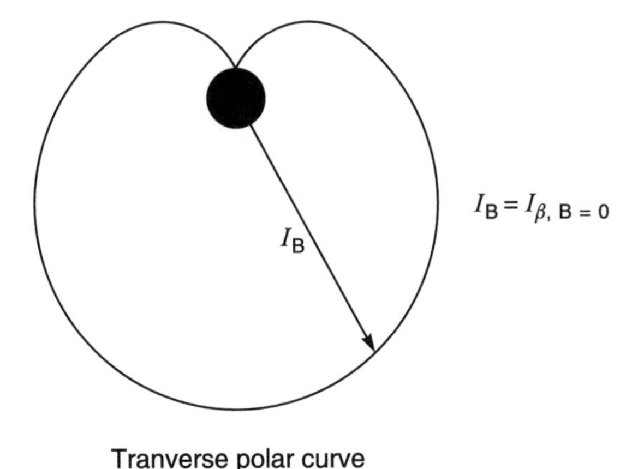

Fig. 3.32 The transverse intensity curve

$$\int_0^{\gamma} \left(\frac{I_\beta}{I_B}\right) \cos\beta \, \mathrm{d}\beta$$

is termed the *aspect factor* and given the symbol AF.

So,

$$E_\mathrm{p} = \frac{I_B}{lh}(AF)$$

Manufacturers sometimes produce tables of AF values for the practical distribution of their luminaires using numerical integration. In the case of a uniformly diffuse line source the aspect factor can be calculated from

$$AF = \int_0^{'} \cos^2 \beta \, \mathrm{d}\beta$$

$$= \frac{1}{2}(\gamma + \sin\gamma \cos\gamma)$$

If the illuminated point is displaced to one side (Figure 3.33) then

$$E_\mathrm{p} = \frac{I_B}{lh} AF \cos B$$

(since the cosine law applies at point P).

With regard to the value of I_B, for uniformly diffusing line sources there are two cases: (1) the flat strip source, and (2) the cylindrical source (Figures 3.34 and 3.35).

(1) $I_B = I_{B(0°)} \cos B$
(2) $I_B = I_{B(0°)}$
$\quad = \text{constant}$

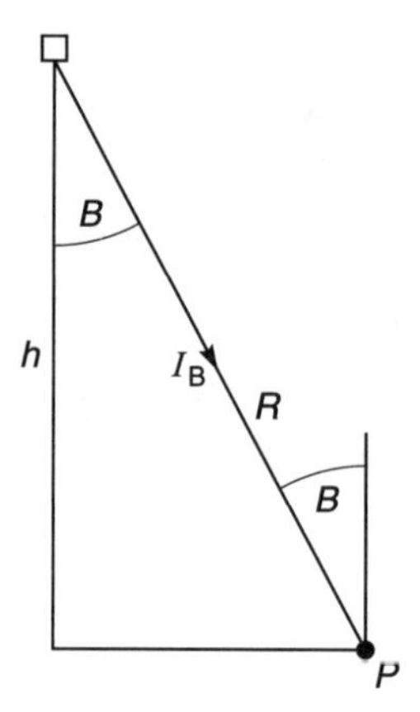

Fig. 3.33 The method fails if the axial intensity curves change shape significantly with the transverse plane angle B

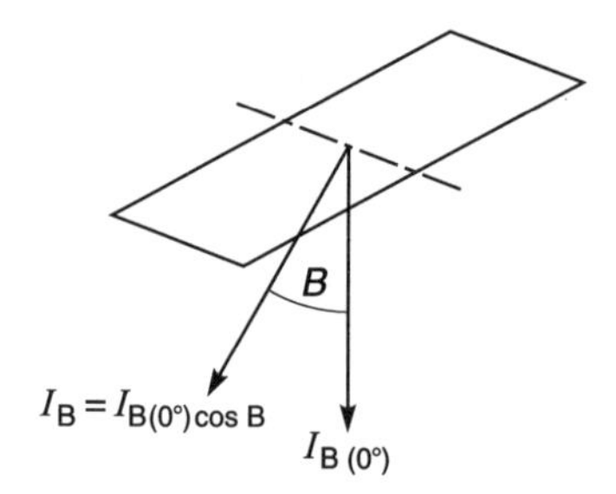

Fig. 3.34 The flat strip line source

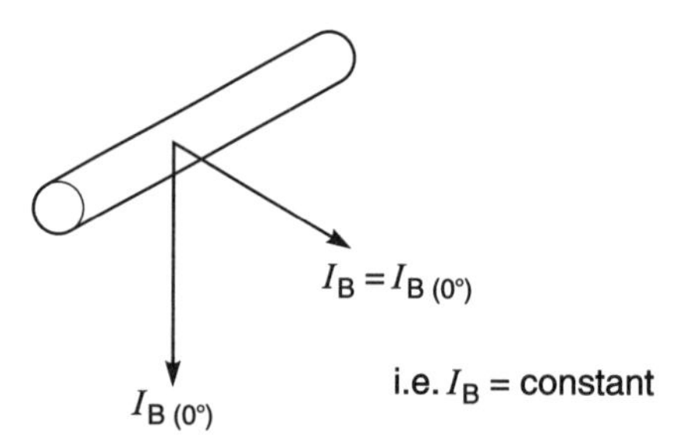

Fig. 3.35 The cylindrical line source

Illuminance on a perpendicular plane parallel to the line source (Figure 3.36)

Since the normal at P has moved through 90° towards the source

$$E_{\mathrm{p(vi)}} = \frac{I_B}{lR} AF \cos\left(\frac{\pi}{2} - B\right)$$

$$= \frac{I_B}{lR} AF \sin B$$

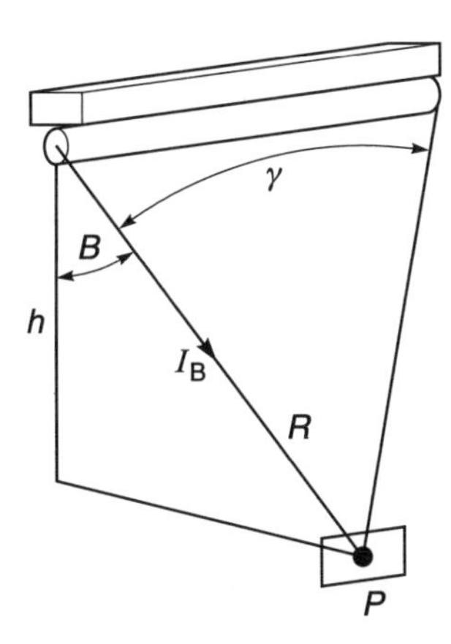

Fig. 3.36 Illuminance on a perpendicular plane parallel to the line source

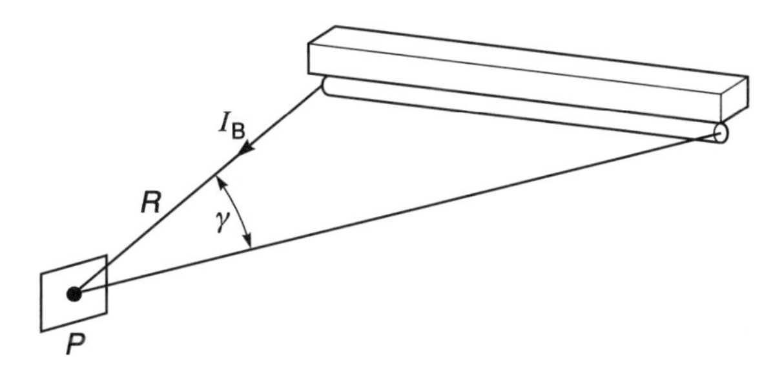

Fig. 3.37 Illuminance on a perpendicular plane normal to the axis of the line source

Illuminance on a perpendicular plane normal to the axis of the source (Figure 3.37)

The derivation is similar to that for the parallel plane except that, since the normal to P is now parallel to the axis of the line source, $\cos\beta$ becomes

$$\cos\left(\frac{\pi}{2}-\beta\right) \text{ or } \sin\beta.$$

$$E_{p(v)} = \frac{I_B}{lR}\int_0^{\gamma}\left(\frac{I_\beta}{I_\beta}\right)\sin\beta\,\mathrm{d}\beta$$

$$= \frac{I_B}{lR}(af)$$

where (af) is the perpendicular plane aspect factor

$$af = \int_0^{\gamma}\left(\frac{I_\beta}{I_B}\right)\sin\beta\,d\beta$$

This time, when the point is displaced from beneath the end of the line source, the cosine law does not operate in the transverse direction since, in effect, the plane of P has been rotated about its normal.

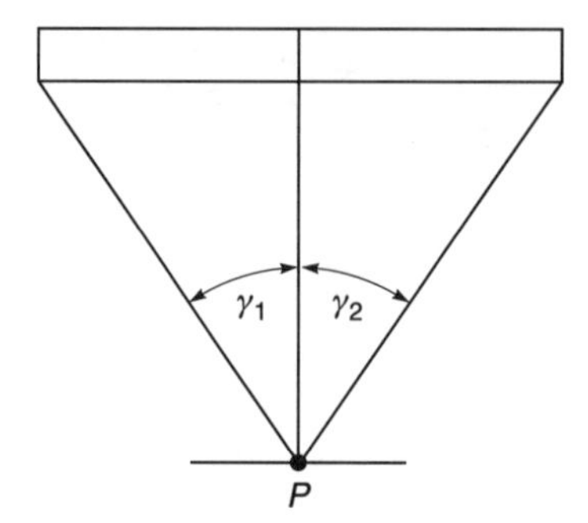

Fig. 3.38 The method of component sources applied to the case where the illuminated point is not opposite one end (beneath the source)

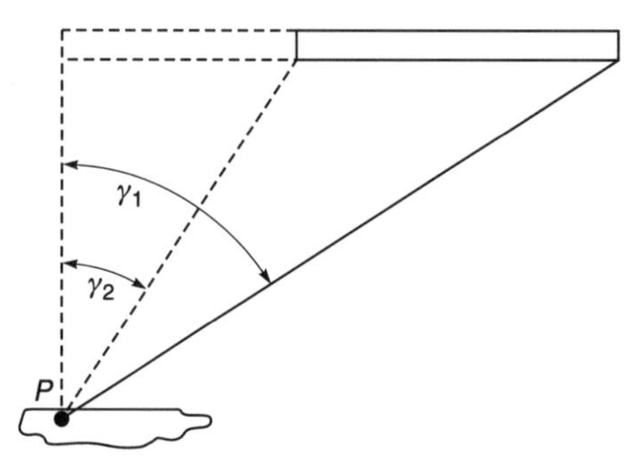

Fig. 3.39 The method of component sources applied to the case where the illuminated point is not opposite one end (beyond the end of the source)

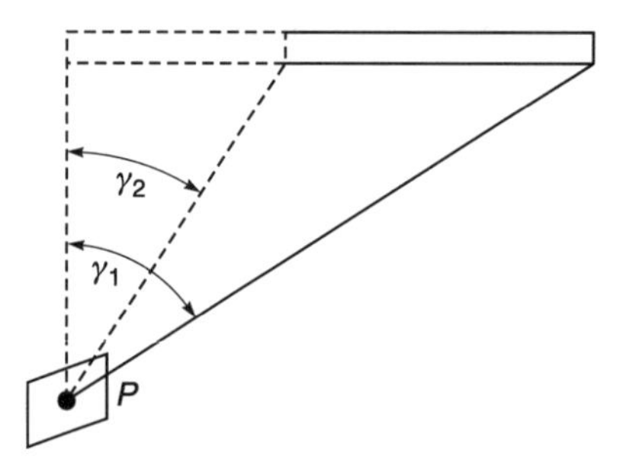

Fig. 3.40 As in Fig. 3.39, but on the perpendicular plane

Illuminance from a line source when the point is not opposite one end

The method of component sources is used to extend the expressions for line sources already developed to other cases (Figures 3.38, 3.39 and 3.40).

$$E_{\mathrm{p}} = E_{\gamma_1} + E_{\gamma_2} = \frac{I_B}{IR}(AF_{\gamma_1} + AF_{\gamma_2}) \cos B$$

$$E_{\text{p}} = E_{\gamma_1} - E_{\gamma_2} = \frac{I_B}{lR}(AF_{\gamma_1} - AF_{\gamma_2}) \cos B$$

$$E_{\text{p(v)}} = E_{(\text{v})\gamma_1} - E_{(\text{v})\gamma_2}$$

$$E_{\text{p(v)}} = \frac{I_B}{lR}(af(\gamma_1) - af(\gamma_2))$$

3.7.4 *ILLUMINANCE FROM VERY LARGE AREA SOURCES*

If we consider an area source that is very large in relation to the distance from the source to a point on a parallel plane, we can use the expression for a cosine power disc source to examine the variation of illuminance with distance (Figure 3.41).

$$E = \frac{2\pi L_{\text{m}}}{n+1}(1 - \cos^{(n+1)}\gamma)$$

From Figure 3.41 we can see that when the source is very large $\gamma \to \pi/2$ and so $\cos \gamma \to 0$. So,

$$E = \frac{2\pi L_{\text{m}}}{n+1} \quad \text{where} \quad L_{\text{m}} = \frac{I_{\text{m}}}{\delta A}$$

The relationship between illuminance E and luminous exitance M is derived as follows:

$$\delta F = 2\pi \int_0^{\pi/2} I_\theta \sin\theta \, d\theta$$

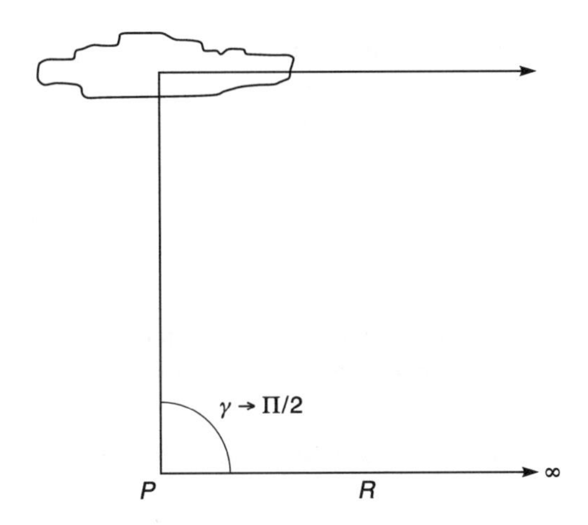

Fig. 3.41 Angle relationships for a very large area source

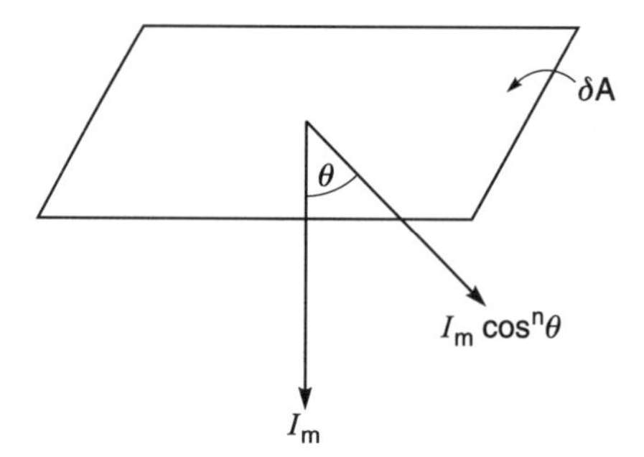

Fig. 3.42 The intensity at angle θ is given by $I_m \cos^n \theta$

We note from Figure 3.42 that

$$I_\theta = I_m \cos^n \theta$$

$$= 2\pi I_m \int_0^{\pi/2} \cos^n \theta \sin\theta \, d\theta$$

$$= \frac{2\pi I_m}{n+1} [-\cos^{(n+1)} \theta]_0^{\pi/2}$$

$$= \frac{2\pi I_m}{n+1} [1.0] = \frac{2\pi L_m \delta A}{n+1}$$

or

$$M = \frac{\delta F}{\delta A} = \frac{2\pi L_{\mathrm{m}}}{n+1}$$

Thus, $E = M$, i.e. the illuminance, is independent of the distance from the source to the illuminated plane.

So, the flux received per unit area equals that emitted by the source per unit area (M) (which is, of course, not a startling result). A similar result would be obtained for a rectangular source.

It is of interest to consider what happens at the boundaries of such a large source. Consider Figures 3.43 and 3.44, which relate to a rectangular source.

At the boundary of the source away from a corner

$$E_{\mathrm{p(edge)}} = \frac{E}{2}$$

since the effect is similar to dividing the source in two and removing one half.

Similarly, at the corner of such a source

$$E_{\mathrm{p(corner)}} = \frac{E}{4}$$

since now three quarters of the source has been removed.

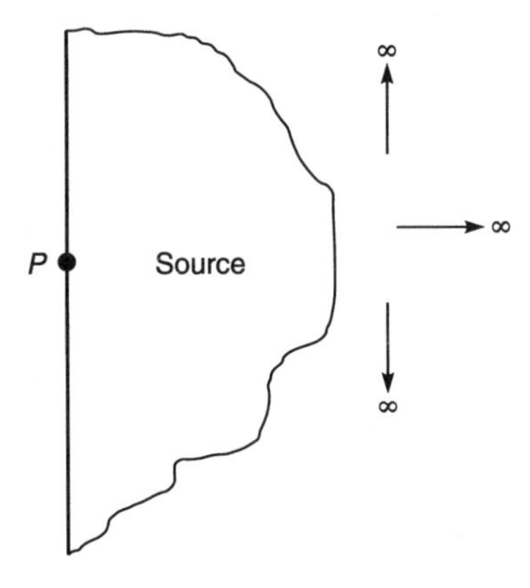

Fig. 3.43 The illuminance under the edge of a very large rectangular area source

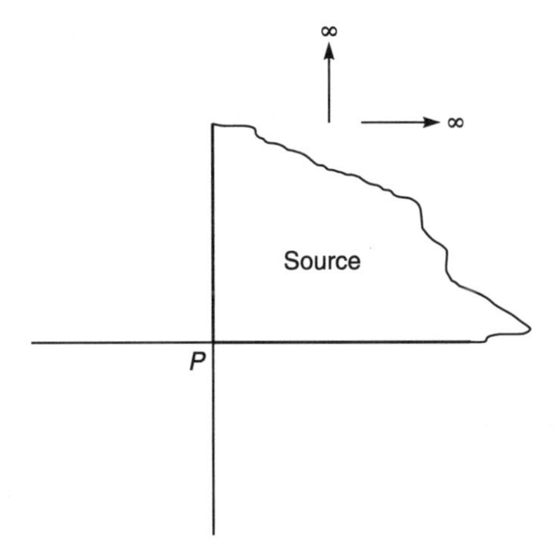

Fig. 3.44 The illuminance under the corner of a very large rectangular source

This means that, although a very large source, such as a luminous ceiling, may provide very good uniformity over most of the area, at the edges the illuminance will fall to half and at the corners to one quarter of the general illuminance.

3.7.5 ILLUMINANCE FROM A VERY LONG LINE SOURCE

Consider Figure 3.45:

$$E_{\mathrm{p}} = \frac{I_B}{lh} 2\left(\frac{1}{2}(\gamma + \sin\gamma\cos\gamma)\right)$$

Let $\gamma = \gamma_1 = \gamma_2$,

$$E_{\mathrm{p}} = \frac{I_B}{lh}\left(\frac{\pi}{2}\right)$$

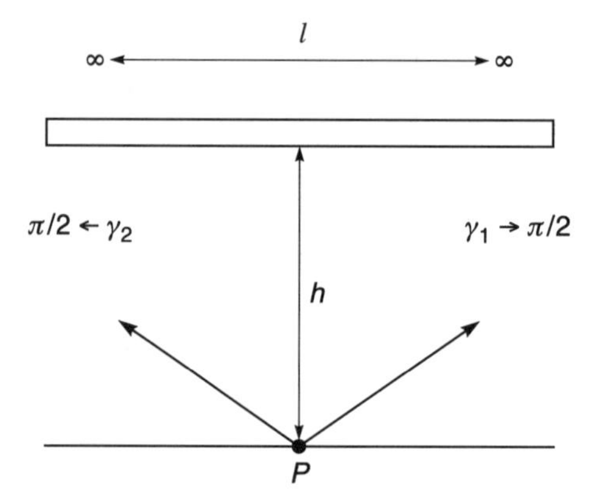

Fig. 3.45 Angle relationships for a very long line source

The illuminance is now given by $\pi/2$ multiplied by intensity per unit length of the source divided by h the distance.

The illuminance is, therefore, inversely proportional to the distance.

Using the same argument as for the boundary of a very large area source, we find the illuminance at the end of a very long line source would tend to be half that beneath the source away from the ends.

$$E_{\mathrm{p}} = \left(\frac{I_B}{l}\right)\frac{\pi}{4h}$$

3.8 Non-planar illuminance

3.8.1 SCALAR ILLUMINANCE (MEAN SPHERICAL ILLUMINANCE FROM A POINT SOURCE)

Scalar illuminance is a measure of the average density of the light flux at the point under consideration. It corresponds to the average illuminance on an infinitesimal sphere located at that point (Figures 3.46, 3.47 and 3.48).

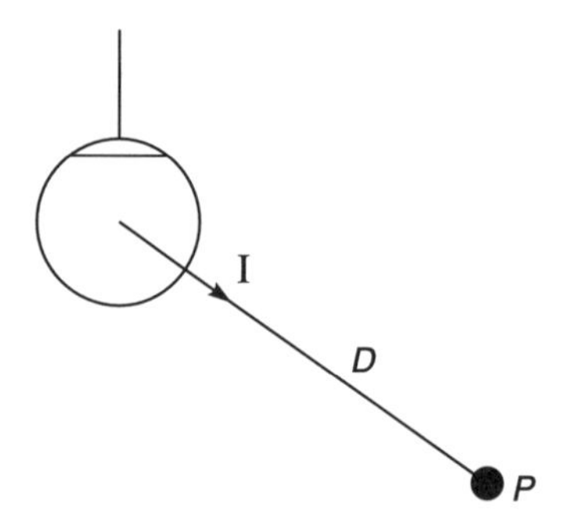

Fig. 3.46 Scalar illuminance at a point from a point source

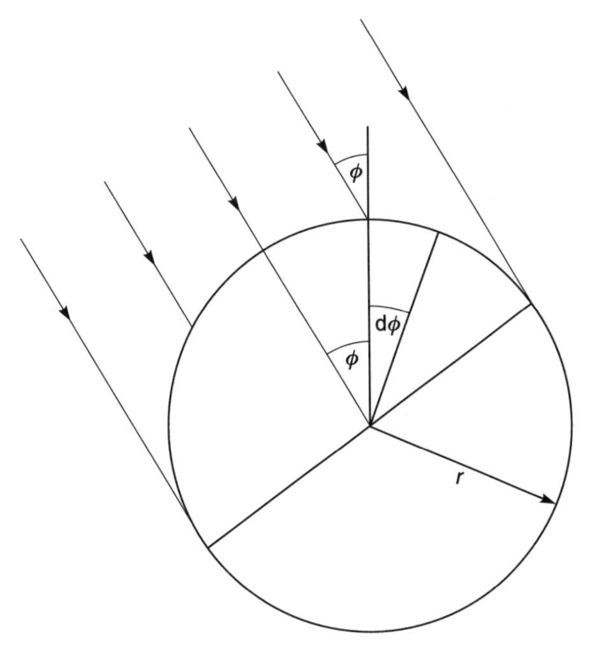

Fig. 3.47 The angular relationships for the calculation of scalar illuminance

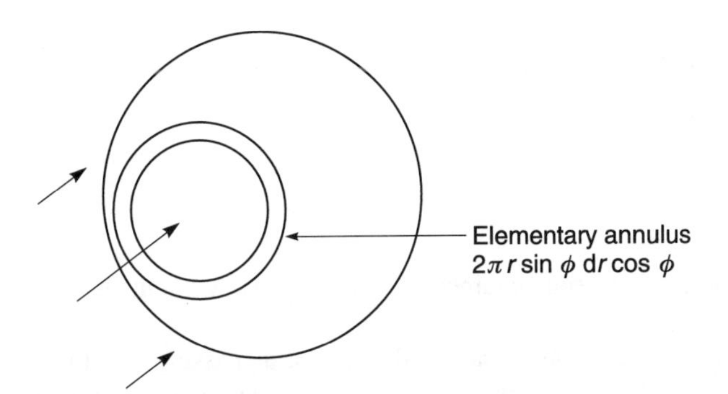

Fig. 3.48 The elementary annulus

The flux received by the illuminated hemisphere

$$\int_{\phi=0}^{\pi/2} \frac{I}{D^2} 2\pi r \sin\phi \; r\mathrm{d}\phi \cos\phi$$

$$= E_{\max}\left[\frac{\sin^2\phi}{2}\right]_0^{\pi/2} 2\pi r^2$$

$$= E_{\max}\pi r^2$$

where

$$E_{\max} = \frac{I}{D^2}$$

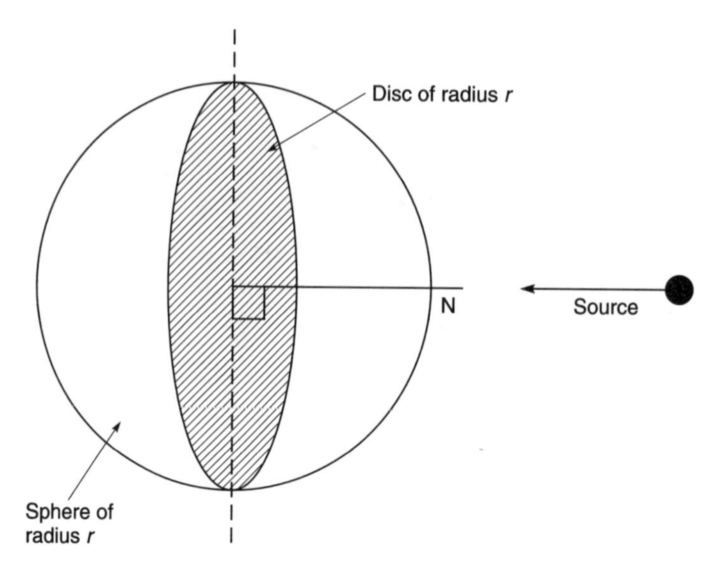

Fig. 3.49 The 'disc' method of calculating the scalar illuminance

Average illuminance of the sphere

$$E_s = E_{max} \frac{\pi r^2}{4\pi r^2}$$

$$= \frac{E_{max}}{4}$$

This result could have been obtained in another way, without the necessity of integrating (Figure 3.49).

The flux received by the sphere can be thought of as passing into the sphere where all of it would be intercepted by a disc of radius r with its normal towards the point source.

Flux received by disc = $E_{max} \times \pi r^2$

$$\text{Average illuminance of the sphere} = E_{max} \left(\frac{\pi r^2}{4\pi r^2} \right)$$

$$= \frac{E_{max}}{4}$$

3.8.2 *CYLINDRICAL ILLUMINANCE (MEAN VERTICAL ILLUMINANCE AT A POINT FROM A POINT SOURCE)*

The cylindrical illuminance is the average illuminance on the curved surface of an infinitesimal cylinder placed at the point of interest. The axis of the cylinder is usually taken as vertical (Figures 3.50, 3.51 and 3.52).

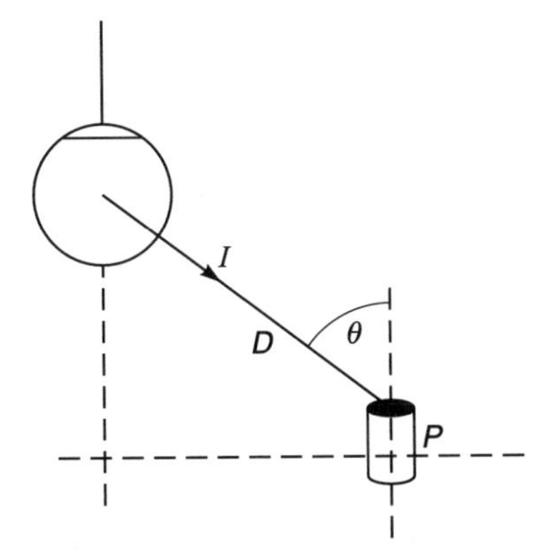

Fig. 3.50 Cylindrical illuminance at a point from a point source

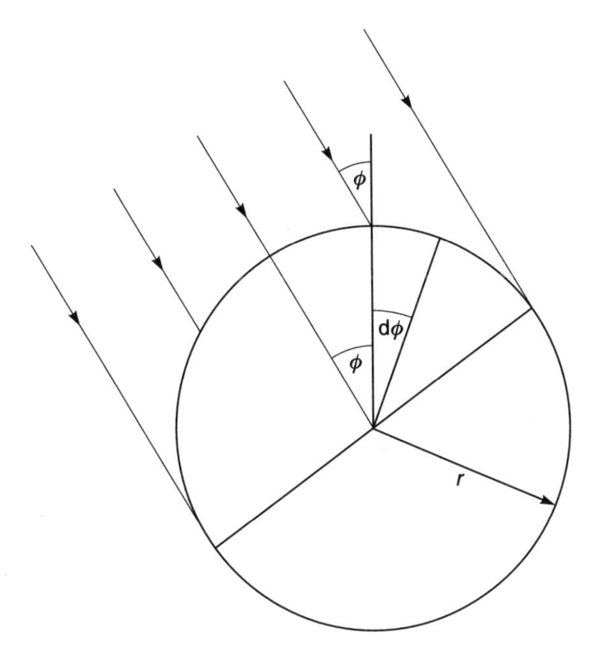

Fig. 3.51 The angular relationships for the calculation of cylindrical illuminance

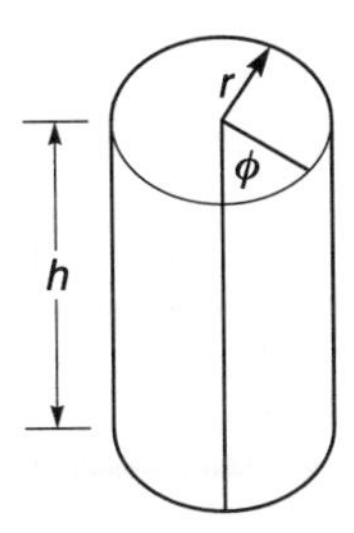

Fig. 3.52 The cylinder is usually considered to be vertical

Flux received over the half cylinder is given by:

$$\int_{-(\pi/2)}^{+(\pi/2)} E_{\text{max}} \sin\theta[\cos\phi \ \mathrm{d}\phi \ rh]$$

where

$$E_{\text{max}} = \frac{I}{D^2}$$

$$= E_{\text{max}} rh \sin\theta[\sin\phi]_{-[\pi/2]}^{+[\pi/2]}$$

$$= 2E_{\text{max}} rh \sin\theta$$

Average illuminance of the whole cylinder

$$E_{\text{cyl}} = \frac{2E_{\text{max}} \ rh \sin\theta}{2\pi rh}$$

$$= E_{\text{max}} \frac{\sin\theta}{\pi}$$

As for scalar illuminance, this result could have been obtained in another way, without the necessity for integration (Figure 3.53).

The flux received by the cylinder could be thought of as passing into the cylinder and being intercepted by a rectangle across the diameter of the cylinder, such that the plane of the normal to the rectangle is also the plane of the direction from the source to the cylinder. The flux reaching the bottom of the cylinder that does not fall on the rectangle is exactly compensated by the flux passing into the top of the cylinder and falling on the rectangle.

Flux received by the rectangle $= E_{\text{max}} \sin\theta h2r$

Average illuminance of cylinder $= E_{\text{max}} \sin\theta \left[\frac{h2r}{2\pi rh}\right]$

$$= E_{\text{max}} \frac{\sin\theta}{\pi}$$

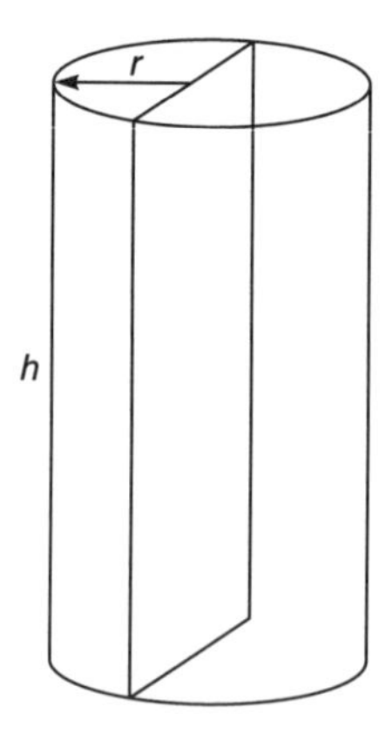

Fig. 3.53 The 'rectangle' method of calculating the cylindrical illuminance

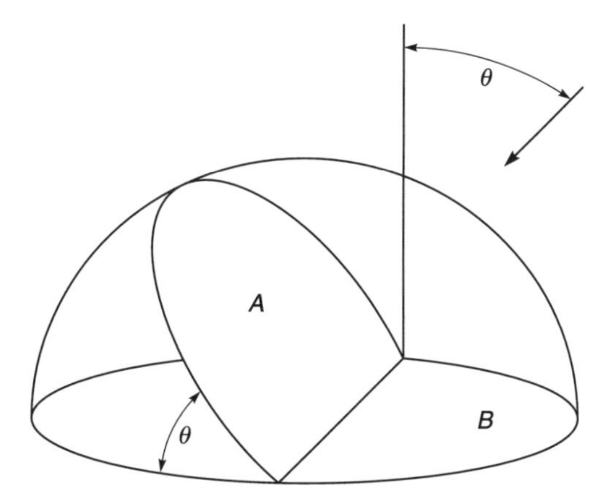

Fig. 3.54 The calculation of hemispherical illuminance

3.8.3 HEMISPHERICAL ILLUMINANCE (FROM A POINT SOURCE)

The second method used in the previous two cases will now be applied to obtain an expression for the hemispherical illuminance at a point.

The hemispherical illuminance is the average illuminance on the curved surface of an infinitesimal hemisphere, which has its base at the reference point (Figure 3.54).

The flux incident on the hemisphere would be equal to that intercepted by the half circular planes *A* and *B* if the flux is passed into the hemisphere.

Flux received by the hemisphere

$$= E_{max} \times \frac{\pi r^2}{2} + E_{max} \cos\theta \frac{\pi r^2}{2}$$

The average illuminance on the hemisphere

$$E_{hem} = E_{max} \frac{\left[\frac{\pi r^2}{2} + \cos\theta \frac{\pi r^2}{2}\right]}{2\pi r^2}$$

$$= \frac{E_{max}}{4}(1 + \cos\theta)$$

3.8.4 SEMI-CYLINDRICAL ILLUMINANCE (FROM A POINT SOURCE)

The second method is also used to obtain an expression for the semi-cylindrical illuminance from a point source (Figure 3.55).

As in the previous case the flux received on the curved surface of the half cylinder is equal to that which would have been intercepted by the rectangles *A* and *B* if the flux passed into the half cylinder. Rectangle *A* is normal to the plane of θ and rectangle *B* occupies half of the diameter of the half cylinder as shown.

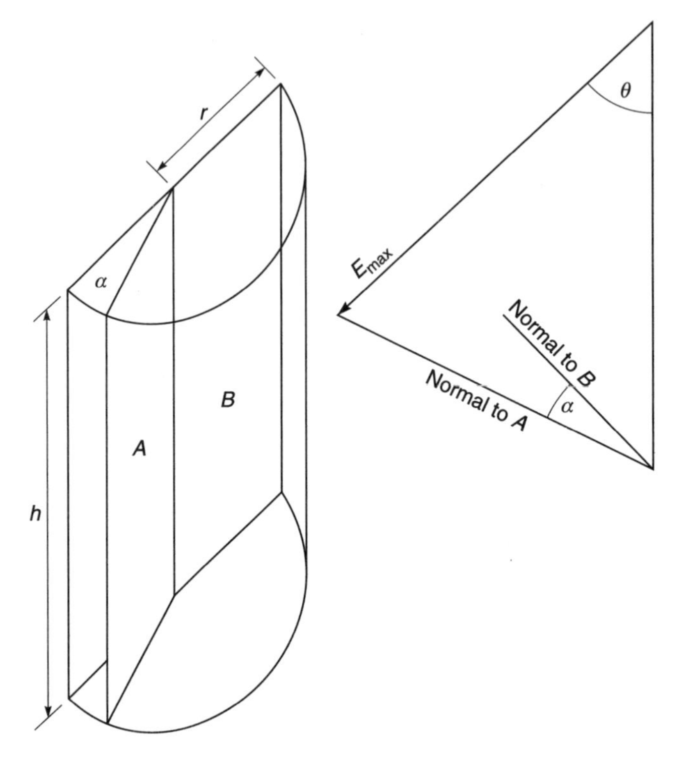

Fig. 3.55 The calculation of semi-cylindrical illuminance

$$\text{Flux received by the half cylinder} = (E_{max} r \times h + E_{max} \cos \alpha\, rh) \sin \theta$$

$$\text{Average illuminance of the semi-cylinder} = E_{max} \sin \theta \left[\frac{hr + \cos \alpha\, rh}{\pi rh} \right]$$

$$E_{sem\text{-}cyl} = E_{max} \frac{\sin \theta}{\pi} [1 + \cos \alpha]$$

3.8.5 CONICAL ILLUMINANCE (FROM A POINT SOURCE)

Another possible measure of the illuminance distribution about a point on a horizontal plane would be the conical illuminance, that is, the average illuminance on an infinitesimal cone placed at that point. The cone could have any angle, but a 45° slope is probably the most useful. An expression for conical illuminance is derived below (Figure 3.56).

Figure 3.56 shows both a plan view of the cone and a side, or elevation, view. On the plan view the projection outline of the top and side of the cone are also shown. The method is to determine the area of this projection in terms of the angle of the cone α and the angle of illuminance θ. The flux received onto this projected area will be the flux received by the cone.

$$\text{Area of triangle } AOB = \frac{1}{2} AB \times AO$$

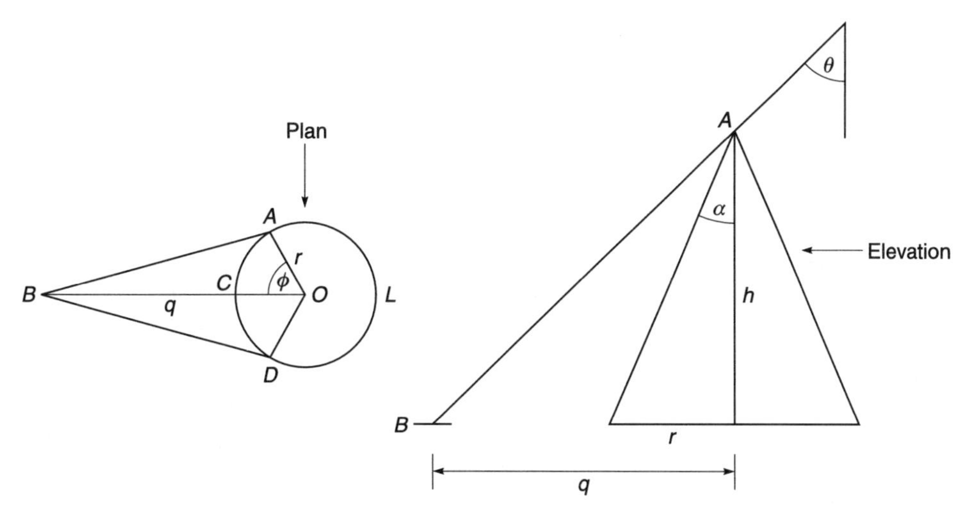

Fig. 3.56 The calculation of conical illuminance

$$= \frac{1}{2} r\sqrt{q^2 - r^2}$$

$$\text{Area } AODC = r^2\phi$$

$$\text{Area } ALDO = \pi r^2 - r^2\phi$$

$$\text{Area } BALD = r\sqrt{q^2 - r^2} + r^2(\pi - \phi)$$

$$\text{Flux received on this projected area } = \frac{I\cos\theta}{d^2}(r\sqrt{q^2 - r^2} + r^2(\pi - \phi))$$

$$\text{Area of cone } = \pi r\sqrt{r^2 + h^2}$$

$$q = h\tan\theta \quad \cos\Phi = \frac{r}{q} = \frac{r}{h\tan\theta}$$

$$E_{\text{con}} = \frac{E_{\max}}{\pi}\cos\theta\left(\frac{r\sqrt{q^2 - r^2} + r^2(\pi - \phi)}{r\sqrt{r^2 + h^2}}\right)$$

$$= \frac{E_{\max}}{\pi}\cos\theta\left(\frac{\sqrt{\tan^2\theta - \tan^2\alpha} + \tan\alpha\left[\pi - \cos^{-1}\left(\frac{\tan\alpha}{\tan\theta}\right)\right]}{\sqrt{1 + \tan^2\alpha}}\right)$$

$$= \frac{E_{\max}}{\pi}\cos\theta\cos\alpha\left(\sqrt{\tan^2\theta - \tan^2\alpha} + \tan\alpha\left[\pi - \cos^{-1}\left(\frac{\tan\alpha}{\tan\theta}\right)\right]\right)$$

For the case when $\alpha = 45°$

$$E_{con} = \frac{E_{max}}{\pi} 0.707 \cos\theta \, [\sqrt{\tan^2\theta - 1} + \pi - \cos^{-1}(\cot\theta)]$$

When $\alpha = 0°$,

$$E_{con} = \frac{E_{max}}{\pi} \sin\theta$$

which is the expression for cylindrical illuminance.

When $\theta = 0°$ the light is incident from directly above the cone and the profile is a disc,

$$E_{con} = \frac{E_{max}\pi r^2}{\pi r \sqrt{r^2 + h^2}} = E_{max} \frac{r}{\sqrt{r^2 + h^2}} = E_{max} \sin\alpha$$

When $\theta \leq \alpha$,

$$E_{con} = E_{max} \cos\theta \sin\alpha$$

When $\theta = 90°$ the formula given above becomes indeterminate since the projected area becomes infinite and the illuminance becomes zero.

In this case the conical illuminance is obtained from the value of the vertical illuminance multiplied by the area of the profile of the cone at this angle (which is a triangle) divided by the surface area of the cone:

$$E_{con} = \frac{I \sin\theta}{d^2} \times \frac{r \times h}{\pi r \sqrt{r^2 + h^2}}$$

$$= \frac{E_{max}}{\pi} \cos\alpha$$

3.9 The scalar product

In certain complex situations, the use of the scalar product from vector analysis allows the equations needed for calculating illuminance to be readily determined.[4] Such situations occur in the lighting of sports stadia, for instance, where the lighting is provided from a number of towers and the illuminance is required on planes facing the camera position, and on the four vertical faces of a cube, which are dealt with in Section 12.3.2. The scalar product provides a very elegant way of tackling this problem.

We need to establish the notation. Vectors are shown in bold type face, for instance ***AB*** or ***s***. A vector of unit magnitude is called a unit vector and three such vectors are used in the later chapters of this book. These are ***n***, which specifies the direction of the normal with respect to the facet or plane containing the illuminated point, ***q***, which specifies the direction of the light source from the illuminated point, and ***e***, which specifies the direction of the illuminance vector.

To signify the direction of the unit vector it may be given a subscript; for example, the normal in the *x* direction is denoted by $\boldsymbol{n}_x$.

The scalar product between two vectors is denoted by a dot. Thus the scalar product of ***a*** and ***b*** is shown as $\boldsymbol{a} \cdot \boldsymbol{b}$. The position of a point in space is expressed by its coordinates relative to the origin in the form (x, y, z), which is sometimes more convenient to use than $x\boldsymbol{n}_x + y\boldsymbol{n}_y + z\boldsymbol{n}_z$.[a]

The modulus or magnitude of a vector is indicated as follows:

$$|\boldsymbol{LM}| = LM$$

3.9.1 DETERMINATION OF THE UNIT VECTOR

Consider a plane facet at the point P given by coordinates (x_p, y_p, z_p). Let the normal to the facet pass through the point S with coordinates (x_s, y_s, z_s), then the distance between the two points is

$$PS = \sqrt{(x_s - x_p)^2 + (y_s - y_p)^2 + (z_s - z_p)^2} \quad (3.3)$$

so

$$\boldsymbol{n}_p = \frac{(x_s - x_p, y_s - y_p, z_s - z_p)}{\sqrt{(x_s - x_p)^2 + (y_s - y_p)^2 + (z_s - z_p)^2}}$$

If the plane facet has its normal parallel to the x axis all the y and z terms are zero. So

$$\frac{x_s - x_p}{\sqrt{(x_s - x_p^2)}} = 1$$

which gives

$$\boldsymbol{n}_x = (1, 0, 0)$$

Similarly,

$$\boldsymbol{n}_y = (0, 1, 0)$$

and

$$\boldsymbol{n}_z = (0, 0, 1)$$

For our purpose it is often convenient to express a vector in terms of its components along the three major axes, thus a vector $\boldsymbol{d}$ could be expressed as follows:

$$\boldsymbol{d} = (\boldsymbol{d}\cdot\boldsymbol{n}_x)\boldsymbol{n}_x + (\boldsymbol{d}\cdot\boldsymbol{n}_y)\boldsymbol{n}_y + (\boldsymbol{d}\cdot\boldsymbol{n}_x)\boldsymbol{n}_x$$

3.9.2 ANGLE BETWEEN TWO VECTORS

If θ is the angle between two vectors $\boldsymbol{LM}$ and $\boldsymbol{PQ}$ then

$$\cos\theta = \frac{\boldsymbol{LM}\cdot\boldsymbol{PQ}}{LM \times PQ} \quad (3.4)$$

Let

$$\boldsymbol{LM} = a\boldsymbol{n}_x + b\boldsymbol{n}_y + c\boldsymbol{n}_z$$

[a] In some texts $\boldsymbol{i}, \boldsymbol{j}$ and $\boldsymbol{k}$ are used instead of $\boldsymbol{n}_x$, $\boldsymbol{n}_y$ and $\boldsymbol{n}_z$.

and

$$\boldsymbol{PQ} = e\boldsymbol{n}_x + f\boldsymbol{n}_y + g\boldsymbol{n}_z$$

then

$$\boldsymbol{LM \cdot PQ} = ae + bf + cg$$

and so from (3.4)

$$\cos\theta = \frac{ae + bf + cg}{\sqrt{(a^2 + b^2 + c^2)(e^2 + f^2 + g^2)}}$$

If $\boldsymbol{LM} = \boldsymbol{n}_x$ so that $a = 1$, $b = 0$ and $c = 0$ then

$$\cos\theta_x = \frac{e}{\sqrt{(e^2 + f^2 + g^2)}}$$

where θ_x denotes the angle between the vector $\boldsymbol{PQ}$ and the x axis. Similarly,

$$\cos\theta_y = \frac{f}{\sqrt{(e^2 + f^2 + g^2)}}$$

and

$$\cos\theta_z = \frac{g}{\sqrt{(e^2 + f^2 + g^2)}}$$

$\cos\theta_x$, $\cos\theta_y$ and $\cos\theta_z$ are known as the direction cosines and the relationship $\cos^2\theta_y + \cos^2\theta_y + \cos^2\theta_z = 1$ is useful.

3.9.3 ILLUMINANCE AT A POINT

$$E_p = \frac{I\cos\theta}{PQ^2}$$

$$= I\frac{\boldsymbol{PQ \cdot n}_p}{|\boldsymbol{PQ}|^2}$$

where

E_p is the illuminance at the point P
I is the luminous intensity directed from the point Q towards P;
θ is the angle of incidence of the light;
$\boldsymbol{n}_p$ is the unit normal to the facet at P.

If $0 < \cos\theta \leq 1$ then the facet is facing the source and receives light. If, on the other hand, $-1 \leq \cos\theta \leq 0$ then the facet does not receive light.

When $PQ > 0$ the sign of $\boldsymbol{PQ \cdot n}_p$ is sufficient to determine whether the facet sees the source:

if $\boldsymbol{PQ}\cdot\boldsymbol{n}_p > 0$ the facet sees the source,
if $\boldsymbol{PQ}\cdot\boldsymbol{n}_p \leq 0$ the facet does not see the source,
in addition if $\boldsymbol{PQ} = 0$ the source and the facet coincide, which cannot happen in practice.

3.9.4 (C, γ) COORDINATES

To calculate the illuminance of a facet it is necessary to know the luminous intensity directed towards the facet. If the (C, γ) system is being used, the values of these two angles have to be determined so that the appropriate value of the luminous intensity can be found from the *I*-table.

Let O be the origin of the (x, y, z) coordinate system shown in Figure 3.57.

Suppose we want to find the luminous intensity in the direction of a point *P* with coordinates (x_p, y_p, z_p) when the luminaire is at (x_L, y_L, z_L). Let the luminaire be aligned with $C = 0°$ along the *x*-axis.

Then from Figure 3.57

$$\cos\gamma = \frac{(\boldsymbol{OP} - \boldsymbol{OL})\cdot(-\boldsymbol{n}_z)}{PL}$$

$$= \frac{[(x_p - x_L)\boldsymbol{n}_x + (y_p - y_L)\boldsymbol{n}_y + (z_p - z_L)\boldsymbol{n}_z]\cdot(-\boldsymbol{n}_z)}{PL}$$

$$= \frac{z_L - z_p}{PL}$$

The negative value of $\boldsymbol{n}_z$ is taken because γ is measured from the downward vertical.

C is found from the equation

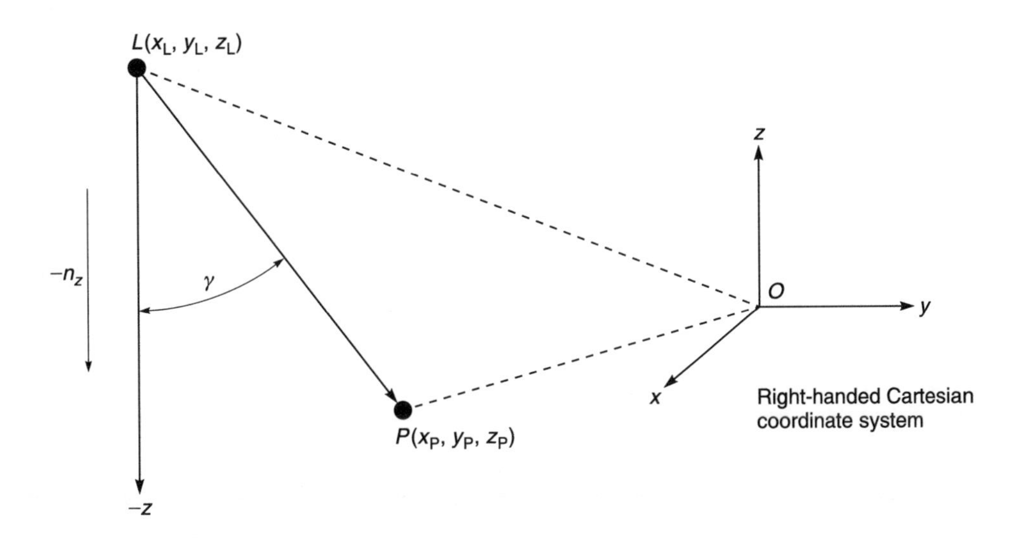

Fig. 3.57 γ in the *x, y, z* coordinate system

$$C = \tan^{-1} \frac{y}{x}$$

The correct quadrant for C is found from the conditions given in Section 2.41, Equation (2.36).

3.10 Examples

1. Point source A produces an illuminance at point P of 200 lux (Fig. 3.58). Calculate the value of the horizontal and vertical components $E_{(-z)}$ and $E_{(-x)}$ at P given that $E_{(-y)} = 0$.

$$E_{(-z)} = 200 \cos 30° = 200 \times 0.866 = 173.2 \text{ lux}$$
$$E_{(-x)} = 200 \sin 30° = 200 \times 0.5 = 100 \text{ lux}$$

2. Given that source B (Fig. 3.58) has the same value of intensity as source A in the direction of point P and is at the same distance from the point, calculate the illuminance produced on the sloping plane at P by source B.

$I_B = E_A D^2 = 200D^2$ since source A is on the normal to the plane of point P.

$$E_B = \frac{I_B \cos 30°}{D^2} = \frac{200D^2 \times 0.866}{D^2} = 173.2 \text{ lux}$$

3. If source C (Fig. 3.58) produces components on the z, y and x axes of the same magnitude as those for source A, calculate the illuminance at point P produced by source C.

   Since the normal to point P slopes away from source C the vertical component is reversed with respect to point P.

$$= 173.2 \times 0.866 + 100 \times (-0.5)$$
$$= 150 - 50$$
$$= 100 \text{ lux}$$

4. Consider the line source shown in Figure 3.6, Section 3.2.2. Given that $\phi = 30°$ and $\beta = 90°$, $E_{(-z)} = 300$ lux and $E_{(-y)} = 100$ lux, calculate the illuminance on the sloping plane at P if $x/h = 0.5$.

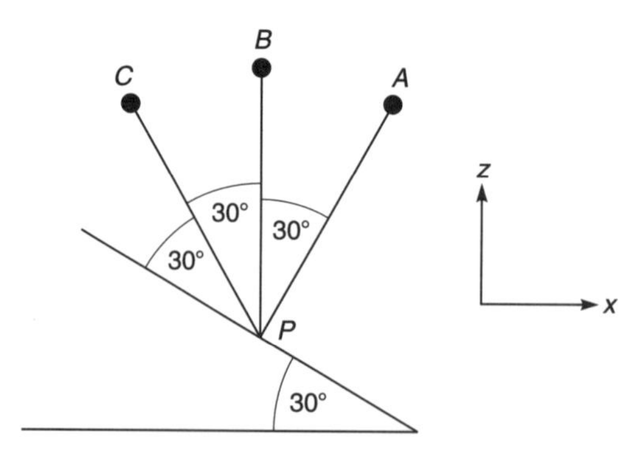

Fig. 3.58 The angles required for examples 1, 2 and 3

$$E_p = E_{(-z)}(\cos\phi + \frac{x}{h}\cos\beta\sin\phi) + E_{(-y)}\sin\beta\sin\phi$$

$$= 300(0.866 + 0.5 \times 0 \times 0.5) + 100 \times 1.0 \times 0.5$$

$$= 309.8 \text{ lux}$$

5. Consider the uniformly diffusing disc source in Figure 3.13, Section 3.6.2. Given that h = 2 m, R = 1.5 m and the source luminance is 150 cd/m^2 calculate the illuminance directly beneath the centre of the disc on the plane parallel to the disc.

$$E_p = \pi L\left(\frac{r^2}{R^2 + h^2}\right)$$

$$= \pi \times 150\left(\frac{1.5^2}{1.5^2 + 2^2}\right)$$

$$= \pi \times 150 \times 0.36$$

$$= 169.6 \text{ lux}$$

If another point, not beneath the centre, on a parallel plane was to receive the same value of illuminance from this source, what condition would need to be fulfilled? The source would have to subtend the same value of the angle α at that point.

6. Consider the rectangular source shown in Figure 3.21. Given that the source is uniformly diffusing and was 3 m wide and 4 m long, calculate the illuminance at a point 2 m below the source on a parallel plane. The point lies on the centre line joining the two 3 m sides and is 1.5 m along this line from one end. The source luminance is 400 cd/m^2.

The source can be considered to consist of four component sources, two of which have dimensions 1.5 m × 1.5 m and the other two 1.5 m × 2.5 m.

(a) The illuminance produced by the 1.5 m × 1.5 m sources,

$$E_{p(1)} = \frac{L}{2}(B_1 \sin A + A_1 \sin B) \quad \text{(see Figure 3.18)}$$

$$B_1 = \tan^{-1}\left(\frac{1.5}{\sqrt{1.5^2 + 2^2}}\right) = 30.96^\circ \text{ or } 0.54 \text{ radians}$$

$$B = \tan^{-1}\left(\frac{1.5}{2.0}\right) = 36.87^\circ$$

$$\sin B = 0.6$$
$$\sin A = 0.6$$
$$A_1 = B_1$$

$$E_{p(1)} = \frac{400}{2}(0.54 \times 0.6 + 0.54 \times 0.6)$$

$= 129.6$ lux

There are two such component sources so $E_{p(i)} = 259.2$ lux.

(b) The angles for the 1.5 m × 2.5 m component sources are:

$$B_1 = \tan^{-1}\left(\frac{2.5}{\sqrt{1.5^2 + 2^2}}\right) = 45° \text{ or } 0.785 \text{ radians}$$

$$B = \tan^{-1}\left(\frac{2.5}{2}\right) = 51.3°$$

$$\sin B = 0.78$$

$$A_1 = \tan^{-1}\left(\frac{1.5}{\sqrt{2.5^2 + 2^2}}\right) = 25.1° \text{ or } 0.438 \text{ radians}$$

$$A = \tan^{-1}\left(\frac{1.5}{2}\right) = 36.87°$$

$$\sin A = 0.6$$

$$E_{p(2)} = \frac{400}{2}(0.785 \times 0.6 + 0.438 \times 0.78)$$

$$= 200(0.471 + 0.342)$$

$$= 162.5 \text{ lux}$$

There are two component sources so $E_{p(2)} = 325$ lux.
The total illuminance at the point is 259.2 + 325 = 584 lux.

7. For a point opposite one end of the source calculate the aspect factor for the parallel plane for a line source 1.8 m long, 1.8 m above the parallel plane, if the axial distribution of the source is given by $I_\beta/I_B = \cos^2 \beta$.

$$AF = \int_0^\gamma \left(\frac{I_\beta}{I_B}\right)\cos\beta \, d\beta$$

$$AF = \int_0^\gamma \cos^2\beta \cos\beta \, d\beta$$

$$= \int_0^\gamma \cos^3\beta \, d\beta$$

$$= \int (1 - \sin^2\beta)\cos\beta \, d\beta$$

$$= \int \cos\beta - \sin^2\beta\cos\beta \, d\beta$$

$$= \left[\sin \beta - \frac{\sin^3 \beta}{3} \right]_0^\gamma$$

$$= \left[0.707 - \frac{0.354}{3} \right]$$

$$= 0.589$$

8. Calculate the following for a point source with an intensity of 1000 cd in the horizontal direction at a horizontal distance of 3 m.

   (a) The scalar illuminance.
   (b) The cylindrical illuminance.

   (a) Scalar illuminance

$$E_s = \frac{E_{max}}{4}$$

$$= \frac{I}{4d^2}$$

$$= \frac{1000}{4 \times 9}$$

$$= 27.8 \text{ lux}$$

   (b) Cylindrical illuminance

$$E_{cyl} = E_{max} \frac{\sin \theta}{\pi}$$

$$= \frac{1000}{9\pi} \sin 90°$$

$$= 35.4 \text{ lux}$$

Problems

1. A light source produces three component illuminances at a particular point; these are $E_{(-x)}$ = 10 lux, $E_{(-y)}$ = 15 lux and $E_{(-z)}$ = 50 lux. Calculate the maximum illuminance that the source can produce at the point in question.

   Answer: [53.2 lux]

2. Consider the line source shown in Figure 3.6. Given that θ =45°, ϕ = 45° and β = 30°, calculate the illuminance on the sloping plane at point *P*. The line source produces the following components of illuminance of point *P*: $E_{(-z)}$ = 400 lux, $E_{(-y)}$ = 100 lux.

   Answer: [563 lux]

3. A 2 m diameter uniformly diffusing disc source is mounted 2 m above a parallel plane. If the disc has a luminance of 400 cd/m^2, calculate the illuminance on the parallel plane directly beneath the centre of the source and also directly beneath the edge of the source.

   Answers: $\begin{bmatrix} \text{centre } 251.3 \text{ lux} \\ \text{edge } 184.1 \text{ lux} \end{bmatrix}$

4. Use the sizes and angles of Example 6 to calculate the illuminance from a similar source; the only difference being that the distribution is cosine cubed. Assume that the maximum luminance (L_m) is 400 cd/m^2.

   Answer: [441.8 lux]

5. Consider the line source in Figure 3.39. Calculate the illuminance at point *P* given the following information. The source is uniformly diffusing and is 2 m long. The intensity in the downward direction is 700 cd. The illuminated point lies 1 m beyond the end of the line source, on a parallel plane 1 m below the source.

   Answer: [46 lux]

6. Calculate the semi-cylindrical illuminance and the hemispherical illuminance at a point from a very small light source having a luminous intensity of 800 cd in all directions.

   The light source is mounted 2 m above the horizontal plane containing the illuminated point and its vertical axis meets the horizontal plane 2 m from the point.

   In the case of the semi-cylindrical illuminance the angle between the vertical plane containing the normal to the diameter of the semi-cylinder and the vertical plane containing the light source is 30°.

   In the case of the hemisphere, its base coincides with the horizontal plane.

   Answers: $\begin{bmatrix} \text{semi-cylindrical illuminance} = 42 \text{ lux} \\ \text{hemispherical illuminance} = 42.7 \text{ lux} \end{bmatrix}$

References

1. Hollindale, S. (1989) *Makers of Mathematics* (Penguin Books) p. 58.
2. Bell, R.I. (1973) A method for the calculation of direct illuminance due to area sources of various distributions, *Lighting Research and Technology*, **5**(2), 99–102.
3. Pierpoint, W. and Hopkins, J. (1984) The derivation of a new area source equation, *Journal of IES*, April, 314–316.
4. Lecocq, J. (1985) Du Produits Scalaire dans les Calculs d'Eclairement, *Lux*, 132, Mars–Avril.

4

Flux Transfer

4.1 Introduction

In Chapter 3 the illuminance at a point from point, line and area sources was considered. In this chapter the luminous flux received by an area from point, line and area sources will be considered and the flux transfer functions, which find application in interreflection calculations, derived.

4.2 Reciprocity

In calculating the flux transfer from one surface to another, it is often convenient to make use of the law of reciprocity. The law states that: if two uniformly diffuse light emitting surfaces illuminate each other, the flux transfer from each surface to the other is the same proportion as their luminous exitance values. Thus, if they have the same value of luminous exitance (M), then the flux transfer will be the same in each direction. This is *General Reciprocity*. What is not always appreciated is that if two emitting surfaces lie in parallel planes, then the limitation is not to uniformly diffuse surfaces but to surfaces emitting light with an intensity distribution that is the same function of angle θ to the intensity at the normal (regardless of the azimuth angle of the plane in which θ lies). This is *Parallel Plane Reciprocity*.

PROOF

1. General Reciprocity (see Figure 4.1)

When the sources have a uniformly diffuse distribution then,

$$I_{\theta_1} = I_{m_1} \cos \theta_1 = \frac{M_1}{\pi} \delta a_1 \cos \theta_1$$

and

$$I_{\theta_2} = I_{m_2} \cos \theta_2 = \frac{M_2}{\pi} \delta a_2 \cos \theta_2$$

The flux transferred from surface 1 to surface 2 (F_{12}) is given by:

$$F_{12} = \int E_2 \, da_2$$

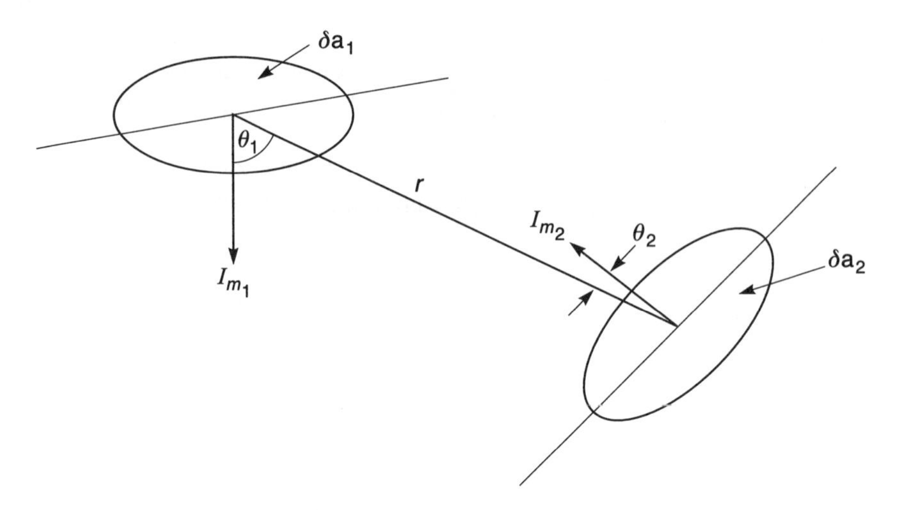

Fig. 4.1 Flux transfer between two surfaces (General Reciprocity)

$$= \frac{1}{\pi} \int_1 \int_2 \frac{M_1 \, da_1 \cos \theta_1 \cos \theta_2 da_2}{r^2}$$

and from surface 2 to surface 1

$$F_{21} = \int E_1 \, da_1$$

$$= \frac{1}{\pi} \int_2 \int_1 \frac{M_2 \, da_2 \cos \theta_2 \cos \theta_1 \, da_1}{r^2}$$

Since the order of integration may be reversed

$$\frac{F_{12}}{M_1} = \frac{F_{21}}{M_2} \quad \text{for any orientation}$$

2. *Parallel Plane Reciprocity* (see Figure 4.2)

$$I_{\theta_1} = I_{m_1} f(\theta) \quad \text{and} \quad I_{\theta_2} = I_{m_2} f(\theta)$$

(Note: since the planes are parallel, $\theta_1 = \theta_2$)

where $I_{m_1} = HM_1 \, da_1$ and $I_{m_2} = HM_2 \, da_2$ and H is a constant relating the luminance normal to the surface to the luminous exitance M.

The flux transferred from surface 1 to surface 2, F_{12} is given by:

$$F_{12} = \int_2 \int_1 \frac{HM_1 \, da_1 f(\theta) \cos \theta \, da_2}{r^2}$$

and

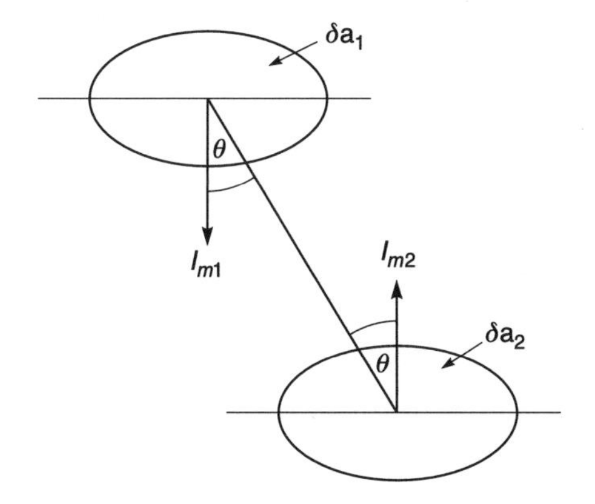

Fig. 4.2 Flux transfer between two surfaces (Parallel Plane Reciprocity)

$$F_{21} = \int_2\int_1 \frac{HM_2\, da_2 f(\theta) \cos\theta\, da_1}{r^2}$$

and as before

$$\frac{F_{12}}{M_1} = \frac{F_{21}}{M_2}$$

Example

Chapter 3, Section 3.7.1, gives the equation for the illuminance at a point beneath the centre of a disc source having a cosine power distribution

$$E_p = \frac{2\pi L_m}{(n+1)}(1 - \cos^{(n+1)}\gamma)$$

Using this formula, the flux received on a small disc of 0.08 m diameter from a large disc source of 4 m diameter, mounted 3.46 m above the centre of the small disc and parallel to it, can be calculated (Figure 4.3). The large disc has a cosine squared intensity distribution and the illuminance over the small disc is regarded as uniform.

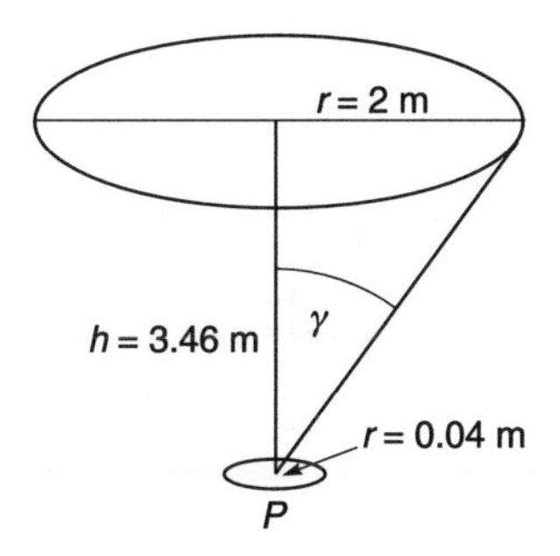

Fig. 4.3 Flux transfer from the large disc to the small disc

The luminous flux (F) received by the small disc is $E_p \times a$ where a is the area of the small disc.

$$F = \frac{2\pi \times a \times L_m}{(n + 1)} (1 - \cos^{(n+1)} \gamma)$$

for $n = 2$:

$$F = \frac{2\pi}{3} \times a \times L_m (1 - \cos^3 \gamma)$$

$$\gamma = \tan^{-1} \frac{2}{3.46} = 30°$$

So,

$$F = \frac{2\pi}{3} \times a \times L_m (1 - 0.125) \tag{4.1}$$

Now a can be calculated and, given L_m, the calculation could be completed.

However, it is of interest to pause at this point and examine the results of reversing the roles of the two discs. Let the small disc become the light source and the large disc be the receiving target (Figure 4.4).

The inverse-square law is now used to calculate the illuminance and hence the flux onto an elementary annulus of the large disc and the resulting expression integrated to obtain the total flux:

$$E = I_m \cos^2 \theta \cos^3 \theta / h^2$$

$$F' = \frac{I_m \cos^2 \theta \cos^3 \theta \, 2\pi r \, dr}{h^2}$$

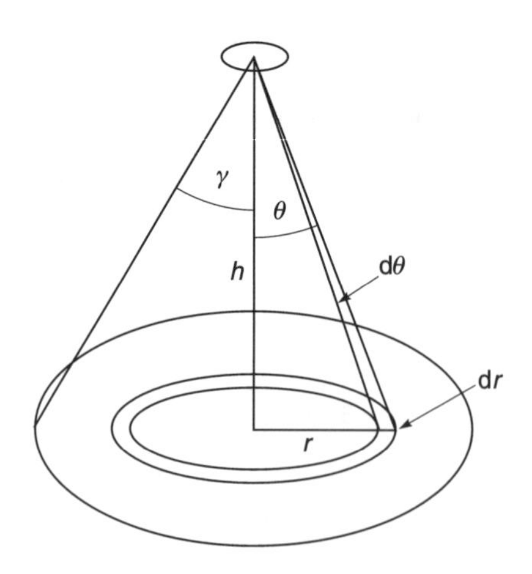

Fig. 4.4 Flux transfer from the small disc to the large disc

$$r = h \tan\theta \quad \text{and} \quad \mathrm{d}r = h \sec^2\theta \, \mathrm{d}\theta$$

where F' is the flux received by the annulus of radius r and width dr.

So,

$$F' = \frac{I_m \cos^2\theta \cos^3\theta \, 2\pi h \tan\theta \, h \sec^2\theta \, \mathrm{d}\theta}{h^2}$$

$$F' = I_m \cos^2\theta \, 2\pi \sin\theta \, \mathrm{d}\theta$$

$$F = 2\pi I_m \int_0^{\gamma} \cos^2\theta \sin\theta \, \mathrm{d}\theta$$

$$\gamma = \tan^{-1} \frac{2}{3.46} = 30°$$

So,

$$F = \frac{2\pi I_m}{3} [-\cos^3\gamma]_0^{30°}$$

$$= \frac{2\pi I_m}{3} [1 - \cos^3 30°]$$

$$= \frac{2\pi I_m}{3} (1 - 0.125)$$

But

$$I_m = L_m \times a$$

giving,

$$F = \frac{2\pi}{3} \times a \times L_m(1 - 0.125) \tag{4.2}$$

Comparing equations (4.1) and (4.2) we see that these are identical and that provided L_m has the same value, the flux transfer will be the same regardless of which disc is the light source.

To complete the calculation, let $L_m = 500$ cd/m^2

$$a = \frac{\pi d^2}{4} = \frac{\pi \times (0.08)^2}{4} = 0.005 \text{ m}^2$$

$$F = \frac{2\pi}{3} \times 0.005 \times 500 \times 0.875$$

$$= 4.58 \text{ lumens}$$

The average illuminance of the small disc would be

4.58/0.005 = 916 lux

whereas that of the larger disc would be

$$4.58/4\pi = 0.364 \text{ lux}$$

This result demonstrates clearly that reciprocity relates to equality of flux transfer and not illuminance.

4.3 Flux transfer from a point source

In the previous section, the flux transfer was calculated from a point source with a cosine squared intensity distribution to a disc directly beneath the source. In this section, the flux transfer from a point source with a *practical* distribution will be calculated to a disc directly beneath the source.

This flux transfer will be calculated in two ways: (1) by means of the Zone Factors introduced in Chapter 1 and (2) by first approximating the practical distribution by means of a cosine power series (as suggested in Chapter 3, Section 3.7) and then (as in the previous section) applying reciprocity to use the formulae for a disc source with a cosine power distribution to each of the terms in the cosine power series.

The purpose of this exercise is to show that, approximating the distribution to a cosine power series can give results that compare favourably with the more commonly used Zone Factor method, and so can be used for cases where Zone Factors cannot be used.

Example

The practical distribution chosen is symmetrical about the vertical axis and is shown in Figure 4.5, and Table 4.1 gives the intensity values at 5° intervals for this distribution.

The dimensions are shown in Figure 4.6.

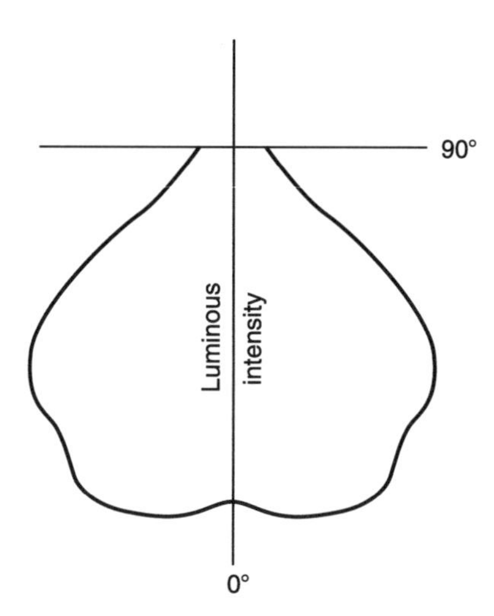

Fig. 4.5 Polar curve for the practical distribution given in Table 4.1

Table 4.1 Intensity values at 5° intervals for the practical distribution used in the example in Section 4.3

Angle from the downward vertical in degrees	Intensity in candelas
0°	1578
5°	1584
10°	1668
15°	1674
20°	1668
25°	1614
30°	1464
35°	1410
40°	1398
45°	1308
50°	978
55°	426
60°	270
65°	204
70°	186
75°	144
80°	120
85°	108
90°	96

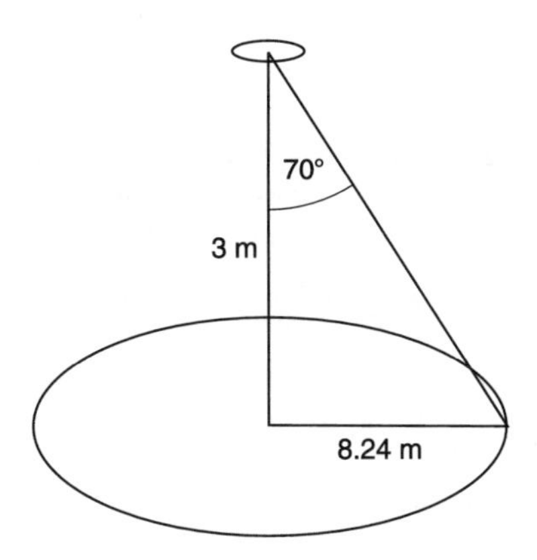

Fig. 4.6 Flux transfer from a small disc source to a large disc for a practical distribution

Method 1. Zone Factors

Angle (deg)	ZF	Mid Zone Intensity (cd)	Lumens
0–10	0.095	1584	150.5
10–20	0.283	1674	473.7
20–30	0.463	1614	747.3
30–40	0.628	1410	885.5
40–50	0.774	1308	1012.4
50–60	0.897	426	382.1
60–70	0.993	204	202.6
			3854.1

From Table 1.1.

Method 2. Cosine power series approximation

Assume that $I_\gamma = A + B\cos\gamma + C\cos^2\gamma + D\cos^3\gamma$

$$\begin{bmatrix} I_{\gamma_1} \\ I_{\gamma_2} \\ I_{\gamma_3} \\ I_{\gamma_4} \end{bmatrix} = \begin{bmatrix} 1 & \cos\gamma_1 & \cos^2\gamma_1 & \cos^3\gamma_1 \\ 1 & \cos\gamma_2 & \cos^2\gamma_2 & \cos^3\gamma_2 \\ 1 & \cos\gamma_3 & \cos^2\gamma_3 & \cos^3\gamma_3 \\ 1 & \cos\gamma_4 & \cos^2\gamma_4 & \cos^3\gamma_4 \end{bmatrix} \times \begin{bmatrix} A \\ B \\ C \\ D \end{bmatrix}$$

where I_γ is the luminous intensity at angle γ to the downward vertical.
So,

$$[I] = [\cos]\ [A]$$

and

$$[A] = [\cos]^{-1}\ [I]$$

Since this is a *flux* calculation, choose values of γ to give equal solid angles between 0° and 90°.
These are $\gamma_1 = 28.96°$, $\gamma_2 = 51.32°$, $\gamma_3 = 67.98°$, γ_4 82.82°.
Obtain the I values by interpolation from Table 4.1:

$$I_{\gamma_1} = 1495.2,\quad I_{\gamma_2} = 832.27,\quad I_{\gamma_3} = 193.27,\quad I_{\gamma_4} = 113.23$$

Inserting the cosine values of γ_1, γ_2, γ_3 and γ_4 we find the matrix to be inverted becomes:

$$\begin{bmatrix} 1.000 & 0.875 & 0.766 & 0.670 \\ 1.000 & 0.625 & 0.391 & 0.244 \\ 1.000 & 0.375 & 0.141 & 0.053 \\ 1.000 & 0.125 & 0.160 & 0.002 \end{bmatrix}^{-1}$$

The inverted matrix is

$$\begin{bmatrix} -0.3008 & 1.2745 & -2.1465 & 2.1728 \\ 3.7474 & -15.2421 & 23.2421 & -11.7474 \\ -11.7895 & 43.3684 & -51.3684 & 19.7895 \\ 10.5263 & -13.5789 & 31.5789 & -10.5263 \end{bmatrix}$$

Multiplying this matrix by the I_γ matrix gives:

$A = 442.146$
$B = 3919.54$
$C = 10776.315$
$D = 5629.675$

By the principle of reciprocity, the cosine power disc source formula may be used to calculate the flux from a point source to a disc source by replacing aL_m in the disc source formula by I_m.

So,

$$\Phi = \frac{2\pi I_m}{(n+1)}(1 - \cos^{(n+1)} \gamma)$$

For $n = 0$, the contribution to the total flux is given by:

$$2\pi I_m(1 - \cos 70°) = 4.134 I_m$$

For $n = 1.0$:

$$\frac{2\pi I_m}{2}(1 - \cos^2 70°) = 2.774 I_m$$

For $n = 2.0$:

$$\frac{2\pi I_m}{3}(1 - \cos^3 70°) = 2.010 I_m$$

For $n = 3.0$:

$$\frac{2\pi I_m}{4}(1 - \cos^4 70°) = 1.55 I_m$$

where the I_m values are given by A, B, C and D.

Flux calculation

$n = 0$, $4.134 \times 442.146 = 1827.8$
$n = 1.0$, $2.774 \times -3919.54 = -10872.8$
$n = 2.0$, $2.010 \times 10776.315 = 21660.39$
$n = 3.0$, $1.55 \times -5629.675 = -8726$

The total flux is therefore 3889.4 lumens.

Comparing this value with that obtained in (1) using Zone Factors

$$\frac{3889.4 - 3854.1}{3854.1} = 0.00916 \quad \text{or} \quad +0.92\%$$

This is a satisfactory level of agreement between the two methods.

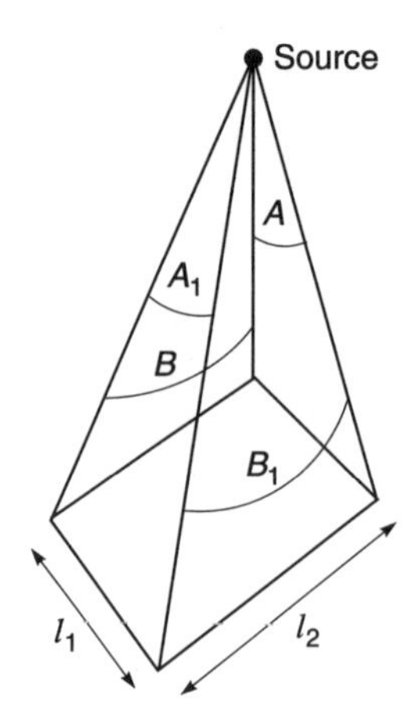

Fig. 4.7(a) Flux transfer from a point source to a rectangle, using cosine powers formulae

4.3.1 FLUX FROM A POINT SOURCE TO A RECTANGULAR AREA

(a) Cosine power series method

The cosine power series approximation demonstrated in the previous section, together with the principle of reciprocity, can be applied to rectangular source calculations, see Figure 4.7(a). It was shown in Section 4.2 that the illuminance formulae for area sources can be converted to flux transfer formulae, from point to area by replacing L_m by I_m. (The assumption is that the area source was illuminating a small area '*a*' which, when reciprocity is applied, can be combined with L_m to give I_m.)

The illuminance equations for cosine powers applied to rectangular sources, as given in Section 3.7.2 relating to Figure 3.28 are restated below, but with L_m replaced by I_m. These then become flux transfer equations related to Figure 4.7(a).

The equation for flux transfer from the point source to area $l_1 \times l_2$ are:

$n = 0$,

$$F_{n=0} = 2I_m \left[\tan^{-1}\left(\tan\left(\frac{A}{2}\right) \tan\left(\frac{B_1}{2}\right)\right) + \tan^{-1}\left(\tan\left(\frac{B}{2}\right) \tan\left(\frac{A_1}{2}\right)\right)\right]$$

$n = 1.0$,

$$F_{n=1.0} = \frac{I_m}{2}(B_1 \sin A + A_1 \sin B)$$

$n = 2.0$,

$$F_{n=2.0} = \frac{I_m}{3} \sin A \cos A \sin B_1 + 2 \tan^1 \left(\tan\left(\frac{A}{2}\right) \tan\left(\frac{B_1}{2}\right)\right)$$

$$+ \sin B \cos B \sin A_1 + 2 \tan^1 \left(\tan\left(\frac{B}{2}\right) \tan\left(\frac{A_1}{2}\right)\right)$$

$n = 3.0$,

$$F_{n=3.0} = \frac{I_m}{4} \sin A \cos^2 A \left(\frac{B_1}{\cos^2 A} + \frac{B_1}{2} + \frac{1}{2} \sin B_1 \cos B_1 \right)$$

$$+ \sin B \cos^2 B \left(\frac{A_1}{\cos^2 B} + \frac{A_1}{2} + \frac{1}{2} \sin A_1 \cos A_1 \right)$$

As in the case of the flux transfer from a point source to a disc, the I_m values are the *A, B, C* and *D* values obtained when the I_γ matrix is used to multiply the inverted cosine matrix.

So,

$$\begin{bmatrix} A \\ B \\ C \\ D \end{bmatrix} = \begin{bmatrix} -0.3008 & 1.2745 & -2.1465 & 2.1728 \\ 3.7474 & -15.2421 & 23.2421 & -11.7474 \\ -11.7895 & 43.3684 & -51.3684 & 19.7895 \\ 10.5263 & -31.5789 & 31.5789 & -10.5263 \end{bmatrix} \begin{bmatrix} I_{\gamma_1} \\ I_{\gamma_2} \\ I_{\gamma_3} \\ I_{\gamma_4} \end{bmatrix}$$

where $\gamma_1 = 28.96°$, $\gamma_2 = 51.32°$, $\gamma_3 = 67.98°$, $\gamma_4 = 82.82°$.

(b) Point by point calculations

With the advent of high speed, high capacity calculators and computers, it has become possible to use point by point methods much more extensively than previously. These methods have the great advantage of dealing with practical intensity distributions without approximation of the intensity distribution (although some error may be introduced by interpolation). The approximation here lies in the number of points for which the calculation is made.

A simple example will be used as an illustration. Consider Figure 4.7(b) where the source is above one corner of the rectangle.

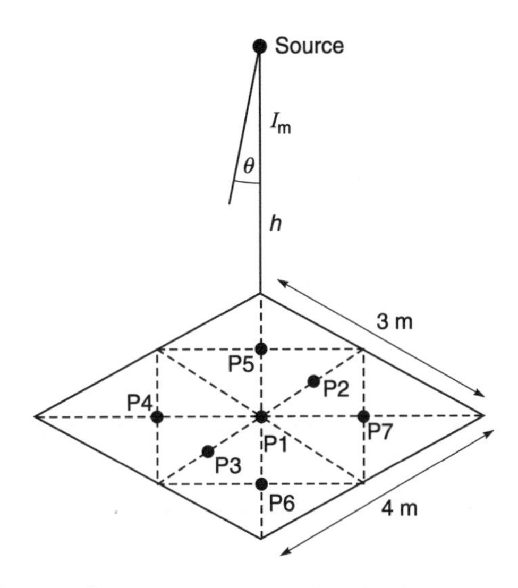

Fig. 4.7(b) Flux transfer from a point source to a rectangle, using inverse-square law calculations

Let

$$I_\theta = I_m \cos\theta \quad \text{and} \quad I_m = 1000 \text{ cd}$$

also let

$$l_1 = 3 \text{ m}, \quad l_2 = 4 \text{ m}, \quad \text{and} \quad h = 2 \text{ m}$$

Calculating the illuminance for one point

$$E_{p1} = \frac{I_\theta \cos^3\theta}{h^2} = \frac{I_m \cos^4\theta}{4}$$

$$\theta = \tan^{-1}\frac{\sqrt{(1.5)^2 + (2)^2}}{2}$$

$$= \tan^{-1} 1.25$$
$$= 51.34°$$

$$E_p = 250 \times \cos^4 51.34°$$
$$= 38.1 \text{ lux}$$

First estimate of flux,

$$E_{p_1} \times l_1 \times l_2 = 38.1 \times 3 \times 4$$
$$= 457 \text{ lumens}$$

Calculating the illuminance for two points P_2 and P_3 gives,

$$E_{p2} = 250 \cos^4\theta$$

$$\theta = \tan^{-1}\frac{\sqrt{(1.5)^2 + (1)^2}}{2} = 42°$$

$$E_{p_2} = 250 \times 0.304$$
$$= 76.1 \text{ lux}$$

$$E_{p3} = 250 \times \cos^4\theta$$

$$\theta = \tan^{-1}\frac{\sqrt{(1.5)^2 + (3)^2}}{2} = 59.2°$$

$$E_{p_3} = 250 \times 0.069$$
$$= 17.2 \text{ lux}$$

Second estimate of flux,

$$\left(\frac{E_{p2} + E_{p3}}{2}\right)(l_1 \times l_2)$$

$$= \frac{93.3}{2} \times 12$$

$$= 559.8 \text{ lumens}$$

Calculating the illuminance for four points P_4, P_5, P_6, P_7 gives,

$$E_{p4} = 250 \times \cos^4 \theta$$

$$\theta = \tan^{-1} \frac{\sqrt{(0.75)^2 + (3)^2}}{2} = 57.1°$$

$$E_{p4} = 250 \times 0.087$$
$$21.75 \text{ lux}$$

$$E_{p5} = 250 \times \cos^4 \theta$$

$$\theta = \tan^{-1} \frac{\sqrt{(0.75)^2 + (1)^2}}{2} = 32°$$

$$E_{p5} = 250 \times 0.517$$
$$= 129.3 \text{ lux}$$

$$E_{p6} = 250 \times \cos^4 \theta$$

$$\theta = \tan^{-1} \frac{\sqrt{(2.25)^2 + (3)^2}}{2} = 61.9°$$

$$E_{p6} = 250 \times 0.049$$
$$= 12.25 \text{ lux}$$

$$E_{p7} = 250 \times \cos^4 \theta$$

$$\theta = \tan^{-1} \frac{\sqrt{(2.25)^2 + (1)^2}}{2} = 50.9°$$

$$E_{p7} = 250 \times 0.158$$
$$= 39.5 \text{ lux}$$

Third estimate of flux

$$\left(\frac{E_{p4} + E_{p5} + E_{p6} + E_{p7}}{4} \right) \times l_1 \times l_2 = \left(\frac{21.75 + 129.3 + 12.25 + 39.5}{4} \right) \times 12$$
$$= 608 \text{ lumens}$$

We can calculate the exact value by using the formula for $I_m \cos \theta$.

So,

$$F = \frac{I_m}{2} (B_1 \sin A + A_1 \sin B)$$

$$B = \tan^{-1} \frac{4}{2} = 63.4°, \qquad A = \tan^{-1} \frac{3}{2} = 56.3°$$

$$B_1 = \tan^{-1}\frac{4}{\sqrt{9+4}} = 48^\circ, \qquad A_1 = \tan^{-1}\frac{3}{\sqrt{16+4}} = 33.87^\circ$$

$$\begin{aligned} F &= 500(0.838 \times 0.832 + 0.591 \times 0.894) \\ &= 500 \times 1.226 \\ &= 613 \text{ lumens} \end{aligned}$$

Difference:

$$\frac{608 - 613}{613} = 0.008 \quad \text{or} \quad 0.8\%$$

From the foregoing it can be seen that, in many cases, the point by point method is the most simple and straightforward approach. However, as in this case, it is useful to be able to calculate an exact value when estimating the number of calculation points required to give a particular level of accuracy.

4.4 Flux transfer from a linear source

In Chapter 3, Section 3.7.3, the aspect factor method of calculating the illuminance at a point from a linear source (such as a fluorescent tubular lamp or luminaire) was introduced. This aspect factor method may be extended and developed into the *K* factor method for calculating flux transfer from a linear source.

Consider Figure 4.8; this shows a sector solid associated with a linear source of length l for a radial distance R from the source. Flux from this source passes through the sector window A, B, C, D. Although the illuminance produced by this flux will vary along the direction parallel to the source, that is in the AB direction, it can be considered constant across the sector window in the AD direction (provided that the mean value of the luminous intensity over the angle $\Delta\theta$ is used in the calculation).

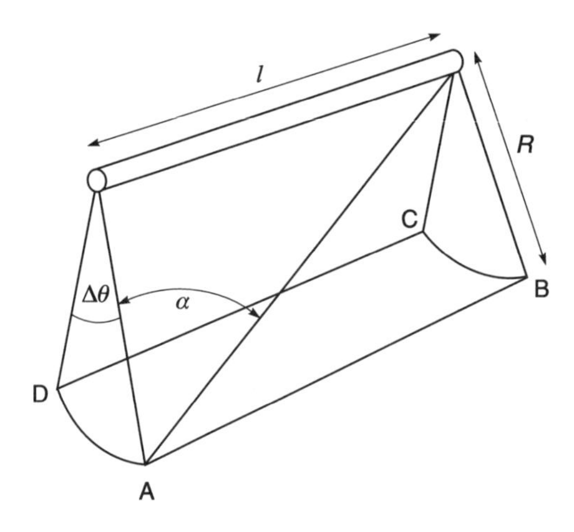

Fig. 4.8 The Sector Solid associated with the K Factor method

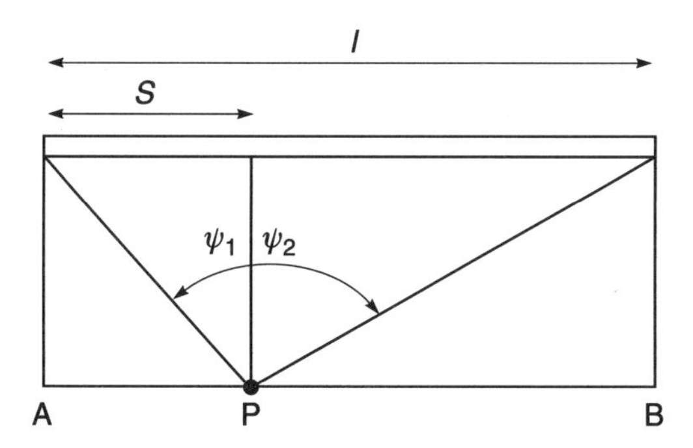

Fig. 4.9 Two component sources producing the axial illuminance at P

In Figure 4.9 it can be seen that the variable illuminance in the axial direction can be considered to be produced by two component sources.

The illuminance at point P will vary with its position along the axis of the sector. This illuminance is made up of a component from a source of length s, subtending angle ψ_1 and a component from a source of length $(l - s)$, subtending angle ψ_2.

The flux through the curved sector window ABCD, in Figure 4.8, is given by:

$$F_{\text{ABCD}} = \int_{s=0}^{s=1} \frac{I}{lR} R\Delta\theta[(AF_{\psi_1}) + (AF_{\psi_2})] \, \mathrm{d}s$$

where I is the mean intensity in the transverse direction and

$$AF_{\psi} = \int_0^{\alpha} f(\psi)\cos\psi \, \mathrm{d}\psi \quad \text{(see Section 3.73)}$$

$$F_{\text{ABCD}} = \frac{I}{l}\Delta\theta \int_0^l [(AF_{\psi_1}) + (AF_{\psi_2})] \, \mathrm{d}s$$

This can be written as:

$$F_{\text{ABCD}} = \frac{I}{l}\Delta\theta \left[\int_0^l (AF_{\psi_1}) \, \mathrm{d}s + \int_0^l (AF_{\psi_2}) \, \mathrm{d}s \right] \tag{4.3}$$

Integrating $\int_0^l (AF_{\psi_1}) \mathrm{d}s$ by parts gives,

$$\int_0^l (AF_{\psi_1}) \, \mathrm{d}s = [s(AF_{\psi_1})]_0^1 - \int_0^l s \, \mathrm{d}(AF_{\psi_1})$$

We note that when

$s = l$, $\psi = \alpha$ (Figure 4.8)

$$\int_0^l (AF_{\psi_1})\,\mathrm{d}s - l(AF_\alpha) - \int_0^l s\,\mathrm{d}\left[\int_0^\alpha f(\psi_1)\cos\psi_1\,\mathrm{d}\psi\right]$$

Now $s = R\tan\psi_1$ giving,

$$\int_0^l (AF_{\psi_1})\,\mathrm{d}s = l(AF_\alpha) - R\int_0^\alpha \tan\psi_1 f(\psi_1)\cos\psi_1\,\mathrm{d}\psi$$

$$= l(AF_\alpha) - R\int_0^\alpha f(\psi_1)\sin\psi_1\,\mathrm{d}\psi$$

From Section 3.7.3,

$$\int_0^\alpha f(\psi_1)\sin\psi_1\,\mathrm{d}\psi = af_\alpha$$

So,

$$\int_0^l (AF_{\psi_1})\,\mathrm{d}s = l(AF_\alpha) - R(af_\alpha)$$

By symmetry,

$$\int_0^l (AF_{\psi_1})\,\mathrm{d}s = \int_0^l (AF_{\psi_2})\,\mathrm{d}s$$

From equation (4.3),

$$F_{\mathrm{ABCD}} = 2I\Delta\theta[(AF_\alpha) - \cot\alpha(af_\alpha)]$$

$$= IK$$

where

$$K = 2\Delta\theta[(AF_\alpha) - \cot\alpha(af_\alpha)] \quad \text{and} \quad \alpha = \cot^{-1}\left(\frac{R}{l}\right)$$

So, for a constant sector angle $\Delta\theta$ (such as 10°), and a given axial distribution, K can be taken as a function of α only.

When $f(\psi)$ is a simple mathematical function such as $\sin\psi$, $\cos\psi$, $\cos^2\psi$ etc., K factors can easily be calculated. Practical distributions can either be approximated to a cosine function or represented by a power series as before, or derived by numerical integration.

Of particular interest is a line source with a cosine axial distribution.

In this case

$$AF_\alpha = \int_0^\psi \cos^2\psi\,\mathrm{d}\psi$$

$$= \frac{1}{2}(\alpha + \sin\alpha\cos\alpha)$$

and

$$af_\alpha = \int_0^\alpha \sin\psi\cos\psi \, d\psi$$

$$= \frac{1}{2}\sin^2\alpha$$

Giving

$$K_\alpha = \Delta\theta(\alpha + \sin\alpha\cos\alpha - \cot\alpha\sin^2\alpha)$$
$$= \Delta\theta\alpha$$

K_α is proportional to α, Figure 4.10 shows the percentage of the total flux emerging through the sector window as a function of aspect angle α.

The sector flux that does not emerge through the sector window passes through the ends of the sector, half through each end.

When K factors are used to calculate the flux incident on a surface parallel to the source, the assumption is made that the value chosen for R for each sector creates errors that cancel out (see Figure 4.11). From this figure it will be seen that the errors created by using a curved sector window operate in opposite senses, that is, one +*ve* and the other –*ve*.

4.4.1 EXAMPLE 1

As an example of the use of K factors, the fraction of the flux emitted by a uniformly diffusing luminous ceiling, and received by the floor, will be calculated by assuming the ceiling consists of three uniformly diffusing strip sources, each located at the centre of a ceiling strip, as shown in Figure 4.12. The length, width and height of the room are equal.

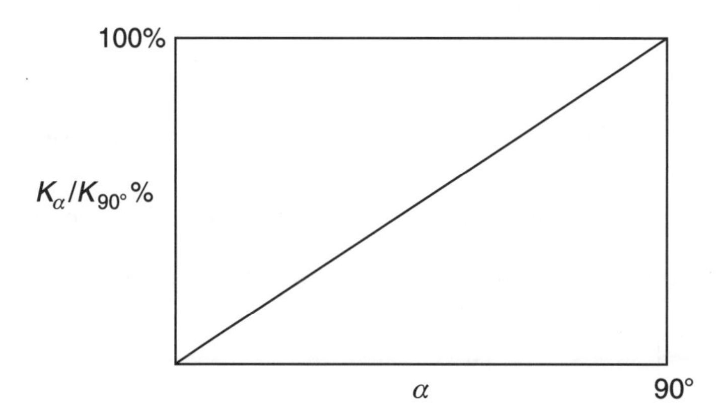

Fig. 4.10 The percentage of total flux emerging through the sector window as a function of aspect angle α

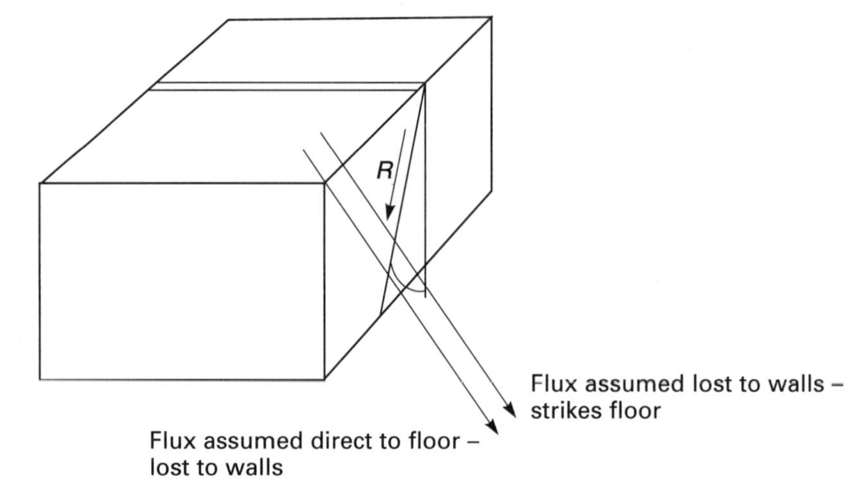

Fig. 4.11 Assumed cancelling errors in the K Factor method

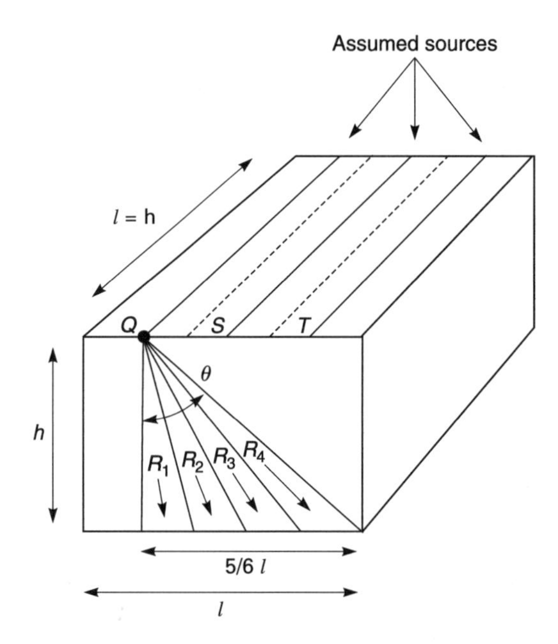

Fig. 4.12 Flux transfer from a uniformly diffusing luminous ceiling, using the K Factor method

From assumed source *Q*, the angle θ to the floor for the right-hand side of the strip is given by

$$\theta_r = \tan^{-1}\left(\frac{\frac{5}{6}l}{l}\right) = 39.8°, \quad \text{say } 40°$$

Assume 4–10° sectors.

The values of R needed to calculate the α values are:

$$R_1 = \frac{h}{\cos 5°} = 1.004h, \qquad R_2 = \frac{h}{\cos 15°} = 1.35h$$

$$R_3 = \frac{h}{\cos 25°} = 1.103h, \qquad R_4 = \frac{h}{\cos 35°} = 1.221h$$

so, since $l = h$,

$$\alpha_1 = \tan^{-1} \frac{h}{1.004h} = 44.89°, \ \alpha_2 = 44°,$$

$$\alpha_3 = 42.2°, \ \alpha_4 = 39.3°$$

$$K_{90°} = \Delta\theta \times \frac{\pi}{2} = 0.1745 \times \frac{\pi}{2} = 0.274$$

So,

$$K = \frac{K_\alpha}{K_{90}}\% \times \frac{0.274}{100}$$

where

$$\frac{K_\alpha}{K_{90}}\% = 1.111 \times \alpha°$$

(From Figure 4.10 where the slope of the graph = 100/90 = 1.111.)

Giving

$$K_1 = 1.111 \times 44.89 \times \frac{0.274}{100}$$

$$= 49.87 \times \frac{0.274}{100} = 0.1366$$

and

$$K_2 = 48.88 \times \frac{0.274}{100} = 0.1339$$

$$K_3 = 46.88 \times \frac{0.274}{100} = 0.1285$$

$$K_4 = 43.66 \times \frac{0.274}{100} = 0.1196$$

The angle from Q to the left-hand side of the floor is:

$$\theta = \tan^{-1}\left(\frac{\frac{1}{6}l}{l}\right) = 9.46^\circ \text{ (less than } 10^\circ)$$

$$R_5 = 1.003h \quad \text{and} \quad \alpha_5 = \tan^{-1}\left(\frac{1}{1.003}\right) = 44.9^\circ$$

$$K_5 = 44.9 \times 1.111 \times \frac{0.274}{100} = 0.1367$$

But true value $K'_5 = 0.946 \times 0.1367$ (only a 9.46° sector) = 0.1293.

Flux from source Q to the floor is therefore:

$I_m \cos 5^\circ \times K_1 + I_m \cos 15^\circ \times K_2 + I_m \cos 25^\circ \times K_3 + I_m \cos 35^\circ \times K_4 + I_m \cos 4.73^\circ \times K'_5$
$= I_m(0.996 \times 0.1366 + 0.966 \times 0.1339 + 0.906 \times 0.1285 + 0.819 \times 0.1196 + 0.996 \times 0.1367) = I_m \times 0.6159$ lumens

Note: 4.73° is the mid-sector angle for the 9.46° sector.

By symmetry, source T gives the same amount of flux to the floor. Source S subtends an angle to the right-hand side of the floor of:

$$\theta = \tan^{-1}\left(\frac{\frac{1}{2}l}{l}\right) = 26.6^\circ$$

From the calculations for source Q:

$K_1 = 0.1366$ 1st 10°

$K_2 = 0.1339$ 2nd 10°

For the remaining 6.6°:

$$R = \frac{h}{\cos 23.3^\circ} = 1.089h$$

$$\alpha = \tan^{-1}\left(\frac{1}{1.089}\right) = 42.6^\circ$$

$$K = 42.6 \times 1.111 \times \frac{0.274}{100} = 0.1296$$

But true value

$K'_3 = 0.66 \times 0.1296$ (only a 6.6° sector) = 0.086

Flux from source S to the right-hand side of the floor is given by:

$I_m \cos 5^\circ \times K_1 + I_m \cos 15^\circ \times K_2 + I_m \cos 23.3^\circ \times K'_3$

$= I_m(0.996 \times 0.1366 + 0.966 \times 0.1339 + 0.918 \times 0.086)$

$= I_m \times 0.3442$

By symmetry, source S contributes the same amount of flux to the left-hand side of the floor.

The total flux to the floor from the three sources Q, S and T is:

$$I_m(2 \times 0.6159 + 2 \times 0.3442) = I_m \times 1.920 \text{ lumens}$$

$$I_m \text{ for each strip of ceiling} = L_m \times l \times \frac{l}{3}$$

$$= \frac{L_m}{3} l^2$$

The total flux emitted from the ceiling is given by:

$$F_c = \pi L_m l^2$$

So the fraction of flux emitted from the ceiling and received on the floor is:

$$\frac{F_{cF}}{F_c} = \frac{\frac{1}{3} L_m l^2 \times 1.920}{\pi L_m l^2}$$

$$= \frac{0.333 \times 1.920}{\pi}$$

$$= 0.2035$$

Zijl in his book *Large Size Perfect Diffusors* has developed a formula for this case (that is, from rectangle to rectangle) for a uniformly diffusing source.

The formula is:

$$\Phi_{\text{ABCD}\rightarrow\text{KLMN}}$$

$$= \frac{2M}{\pi}\left[a\sqrt{(b^2+h^2)}\tan^{-1}\left(\frac{a}{\sqrt{b^2+h^2}}\right) + b\sqrt{a^2+h^2} \times \right.$$

$$\tan^{-1}\left(\frac{b}{\sqrt{a^2+h^2}}\right) - ah\tan^{-1}\left(\frac{a}{h}\right) - bh\tan^{-1}\left(\frac{b}{h}\right)$$

$$\left. -\frac{1}{2}\log_e \frac{(a^2+b^2+h^2)h^2}{(a^2+h^2)\ (b^2+h^2)} \right]$$

where a, b and h are the length, breadth and height of the room respectively.

Note: ABCD is the emitting rectangle and KLMN the receiving rectangle. M is the luminous exitance.

Let $a = b = h = 1.0$,

Flux received

$$= \frac{2M}{\pi}\left[\sqrt{2}\tan^{-1}\left(\frac{1}{\sqrt{2}}\right) + \sqrt{2}\tan^{-1}\left(\frac{1}{\sqrt{2}}\right)\right.$$

$$\left. - \tan^{-1}(1.0) - \tan^{-1}(1.0) - \frac{1}{2}\log\frac{(1+1+1)}{(1+1)\ (1+1)}\right]$$

$$= M[0.2003]$$

Fraction of emitted flux received

$$= \frac{M[0.2003]}{M}$$

$$= 0.2003$$

Difference

$$\frac{0.2035 - 0.2003}{0.2003} = 0.016 \quad \text{or} \quad 1.6\%$$

4.4.2 EXAMPLE 2

The second example relates to the calculation of ceiling illuminance in a practical lighting design.

A banking hall has a 4 m radius semi-cylindrical ceiling as shown in Figures 4.13(a), (b) and (c).

The ceiling is to be lit by a continuous row of fluorescent tubes mounted on a cornice running the 15 m length of the hall.

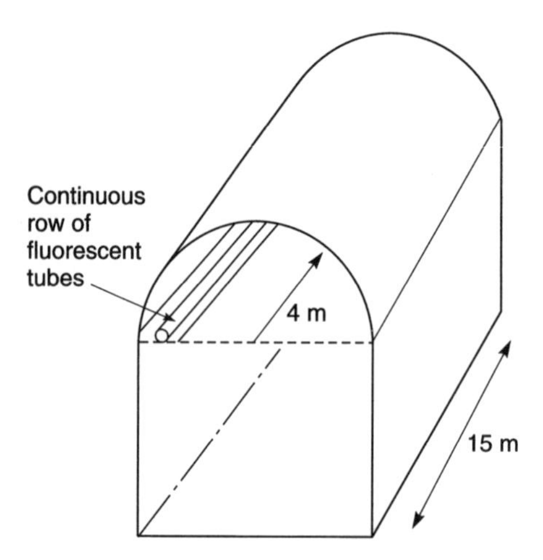

Fig. 4.13(a) The position of the cornice lighting relative to the curved ceiling

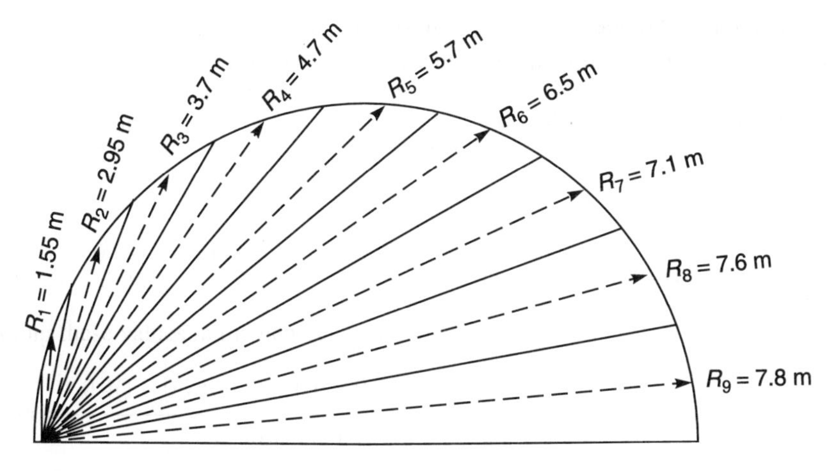

Fig. 4.13(b) The radial distances required for the calculation of the K Factors

In Figure 4.13(b) nine, 10°, sectors have been drawn (solid lines) covering the area lit by one row of fluorescent tubes. For the purpose of this example the axial distribution of the lamps is taken to be a cosine and the transverse intensity as constant.

In Table 4.2, the values of R have been scaled from the drawing (Figure 4.13(b)) as they would be most conveniently obtained in practice.

As before,

$$K = \frac{K_\alpha}{K_{90}}\% \times \frac{0.274}{100}$$

$$= 1.111 \times \alpha^\circ \times \frac{0.274}{100}$$

$$= 3.04 \times 10^{-3} \times \alpha^\circ$$

The K values are summed because the transverse intensity is assumed constant.

Table 4.2 Measured R values and the corresponding K values for the example in Section 4.5.2

	$R_{(M)}$	1/R	α°	K
(1)	1.55	9.677	84.1	0.256
(2)	2.95	5.085	78.9	0.240
(3)	3.7	4.054	76.1	0.231
(4)	4.7	3.191	72.6	0.221
(5)	5.7	2.632	69.2	0.210
(6)	6.5	2.308	66.6	0.202
(7)	7.1	2.113	64.7	0.197
(8)	7.6	1.974	63.1	0.192
(9)	7.8	1.923	62.5	0.190
				1.939

Consequently, the flux to the ceiling is given by:

$$F = 1.939I \text{ lumens} \quad \text{or} \quad 1.94I \text{ lumens}$$

The average illuminance on the ceiling is,

$$= \frac{1.94I}{\pi r l} = \frac{1.94I}{\pi \times 4 \times 15} = 0.01029I \text{ lux}$$

The longest hot cathode fluorescent tube available would be 2.4 m and six of these would meet the length requirement of 15 m with an acceptable tolerance (14.4 m).

A typical initial lumen output of a warm white lamp would be 9500. The total output of the row of six lamps would be 6 × 9500 lumens. We can use this value to calculate I.

$$F = 36 \times K_{90} \times I$$

So,

$$I = \frac{F}{36 \times \Delta\theta \times \frac{\pi}{2}}$$

$$= \frac{9500 \times 6}{36 \times \frac{2\pi}{36} \times \frac{\pi}{2}}$$

$$= 5775 \text{ cd}$$

This gives

$$E_{av} = 0.01029 \times 5775$$
$$= 59 \text{ lux}$$

This is the average illuminance of the curved area of the ceiling. The flux in each sector that does not fall on the ceiling in this way is lost to the side walls above the cornice height, Figure 4.13(c).

The total flux emitted by the six lamps is 9500 × 6. This is emitted over 36, 10° sectors. Nine sectors were used in calculation and so the total amount of useful flux emitted into the ceiling cavity was

$$9500 \times 6 \times \frac{9}{36} = 14250 \text{ lumens}$$

The flux reaching the curved surface of the ceiling was

$$1.94 \times 5775 = 11203$$

The flux to each end is, therefore,

$$\frac{14250 - 11203}{2} = 1524 \text{ lumens}$$

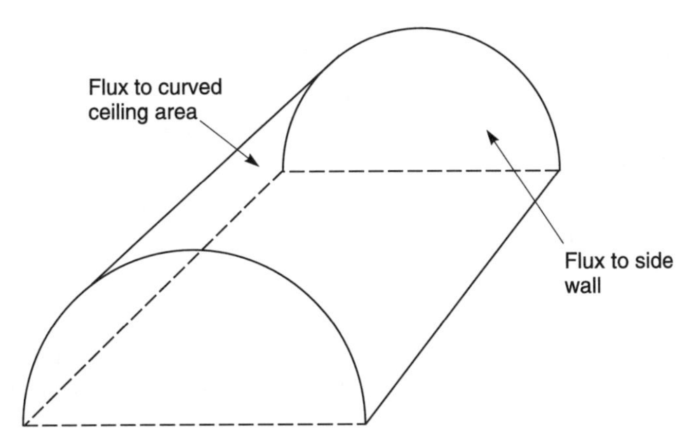

Fig. 4.13(c) The flux passing through the ends of the sectors is incident on the side walls of the ceiling vault

The area of each end wall is

$$\frac{\pi \times 16}{2} \quad \text{or} \quad 25 \text{ m}^2$$

The average illuminance of the end walls is, therefore,

$$\frac{1524}{25} \quad \text{or} \quad 61 \text{ lux}$$

Note: Three practical matters should be kept in mind.

(1) The assumption has been made that the lamps are uniformly diffusing; that is, the axial polar curve is a tangential circle. In practice, fluorescent tubes give a lower intensity at angles approaching 90° and this affects the relationship between the lamp output in lumens and the maximum intensity in candelas.

Photometric measurements show that a typical fluorescent lamp gives:

$$F = 0.925I \quad \text{or} \quad I = 1.08F$$

We have assumed that a 10° sector gives

$$\left(0.1745 \times \frac{\pi}{2}\right) I$$

which, for 36 sectors, gives

$$F = 0.987I \quad \text{or} \quad I = 1.013F$$

So the intensities achieved for a given lumen output will be slightly higher than we have assumed.

However, the aspect factor values from which the K values are calculated will be slightly lower for the practical distribution and this compensates, so that the overall error is less than 5%.

(2) At this stage, even though we are considering a practical problem, no allowance has been made for deterioration of the installation with time. This is dealt with later.
(3) The illuminance of the ceiling would be very uneven across the vault, with only one row of lamps on one of the cornices. In practice, both cornices would probably be used, doubling the average illuminance.

4.4.3 EXTENSION TO THE CASE WHERE THE SECTOR WINDOW IS LONGER THAN THE LIGHT SOURCE

In some practical situations, the light source, although linear, does not span the entire distance between the walls. The *K* factor method can be extended to meet such a case.

Below, the formula is derived for the case where the line source is shorter than the sector window length required to fit the area receiving the flux.

In the treatment given below, the distances from the walls to the end of the line source are assumed to be equal, but the same approach can be used to deal with unequal distances. Although, in Figure 4.14, all sections have equal length, it is only opposite sections that must be

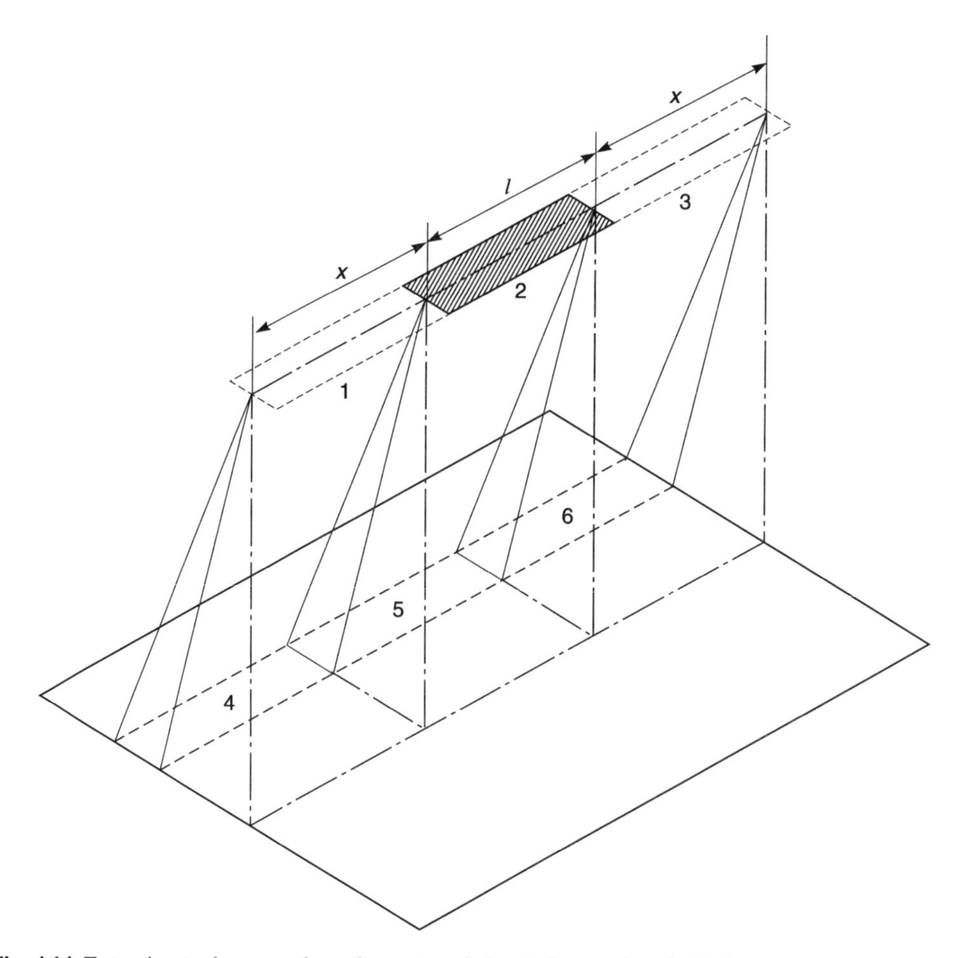

Fig. 4.14 Extension to the case where the sector window is longer than the light source

equal; that is, section 1 equals section 4, section 2 equals section 5 and section 3 equals section 6.

In Figure 4.14, section 2 is the real source and section 1 represents an imaginary source. Let the imaginary source have an intensity distribution identical to the real source and let the luminous exitance of the imaginary source be equal to that of the real source. We wish to find the flux from section 2 to sections 4, 5 and 6; that is, $F_{2(4,5,6)}$.

The flux from the real source (section 2) is found as follows:

$$F_{(1,2)(4,5)} - F_{14} = F_{2(4,5)} + F_{15}$$

where F is the flux and the first subscript indicates the sending surface and the second subscript the receiving surface.

Noting that section 1 has the same length as section 6 and section 2 has the same length as section 5, and applying the principle of reciprocity and by assuming section 5 to be an imaginary light source with the same characteristics as those already defined, we can say:

$$\frac{F_{15}}{M} = \frac{F_{51}}{M} = \frac{F_{26}}{M} \quad \text{giving} \quad F_{15} = F_{26}$$

So,

$$F_{(1,2)(4,5)} - F_{14} = F_{2(4,5,6)}$$

Since the sources all have the same value of luminous exitance then the intensity for the length of source represented by section (1 + 2) is given by

$$\left(\frac{I_\theta}{l}\right) \times (x + l)$$

and the intensity for the length of source represented by section 1 is given by

$$\left(\frac{I_\theta}{l}\right) \times x$$

Giving

$$F_{2(4,5,6)} = \frac{I_\theta}{l}\,[(x + l)K_{(1,2)} - xK_1]$$

where

$K_{(1,2)}$ is the value of K for

$$\alpha = \tan^{-1}\left(\frac{x + l}{R}\right)$$

and K_1 is the value of K for

$$\alpha = \tan^{-1}\left(\frac{x}{R}\right)$$

4.5 Flux transfer between opposite parallel rectangular surfaces

4.5.1 INTRODUCTION

Such a surface could be a ceiling reflecting light to the floor or a light emitting ceiling of prismatic or louvred panels illuminating the floor or work plane.

Although Zijl[1] produced rigorous solutions for the transfer of flux between parallel surfaces in the case of uniform diffusion, Croft[2] produced a somewhat simpler and more versatile solution by developing zonal multipliers for continuous distributions that could be applied to practical distributions as well as to mathematically defined distributions.

4.5.2 ZONAL MULTIPLIERS FOR AREA SOURCES

A luminous surface can be considered to be an array of infinitely small sources with each luminous element touching the other similar elements. Each of these elements will have the same value of luminous flux in a given angular zone (see Chapter 1 for the calculation of zonal flux, Section 1.7.1). The zonal multiplier is the fraction of this flux that lies within the boundary of the illuminated rectangular surface. The value of this multiplier depends upon the position of the element in the area source array; that is, how near it is to the boundary. Consider Figure 4.15(a). This shows some of the parameters needed in the calculation.

Because of symmetry, if the zonal multipliers for one quadrant of a room are determined, the zonal multipliers for the other quadrants will have the same value. Thus, we need only consider one quadrant in our calculation. The shaded area of the zone represents the flux fraction lost to the walls.

4.5.3 ZONAL MULTIPLIERS FOR RECTANGULAR ROOMS

Consider Figure 4.15(b).

Four areas are shown in this figure and these need to be considered in developing the zonal multipliers. The areas are as follows.

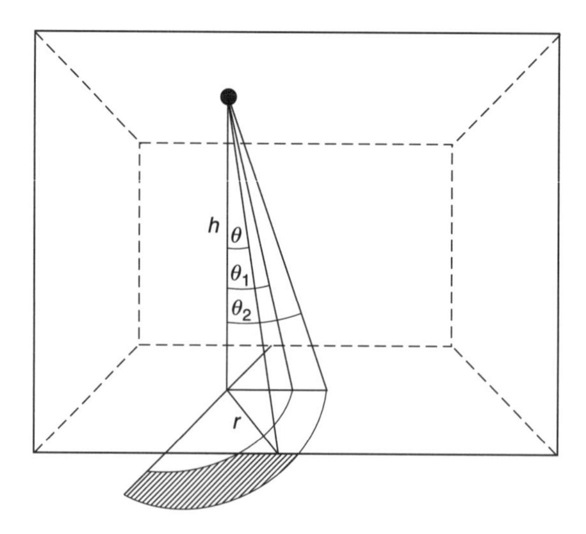

Fig. 4.15(a) The zone related to a quadrant

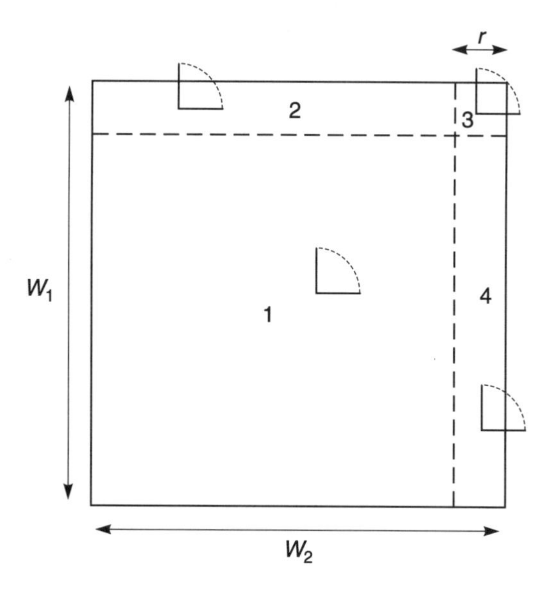

Fig. 4.15(b) The four areas to be considered for the zonal multiplier calculations

Area 1, where the zonal multiplier arc for the source falls entirely within the boundary of the room. This gives a zonal multiplier of unity.
Area 2, where the zonal multiplier arc cuts the boundary in one direction only.
Area 3, here the zonal multiplier arc cuts the boundary in two directions.
Area 4, this is similar to area 2 but may be of a different length.

For a given angular zone, the size of each area is determined by the radius of the zonal multiplier arc, that is $h \tan \theta$.

As an example, consider area 3. Here, the sides of the square (w) to which the zonal multiplier applies are equal to the radius (r) of the zonal multiplier arc.

Thus, when $r = w$ only area 3 exists.

There is, of course, a fifth area where the zonal multiplier arc lies outside the rectangle and the zonal multiplier is zero.

Therefore, it is possible to obtain general zonal multipliers for each type of area and then weight them according to the area of the quadrant to which they apply, for a given value of arc radius r.

Area 1. The zonal multiplier is unity
Areas 2 and 4. Consider Figure 4.15(c)

It is convenient to work in units of r, so, $r = 1.0$. This does not affect the outcome of the calculations.

The zonal multiplier is the fraction of all the quadrant arcs that lie within the area considered.

The length of each arc is equal to the radius multiplied by the angle in radians, but since the radius is 1.0 the arc equals the angle.

$$ZM = \frac{\Sigma \text{ of actual arcs}}{\Sigma \text{ of all quadrant arcs}}$$

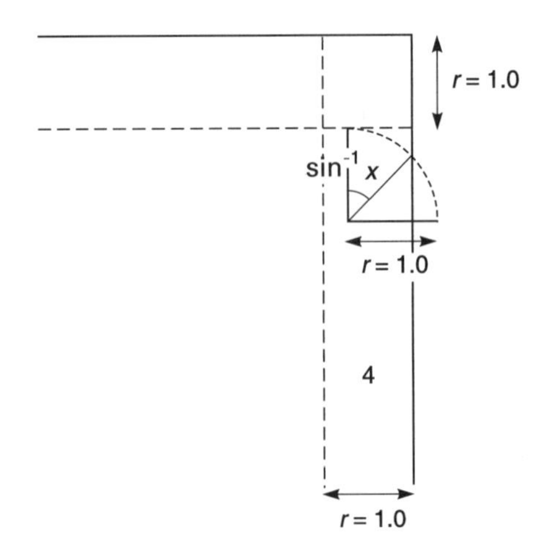

Fig. 4.15(c) Zones 2 and 4, where the arc cuts the boundary in one direction only

$$ZM \int_0^{1.0} \frac{\pi}{2} \, dx = \int_0^{1.0} \sin^{-1} x \, dx$$

$$ZM = \frac{2}{\pi} \int_0^{1.0} \sin^{-1} x \, dx$$

$$= \frac{2}{\pi} \left[x \sin^{-1} x - \int_0^{1.0} \frac{x}{\sqrt{(1-x^2)}} \, dx \right]$$

$$ZM = \frac{2}{\pi} [x \sin^{-1} x + \sqrt{(1-x^2)}]_0^{1.0}$$

$$= 1 - \frac{2}{\pi}$$

$$= 0.363$$

This is the zonal multiplier for areas 2 *and* 4.

Area 3. Consider Figure 4.15(d). Here the arc cuts the boundary at two points. In this case,

$$ZM_3 \int_0^{1.0} \int_0^{1.0} \frac{\pi}{2} \, dx \, dy = \int_0^{1.0} \int_{\sqrt{1-y^2}}^{1.0} \left(\frac{\pi}{2} - \cos^{-1} x - \cos^{-1} y \right) dx \, dy$$

Then,

$$ZM_3 = \frac{2}{\pi} \int_0^{1.0} \int_{\sqrt{1-y^2}}^{1.0} (\sin^{-1} x - \cos^{-1} y) \, dx \, dy$$

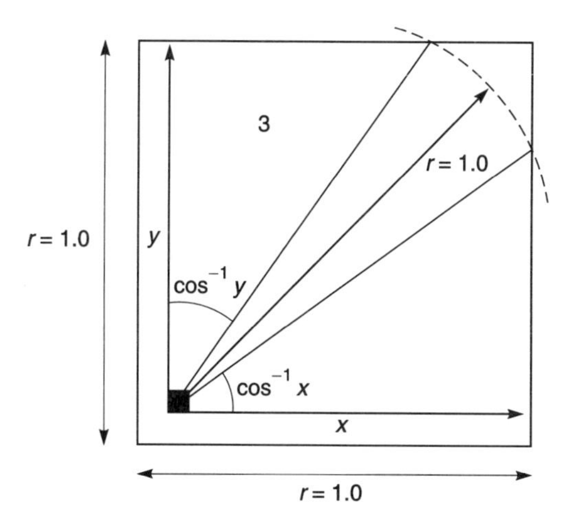

Fig. 4.15(d) Zone 3, where the arc cuts the boundary in two directions

$$\int \sin^{-1} x \ \mathrm{d}x = x \sin^{-1} x - \int \frac{x}{\sqrt{(1-x^{2})}}$$

$$= x \sin^{-1} x + \sqrt{(1-x^2)}$$

Also,

$$\int \cos^{-1} y \ \mathrm{d}x = x \cos^{-1} y$$

So

$$\int_{\sqrt{1-y^2}}^{1.0} (\sin^{-1} x - \cos^{-1} y) \ \mathrm{d}x \ \mathrm{d}y$$

$$= [x \sin^{-1} x + \sqrt{(1-x^2)} - x \cos^{-1} y]_{\sqrt{1-y^2}}^{1.0}$$

$$= \frac{\pi}{2} - \cos^{-1} y - y$$

$$\int_0^{1.0} \left(\frac{\pi}{2} - \cos^{-1} y - y \right) \ \mathrm{d}y$$

$$= \int_0^{1.0} (\sin^{-1} y - y) \ \mathrm{d}y$$

$$= \left[y \sin^{-1} y + \sqrt{1-y^2} - \frac{y^2}{2} \right]_0^{1.0}$$

$$= \frac{\pi}{2} - \frac{3}{2}$$

Thus,

$$ZM_3 = \frac{2}{\pi}\left(\frac{\pi}{2} - \frac{3}{2}\right)$$

$$= 1 - \frac{3}{\pi}$$

$$= 0.045$$

Thus, the zonal multipliers are:

Area type 1	$ZM_1 = 1.0$
Area types 2 and 4	$ZM_2 = ZM_4 = 0.363$
Area type 3	$ZM_3 = 0.045$

The mean zonal multiplier for the whole rectangular area is obtained by weighting each of these zonal multipliers according to the part of the area to which it applies. The results are added together and then divided by the total area.

This formula can then be used to produce a zonal multiplier value for each zone of the area light source flux distribution and so the flux transfer between the two parallel surfaces may be obtained. This is usually the flux transfer between the ceiling and the working plane within a room.

The weighting value for area 1 is given by $(w_1 - r)(w_2 - r)$.
The weighting value for area 2 is given by $r(w_2 - r)$.
The weighting value for area 3 is given by r^2.
The weighting value for area 4 is given by $r(w_1 - r)$.
The formula for the required mean zonal multiplier is:

$$ZM = \frac{0.045r^2 + 0.363[r(w_2 - r) + r(w_1 - r)] + 1.0[(w_2 - r)(w_1 - r)]}{w_1 \times w_2}$$

$$= \frac{0.045r^2 + 0.363[r(w_1 + w_2) - 2r^2] + [w_1w_2 - r(w_1 + w_2) + r^2]}{w_1 \times w_2}$$

Substituting $h \tan \theta$ for r, we obtain

$$ZM = 0.045(h \tan \theta)^2 + 0.363[h \tan \theta(w_1 + w_2)$$

$$\frac{- 2(h \tan \theta)^2] + [w_2w_1 - h \tan \theta(w_2 + w_1) + (h \tan \theta)^2]}{w_1 \times w_2}$$

For a rectangular room the ratio of the horizontal surface area to the vertical surface area is called the room index (RI) and this can be used to simplify the above formula

$$RI = \frac{2 \times w_1 \times w_2}{2h(w_1 + w_2)} = \frac{w_1 \times w_2}{h(w_1 + w_2)}$$

Substituting and simplifying gives:

$$ZM = 0.318 \left(\frac{h}{w_1} \times \frac{h}{w_2} \right) \tan^2 \theta - 0.637 \left(\frac{\tan \theta}{RI} \right) + 1$$

In the special case of a square room as $w_1 = w_2$,

$$ZM = 0.080 \left(\frac{\tan \theta}{RI} \right)^2 - 0.637 \left(\frac{\tan \theta}{RI} \right) + 1$$

Example 1

Calculate 10° zonal multipliers for $w_1/h = 8$ and $w_2/h = 3.0$ and hence determine the fraction of flux transferred from a uniformly diffusing luminous ceiling to the floor

$$RI = \frac{8 \times 3}{8 + 3} = 2.18$$

giving

$$ZM = 0.318 \left(\frac{1}{8} \times \frac{1}{3} \right) \tan^2 \theta - 0.637 \frac{\tan \theta}{2.18} + 1$$

$$= 0.01325 \tan^2 \theta - 0.292 \tan \theta + 1$$

Once $h \tan \theta$ is greater than $\sqrt{w_1^2 + w_2^2}$ then the zonal multiplier is zero. In the above example, this occurs at 85° and so for this angle the zonal multiplier is zero (Table 4.3).

Table 4.3 Tabulated calculation of the 10° interval zonal multipliers for Example 1

Mid-zone value $\theta°$	$\tan \theta$	$\tan^2 \theta$	$0.01325 \tan^2 \theta$	$0.292 \tan \theta$	*ZM*
5	0.0875	7.65×10^{-3}	1.014×10^{-4}	0.0255	0.974
15	0.268	0.0718	9.514×10^{-4}	0.0783	0.922
25	0.466	0.217	2.875×10^{-3}	0.1364	0.864
35	0.700	0.490	6.49×10^{-3}	0.204	0.802
45	1.000	1.000	0.01325	0.292	0.721
55	1.428	2.040	0.0270	0.417	0.610
65	2.145	4.599	0.0609	0.626	0.435
75	3.732	13.928	0.1845	1.0897	0.095
85	11.430				0.000

To calculate the fraction of flux transferred from the ceiling to the floor, the fraction of the total downward flux emitted in each zone must first be determined. This is easily calculated for distributions of the form $I_\theta = I_m \cos^n \theta$,

$$\text{Zonal fraction} = \frac{\text{Flux emitted in the zone}}{\text{Total downward flux}}$$

$$= \frac{2\pi \int_{\theta_1}^{\theta_2} I_\theta \sin\theta \ \mathrm{d}\theta}{2\pi \int_{0}^{\pi/2} I_\theta \sin\theta \ \mathrm{d}\theta}$$

When $I_\theta = I_m \cos^n \theta$, this becomes

$$ZFR = \cos^{(n+1)} \theta - \cos^{(n+1)} \theta_2$$

In all cases the fraction of flux to the floor or working plane

$$= \sum_{0}^{\pi/2} ZM.ZFR$$

In Table 4.4 the zonal fractions are calculated for a cosine distribution and multiplied by the appropriate zonal multipliers (Table 4.5) to obtain the required value for the fraction of flux transferred from the ceiling to the floor.

The fraction of flux transferred from the 'ceiling' to the 'floor' is called the Surface Distribution Factor, D_S. In this case, $D_S = 0.646$.

Zijl's formula for a uniformly diffusing rectangular source, can be used to calculate a precise value for D_S for a room with $w_1/h = 8$ and $w_2/h = 3.0$.

Table 4.4 Tabulated calculation of the zonal fractions for a cosine distribution of 10° zones

Zone angle in degrees	$\cos\theta_1$	$\cos\theta_2$	$\cos^2\theta_1$	$\cos^2\theta_2$	$ZFR = \cos^2\theta_1 - \cos^2\theta_2$
0–10	1	0.985	1	0.9698	= 0.0302
10–20	0.985	0.9397	0.9698	0.883	= 0.0868
20–30	0.9397	0.866	0.883	0.75	= 0.133
30–40	0.866	0.766	0.75	0.587	= 0.163
40–50	0.766	0.643	0.587	0.413	= 0.174
50–60	0.643	0.5	0.413	0.25	= 0.163
60–70	0.5	0.342	0.25	0.117	= 0.133
70–80	0.342	0.1736	0.117	0.030	= 0.087
80–90	0.1736	0	0.030	0	= 0.030
					1.000

Table 4.5 Tabulated calculation of $\sum ZM.ZF$ to obtain the surface distribution factor in Example 1

	ZM	*ZFR*	*ZM,ZF*
0–10	0.974	0.0302	0.0294
10–20	0.922	0.0868	0.0800
20–30	0.864	0.133	0.1149
30–40	0.802	0.163	0.1307
40–50	0.721	0.174	0.1255
50–60	0.610	0.163	0.0994
60–70	0.435	0.133	0.0579
70–80	0.095	0.087	0.0083
80–90	0	0.030	0.000
		Total	0.646

$$\Phi_{(\text{ABCD} \rightarrow \text{KLMN})} = \frac{2M}{\pi}\left[a\sqrt{(b^2+h^2)}\tan^{-1}\left(\frac{a}{\sqrt{b^2+h^2}}\right) + b\sqrt{(a^2+h^2)}\right.$$

$$\times\tan^{-1}\left(\frac{b}{\sqrt{a^2+h^2}}\right) - ah\tan^{-1}\left(\frac{a}{h}\right) - bh\tan^{-1}\left(\frac{b}{h}\right)$$

$$\left. -\frac{1}{2}\log_e\frac{(a^2+b^2+h^2)h^2}{(a^2+h^2)(b^2+h^2)}\right]$$

Let $h = 1$, $a = 8$ and $b = 3$,

$$\Phi = \frac{2M}{\pi}\left[8\sqrt{3^2+1^2}\tan^{-1}\left(\frac{8}{\sqrt{3^2+1^2}}\right) + 3\sqrt{8^2+1^2}\tan^{-1}\left(\frac{3}{\sqrt{8^2+1^2}}\right)\right.$$

$$\left. -\ 8\times1\tan^{-1}\left(\frac{8}{1}\right) - 3\times1\tan^{-1}\left(\frac{3}{1}\right) - \frac{1}{2}\log_e\frac{(8^2+3^2+1^2)}{(8^2+1^2)(3^2+1^2)}\right]$$

$$= \frac{2M}{\pi}[25.3\tan^{-1}(2.53) + 24.187\tan^{-1}(0.372) - 8\tan^{-1}(8) - 3\tan^{-1}(3) - \frac{1}{2}\log_e 0.9867]$$

$$= \frac{2M}{\pi}[30.22 + 8.61 - 11.572 - 3.747 - \frac{1}{2}\log_e 0.9867]$$

$$= \frac{2M}{\pi}[23.511 + 0.0067]$$

$$= \frac{23.50\times 2M}{\pi} = 14.96M$$

Fraction of emitted flux received

$$= \frac{14.96M}{M} = 14.96$$

$$D_S = \frac{14.96}{8 \times 3} = 0.624$$

Difference,

$$\frac{0.646 - 0.624}{0.624} = 0.035 \quad \text{or} \quad 3.5\%$$

Thus, the value of 0.646 obtained by zonal multipliers compares favourably with that obtained using Zijl's formula.

The next example shows that the agreement between the zonal multiplier value and the value calculated using Zijl's formula is much closer when square rooms are considered.

Example 2

Let the room have the same room index as the previous example, but let it be a square room, then

$$ZM = 0.080 \left(\frac{\tan\theta}{RI} \right)^2 - 0.637 \left(\frac{\tan\theta}{RI} \right) + 1$$

$$RI = 2.18$$

$$ZM = 0.0168 \tan^2\theta - 0.292 \tan\theta + 1$$

The zonal multiplier is calculated in Table 4.6 and the surface distribution factor (D_S) in Table 4.7.

So, $D_S = 0.655$.

Table 4.6 Tabulated calculation of the zonal multipliers for Example 2

Mid-zone angle ($\theta°$)	$\tan\theta$	$\tan^2\theta$	$0.0168 \tan^2\theta$	$0.292 \tan\theta$	ZM
5	0.0875	7.65×10^{-3}	1.285×10^{-4}	0.0255	0.974
15	0.268	0.0718	1.206×10^{-3}	0.0786	0.923
25	0.466	0.217	3.72×10^{-3}	0.1364	0.867
35	0.700	0.490	8.23×10^{-3}	0.204	0.804
45	1.000	1.000	0.0168	0.292	0.725
55	1.428	2.040	0.0343	0.417	0.617
65	2.145	4.599	0.0773	0.626	0.451
75	3.732	13.928	0.234	1.089	0.144
85	11.43	–	–	–	0.000

(Arc outside rectangle)

Table 4.7 Tabulated calculation of the distribution factor for Example 2

Zone	*ZM*	*ZF*	*ZM.ZF*
0–10	0.974	0.0302	0.0294
10–20	0.922	0.0865	0.080
20–30	0.867	0.133	0.115
30–40	0.804	0.163	0.131
40–50	0.725	0.174	0.126
50–60	0.617	0.163	0.101
60–70	0.451	0.133	0.060
70–80	0.144	0.087	0.013
			0.655

Now let us compare this value with that obtained using Zijl's formula:

$$RI = 2.18 = \frac{w}{2h}, \text{ let } h = 1.0 \text{ so } w = 4.36 = a = b$$

$$\Phi = \frac{2M}{\pi}[4.36\sqrt{(4.36^2+1^2)}\tan^{-1}\left(\frac{4.36}{\sqrt{(4.36^2+1^2)}}\right) + 4.36\sqrt{(4.36^2+1^2)}$$

$$\times\tan^{-1}\left(\frac{4.36}{\sqrt{(4.36^2+1^2)}}\right) - 4.36\tan^{-1}\left(\frac{4.36}{1}\right) - 4.36\tan^{-1}\left(\frac{4.36}{1}\right)$$

$$-\frac{1}{2}\log_e\frac{(4.36^2+4.36^2+1^2)1^2}{(4.36^2+1^2)(4.36^2+1^2)}$$

$$= 12.48M$$

$$D_s = \frac{12.48}{4.36^2 M} = 0.656$$

This value agrees with that obtained by zonal multipliers.

O'Brien[3] calculated the illuminance and luminous emittance (exitance) for 35 000 rooms and published the results for 160. He points out that square and infinitely long rooms of the same room index exhibit differences in luminous characteristics that are generally less than 10%. (See also the comment at the end of Section 5.4.5.)

The advantage of Croft's method of zonal multipliers for continuous distributions is that it may be applied to any distribution that can be taken as symmetrical about the vertical axis or averaged to an equivalent symmetrical distribution. That is, it can be applied directly to practical distributions and not just to the uniformly diffuse case.

An important advantage compared to a point by point calculation is that, because the mid-zone angles are used, the values from I-tables may be used in the calculation of zonal flux without interpolation.

4.5.4 AVERAGE HORIZONTAL SCALAR AND CYLINDRICAL ILLUMINANCE USING THE ZONAL MULTIPLIER METHOD

The zonal multipliers developed in the previous sections enable the direct flux to the horizontal plane to be calculated and hence the average horizontal plane illuminance (E_h). In this section it is shown how these multipliers may be modified to enable the average direct scalar or cylindrical illuminances, as well as horizontal illuminance, to be calculated.

For any rectangular horizontal area A, the average value of

$$E_{\mathrm{h}} = \frac{\Sigma\, ZM \times ZF \times I_\theta}{A}$$

where ZM is the zonal multiplier for the zone for which I_θ is the mid-zone intensity and ZF is the zone factor.

In Section 3.8.1 it was shown that the scalar illuminance from a point source at point is given by

$$E_{\mathrm{s}} = \frac{E_{\mathrm{max}}}{4}$$

If the light is incident at an angle θ at a point on the horizontal plane then we may write

$$E_{\mathrm{s}} = \frac{E_{\mathrm{h}}}{4 \cos \theta}$$

where E_{h} is the horizontal illuminance at that point.

If the point under consideration is an average point on the horizontal area then the average value of

$$E_{\mathrm{s}} = \frac{\Sigma\, ZM \times ZF \times I_\theta}{4 \times A \times \cos \theta}$$

where θ is the mid-zone angle for each zone.

It is convenient to associate $1/(4 \cos \theta)$ with the zonal multiplier ZM and so develop a zonal multiplier related to the average scalar illuminance.

Thus, the average value of E_{s} is given by

$$E_{\mathrm{s}} = \frac{\Sigma\, ZM_{\mathrm{s}} \times ZF \times I_\theta}{A}$$

where $ZM_{\mathrm{s}} = ZM/(4 \cos \theta)$.

In a similar way, zonal multipliers for mean vertical or cylindrical illuminance may be developed.

In Section 3.8.2 it was shown that cylindrical illuminance from a point source at a point is given by:

$$E_{\mathrm{cyl}} = E_{\mathrm{max}} \frac{\sin \theta}{\pi}$$

Applying the same reasoning as for scalar illuminance gives:

$$E_{cyl} = \frac{\Sigma\, ZM_{cyl} \times ZF \times I_\theta}{A}$$

where

$$ZM_{cyl} = ZM \frac{\tan \theta}{\pi}$$

Example

Calculate the average horizontal, scalar and cylindrical illuminances for the working plane of a room 16 m long by 6 m wide by 2.8 m high. Assume the room has a luminous ceiling with a luminous exitance of 500 lumens per square metre and a relative luminous intensity distribution given in the tabulated calculation in Table 4.8.

The zonal fractions calculated in Table 4.8, when multiplied by the zonal multipliers, enable a surface distribution factor to be calculated. The surface distribution factor multiplied by the total flux emitted from the ceiling gives the flux incident at the working plane, and when divided by the area of the working plane gives the average illuminance.

Assuming a working plane at 0.8 m above the floor gives $h = 2$ m for the room index calculation,

$$RI = \frac{16 \times 6}{2(16 + 6)} = 2.18$$

Horizontal plane zonal multipliers have already been calculated for these room proportions and this value of room index (see Example 1 of this chapter). The distribution factor for the horizontal plane illuminance is given by $D_S = \Sigma\, ZM \times ZFR$ (see Table 4.9).

Average value of

$$E_h = \frac{D_S \times M \times A}{A}$$

$$= D_S \times M$$
$$= 0.727 \times 500$$
$$= 363.5 \text{ lux}$$

Table 4.8 Tabulated calculation of the zonal fractions for the intensity distribution used for the example

θ°	Relative Intensity	Zone Factor	Relative Zonal Flux	Zonal Fraction
5	1.000	0.095	0.095	0.036
15	1.057	0.283	0.299	0.115
25	1.019	0.463	0.472	0.181
35	0.890	0.628	0.559	0.215
45	0.826	0.774	0.639	0.245
55	0.269	0.897	0.241	0.093
65	0.129	0.993	0.128	0.049
75	0.091	1.058	0.096	0.037
85	0.068	1.091	0.074	0.028
			Total 2.603	

Table 4.9 Tabulated calculation of the distribution factor for the example

$\theta°$	ZM	ZFR	$ZM \times ZFR$
5	0.974	0.036	0.035
15	0.922	0.115	0.106
25	0.864	0.181	0.156
35	0.802	0.215	0.172
45	0.721	0.245	0.177
55	0.610	0.093	0.057
65	0.435	0.049	0.021
75	0.095	0.037	0.003
85	0.000	0.028	0.000
			$D_S = 0.727$

Table 4.10 Tabulated calculation of the scalar distribution factor for the example

$\theta°$	$ZM_s \left(= \frac{ZM}{4 \cos \theta}\right)$	ZFR	$ZM_s \times ZRF$
5	0.244	0.036	0.009
15	0.239	0.115	0.027
25	0.238	0.181	0.052
35	0.244	0.215	0.052
45	0.255	0.245	0.062
55	0.266	0.093	0.025
65	0.257	0.049	0.013
75	0.092	0.037	0.003
85	0.000	0.028	0.000
		$D_{S(scalar)}$	0.243

The distribution factor for scalar illuminance is given by: $D_{S(scalar)} = \Sigma\, ZM_s \times ZFR$ (see Table 4.10).

Average value of

$$E_s = \frac{D_{S(scalar)} \times M \times A}{A}$$

$$= D_{S(scalar)} \times M$$
$$= 0.243 \times 500$$
$$= 121.5 \text{ lux}$$

The distribution for cylindrical illuminance is given by: $D_{S(cylindrical)} = \Sigma\, ZM_{cyl} \times ZFR$ (see Table 4.11).

Table 4.11 Tabulated calculation of the cylindrical distribution factor for the example in section 4.5.4

$\theta°$	$ZM_{cyl}\left(=\dfrac{ZM \tan\theta}{\pi}\right)$	ZFR	$ZM_{cyl} \times ZRF$
5	0.027	0.036	0.000
15	0.079	0.115	0.009
25	0.128	0.181	0.023
35	0.179	0.215	0.039
45	0.229	0.245	0.056
55	0.227	0.093	0.021
65	0.297	0.049	0.014
75	0.113	0.037	0.004
85	0.000	0.028	0.000
		$D_{S(cylindrical)}$	0.166

Average value of

$$E_{cyl} = \frac{D_{S(cylindrical)} \times M \times A}{A}$$

$$= D_{S(cylindrical)} \times M$$
$$= 0.166 \times 500$$
$$= 83 \text{ lux}$$

The required values are therefore:

$E_h = 363.5$ lux
$E_s = 121.5$ lux
$E_{cyl} = 83$ lux

4.6 Flux transfer to a vertical surface

4.6.1 FLUX TRANSFER TO VERTICAL SURFACES WITHIN A SQUARE ROOM, WHERE $w_1 = w_2$ (Figure 4.16)

The flux transferred from (say) the ceiling to the walls can be obtained by subtracting the flux received by the floor from the total flux emitted.

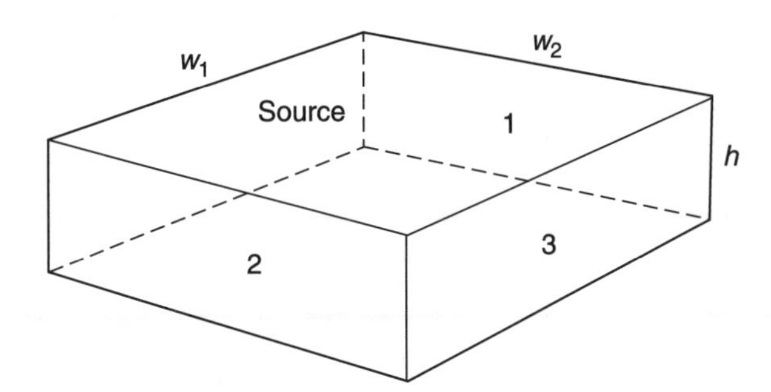

Fig. 4.16 Flux transfer to vertical surfaces within a square room, that is the walls (Surface 3)

So,

Flux to walls = total flux – flux to floor

or, in terms of the proportion of flux transferred,

Fraction of total
flux received by walls = 1 – (Fraction of flux received by floor)
= $1 - D_S$

In the example for a square room calculated earlier, the fraction of total flux received by the floor D_S was 0.655.

So, in that case,

Fraction of flux to walls = 1 – 0.655
= 0.345

This is for all four walls. The fraction for an individual wall is

0.345 ÷ 4
= 0.0863

4.7 Flux transfer within a cylindrical enclosure

Before moving on, it is worthwhile demonstrating how the Principle of Equivalence together with the geometrical properties of a sphere enable the flux transfer within a cylindrical enclosure to be obtained in a simple way (see Figures 4.17(a) and (b)).

The flux transfer between the two parallel discs that form the ends of the cylinder can be calculated as follows. Consider Figure 4.17(b). Here the cylinder is shown enclosed in a sphere.

Using the Principle of Equivalence (Chapter 3, Section 3.3) the illuminance at all points on the inside of the sphere produced by the disc AB forming the end of the cylinder must be the same as that which the spherical cap ABC would produce if it had the same value of luminance.

It was also shown that the direct illuminance on the inside of a sphere from any luminous element of the sphere has the same value at all points within the sphere. It follows that the illuminance on the inside of the spherical cap DEF must have the same value at all points. The flux passing through the disc DF (the opposite end of the cylinder), and which is therefore equal to the flux received by the disc, must be

$$F_2 = E \times (\text{Area of spherical cap DEF})$$

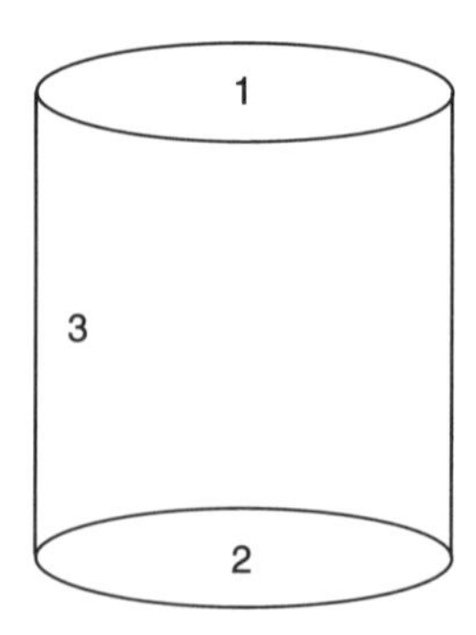

Fig. 4.17(a) Flux transfer within a cylindrical enclosure

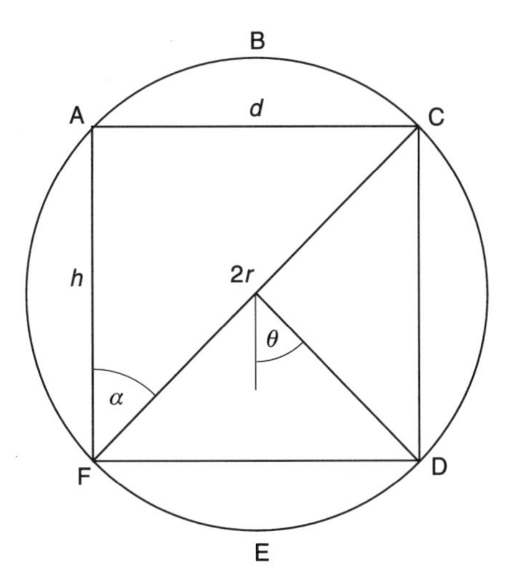

Fig. 4.17(b) The cylinder enclosed in a sphere

The illuminance at point F is given by the disc source formula

$$E = \frac{\pi L}{2}(1 - \cos\alpha)$$
$$= \frac{\pi L}{2}\left(1 - \frac{h}{\sqrt{h^2 + d^2}}\right)$$

and since E is constant over the inside of the cap DEF

$$F_{12} = \frac{\pi L}{2}\left[1 - \frac{h}{\sqrt{h^2 + d^2}}\right] \times \text{area of spherical cap DEF}$$

The area of the spherical cap is equal to the solid angle subtended by the cap at the centre of the sphere multiplied by the square of the sphere radius (Chapter 1, Section 1.1).

So,

$$\text{Area of cap} = 2\pi(1 - \cos\theta) \times r^2$$

In the case chosen, angle θ equals angle α, so that

$$\text{Area of cap} = 2\pi\left(1 - \frac{h}{\sqrt{h^2 + d^2}}\right)r^2$$

Also,

$$r^2 = \frac{h^2 + d^2}{4}$$

Giving

$$\text{Area of cap} = 2\pi\left(1 - \frac{h}{\sqrt{h^2 + d^2}}\right)\frac{h^2 + d^2}{4}$$

So,

$$F_{12} = \frac{\pi L}{2}\left(1 - \frac{h}{\sqrt{h^2 + d^2}}\right)\left(2\pi\left(1 - \frac{h}{\sqrt{h^2 + d^2}}\right)\frac{h^2 + d^2}{4}\right)$$

$$= \pi^2 L\left(1 - \frac{h}{\sqrt{h^2 + d^2}}\right)^2\left(\frac{h^2 + d^2}{4}\right)$$

Although this expression could be further developed, in its present form it is quite convenient for calculation.

The total flux emitted by surface 1 (the top disc of the cylinder)

$$= \frac{\pi^2 d^2 \times L}{4}$$

The fraction of flux emitted from surface 1 received by surface 2,

$$D_{S'} = \frac{4}{d^2}\left(1 - \frac{h}{\sqrt{h^2 + d^2}}\right)^2\left(\frac{h^2 + d^2}{4}\right)$$

$$= \left(\frac{h^2}{d^2} + 1\right)\left(1 - \frac{h}{\sqrt{h^2 + d^2}}\right)^2$$

$$= \left(\left(\frac{h}{d}\right)^2 + 1\right)\left(1 - \frac{\frac{h}{d}}{\sqrt{\left(\frac{h}{d}\right)^2 + 1}}\right)^2$$

Example

Let us calculate the flux received by disc 2 from disc 1, given the following information:

$h = 10$ m, $d = 10$ m and $L = 200$ cd/m^2

$$F_{12} = \pi^2 L\left(1 - \frac{h}{\sqrt{h^2 + d^2}}\right)^2\left(\frac{h^2 + d^2}{4}\right)$$

$$= \pi^2 \times 200\left(1 - \frac{10}{\sqrt{100 + 100}}\right)^2\left(\frac{100 + 100}{4}\right)$$

$$= 200\pi^2(1 - 0.707)^2(50)$$

$$= 8473 \text{ lumens}$$

Flux emitted by disc

$$= \frac{\pi d^2}{4} \times \pi L = \pi^2 \times 25 \times 200 = 49348 \text{ lumens}$$

Fraction of flux emitted from surface 1 received by surface 2

$$D_S = \frac{8473}{49348}$$

$$= 0.1717$$

The flux transferred to the vertical surface or wall (3) of the cylinder can be obtained by subtraction:

Flux to wall	= 49348 – 8473
	= 40875 lumens
Fraction of flux transferred to wall	= 1 – 0.1717
	= 0.8283

4.7.1 *FLUX TRANSFER FUNCTION*

A convenient form of notation to represent the fraction of flux transferred from one surface to another is the flux transfer function. Consider the cylinder used in the previous section (Figures 4.17(a) and (b)).

Denoting the 'ceiling' of the cylinder by 1, the 'floor' by 2 and the 'wall' by 3 enables us to write

f_{12} = fraction of flux transferred from surface 1 to surface 2
f_{13} = fraction of flux transferred from surface 1 to surface 3
f_{21} = fraction of flux transferred from surface 2 to surface 1
f_{33} = fraction of flux transferred from surface 3 to surface 3 (since surface 3 can 'see' itself)

4.7.2 *TRANSFER FUNCTIONS FOR UNIFORMLY DIFFUSE EMISSION*

The transfer functions, or form factors as they are often called, may be extended to relate to rectangular rooms. In the simplest case, the four walls are treated as one surface. This means that the room is considered to consist of three surfaces; the ceiling (1), the floor (2) and the walls (3) (see Figure 4.18). Also, if the emitting surface is not only emitting with a uniform value of luminous exitance over its surface, but also with a uniformly diffuse distribution, then the Principle of Reciprocity can be used to produce further transfer functions.

(1) Denotes the top surface or 'ceiling'
(2) Denotes the bottom surface or 'floor'
(3) Denotes the vertical surface area or 'walls'

$$f_{13} = 1 - f_{12}$$

by symmetry

$$f_{21} = f_{12}$$

and

$$f_{23} = f_{13}$$

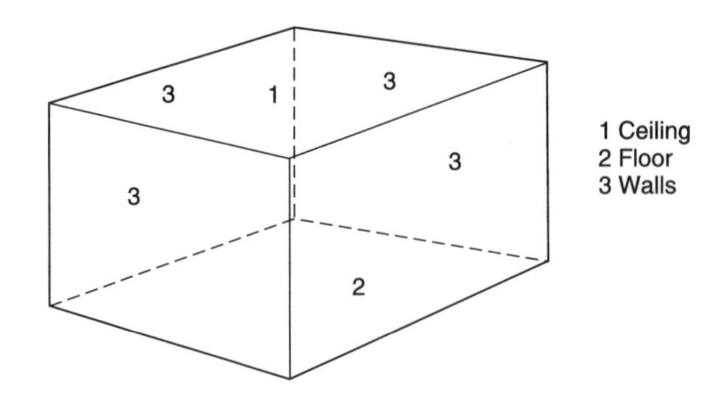

Fig. 4.18 The three surface room for simple transfer function (form factor) calculations

Applying the Principle of Reciprocity we may write:

$$\frac{\text{Flux received by 1 from 3}}{\text{Luminous exitance of 3}} = \frac{\text{Flux received by 3 from 1}}{\text{Luminous exitance of 1}}$$

$$= \frac{M_3 S_3 f_{31}}{M_3} = \frac{M_1 S_1 f_{13}}{M_1}$$

So,

$$f_{31} = f_{13} \frac{S_1}{S_3}$$

Also,

$$f_{33} = 1 - f_{31} - f_{32}$$
$$= 1 - 2f_{31}$$

4.7.3 AN ALTERNATIVE MEANING FOR THE TRANSFER FUNCTION

By definition:

$$f_{ab} = \frac{\text{Flux received by surface } b}{\text{Flux emitted by surface } a}$$

So,

$$f_{ab} = \frac{S_b E_b}{S_a M_a} \quad \text{and} \quad f_{ba} = \frac{S_a E_a}{S_b M_b}$$

Applying the Principle of Reciprocity, we obtain

$$\frac{S_b E_b}{M_a} = \frac{S_a E_a}{M_b}$$

giving,

$$\frac{S_a E_a}{S_a M_b} = f_{ab}$$

$$= \frac{E_a}{M_b}$$

So, f_{ab} can also be defined as the

Illuminance of surface '*a*' divided
by the luminous exitance of surface '*b*'

To sum up, when the transfer functions are considered as ratios between illuminance and luminous exitance, the first subscript refers to the receiving surface, but when they are taken as the ratio of two fluxes the first subscript refers to the emitting surface.

A consequence of the above analysis is that the relationship between f_{ab} and f_{ba} is also established, since

$$f_{ba} = \frac{S_a E_a}{S_b M_b}$$

$$= \frac{S_a}{S_b} f_{ab}$$

4.7.4 FLUX TRANSFER BETWEEN INFINITE PLANES

The fraction of flux emitted from a disc source to a parallel disc with the same axis and the same diameter is given by its distribution factor, D_S.

In this case

$$D_S = \left(\left(\frac{h}{d}\right)^2 + 1\right)\left(1 - \frac{\left(\frac{h}{d}\right)}{\sqrt{\left(\frac{h}{d}\right)^2 + 1}}\right)^2$$

As d is continually increased, h/d tends to zero and, for two parallel infinite surfaces, $D_S = 1.0$.

The direct illuminance of one surface by the other is therefore equal to the luminous exitance of the source surface M (lumens per metre squared).

Although the above treatment is for a uniformly diffuse source the result holds true for other distributions, since if the planes are *infinite* then distance between them may be neglected; that is, all the flux from one is incident on the other.

4.8 Cavities

Consider flux transfer from 'surface' 1 to 'surface' 2 (Figure 4.19).

The light passing into the aperture 'surface' 1 is all incident upon 'surface' 2, the wall of the cavity, so that $f_{12} = 1.0$.

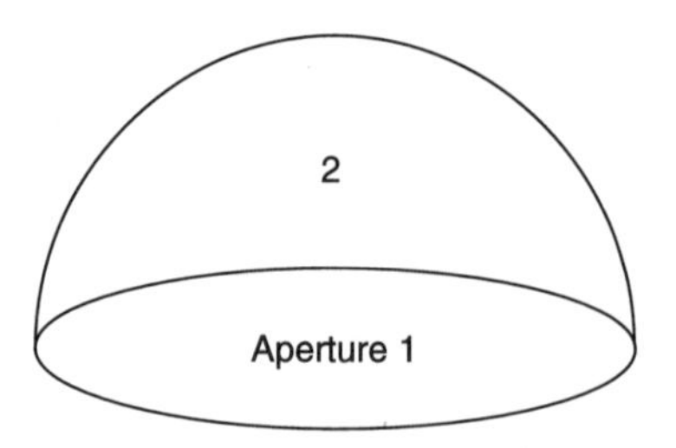

Fig. 4.19 Flux transfer within a cavity

Now consider flux transfer from the interior of the cavity to the aperture; that is, from 'surface' 2 to 'surface' 1.

If we apply the Principle of Equivalence – that two uniformly diffuse sources will produce the same value of illuminance at a point *P*, if they are of equal luminance and appear to have the same boundary when viewed from *P* – then it may be concluded that the result will be the same as if the aperture was placed in front of a uniformly diffusing source of infinite size.

Hence, the illuminance over the aperture would be equal to the luminous exitance of the interior of the cavity; that is 'surface' 2.

So, flux received at the aperture 'surface' 1

$$= M_2 S_1$$

Total flux emitted by 'surface' 2

$$= M_2 S_2$$

giving

$$f_{21} = \frac{S_1}{S_2}$$

Flux transferred across the cavity; that is from 2 to 2 (since 2 can 'see' itself) is obtained by subtracting the aperture transfer function f_{21} from unity

So,

$$f_{22} = 1 - \frac{S_1}{S_2}$$

References

1. Zijl, H. (1960) *Large-size Perfect Diffusors* (Philips Technical Library), p. 52.
2. Croft, R. (1959) *Average Illumination on the Vertical and Calculation of Zonal Multipliers for Continuous Distributions,* Monograph No. 2 (London: IES) p. 8.
3. O'Brien, P.F. (1959) *Lighting Calculations for Thirty-Five Thousand Rooms* (IES). (NY).

5

Interreflected Light

5.1 Introduction

The previous four chapters have dealt with direct illuminance and flux transfer. This chapter deals with interreflected components of illuminance and flux transfer.

When light is emitted from a lamp or luminaire it produces direct illuminance on the surfaces that intercept the light. If the receiving surface is part of an array of surfaces, e.g. the floor, ceiling and walls of a room, then some of the light will undergo multiple reflections until it is all absorbed or transmitted out of the space.

For many years, this aspect of interreflected light was found difficult to deal with and, until the 1950s, most calculations of average illuminance, in practical rooms, were based on tables produced from experimental measurements in a large model room carried out by Harrison and Anderson in the USA during the 1914–18 War.

The difficulties encountered had two elements:

(1) the accurate calculation of the direct illuminances of the room surfaces for the many different types of luminaire;
(2) the interreflection calculations themselves.

The earlier chapters have dealt with the calculation of the direct illuminances, which can, if necessary, be calculated by dividing all the surfaces into a series of elements and calculating the direct illuminance on each element using the inverse-square law. Where necessary, large luminaires are divided into elements small enough for the inverse-square law to be applied. This approach, although fundamental, relies on the advent of powerful computing facilities that have become available in recent times.

Once the direct illuminances are known then the interreflection calculations can be carried out in a number of ways.

(1) By an energy balance approach (based on the Conservation of Energy principle), where the equations are written to represent the steady state where all the lamp flux is accounted for by absorption or transmission out of the space. The equations are then solved simultaneously to give the required illuminance values. This is the *radiosity method.*
(2) By an iterative method where the reflected flux is traced through successive reflections until further calculations are found to make no significant change to the illuminance values.

It is the radiosity method that is developed in this chapter.[1]

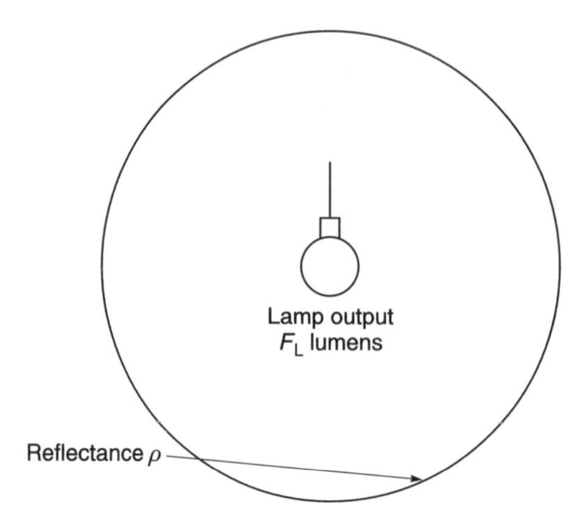

Fig. 5.1 Spherical enclosure containing a light source (integrating sphere)

5.2 Radiosity

The simplest example of the radiosity method is that of a spherical enclosure containing a light source. A practical example of such an enclosure would be an integrating sphere used for comparing the light output of light sources, see Figure 5.1.

A single equation can be written representing the energy balance for this situation.

Total incident flux (F) on the sphere interior	=	The direct flux from the lamp (F_L)	+	The reflected flux

$$F = F_L + F \times \rho$$

$$F(1 - \rho) = F_L$$

$$F = \frac{F_L}{1 - \rho}$$

The average illuminance on the inside of the sphere,

$$E_{av} = \frac{F_L}{A(1 - \rho)}$$

where A is the interior surface area.

The interreflected or indirect component is given by:

$$E_{ind} = E_{av} - E_{direct}$$

$$= \frac{F_L}{A(1 - \rho)} - \frac{F_L}{A}$$

$$= \frac{F_L}{A}\left(\frac{\rho}{1 - \rho}\right)$$

$$= E_{\text{direct}}\left(\frac{\rho}{1-\rho}\right)$$

Example

(a) Let the direct illuminance within a sphere be 500 lux and let the surface reflectance be 0.8. Calculate the final illuminance.

$$E_{\text{av}} = \frac{E_{\text{direct}}}{1-\rho}$$

$$= \frac{500}{1-0.8}$$

$$= 2500 \text{ lux}$$

Notice that the size of the sphere does not appear in this calculation. That is because it has already been taken into account when calculating E_{direct}.

(b) What would be the effect of changing the sphere reflectance to 0.5?

$$E_{\text{av}} = \frac{500}{1-0.5}$$

$$= 1000 \text{ lux}$$

(c) What would be the effect of changing the sphere reflectance to 0.9?

$$E_{\text{av}} = \frac{500}{1-0.9}$$

$$= 5000 \text{ lux}$$

Thus, the original 500 lux in direct illuminance is multiplied by 10.

Comment

In practice, such high values of reflectance are avoided in integrating spheres, since small changes in reflectance over time will have a large effect on the measured illuminance.

5.3 Luminaires

The above type of calculations can be used as a basis for estimating the light output ratio of a luminaire.

Consider the simple opal glass spherical luminaire shown in Figure 5.2.

Let the glass of the sphere have an internal reflectance ρ and a transmittance τ. Then the total flux incident on the inside of the sphere is:

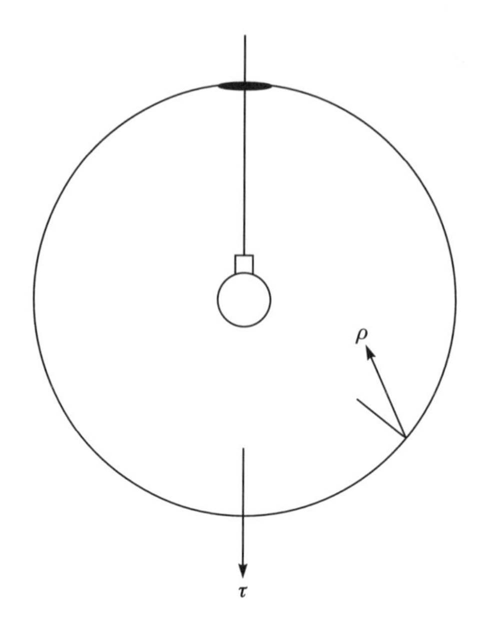

Fig. 5.2 Opal glass spherical luminaire

$$F = \frac{F_L}{(1 - \rho)}$$

where F_L is the lamp flux.

The flux emitted from the sphere surface is:

$$F_{out} = \frac{\tau F_L}{(1 - \rho)}$$

The light output ratio is:

$$LOR = \frac{F_{out}}{F_L} = \frac{\tau}{(1 - \rho)}$$

Example

Let the sphere internal reflectance be 0.4 and the transmittance of the opal glass be 0.5. Calculate the *LOR*:

$$LOR = \frac{0.5}{1 - 0.4}$$

$$= 0.83 \quad \text{or} \quad 83\%$$

In practice, there would be additional losses at the luminaire support point and some absorption by the lamp and its holder, but a value of about 80% may be achieved.

A general expression for the *LOR* of luminaires can be developed providing that the following conditions are observed:

(1) the flux distribution over each surface is assumed to be uniform;
(2) the reflectance or transmittance of each surface is assumed to be uniform and to give uniform diffusion;
(3) the surfaces do not have re-entrants that would trap light and hence alter the effective reflectance of the surface.

Consider the luminaire shown in Figure 5.3.

This general model for a luminaire consists of an aperture (1) which connects two cavities (2) and (3).

In Section 4.4 the flux transfer functions (form factors) were derived for a cavity.

The fraction of flux transferred from each cavity through the joining aperture is:

for the top cavity $f_{31} = \dfrac{\text{area } S_1}{\text{area } S_3}$ and for the bottom cavity $f_{21} = \dfrac{\text{area } S_1}{\text{area } S_2}$

The flux exchange within a cavity is given by:

$$f_{22} = 1 - \frac{S_1}{S_2} \quad \text{and} \quad f_{33} = 1 - \frac{S_1}{S_3}$$

Since all the flux from 3 passing through aperture 1 reaches surface 2 then

$$f_{32} = f_{31} = \frac{S_1}{S_3}$$

and similarly $f_{23} = f_{21}$.

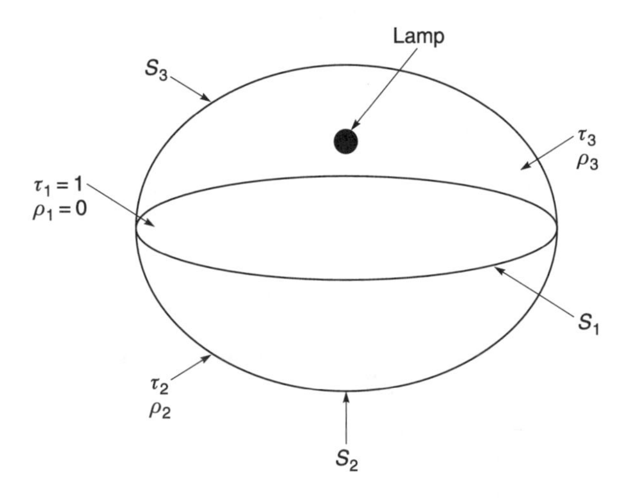

Fig. 5.3 The general luminaire model

The energy balance equations for surfaces 2 and 3 are given below:

$$F_2 = F_3\rho_3 f_{32} + F_2\rho_2 f_{22} + D_2 F_L$$
$$F_3 = F_2\rho_2 f_{23} + F_3\rho_3 f_{33} + D_3 F_L$$

where F_2 and F_3 are the fluxes incident on surfaces 2 and 3 respectively. D_2 and D_3 are the fractions of the lamp flux received by surfaces 2 and 3 directly. ρ_2 and ρ_3 are the reflectances of surfaces 2 and 3.

Solving these equations simultaneously gives

$$\frac{F_2}{F_L} = \frac{D_2(1-\rho_3 f_{33}) + D_3\rho_3 f_{32}}{(1-\rho_2 f_{22})(1-\rho_3 f_{33}) - \rho_2\rho_3 f_{23} f_{32}}$$

$$\frac{F_3}{F_L} = \frac{D_3(1-\rho_2 f_{22}) + D_2\rho_2 f_{23}}{(1-\rho_3 f_{33})(1-\rho_2 f_{22}) - \rho_2\rho_3 f_{23} f_{32}}$$

The upward light output ratio,

$$ULOR = \frac{F_3}{F_L} \times \tau_3$$

where τ_3 is the transmittance of surface 3, and the downward light output ratio,

$$DLOR = \frac{F_2}{F_L} \times \tau_2$$

where τ_2 is the transmittance of surface 2.

The $LOR = ULOR + DLOR$.

The general expression for LOR is:

$$LOR = \frac{\tau_2[D_2(1-\rho_3 f_{33}) + D_3\rho_3 f_{32}] + \tau_3[D_3(1-\rho_2 f_{22}) + D_2\rho_2 f_{23}]}{(1-\rho_2 f_{22})(1-\rho_3 f_{33}) - \rho_2\rho_3 f_{23} f_{32}}$$

Examples

(1) Calculate the *LOR* for the luminaire shown in Figure 5.4(a).

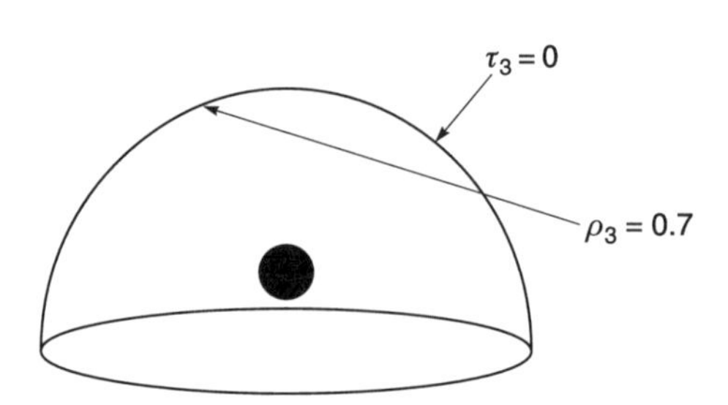

Fig. 5.4(a) Spun metal reflector with uniformly diffusing internal finish

This is a spun metal reflector with a uniformly diffusing internal finish of reflectance 0.7.

Assume 40% of the lamp flux passes directly through the reflector aperture and that the ratio of aperture area to reflector surface area is 0.5.

The essential data are:

$$\tau_3 = 0 \quad \rho_3 = 0.7 \quad \tau_2 = \tau_1 = 1.0$$

$$\rho_2 = \rho_1 = 0 \quad \frac{S_1}{S_3} = 0.5 = f_{31} = f_{32} \quad f_{33} = 1 - \frac{S_1}{S_3} = 0.5$$

Substituting these values in the general equation gives:

$$LOR = \frac{0.4(1 - 0.7 \times 0.5) + 0.6 \times 0.7 \times 0.5 + 0}{1 - 0.7 \times 0.5}$$

$$= 0.72 \quad \text{or} \quad 72\%$$

(2) Calculate the *LOR* for the luminaire shown in Figure 5.4(b).

Assume the area ratios are

$$\frac{S_1}{S_3} = 0.5, \qquad \frac{S_1}{S_2} = 0.4,$$

and the lamp flux fractions are $D_2 = 0.45$, $D_3 = 0.55$.

The essential data are:

$$\tau_3 = 0, \quad \rho_3 = 0.7, \quad \tau_2 = 0.5, \quad \rho_2 = 0.4$$

$$f_{32} = \frac{S_1}{S_3} = 0.5, \quad f_{23} = \frac{S_1}{S_2} = 0.4, \quad f_{33} = 1 - 0.5 = 0.5, \quad f_{22} = 1 - 0.4 = 0.6$$

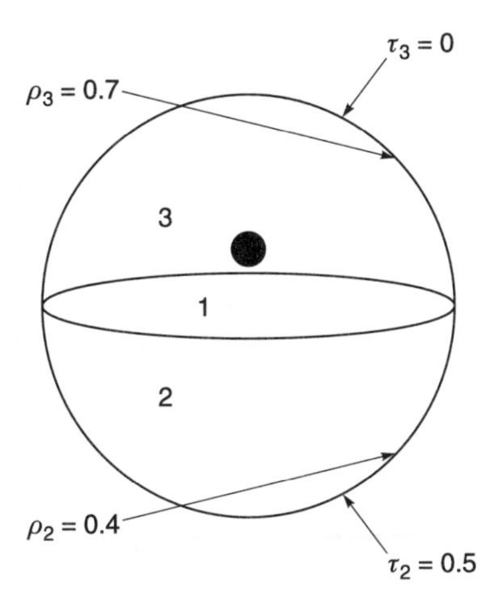

Fig. 5.4(b) Example 2, a reflector as in Fig. 5.4(a) but with a translucent base

Substituting in the general equation gives:

$$LOR = \frac{0.5[0.45(1 - 0.7 \times 0.5) + 0.55 \times 0.7 \times 0.5] + 0}{(1 - 0.4 \times 0.6)(1 - 0.7 \times 0.5) - 0.4 \times 0.7 \times 0.4 \times 0.5}$$

$$= \frac{0.5(0.293 + 0.193)}{0.76 \times 0.65 - 0.056}$$

$$= 0.555 \quad \text{or} \quad 55.5\%$$

(3) Calculate the *LOR* for the luminaire shown in Figure 5.4(c).
In this case S_1 and S_2 are the same area so,

$$\frac{S_1}{S_2} = 1.0 = f_{23}$$

assume

$$\frac{S_1}{S_3} = 0.5 = f_{32}$$

and

$$D_2 = 0.4, \quad D_3 = 0.6$$

The essential data are:

$$\tau_3 = 0, \quad \rho_3 = 0.7, \quad \tau_2 = 0.6, \quad \rho_2 = 0.17$$

$$f_{32} = 0.5, \quad f_{23} = 1.0, \quad f_{33} = 1 - 0.5 = 0.5, \quad f_{22} = 0$$

Substituting in the general equation gives:

$$LOR = \frac{0.6[0.4(1 - 0.7 \times 0.5) + 0.6 \times 0.7 \times 0.5]}{(1 - 0.7 \times 0.5) - 0.17 \times 0.7 \times 1.0 \times 0.5}$$

$$= \frac{0.366}{0.65 - 0.06}$$

$$= 0.62 \quad \text{or} \quad 62\%$$

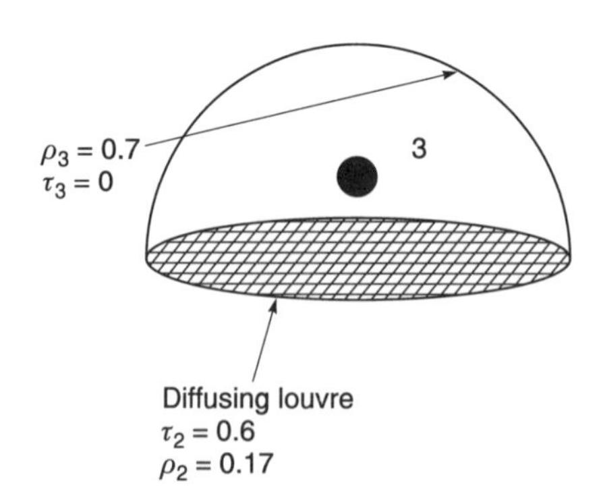

Fig. 5.4(c) Example 3, a reflector as in Fig. 5.4(a) but with a louvred base

5.4 Louvres

In example (3) above, the transmittance and reflectance of a diffusing louvre were used. These may be calculated by using a similar method to that for luminaires.

Consider the louvre cell shown in Figure 5.5.

This consists of the louvre walls (3) and the top and bottom apertures (1) and (2) respectively. The assumption is made that aperture (1) is uniformly illuminated from above and is equivalent to a uniformly diffuse area source filling the aperture.

The energy balance equation is:

Flux received by (2) = Flux received from (1) + Flux received from (3)

$$F_2 = F_1 f_{12} + F_3 \rho_3 f_{32}$$

Flux received by (3):

$$F_3 = F_1 f_{13} + F_3 \rho_3 f_{33}$$

$$F_3 - F_3 \rho_3 f_{33} = F_1 f_{13}$$

$$F_3 = \frac{F_1 f_{13}}{(1 - \rho_3 f_{33})}$$

So,

$$F_2 = F_1 f_{12} + \frac{F_1 f_{13} \rho_3 f_{32}}{(1 - \rho_3 f_{33})}$$

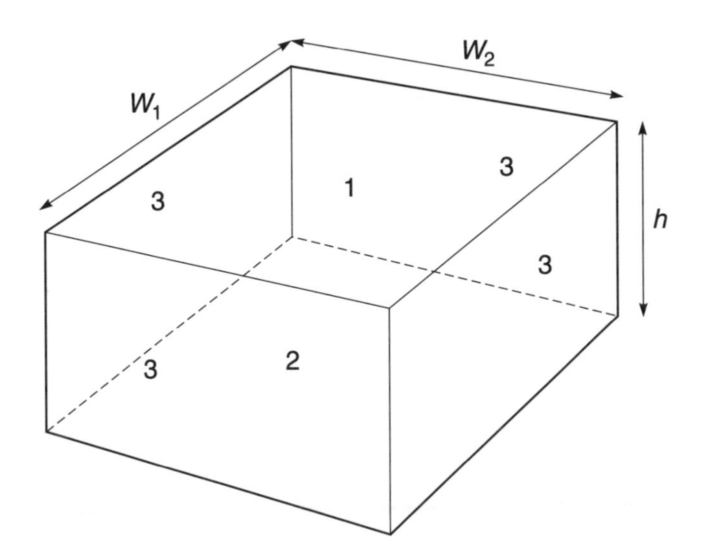

Fig. 5.5 The rectangular louvre cell

Transmittance

$$\frac{F_2}{F_1} = f_{12} + \frac{f_{13} f_{32} \rho_3}{(1 - \rho_3 f_{33})}$$

$$\tau = f_{12} + \frac{f_{13} f_{32}}{\left(\dfrac{1}{\rho_3} - f_{33}\right)} \tag{5.1}$$

Now f_{12} represents the fraction of flux transmitted directly through the louvre without reflection and so the other term must represent the fraction of flux transmitted by reflection. For a uniformly diffusing louvre the same fraction of the flux received by (3) would be reflected upwards and downwards and so the second term represents the equivalent reflectance of the louvre.

Thus,

$$\rho = \frac{f_{13}\ f_{32}}{\left(\dfrac{1}{\rho_3} - f_{33}\right)} \tag{5.2}$$

This expression holds as long as the assumption of uniform illuminance on surface (3) can be justified. Non-uniformity begins to have a serious effect on the value of ρ when the ratio of the area of louvre aperture to louvre wall area is less that 0.5.

From the results obtained for rectangular enclosures in Section 4.5.2,

$$f_{12} = D_S \quad \text{and} \quad f_{13} = 1 - D_S$$

$$f_{32} = f_{31} = f_{13} \frac{S_1}{S_3}$$

where S_1 is the area of aperture (1) and S_3 is the area of the louvre walls.

$$f_{33} = 1 - 2f_{32}$$

giving (by substitution in Equation (5.1) above)

$$\tau = D_S + \frac{(1 - D_S)(1 - D_S)\dfrac{S_1}{S_3}}{\dfrac{1}{\rho_3} - \left(1 - 2(1 - D_S)\dfrac{S_1}{S_3}\right)}$$

and (by substitution in Equation (5.2) above)

$$\rho = \frac{(1 - D_S)(1 - D_S)\dfrac{S_1}{S_3}}{\dfrac{1}{\rho_3} - \left(1 - 2(1 - D_S)\dfrac{S_1}{S_3}\right)}$$

by substitution in Equation 5.2 above.

$$\tau = D_S + \frac{(1 - D_S)^2 \dfrac{S_1}{S_3}}{\dfrac{1}{\rho_3} - \left(1 - 2(1 - D_S)\dfrac{S_1}{S_3}\right)} \tag{5.3}$$

$$\rho = \frac{(1 - D_S)^2 \dfrac{S_1}{S_3}}{\dfrac{1}{\rho_3} - \left(1 - 2(1 - D_S)\dfrac{S_1}{S_3}\right)} \tag{5.4}$$

where, for a rectangular louvre, $S_1 = W_1 \times W_2$ and $S_3 = 2h(W_1 + W_2)$.

If the thickness of the cell wall is not negligible then an adjustment should be made,

$$\tau = \tau\,(\text{louvre cell}) \times \frac{\text{total area of louvre apertures}}{\text{total horizontal area of louvre}}$$

and

$$\rho = \rho\,(\text{louvre cell}) \times \frac{\text{total area of louvre apertures}}{\text{total horizontal area of louvre}} +$$

$$\rho\,(\text{louvre material}) \times \left(1 - \frac{\text{total area of louvre apertures}}{\text{total horizontal area of louvre}}\right)$$

5.4.1 *TRANSLUCENT LOUVRES*

The formulae obtained for opaque rectangular louvres can be used for translucent louvres, if the transmittance of the louvre cell wall is added to its reflectance. This can be done because the flux emerging from the wall of one louvre cell by transmission from another cell is balanced by a similar amount lost by transmission to adjacent cells. The result is the same as an increase in the reflectance of all cells except those at the extreme edge of the louvre by the amount of the louvre wall transmittance.

5.4.2 *PRACTICAL RESULTS*

The authors have investigated the accuracy with which the performance of practical louvres may be predicted using the method developed above.[2] The general conclusion is that the accuracy is of the order of 5%.

As already mentioned, the method relies on the assumption of a uniform distribution of flux on the louvre walls. If accurate calculations are required for deep louvres; that is, when the height of the louvre exceeds its width, then the wall must be divided into two or more sections so that uniform flux distribution may be assumed for each section.

Table 5.1 Surface Distribution Factors

	Room Index (*RI*)								
	0.75	1.0	1.25	1.5	2.0	2.5	3.0	4.0	5.0
D_S	0.320	0.415	0.491	0.547	0.632	0.691	0.733	0.788	0.823

Examples

(1) Calculate the reflectance and transmittance of a square louvre of width 30 mm and depth 15 mm. The reflectance of the louvre material is 0.7. The thickness of the louvre material may be disregarded.

The ratio of aperture area to cell wall area

$$\frac{S_1}{S_3} = \frac{30 \times 30}{2 \times 15(30 + 30)} = 0.5$$

Values of D_S are tabulated against room index (*RI*) in Table 5.1, where

$$RI = \frac{2S_1}{S_3} = 1.0$$

in this case, giving,

$$D_S = 0.415$$

$$\tau = D_S + \frac{(1 - D_S)^2 \dfrac{S_1}{S_3}}{\dfrac{1}{\rho_3} - \left(1 - 2(1 - D_S)\dfrac{S_1}{S_3}\right)}$$

$$= 0.415 + \frac{(1 - 0.415)^2 \times 0.5}{\dfrac{1}{0.7} - [1 - (1 - 0.415) \times 1]}$$

$$= 0.415 + 0.1687$$

(since $\tau = D_S + \rho$)

$\tau = 0.58$ or 58%
$\rho = 0.169$ or 17%

(2) Calculate the transmittance and reflectance for a louvre of similar size to that in Example (1), but where the reflectance of the louvre material is 0.6 and the transmittance of the louvre wall is 0.3. In this case the transmittance and reflectance of the louvre material are added together to give $\rho_3 = 0.6 + 0.3 = 0.9$:

$$\tau = 0.415 + \frac{(1 - 0.415)^2 \times 0.5}{\dfrac{1}{0.9} - [1 - (1 - 0.415) \times 1]}$$

$= 0.415 + 0.246$
$= 0.66$ or 66%
$\rho = 0.246$ or 25%

(since $\tau = D_S + \rho$).

5.4.3 CLASSIFICATION OF ROOMS

Consider a square room where the height is much greater than the width. Let there be a luminaire with a general downward distribution mounted at the centre of the ceiling, see Figure 5.6(a).

Let the angle across the width of the room subtended by the luminaire be α.

Consider two such identical rooms placed side by side, Figure 5.6(b).

In each room the angle subtended at the floor by the luminaires will be α.

Let the dividing wall between the two rooms be removed, Figure 5.6(c).

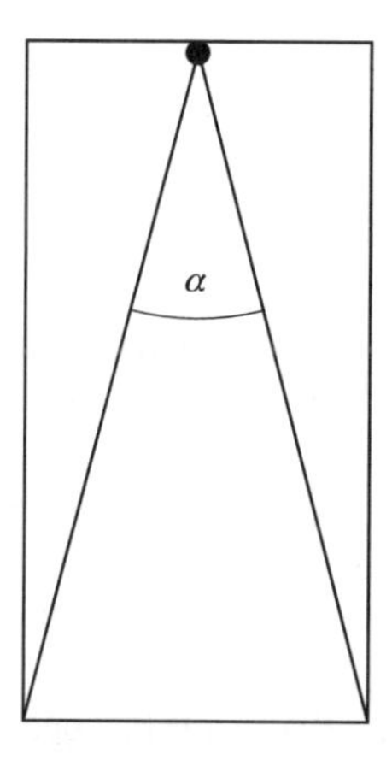

Fig. 5.6(a) Cross-section of a tall square room

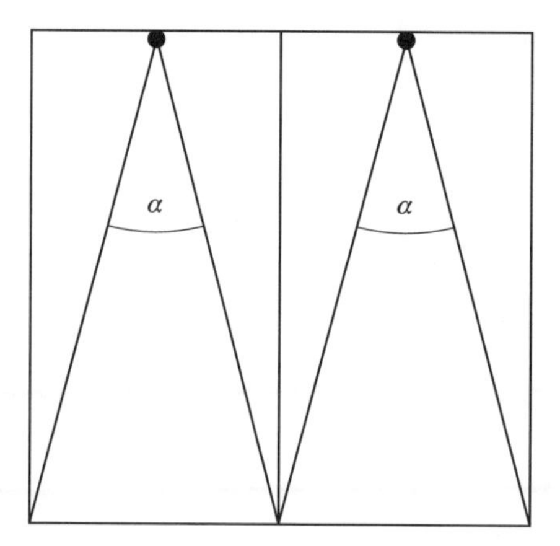

Fig. 5.6(b) Two rooms side by side

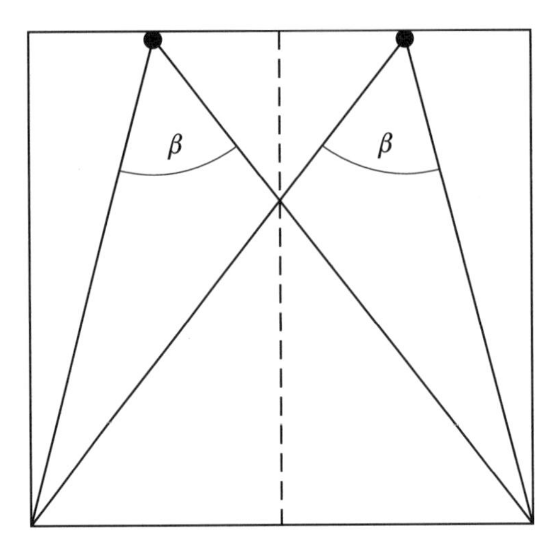

Fig. 5.6(c) The two rooms with the dividing wall removed

Each luminaire now subtends a much larger angle β at the floor. The flux per unit area released into the two rooms side by side is still the same, but with the partition removed more of that flux reaches the floor directly.

Clearly, when we removed the dividing wall the ratio of ceiling area to wall area in the enlarged room was increased and this has been followed by an increase in the average direct illuminance of the floor or working plane.

This increase of efficiency in the transfer of flux from ceiling-mounted sources to the floor is the basis of classifying rooms according to the ratio of horizontal area to vertical area and it has been found that, all other things being equal, rooms with similar ratios of horizontal to vertical area have similar utilization factors, see Section 4.5.3.

This ratio is called the Room Index (*RI*) and it has been introduced in Chapter 4 to simplify the zonal multiplier formulae.

$$RI = \frac{\text{area of ceiling + area of floor}}{\text{area of walls}}$$

$$= \frac{2 \times W_1 \times W_2}{2 \times h(W_1 + W_2)}$$

$$= \frac{W_1 W_2}{h(W_1 + W_2)}$$

which, for a square room, becomes

$$RI = \frac{W}{2h}$$

The Room Index formula is often expressed in terms of the room length (L) and the room width (W). This gives,

$$RI = \frac{L \times W}{h(L + W)}$$

Examples

(1) Calculate the Room Index for a room 6 m long by 4 m wide and 3 m high.

$$RI = \frac{6 \times 4}{3(6 + 4)} = 0.8$$

(2) Calculate the Room Index for a square room 10 m wide and 3 m high.

$$RI = \frac{10}{2 \times 3} = 1.67$$

(3) A square room has a Room Index of 5 and a height of 2.8 m. Calculate its width.

$$RI = \frac{W}{2h}$$

$$5 = \frac{W}{2 \times 2.8}$$

$$W = 5 \times 2 \times 2.8$$
$$= 28 \text{ m}$$

5.5 Interreflections in rooms

When the direct flux received by each room surface is known, then the total surface flux to each surface, including interreflected flux, may be obtained using the same energy balance or flux balance methods already used for luminaires and louvres.

If we consider the room to consist of three surfaces, namely: ceiling (1), floor (2) and walls (3) (see Figure 5.7).

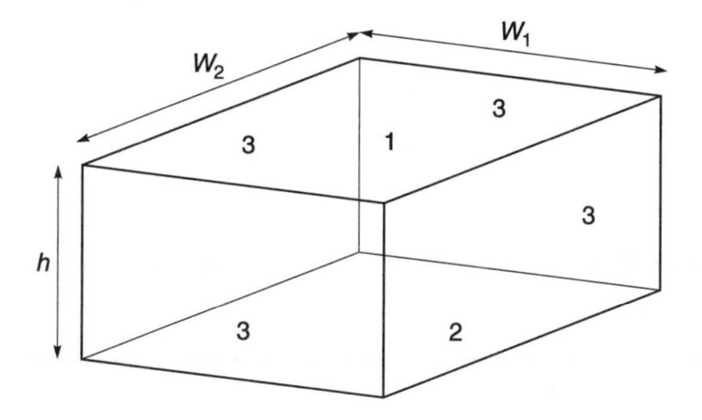

Fig. 5.7 The three surface room for interreflection calculations

Then three energy balance equations can be set up to determine the final illuminance, including the interreflected component.

The following terms are used in setting up the equations:

ρ_1 etc., the surface reflectances
f_{12} etc., the flux transfer functions already introduced
F_1 the total flux received by the ceiling
F_2 the total flux received by the floor
F_3 the total flux received by the walls
F_L the total lamp flux of the installation
$DF(1)$ the ratio of the direct flux received by surface (1) to the total lamp flux of the installation. (*DF* is an abbreviation of distribution factor.)
$DF(2)$ as above, but for the floor
$DF(3)$ as above, but for the walls
$UF(1)$ the ratio of the total flux received by surface (1) to the total lamp flux of the installation. (*UF* is an abbreviation of utilization factor.)
$UF(2)$ as above, but for the floor
$UF(3)$ as above, but for the walls

The equations are:

Flux to ceiling = Direct Flux + Flux received from the floor + Flux received from walls

$$F_1 = DF(1)F_L + F_2\,\rho_2 f_{21} + F_3\,\rho_3 f_{31}$$

Flux to floor = Direct Flux + Flux received from the ceiling + Flux received from the walls

$$F_2 = DF(2)F_L + F_1\,\rho_1 f_{12} + F_3\,\rho_3 f_{32}$$

Flux to walls = Direct Flux + Flux received from the ceiling + Flux received from the floor + Flux received from the walls

(As the walls are being treated as single surface that can 'see' itself.)

$$F_3 = DF(3)F_L + F_1\,\rho_1 f_{13} + F_2\,\rho_2 f_{23} + F_3\,\rho_3 f_{33}$$

Dividing through by F_L we find that the equations may be written as,

$$UF(1) = DF(1) + UF(2)\rho_2 f_{21} + UF(3)\rho_3 f_{31}$$

$$UF(2) = DF(2) + UF(1)\rho_1 f_{12} + UF(3)\rho_3 f_{32}$$

$$UF(3) = DF(3) + UF(1)\rho_1 f_{13} + UF(2)\rho_2 f_{23} + UF(3)\,\rho_3 f_{33}$$

or

$$DF(1) = UF(1) - UF(2)\rho_2 f_{21} - UF(3)\rho_3 f_{31}$$

$$DF(2) = -\,UF(1)\rho_1 f_{12} + UF(2) - UF(3)\rho_3 f_{32}$$

$$DF(3) = -\,UF(1)\rho_1 f_{13} - UF(2)\rho_2 f_{23} + UF(3)(1 - \rho_3 f_{33})$$

Before expressing these equations in matrix form we should note the following relationships:

Because of symmetry $f_{12} = f_{21}$ and $f_{13} = f_{23}$, also,

$$f_{31} = f_{13}\left(\frac{\text{area of ceiling}}{\text{area of walls}}\right)$$

giving

$$f_{31} = \frac{(RI)}{2} f_{13} \quad \text{where } f_{13} = (1 - f_{12})$$

and

$$f_{33} = (1 - f_{31} - f_{32}) = (1 - 2f_{31}) = (1 - (RI)f_{13})$$

giving

$$\begin{bmatrix} DF(1) \\ DF(2) \\ DF(3) \end{bmatrix} = \begin{bmatrix} 1 & -\rho_2 f_{12} & -\rho_3 f_{31} \\ -\rho_1 f_{12} & 1 & -\rho_3 f_{31} \\ -\rho_1(1 - f_{12}) & -\rho_2(1 - f_{12}) & 1 - \rho_3 + 2\rho_3 f_{31} \end{bmatrix} \begin{bmatrix} UF(1) \\ UF(2) \\ UF(3) \end{bmatrix}$$

So,

$$[DF] = [f][UF]$$

or

$$[UF] = [f]^{-1}[DF]$$

Thus, multiplying the inverted $[f]$ matrix by the $[DF]$ matrix, gives the ratio of the final flux on each surface to the total lamp flux for the installation; that is, the utilization factors (UF).

It is not too difficult to invert the f matrix as it stands; that is, in terms of symbols, but the resultant formulae are complex.

If the calculation is to be done for a specific case it is much simpler to give values to the terms in the matrix so that it consists of single numbers and then invert it.

The surface distribution factors (D_S) required for the calculation of the flux transfer factors may be derived by the method given in Section 4.5.2 or obtained from Table 5.1.

Example

Calculate the utilization factors for an installation given the following data:

$$RI = 1.0, \quad \rho_1 = 0.7, \quad \rho_2 = 0.3, \quad \rho_3 = 0.5$$

$$DF(1) = 0.2, \quad DF(2) = 0.4, \quad DF(3) = 0.4$$

From Table 5.1 we find $D_S = 0.415 = f_{12}$

$$\begin{bmatrix} UF(1) \\ UF(2) \\ UF(3) \end{bmatrix} = \begin{bmatrix} 1 & -\rho_2 f_{12} & -\rho_3(1 - f_{12})\frac{RI}{2} \\ -\rho_1 f_{12} & 1 & -\rho_3(1 - f_{12})\frac{RI}{2} \\ -\rho_1(1 - f_{12}) & -\rho_2(1 - f_{12}) & (1 - \rho_3(1 - (1 - f_{12})RI)) \end{bmatrix}^{-1} \begin{bmatrix} DF(1) \\ DF(2) \\ DF(3) \end{bmatrix}$$

Inserting numerical values into the f matrix to be inverted, gives

$$[f] = \begin{bmatrix} 1 & -0.1245 & -0.1463 \\ -0.2905 & 1 & -0.1463 \\ -0.4095 & -0.1755 & 0.7925 \end{bmatrix}^{-1}$$

The inversion requires the following steps:

Step 1. Find the determinant of the matrix $|f|$.
Step 2. Transpose the matrix $[f_t]$.
Step 3. Replace each term of the transposed matrix by its minor.
Step 4. Divide each term in the new matrix by the determinant of matrix $[f]$.
Step 5. Multiply each term by (+1) or (–1) according to the following pattern

$$\begin{matrix} + & - & + \\ - & + & - \\ + & - & + \end{matrix}$$

Step 1. The determinant

$$\begin{aligned} \Delta &= (1 \times 1 \times 0.7925) + (-0.1245)(-0.1463)(0.4095) \\ &\quad + (-0.1463)(-0.2905)(-0.1755) - (1) \times (-0.1463)(-0.1755) \\ &\quad - (0.1245)(-0.2905)(0.7925) - (-0.1463)(1)(-0.4095) \\ &= 0.663 \end{aligned}$$

Step 2. Transpose of the matrix (exchange rows and columns)

$$\begin{bmatrix} 1 & -0.2905 & -0.4095 \\ -0.1245 & 1 & -0.1755 \\ -0.1463 & -0.1463 & 0.7925 \end{bmatrix}$$

Step 3. Replace each by its minor (the matrix left when the row and column containing the term is removed)

For example: the minor of the term in the first row and first column, that is 1, 1 is:

$$\begin{bmatrix} 1 & -0.1755 \\ -0.1463 & 0.7925 \end{bmatrix}$$

$$\begin{aligned} &= (1)(0.7925) - (-0.1755)(-0.1463) \\ &= 0.767 \end{aligned}$$

Step 4. Divide by the determinant obtained in 1,

$$\frac{0.767}{0.663} = 1.156$$

Step 5. Multiply by (+1)

$$(+1)(1.156) = 1.156$$

Repeating this process for all terms gives the inverted matrix $[f]^{-1}$, we find,

$$\begin{bmatrix} UF(1) \\ UF(2) \\ UF(3) \end{bmatrix} = \begin{bmatrix} 1.156 & 0.187 & 0.248 \\ 0.437 & 1.105 & 0.285 \\ 0.694 & 0.342 & 1.453 \end{bmatrix} \begin{bmatrix} DF(1) \\ DF(2) \\ DF(3) \end{bmatrix}$$

So,

$$\begin{aligned} UF(1) &= 1.156DF(1) + 0.187DF(2) + 0.248DF(3) \\ &= 1.156 \times 0.2 + 0.187 \times 0.4 + 0.248 \times 0.4 \\ &= 0.405 \end{aligned}$$

$$UF(2) = 0.437 \times 0.2 + 1.105 \times 0.4 + 0.285 \times 0.4$$
$$= 0.643$$

$$UF(3) = 0.694 \times 0.2 + 0.342 \times 0.4 + 1.453 \times 0.4$$
$$= 0.857$$

The values of $UF(2)$ are sometimes published by manufacturers relating to their indoor luminaires for a complete range of room index values and a range of reflectances for ceilings, walls and floor.

The numerical values in the above matrix, that is 1.156, etc., are termed *Transfer Factors* and the UF formulae written as:

$$UF(1) = TF(1, 1)DF(1) + TF(2, 1)DF(2) + TF(3, 1)DF(3), \text{ etc.}$$

A complete set of Transfer Factors was published by Bean and Bell[3] in 1976. This method was subsequently used as the basis for Technical Memoranda TM5 (London: Chartered Institution of Building Services Engineers, 1980).

The CIE produced an alternative method, published in 1982.[4] This method relied on approximating the flux distribution of the luminaire to a cosine power series, but the interreflection calculations were also based on the same energy balance approach.

5.5.1 TUNNELS

The utilization factors for tunnels can be considered as a special case of the calculations for rooms. The matrix to be inverted is the same, since a tunnel can still be considered as consisting of three surfaces; ceiling, floor and walls. The main problem is finding the surface distribution factor for a 'room' that can be considered to be infinitely long.

One approach would be to divide the ceiling into three infinitely long strips and to use the K factor method for linear sources that was developed in Section 4.4. The assumption of an infinitely long tunnel greatly simplifies this method, since the K values are all the same and therefore, since the surface distribution factor is a ratio, cancel out.

As an example, we will calculate the surface distribution factor for a long tunnel where the width is twice the height, see Figure 5.8.

The ceiling of the tunnel is assumed to give uniformly diffusing reflectance, so that $I_\theta = I_m \cos\theta$. To apply the K factor method we assume that the ceiling is divided into three strips (the more strips the greater the accuracy of the approximation). Each strip is treated as a line source located at the centre point of the strip. Examining Figure 5.8 we see that the flux to the floor from source 3 will have the same value as that from source 1; also that the flux from source 2 will be given by twice that emitted to the floor over angle θ_3.

Consider source 2. Using the normal K factor method, the fraction of the flux emitted by the line source (representing the strip) in the 0° to 90° sector, which reaches the floor, is given by:

$$\frac{I_m \sum_{0}^{\theta_3} \cos\theta\, K}{I_m \sum_{0}^{90^\circ} \cos\theta\, K}$$

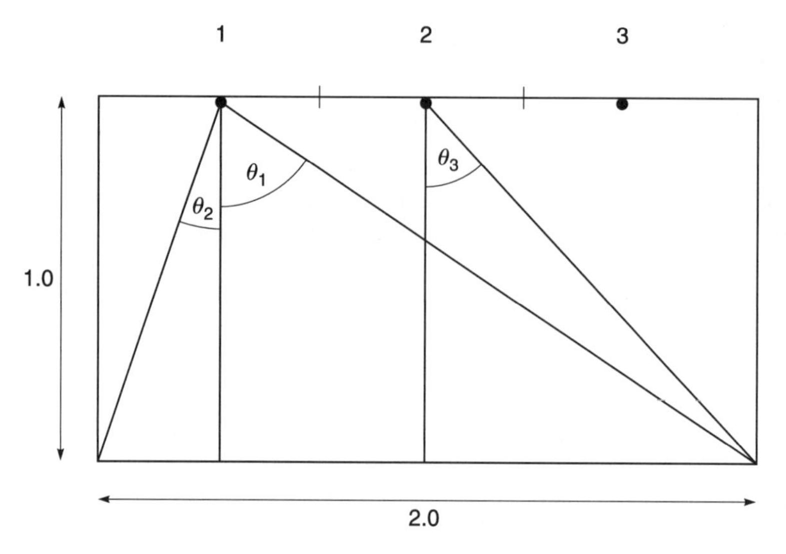

Fig. 5.8 Surface distribution factor for a tunnel using the *K* factor method

However, since in the case of an infinitely long room or tunnel *K* is constant we get

$$\frac{\sum_{0}^{\theta_3} \cos\theta}{\sum_{0}^{90^\circ} \cos\theta}$$

When *K* varies, small sectors such as 10° are chosen, but in this case *K* has cancelled so we can assume infinitely small sectors and integrate.

Consequently, the fraction of the 0–90° flux that reaches the floor is given by:

$$\frac{\int_0^{\theta_3} \cos\theta}{\int_0^{90^\circ} \cos\theta} = \frac{[-\sin\theta]_0^{\theta_3}}{[-\sin\theta]_0^{90^\circ}} = \sin\theta_3$$

Now $\theta_3 = 45°$ so the fraction of 0 – 90° reaching the floor is sin 45° = 0.707.

For the two sides of the line source, the total would be 1.5 from an output of 2 units (e.g. each side of the source).

Now consider the case of source 1:

$$\theta_1 = \tan^{-1}\left(\frac{5}{6} \times \frac{2}{1}\right) = 59°$$

$$\begin{aligned} \sin 59° &= 0.857 \\ &= \text{fraction of } 0 - 90° \text{ flux received on floor over angle } \theta_1 \end{aligned}$$

$$\theta_2 = \tan^{-1}\left(\frac{1}{6} \times \frac{2}{1}\right) = 18.4°$$

sin 18.4° = 0.316

= fraction of 0 – 90° flux received on floor over angle θ_2

Total from source 1 is 0.857 + 0.316 = 1.173 (from two units of output).

Source 3 gives a similar result.

Considering each side of the three line sources the total units of output are 6.

Adding the results for the three line sources we arrive at

$$\frac{1.5 + 1.173 + 1.173}{6} = 0.641$$

So, $D_S = 0.641$.

If we use Zijl's formula (Chapter 4) to calculate D_S for a tunnel $h = 1.0$, $w = 2.0$, $l = 100$ we arrive at a value of 0.614 – a difference of 4%.

The difference is because the *K* factor method assumes that the ceiling strips can be represented by line sources located at the centre of each strip and this reduces the flux lost to the walls, hence the higher value of D_S.

It is worth noting that if we calculate the Room Index for those values used above in Zijl's formula and consult Table 5.1 we obtain an interesting result.

$$RI = \frac{W_1 W_2}{h(W_1 + W_2)} = \frac{2 \times 100}{1(2 + 100)} = 1.96$$

taking $RI = 2.0$ gives a value of $D_S = 0.632$ – a difference of 3%.

If we have taken the Room Index for the tunnel as simply W/h then we would have a value of 2.0 and $D_S = 0.632$ as before. Examination of the Room Index formula shows that for very long tunnels

$$RI \rightarrow \frac{W}{h}$$

Thus, if the *RI* value is taken as the ratio of W/h then the D_S value for a square room of similar room index applies to a good approximation (e.g. 2% for narrow tunnels to 5% for wide tunnels).

5.5.2 CHANGE OF REFLECTANCE

In Section 5.5 the matrix used to obtain utilization factors was given. An example was also given of the matrix inversion required, but for simplicity this was for a specific case. If the process is carried out using only symbols, an interesting fact is noticed. When the minors for surface 2 (the floor) are obtained it is found that they do not contain ρ_2, e.g. minor 2, 1 is,

$$\begin{vmatrix} -\rho_1 f_{21} & -\rho_1 f_{13} \\ -\rho_3 f_{31} & 1 - \rho_3 + 2\rho_3 f_{31} \end{vmatrix}$$

(see Step 3 in the example in Section 5.4.4).

Minors 2, 2 and 2, 3 also do not contain ρ_2. The determinant by which the minors have to be divided to obtain the transfer factors of necessity contains *all* the reflectances.

Thus, in changing the reflectance of the floor, only the determinant is changed. This means that for a given room index with fixed wall and ceiling reflectances the effect on the utilization factor of varying the floor reflectance will be the same for generally diffusing lighting as for any light distribution. Thus, a single table of correction factors can be produced, in terms of room reflectances and room index, to allow for different values of floor reflectance for use with the utilization factor tables for all the different types of lighting unit.

5.5.3 CAVITIES

When light enters a cavity, the amount of that light reflected back out of the cavity depends not only upon the surface reflectance of the cavity but also upon its shape, see Figure 5.9.

The equivalent reflectance of the cavity is given by F_2/Fo_1 where F_2 is the flux from surface 1 passing through aperture 2 and Fo_1 is the initial flux entering the cavity and falling on surface 1.

$$F_1 = Fo_1 + F_1 \rho_1 f_{11} \quad \text{or} \quad F_1 = \frac{Fo_1}{1 - \rho_1 f_{11}}$$

$$F_2 = F_1 \rho_1 f_{12}$$

giving

$$F_2 = \frac{Fo_1 \rho f_{12}}{1 - \rho_1 f_{11}}$$

$$\rho_{eq} = \frac{\rho_1 f_{12}}{1 - \rho_1 f_{11}}$$

where ρ_{eq} is the equivalent reflectance of the cavity.

f_{12} and f_{11} have been determined in Section 4.9.1. (Note: the surfaces have been numbered differently because only two surfaces are considered, not three, and here surface 1 combines 1 and 3 in that previous treatment.)

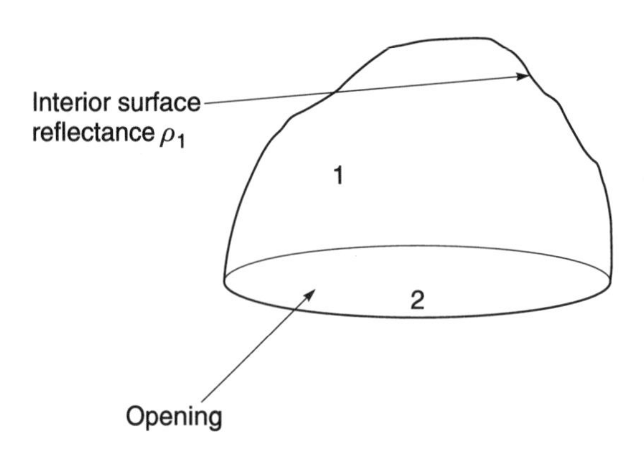

Fig. 5.9 Equivalent reflectance of a cavity

$$f_{12} = \frac{S_2}{S_1} \qquad f_{11} = 1 - \frac{S_2}{S_1}$$

$$\rho_{eq} = \frac{\rho_1 \dfrac{S_2}{S_1}}{1 - \rho_1 \left(1 - \dfrac{S_2}{S_1}\right)}$$

$$= \frac{\rho_1 S_2}{\rho_1 S_2 + S_1(1 - \rho_1)}$$

Example

Calculate the reflectance of a hemispherical cavity of uniformly diffuse surface reflectance 0.7.

$$S_1 = 2\pi r^2 \qquad S_2 = \pi r^2$$

So,

$$\frac{S_2}{S_1} = \frac{1}{2}$$

$$\rho_{eq} = \frac{0.7 \times 0.5}{1 - 0.7(1 - 0.5)}$$

$$= 0.54$$

5.5.4 THE CAVITY METHOD FOR UTILIZATION FACTORS

In the preceding sections the calculations have generally assumed that the luminaires are mounted on the ceiling. In practice, the luminaires are often suspended from the ceiling and the illuminances required at desk or bench level. This, and not the floor, is called *the working plane*. The utilization factor calculations for suspended luminaires and for a working plane above the floor may be simplified by adopting the cavity method. The cavity method of utilization factors is illustrated in Figure 5.10.

The room is divided horizontally into three sections (1) the ceiling cavity, (2) the floor cavity and (3) the room cavity.

The methods previously described can be applied to obtain the utilization factors for the room cavity provided that reflectances are ascribed to the ceiling and floor cavities.

Since the floor cavity may contain furniture that may have any reflectance and may be moved about or changed at any time it is usual to estimate the floor cavity reflectance and to ascribe what is considered to be an appropriate value, taking into account the actual floor reflectances. The value is usually chosen from 10%, 20% or 30%.

The ceiling cavity reflectance can be calculated by using the simplified formula developed in Section 5.5.3 or by the more accurate method where the ceiling cavity is treated as a room of height h_s which has a 'floor' reflectance of zero.

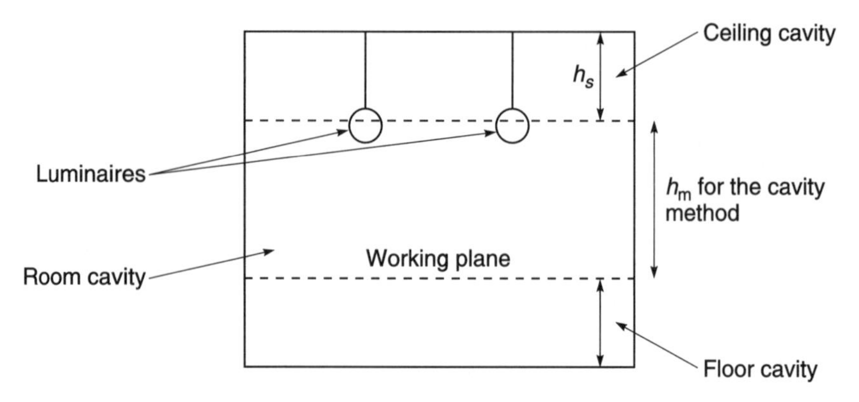

Fig. 5.10 The cavity method for utilization factors

Example 1

$$\rho_{eq} = \frac{\rho_1 \dfrac{S_2}{S_1}}{1 - \rho_1 \left(1 - \dfrac{S_2}{S_1}\right)}$$

Let the suspension distance of the luminaires from the ceiling be $h_s = 1.0$ m and let the room be 8 m × 8 m. Assume ceiling reflectance 0.7 and upper wall reflectance 0.3.

$$S_2 = 8 \times 8 = 64$$

$$S_1 = 64 + 4 \times 8 \times 1 = 96$$

$$\frac{S_2}{S_1} = \frac{64}{96} = 0.67$$

$$\rho_1 = \text{average reflectance of } S$$

$$= \frac{0.7 \times 64 + 0.3 \times 32}{96}$$

$$= 0.57$$

$$\rho_{eq} = \frac{0.57 \times 0.67}{1 - 0.57(1 - 0.67)}$$

$$= 0.468$$

Example 2 (for the same ceiling cavity as above)

In this second example the 'floor' of the cavity is the aperture of the cavity and the aperture is assumed to have a uniform distribution of incoming flux, which is then evenly distributed over the ceiling and upper walls above the level of the luminaires. In this example the reflectances are not averaged. $UF(2)$ is the ratio of flux received by the surface to the lamp or input flux which, in this case, is the flux entering the cavity. $UF(2)$ is, therefore, the equivalent reflectance of the cavity (ρ_{eq}).

The required surface distribution factor for this example can be obtained from Table 5.1 by calculating a cavity index and treating it as a room index when entering the table. The cavity index formula is the room index formula with the luminaire mounting height (h_m) replaced by the suspension distance from the ceiling (h_s).

$$CI = \frac{W_1 \times W_2}{h_s(W_1 + W_2)}$$

which for a square room becomes

$$CI = \frac{W}{2h_s}$$

Obviously, if a table for CI values above 5.0 is required the additional values of D_S would have to be calculated, using the zonal multiplier method or the Zijl formula.

Example

For this calculation, we will assume the same dimensions as for Example 1.

$$\text{Cavity Index (Room Index)} = \frac{8}{2 \times 1} = 4$$

$$\rho_1 = 0.7 \quad \text{and} \quad \rho_3 = 0.3, \quad \rho_2 = 0$$

The surface distribution factor D_S for $CI(RI) = 4.0$ is 0.788, that is $f_{12} = 0.788$.

Inserting the above values into the matrix given on page 185 gives,

$$\begin{vmatrix} 1 & 0 & -0.127 \\ -0.552 & 1 & -0.127 \\ -0.148 & 0 & 0.894 \end{vmatrix}$$

The procedure is then that described on page 186 for inverting the matrix.

However, since we are only interested in $UF(2)$ only the minors for column/row 2,1 2,2 and 2,3 need to be considered. Also, since $DF(2) = 0$, 2, 2 is not required. It is, therefore, only necessary to calculate the determinant and two transfer factors.

1. The determinant

$$\begin{aligned} \Delta &= 0.894 + 0 + 0 - 0 - 0 - 0.0188 \\ &= 0.875 \end{aligned}$$

2. Minor 2, 1

$$\begin{vmatrix} -0.552 & -0.148 \\ -0.127 & 0.894 \end{vmatrix}$$

$= -0.493 - 0.0188 = -0.512$
multiply by -1
$= 0.512$

Transfer factor from cavity 'ceiling' to cavity aperture ('floor') = 0.512/0.875 = 0.585.

3. Minor 2, 3

$$\begin{vmatrix} 1 & -0.552 \\ -0.127 & -0.127 \end{vmatrix}$$

$= -0.127 - 0.701 = -0.197$
multiply by -1
$= 0.197$

Transfer factor from cavity 'walls' to cavity aperture

$$= \frac{0.197}{0.875} = 0.225$$

$$UF(2) = \rho_{eq} = 0.585DF(1) + 0.225DF(2)$$

Because the aperture is the source of light and is treated as a uniformly diffusing area source, $DF(1) = f_{12} = 0.788$ and $DF(3) = f_{13} = (1 - 0.788) = 0.212$.

So,

$$\begin{aligned}\rho_{eq} &= 0.585 \times 0.788 + 0.225 \times 0.212 \\ &= 0.461 + 0.0477 \\ &= 0.509\end{aligned}$$

Comparing the two examples we find

Example 1. $\rho_{eq} = 0.468$
Example 2. $\rho_{eq} = 0.509$

Taking the percentage of the more accurate calculation (that is, three surfaces instead of two)

$$\frac{0.468 - 0.509}{0.509} = -0.08 \quad \text{or} \quad 8\%$$

Consider again the formula used in Example 1 above, that is

$$\rho_{eq} = \rho_1 \frac{\dfrac{S_2}{S_1}}{1 - \rho_1 \left(- \dfrac{S_2}{S_1} \right)}$$

Multiply by S_1 top and bottom,

$$\rho_{eq} = \frac{\rho_1 S_2}{S_1 - \rho_1(S_1 - S_2)}$$

Expand by adding $+S_2$ and $-S_2$ to the denominator

$$\rho_{eq} = \frac{\rho_1 S_2}{S_1 - S_2 + S_2 - \rho_1 S_1 + \rho_1 S_2}$$

$$= \frac{\rho_1 S_2}{S_2 + (S_1 - S_2)(1 - \rho_1)}$$

Multiply top and bottom by

$$\frac{2}{(S_1 - S_2)}$$

Since *CI*, the Cavity Index for a rectangular cavity

$$= \frac{2S_2}{S_1 - S_2}$$

then

$$\rho_{eq} = \frac{CI\rho_1}{CI + 2(1 - \rho_1)}$$

which is the formula given in CIBSE Memorandum TM5.

5.5.5 REFLECTED COMPONENT OF SCALAR AND CYLINDRICAL ILLUMINANCE

In Section 4.5.3 the zonal multipliers for continuous distributions such as luminous ceilings, or for calculating distribution factors from rectangular area sources such as a ceiling reflecting light, were further developed to allow for the calculation of average scalar or cylindrical illuminance from such sources onto the floor or working plane (introduced in Sections 3.8.1 and 3.8.2).

If the interreflection method introduced in Section 5.5 is used, the final luminances of the ceiling, floor and walls may be calculated and the zonal multipliers for scalar or cylindrical illuminance used to calculate the reflected components of these illuminances on the working plane from the ceiling. It is then possible also to calculate the contributions to these illuminances from the working plane cavity and from the walls.

Because the modified zonal multipliers developed in Section 4.5.3 related to illuminance rather than simply to flux transfer Croft[5] used the term *inter-illumination factor* for the distribution factor for scalar and cylindrical illuminances. In keeping with present practice, we will use the term inter-illuminance factor.

In Section 4.5.3 it will be seen that the scalar illuminance is given by:

$$E_s = D_{s(scalar)} \times M$$

where M is the luminous exitance in lumens per square metre.

It is convenient at this point to designate $D_{s(scalar)}$ as one of the inter-illuminance factors (*Ilf*). Thus,

$$Ilf_{s(c)} = D_{s(scalar)} \text{ (ceiling to floor or working plane)}$$

So, the reflected component of scalar illuminance from the ceiling onto the working plane is given by:

$$E_s = Ilf_{s(c)} \times M$$

where M is the luminous exitance of the ceiling.

The inter-illuminance factor for the working plane or floor cavity, where $h = 0$, is obtained in a simple manner. Since the scalar illuminance is the average illuminance on an infinitesimal sphere at the point under consideration, the working plane is considered to be covered in such spheres. To each of these spheres the working plane becomes a light source of infinite extent (provided its reflectance is not zero).

If we use the principle of equivalence introduced in Section 3.5 the infinite plane can be replaced by a hemisphere of similar luminance (Figure 5.11). This hemisphere produces the same illuminance on the surface of the sphere as the infinite plane.

We know that if the infinitesimal sphere was completely enclosed in a sphere of similar luminance to that of the infinite plane, then the infinitesimal sphere illuminance would be given by:

$$\begin{aligned} E_s &= \pi L_f \\ &= M_f \end{aligned}$$

where f indicates the floor cavity or working plane value.

However, in our case, the top half of the sphere is missing, so

$$E_s = \frac{M_f}{2}$$

This gives $Ilf_{s(f)} = 0.5$.

Two of the three inter-illuminance factors have now been determined. The third factor is that for the walls, which is $Ilf_{s(w)}$.

This is obtained by considering the situation where the ceiling, floors or working plane and the walls are all of unit luminous exitance. The inter-illuminance factors required in this case

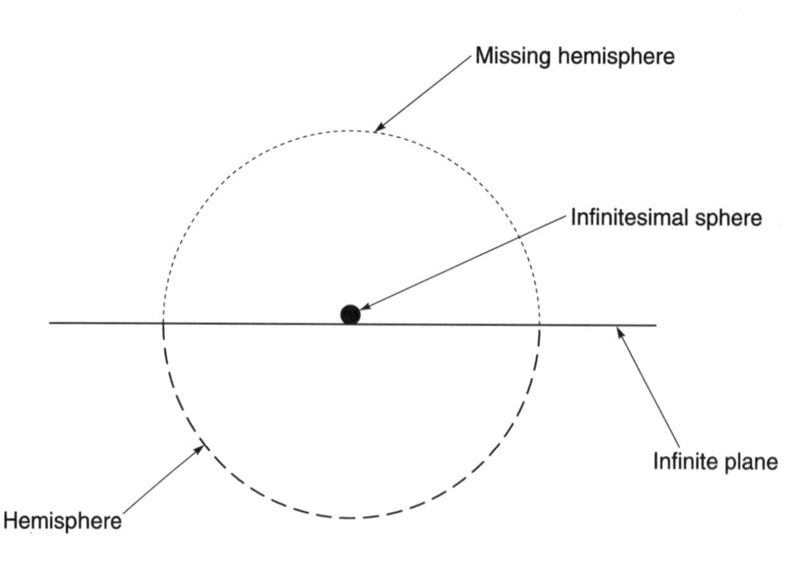

Fig. 5.11 Reflected component of scalar illuminance due to the floor cavity

will be those for any other set of illuminances and we already know that, by the principle of equivalence, the result for this case is the same as surrounding the infinitesimal sphere with an enclosing sphere of unit luminous exitance, that is:

$$E_s = 1$$

So,

$$E_s = Ilf_{s(c)} \times 1 + Ilf_{s(f)} \times 1 + Ilf_{s(w)} \times 1 = 1$$

Thus,

$$Ilf_{s(w)} = 1 - Ilf_{s(c)} - Ilf_{s(f)}$$

Or,

$$Ilf_{s(w)} = 0.5 - Ilf_{s(c)} = 0.5 - D_{s(scalar)}$$

giving

$$Ilf_{s(c)} = D_{s(scalar)}$$
$$Ilf_{s(f)} = 0.5$$
$$Ilf_{s(w)} = 0.5 - D_{s(scalar)}$$

Thus,

$$E_s = Ilf_{s(c)}M_c + Ilf_{s(f)}M_f + Ilf_{s(w)}M_w$$

A similar argument can be used in the case of cylindrical illuminance, so that,

$$E_c = Ilf_{c(c)} \times M_c + Ilf_{c(f)} \times M_f + Ilf_{c(w)} \times M_w$$

where

$$Ilf_{c(c)} = D_{s(cylindrical)}$$
$$Ilf_{c(f)} = 0.5$$
$$Ilf_{c(w)} = 0.5 - D_{s(cylindrical)}$$

Example

Calculate the reflected components of scalar and cylindrical illuminances using the data given and calculated in the example of utilization factor calculations in Section 5.5.

The data given are $RI = 1.0$, $\rho_1 = 0.7$, $\rho_2 = 0.3$, $\rho_3 = 0.5$, and the calculated UF values are:

$$UF(1) = 0.405, \quad UF(2) = 0.643, \quad UF(3) = 0.857$$

Assume $h_m = 1.6$ m and that the room is square,

$$RI = 1.0 = \frac{W}{2 \times 1.6} \quad \text{then} \quad W = 3.2 \text{ m}$$

Area of working plane

$$= 3.2^2$$
$$= 10.24 \text{ m}^2$$

Assume lamp output in the room is 7400 lumens and ignore depreciation allowances

$$E_2 = \frac{7400 \times 0.643}{10.24} = 465 \text{ lux}$$

So, $M_2 = 465 \times 0.3 = 139.4$ lumens per square metre

$$E_1 = \frac{7400 \times 0.805}{10.24} = 292.7 \text{ lux}$$

So, $M_1 = 292.7 \times 0.7 = 204.9$ lumens per square metre

$$E_3 = \frac{7400 \times 0.857}{20.48} = 309.7 \text{ lux}$$

So, $M_3 = 309.7 \times 0.5 = 154.8$ lumens per square metre.

The inter-illuminance factors

$$Ilf_{s(c)} = D_{S(scalar)} \quad \text{and} \quad Ilf_{c(c)} = D_{S(cylindrical)}$$

To calculate $D_{S(scalar)}$ or $D_{S(cylindrical)}$ we use zonal multipliers for $RI = 1.0$.

The zonal multipliers for horizontal illuminance are given by:

$$ZM = 0.080 \left(\frac{\tan\theta}{RI}\right)^2 - 0.637 \left(\frac{\tan\theta}{RI}\right) + 1$$

(see Section 4.5.2).

To convert from horizontal illuminance to scalar illuminance we divide each multiplier by $4\cos\theta$.

To convert from horizontal illuminance to cylindrical illuminance we divide each multiplier by π and multiply by $\tan\theta$ (see Chapter 3).

Much of the work has already been done in Chapter 4 (Example 2 of Section 4.5.3) where zonal multipliers for a square room of $RI = 2.18$ were calculated.

In that example, values of

$$0.080 \left(\frac{\tan\theta}{2.18}\right)^2 \quad \text{and} \quad 0.637 \left(\frac{\tan\theta}{2.18}\right)$$

were calculated. In this example the $RI = 1.0$.

Therefore, if we multiply one set of values by 2.18^2 and the other by 2.18 we will have the values required for our example, see Table 5.2.

We now multiply each zonal multiplier by the appropriate factor to convert to ZM_s or ZM_{cyl} and then multiply by the zonal fractions. Summation produces the required values of $D_{S(scalar)}$ and $D_{s(cylindrical)}$, see Table 5.3.

The reflected component of

$$\begin{aligned} E_s &= 0.234 \times 204.9 + 0.5 \times 139.4 + (1 - 0.234) \times 154.8 \\ &= 47.95 + 69.7 + 118.57 \\ &= 236 \text{ lux} \end{aligned}$$

Table 5.2 Adaptation of the zonal multiplier calculation from Table 4.6 to a new room index value of 1.0 for the example

$$ZM = 0.080\left(\frac{\tan^2\theta}{RI}\right) - 0.637\left(\frac{\tan\theta}{RI}\right) + 1$$

Mid-zone angle θ	$0.08\tan^2\theta$	$0.637\tan\theta$	ZM
5	6.106×10^{-4}	0.0556	0.944
15	5.731×10^{-3}	0.1713	0.835
25	0.0177	0.2973	0.720
35	0.0391	0.4447	0.593
45	0.0798	0.6366	0.443
55	0.1630	0.9091	0.253
65	0.3673	1.3647	0.020
75	1.1120	2.3740	0.000
85			0

Table 5.3 Tabulated calculation of scalar and cylindrical distribution factors for the example

θ	$ZM_s = \frac{ZM}{4\cos\theta}$	$ZM_{cyl} = \frac{ZM\tan\theta}{\pi}$	ZFR	$ZM_s \times ZFR$	$ZM_{cyl} \times ZFR$
5	0.237	0.026	0.036	0.085	0.001
15	0.216	0.071	0.115	0.025	0.008
25	0.199	0.107	0.181	0.036	0.019
35	0.181	0.132	0.215	0.039	0.028
45	0.157	0.141	0.245	0.038	0.035
55	0.110	0.115	0.093	0.010	0.011
65	0.012	0.014	0.049	0.001	0.001
75	0.000	0.000	0.037	0.000	0.000
85			0.028	0	0
			Total	0.234 $= D_{s(scalar)}$	0.103 $= D_{s(cylindrical)}$

The reflected component of

$$\begin{aligned} E_{cyl} &= 0.103 \times 204.9 + 0.5 \times 139.4 + (1 - 0.103) \times 154.8 \\ &= 21.1 + 69.7 + 138.9 \\ &= 230 \text{ lux} \end{aligned}$$

These are only the interreflected components of E_s and E_c.

The direct components from the luminaires must also be calculated. The formulae for point source calculation of E_s and E_{cyl} were given in Section 3.9.3.

References

1. Cuttle, C. (1991) Sumpner's principle: a discussion, *Lighting Research and Technology*, **23**(2), 99.
2. Bean, A. R. and Simons, R. H. (1965) *Monograph No. 9* (IES London).
3. Bean, A. R. and Bell, R. I. (1976) Calculation of utilisation factors, *Lighting Research and Technology*, **8**(4), 200–210.
4. (1982) Calculations for Interior Lighting Applied Method, CIE No. 52 (TC 1.5).
5. Croft, R. (1959) *Average Illumination on the Vertical and Calculation of Zonal Multipliers for Continuous Distributions*, Monograph No. 2 (London: IES) p. 8.

6
Optical Design

6.1 Introduction

In this chapter the various methods by which light from a source can be controlled are described and discussed. The challenge is to produce a design that is the cheapest possible, has the best possible light distribution, involves the least possible loss of light, is of pleasing appearance, and does not involve complicated manufacturing processes. Some of these requirements are nearly always conflicting, some – for example the best possible light distribution or good appearance – may be matters of opinion, but they do provide the questions the designer has to keep asking.

6.2 Approaches to optical design

Although, in this chapter, the theoretical aspects of optical design will be described, optical design as applied to luminaires is a subject with a large practical element. This is because, even with the comprehensive and powerful computer programs available today, it is still not possible to predict fully the optical performance of the manufactured luminaire. There are many reasons for this. One is manufacturing tolerances, which vary according to the process used. For instance, extrusion of plastics cannot be expected to give good accuracy for prism banks. Injection moulding is much better but even here there will be some rounding of prism edges and there may be some shrinking of the material after cooling so that surfaces are not exactly flat. Small inaccuracies in specular surfaces will produce large discrepancies in light distribution. The reflection properties of semi-specular surfaces are difficult to control over the life of the manufacturing tools, or from batch to batch. The light source itself may be subject to luminance variation over its surface, as will be described in the next section.

All these matters, and there are others, mean that – as far as is possible – a mock-up of the final design should be photometrically tested to ensure that the design meets the requirements. Judgement may have to be exercised at this stage to assess whether the performance has sufficient margin to allow for manufacturing tolerances and other imponderables.

Optical design, then, is not just a matter of sitting down and calculating the layout and profile of light control elements, it includes the assessment of the design to meet the requirements when it is manufactured. This is a significant responsibility for the designer because, if the design is not satisfactory, tools may have to be altered, which can be expensive, and time will be lost, which may be even more expensive, since it may delay the launch of a product. Making mock-ups that will truly represent the performance of the finished product is then an important aspect of the optical design process. Making these has been helped in recent years by the use of three-dimensional CAD drafting.[1] The data file available from this can be used to make a mould from which a reflector can be produced by hydro-forming or vacuum forming a suitable plastics material, which is subsequently aluminized. Linear prisms can be cut directly by purpose-made cutters, but mock-ups of prismatic refractor bowls with asymmetrical double curvature are difficult to make. If possible, the design should be tested before the moulding tools are hardened.

Even when the required light distribution is known it is rarely possible to work back to a unique solution. The designer has to rely on background knowledge and experience to decide on the general form of the light control system to be used, and then design the control elements. This may involve using parts of previous designs whose performance is known, which may be quicker and more certain than starting from scratch. He or she can then use the calculation methods outlined in this chapter to predict the performance of the proposed design. Modifications may have to be made to optimize the design. In fact the process is often iterative.

6.3 The light source

The starting point of optical control is the light source. In this section, we will briefly review various types of light source from the point of view of their suitability for optical control.

The fluorescent lamp – In its commonest manifestation this is a diffusing cylinder. The distribution of luminance over its surface can be regarded as even except towards the ends where there is a fall off. Its size is usually substantial with respect to the optical control system so that it may obstruct and absorb some of the light redirected by reflectors. In multi-lamp luminaires the lamps may obstruct each other. This also applies to the limbs of compact fluorescent lamps (CFLs). One property that may have a large effect on the efficacy of the lamp is the variation of light output with temperature. It is a physical property of fluorescent lamps that their light output varies with temperature (Figure 6.1). Usually, the maximum light output occurs at 25°C. This

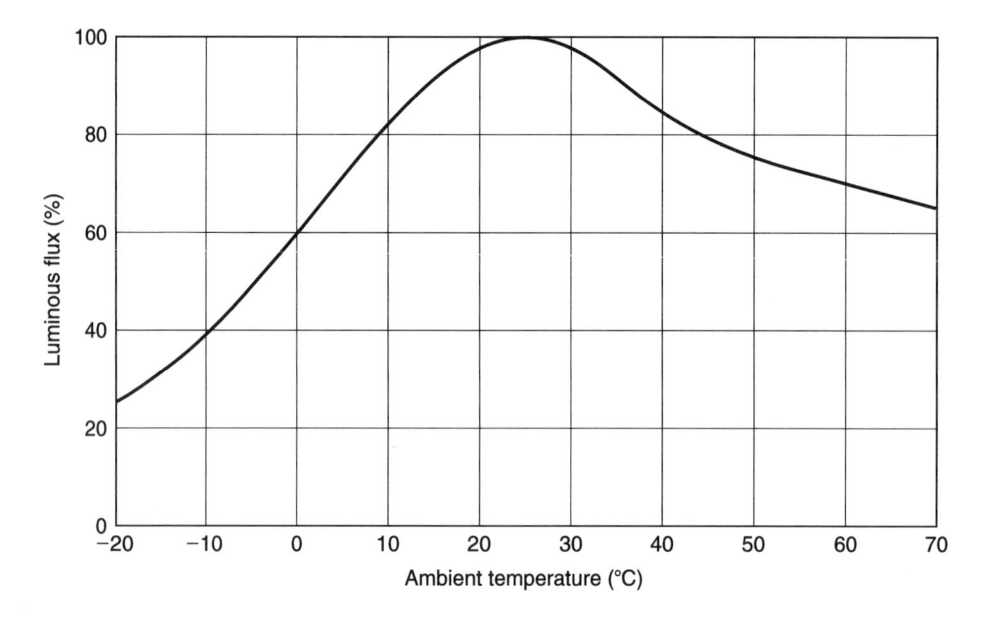

Fig. 6.1 Variation of light output of a fluorescent lamp with ambient temperature

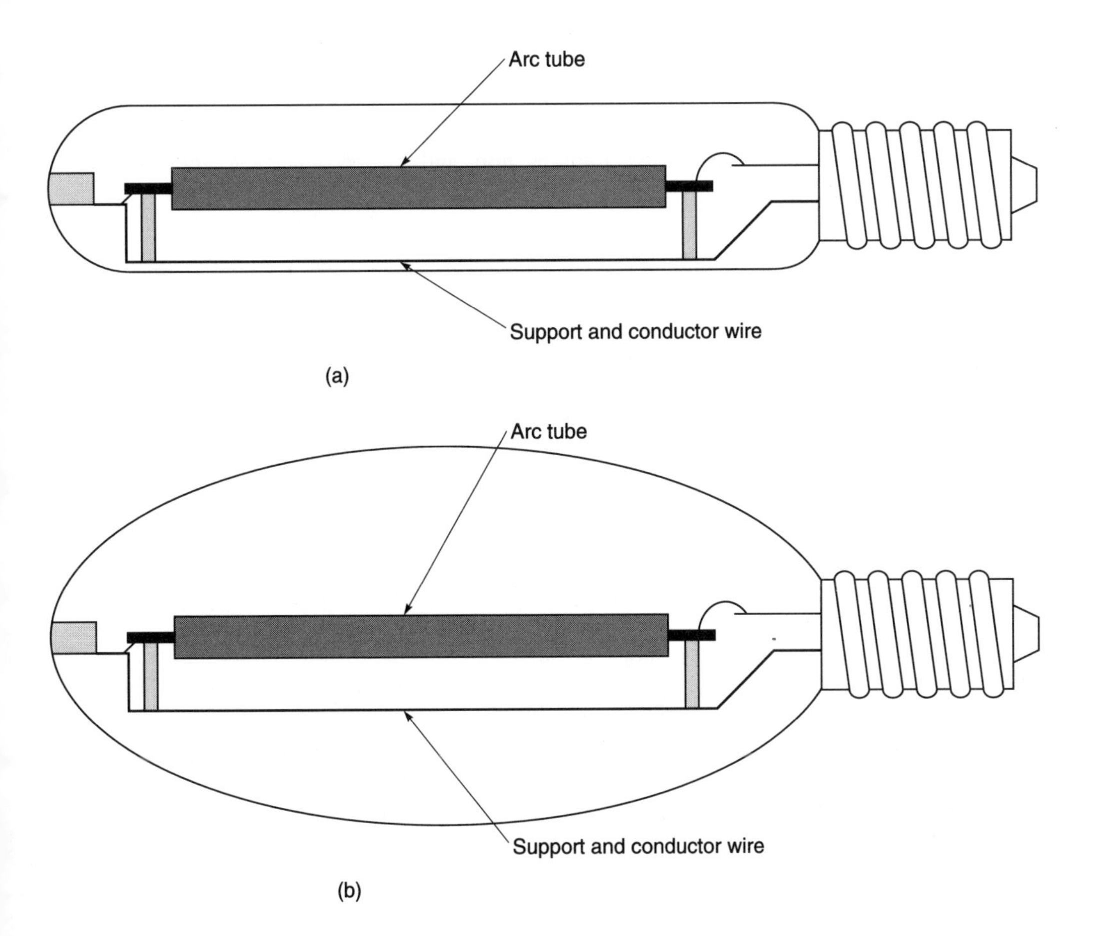

Fig. 6.2 HID discharge lamps with cylindrical and elliptical envelopes: (a) cylindrical envelope; (b) elliptical envelope

means that when the lamp is enclosed in a luminaire, its light output will drop. This becomes more pronounced in enclosed or multi-lamp luminaires.

The high intensity discharge lamp (HID lamp) – Figure 6.2 shows the main elements of the HID lamp in so far as they affect optical behaviour. The arc tube, which contains the arc, may be clear, as in the high pressure mercury vapour lamp, or diffuse, as in the high pressure sodium lamp. The arc has a central core of very high luminance and, owing to convection currents, has an upwards curvature. In some discharge lamps different parts of the arc, from top to bottom, have different colours, which become apparent when the arc tube is focused to form an image on a surface. The support and conductor wire for the arc tube, even though it is thin, may throw a shadow on the optical control system producing some asymmetry in the light distribution. Most lamps have elliptical glass envelopes although some are tubular. In some elliptical types the glass envelope may be coated with a diffusing or fluorescent powder. In the clear glass types the envelope has little effect on the light distribution, except in one case. This occurs in reflector systems in directions where the edge of the envelope is in line with the image of the arc tube in the reflector. This causes a dark line that reduces the luminous intensity. In the coated types the envelope is usually large with respect to the optical system, which may cause considerable obstruction to the reflected light. Also, the arc tube will usually be visible through the coating, which means that the lamp cannot be regarded as a source of even luminance. Sometimes there is a variation in the thickness of the coating on a lamp and between lamps, nominally of the same type, giving cause for variation in performance.

The tungsten filament lamp – In this category we include the tungsten halogen lamp. In the ordinary GLS lamp the filament is usually too loosely coiled for precise optical control to be achieved. With low voltage lamps the filament is thicker and more closely coiled so allowing good optical control to be achieved, as in car headlights. Tungsten halogen lamps generally have compact filaments making them good for optical control. The filament may be a linear coil that has to be used horizontally. This is ideal for floodlights, where control is generally required in vertical planes. The configuration of the filament can often be suited to the application.

A full description of the various lamp types is given in the book *Lamps and Lighting* listed in the Bibliography at the end of this chapter.

6.4 General principles

The approach to an optical design problem will depend on the form in which it is stated. Usually a certain illuminance has to be achieved on a specified surface. From this it is possible to work back and find the required luminous intensity distribution. However, the answer might not be unique, particularly when a number of luminaires provide the illuminance at any given point. Moreover, the demand may be for a light control system that gives the optimum performance in a range of situations, such as rooms of differing reflectances and sizes. In main road lighting, the requirements are in terms of road luminance, which makes working back to an optical design very difficult indeed.

There may be additional requirements to be met. These may be for the control of discomfort or disability glare, or they may be related to appearance.

Once the photometric requirements have been decided it is possible to start on the design of the optical system. There are a number of approaches available that enable the feasibility of a project to be assessed and which give an insight into the kind of optical control that will be needed. These will now be described.

6.4.1 LUMINOUS FLUX AVAILABILITY

In this technique, the luminous flux required is estimated and a source of suitable output chosen. It can also be used to find which zones have an excess of flux and which zones are deficient in luminous flux with a view to deciding how the luminous flux is to be redistributed. It is useful in the design of reflector contours and an example will be worked out when we deal with them.

6.4.2 LUMINANCE OF LIGHT SOURCE AND LUMINAIRE

There are two ways in which the consideration of luminance can be invoked in optical design, namely

(a) The luminous intensity in any direction of a light control system is equal to the projected area of its flashed or bright parts multiplied by their luminances. If the light control system is divided into m parts, each substantially of even luminance, the resultant luminous intensity can be expressed mathematically as:

$$I = \sum_{k=1}^{m} A_k L_k \tag{6.1}$$

where
A_k is the area of a part (numbered k) of the optical system of substantially uniform luminance,
L_k is the luminance of that part,
Σ signifies the sum of the products.

(b) If a reflector is specular, the luminance of any part of the reflector is equal to the luminance of the source of light reflected in that part multiplied by the reflectance of the reflector. Similarly, the luminance of any part of a flashed lens or any refractor system is the luminance of the source facing that part multiplied by the transmittance of the materials of the optical system.

These two facts often enable the feasibility of a problem to be assessed quickly. For example, consider the kind of problem where the opening through which the light can emerge is limited in size. This occurs in airfield lighting. The projection of the runway lights above the surface of the runway is limited so that they do not cause damage to the tyres of landing aircraft. Suppose the maximum window size allowed is 625 mm^2 and we have to produce a peak luminous intensity of 15 000 candelas with a source of luminance 13 cd/mm^2. If the transmittance of the refractor system is 0.70 and we assume that it can be completely flashed, then the luminous intensity obtainable will be $625 \times 13 \times 0.70 = 5688$ candelas, which is well below the required peak. Another example occurs with fibre optics. The luminance of the light emitting ends of the fibres cannot exceed the luminance of the source, which provides a very quick way of calculating the maximum luminous flux that can be transmitted by the fibres or bundle of fibres.

The uninitiated often put forward arguments for trying batteries of lenses and reflectors to concentrate the light and so increase the luminance. As stated above this is erroneous. Such arguments are commonly founded on considering a point source of light, which does not exist in nature and would have infinite luminance. However, what supplementary reflectors behind a lens system, for instance, can do is increase the spread of light and help to increase the flashed area if there are unflashed areas.

6.4.3 MAGNIFICATION OF SOURCE

In this method, the magnification of the area of the source in a direction is calculated. The luminous intensity in that direction will then be the luminous intensity of the source in the direction of the optical system multiplied by the magnification and the losses in the system. This method is only applicable when the source is small compared with the size of the control system so the image or images are complete.

6.4.4 RAY TRACING

Ray tracing is used for locating the images in lens and prism systems. Until recently it has not commonly been used for predicting the luminous intensity distribution of optical systems, because it is tedious to apply. However, computers make this a feasible process.

6.5 Reflector systems

In this section we will describe how use can be made of specular or highly polished metallic surfaces in the control of light. Such surfaces are used when precise or fairly precise control of light is required as in floodlights, spotlights and road lighting luminaires.

Light reflected from a specular surface obeys the following laws (Figure 6.3).

(1) The incident ray of light, the normal to the reflecting surface at the point of incidence, and the reflected ray are all in the same plane.
(2) The angle of reflection (r) is equal to the angle of incidence (i).

The position of an image can be determined by using these laws. In Figure 6.4, P is an object, the image of which is P'. Two rays are shown reflected at the points L and M.

Since $i = r$, $\angle PLT = \angle P'LT = 90° - r$. Similarly, $\angle PMT = \angle P'MT$. Therefore, the triangles PLM and $P'LM$ are congruent, and PT is equal to $P'T$. Hence the rays of light appear to emanate from an image that is on the normal produced (that is, at P') and is as far behind the mirror as the object is in front of it. This is referred to as a virtual image as it only appears to exist and cannot be formed on a screen.

It can also be shown that if a plane mirror is rotated through an angle, the reflected rays will be rotated twice that angle. This has the effect of magnifying inaccuracies in mirrors whether they be plane or contoured, since contoured mirrors can be regarded as being made up of infinitesimal plane mirrors.

Figure 6.5 shows what happens when a mirror is rotated through an angle θ. The angle of incidence i is increased to $i + \theta$, so the angle of reflection is likewise increased to $r + \theta$. Whereas before the mirror was rotated, the angle between the incident and reflected rays was $i + r$, after rotation it is $i + r + 2\theta$. So the reflected ray is turned through an extra 2θ.

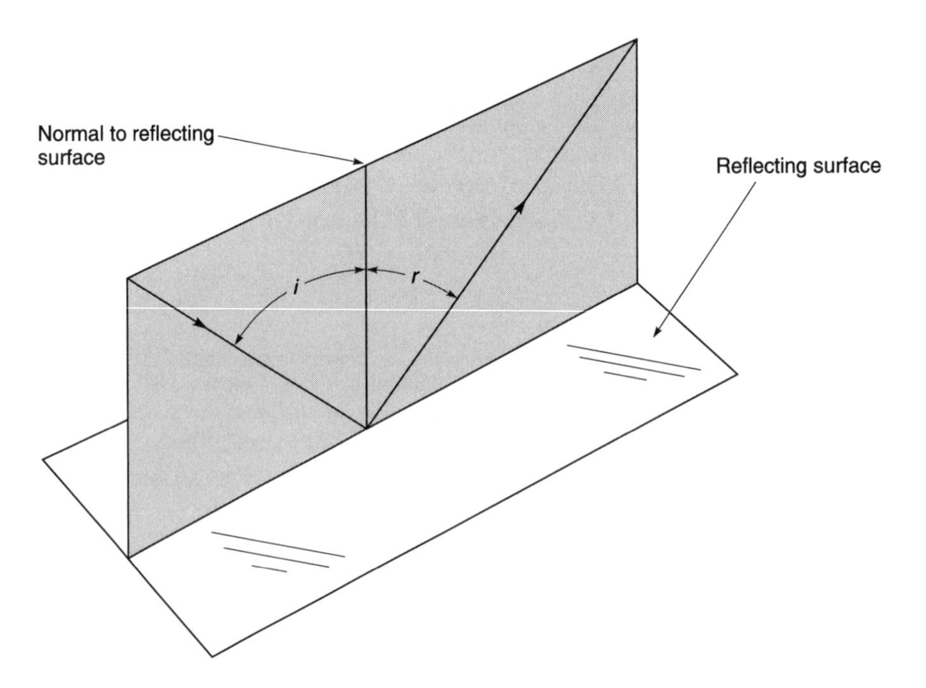

Fig. 6.3 Reflection at a surface

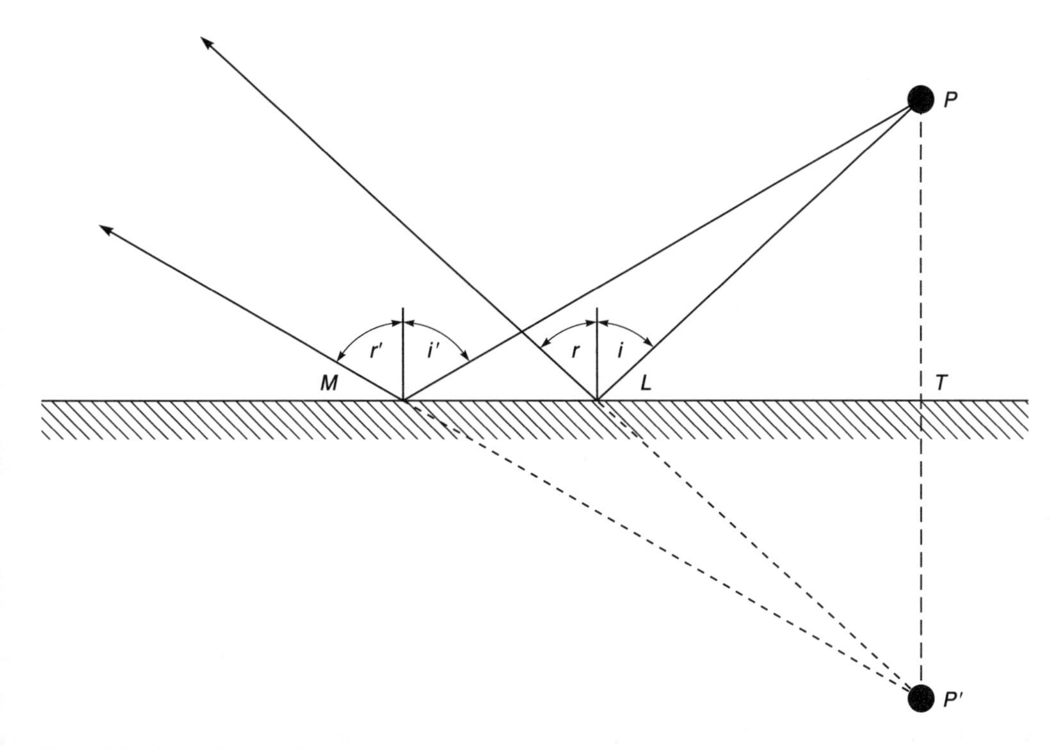

Fig. 6.4 Position of image in mirror

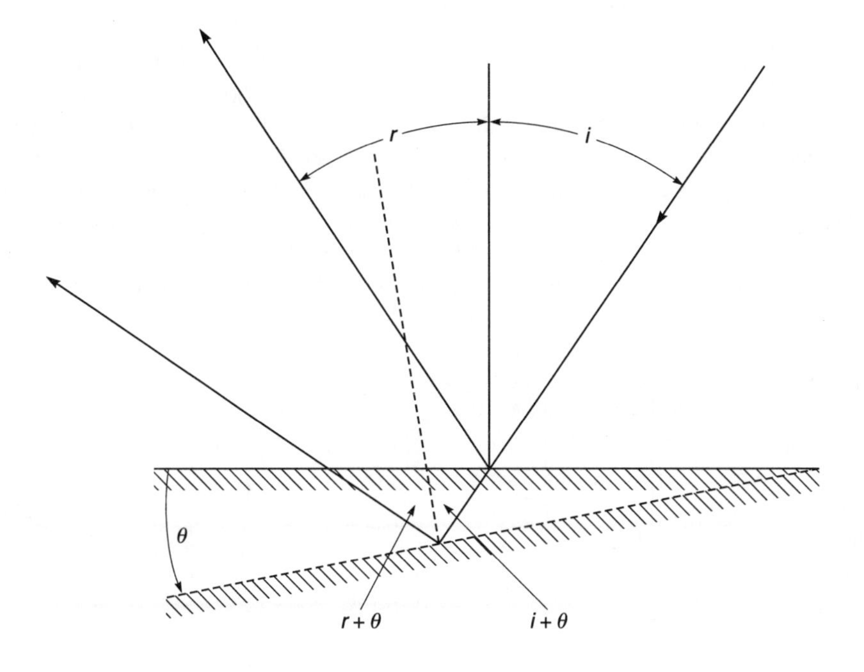

Fig. 6.5 Rotation of a mirror

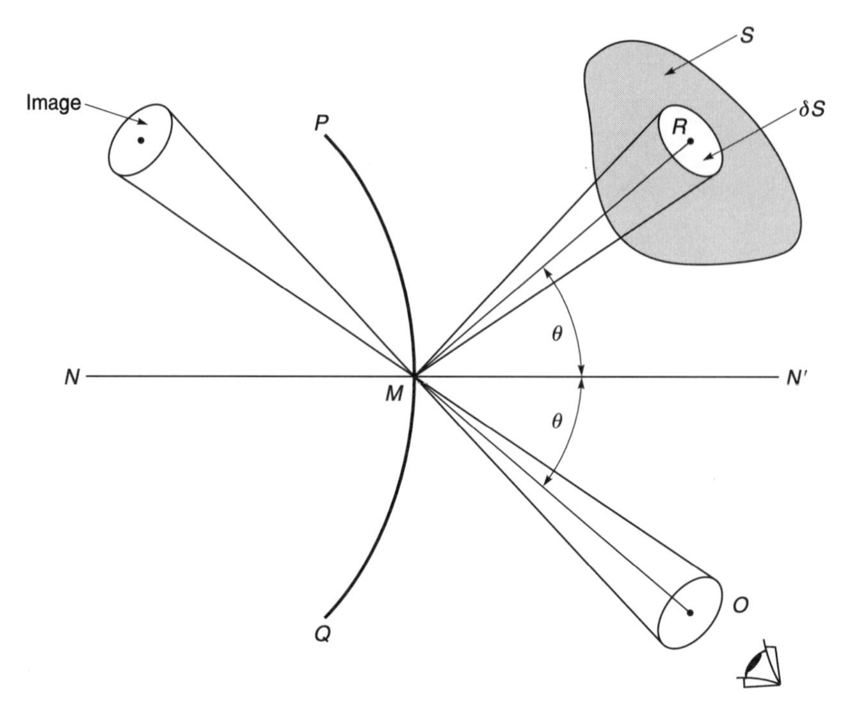

Fig. 6.6 Luminance of image

6.5.1 LUMINANCE OF IMAGE

The luminance of the image has to be known in order to calculate the luminous intensity to be expected from the flashed area of a mirror. As previously stated, this is equal to the luminance of the source multiplied by the reflectance. It is not obvious that this law is true, particularly for a curved surface, so we now give a proof of it.

In Figure 6.6, *PQ* is a curved specular mirror. *S* is the source. *O* is the eye of the observer, who is looking at the light rays reflected from *M*. *NMN′* is the normal at *M*. The axes of the cones of light which are incident and reflected at *M* make an angle θ with *MN′*.

Consider a small element δS located at *R*. Let it have a projected area δS and a luminance *L* in the direction *RM*.

Then the luminous intensity of δS is $L\delta S$.

Therefore, the illuminance on *M* is $L\delta S/d^2 \cos\theta$, where $d = RM$.

The luminous flux falling on an area δA at *M* is $L\delta S/d^2 \cos\theta\delta A$.

The reflectance ρ is defined as the reflected luminous flux divided by the incident luminous flux. Hence, the luminous flux leaving element δA is $L\delta S/d^2 \cos\theta\rho\delta A$.

Now, the solid angle of the incident cone of light is equal to $\delta S/d^2$, and this is equal to the solid angle of the reflected cone, since all angles of incidence and reflection are equal. Therefore

$$\text{luminous intensity of } \delta A = L \cos\theta\rho\delta A \qquad (6.2)$$

since luminous intensity equals luminous flux divided by solid angle.

The apparent or projected area of element δA in the direction of the eye is $\cos\theta\delta A$, and since luminance equals luminous intensity divided by projected area

$$\text{luminance of } \delta A = \rho L \tag{6.3}$$

Thus, the proof hinges on the fact that the solid angle contained by the reflected cone of light equals the solid angle contained by the incident cone. This proof can be repeated for every part of a large source and every part on a mirror from which the source is visible.

Using this relation we can now derive an expression for the luminous intensity of the image in terms of its magnification.

Let the projected area of the object in the direction of view be A_1 and that of the image be A_2, also let the corresponding luminous intensities be I_1 and I_2, and the luminances be L_1 and L_2.

From (6.3), the luminance L_2 of the image is ρL_2, and since

$$L_1 = \frac{I_1}{A_1} \quad \text{and} \quad L_2 = \frac{I_2}{A_2} \tag{6.4}$$

it follows that

$$\frac{I_2}{I_1} = \rho \frac{A_2}{A_1} \tag{6.5}$$

If the whole of the image is visible, A_2/A_1 is the area magnification of the optical system, so we can say in this special case that the luminous intensity is proportional to the magnification of the optical system.

6.5.2 *LUMINOUS INTENSITY DISTRIBUTION FROM A PLANE MIRROR*

Figure 6.7 shows schematically how the luminous intensity distribution from a plane mirror can be derived. For simplicity, the source is assumed to be a flat uniform diffuser. Its image is positioned as shown and the mirror is large enough for the whole of the image to be seen in the direction of the normal to the mirror.

In Figure 6.7(a) the luminous intensity is that of the source multiplied by the reflectance of the mirror. In Figure 6.7(b) the edge of the image coincides with edge of the mirror and to this point the luminous intensity decreases as $\cos\theta$. Past this point the decrease is much more rapid because only part of the source is visible, as in Figure 6.7(c). In Figure 6.7(d) the luminous intensity is zero. The actual rate of run-back of the luminous intensity can be calculated from the size of the mirror and the position of the image. The mirror acts as a pupil through which the image can be seen. This is the *exit pupil,* which is a useful concept for calculating the luminous intensity distribution for more complicated problems.

6.5.3 *INCLINED PLANE MIRRORS*

When two or more mirrors are inclined towards each other, multiple images are formed and, in certain designs, such as faceted reflectors, it is useful to be able to predict how the light is reflected between them.

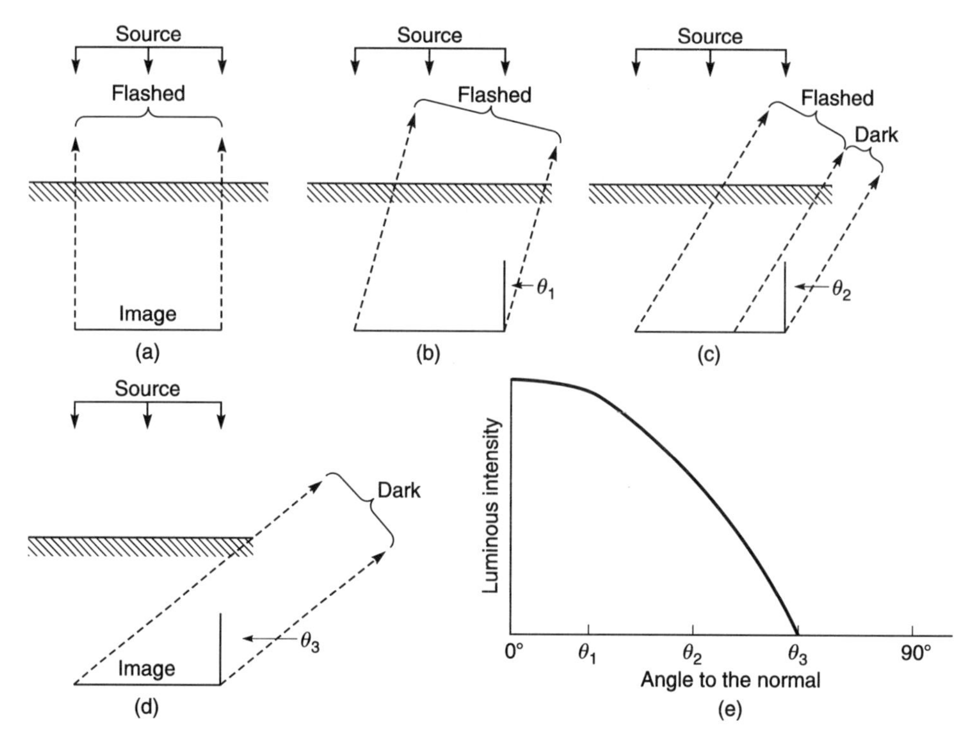

Fig. 6.7 Light distribution from a plane mirror

In Figure 6.8, *AO* and *BO* are two inclined mirrors. An object is situated at *S* and the observer is at *T*. The problem is to locate the images of *S* when these are viewed in a plane at right angles to the line of intersection of the mirrors. The first image in the figure is produced by *AO* at I_1 located so that *SU* equals UI_1 and such that SI_1 is perpendicular to *AO*. To locate the second image, *BO* is extended back as shown. The second image is then located at I_2 such that I_1X is equal to XI_2 and that I_1I_2 is perpendicular to *BX*. No image of I_2 by *AO* extended is formed because this image would lie behind both mirrors. An image of *S* is formed in *BO* at I_3, and in *AO* extended at I_4. For clarity the construction lines are not shown. All these images are formed on a circle centre *O* passing through *S*. This is because the construction for the formation of the images makes *OS* equal to OI_1, OI_1 equal to OI_2, and so on. *SVT* and *SWZT* are the traces of the rays reaching the observer's eye from the first two images.

The number of images formed depends on the angle between the mirrors, the smaller this is the greater the number of images.

6.5.4 CIRCULAR SECTION MIRRORS

Mirrors of circular section may be developed as parts of cylinders or parts of spheres. Part cylinders or troughs are typically used for linear sources, with the axis of the lamp lying in the axis of the part cylinder, whereas part spheres are typically used for light sources which approximate to a point.

Figure 6.9 shows two possibilities for using a circular or concave mirror. In both of these *O*

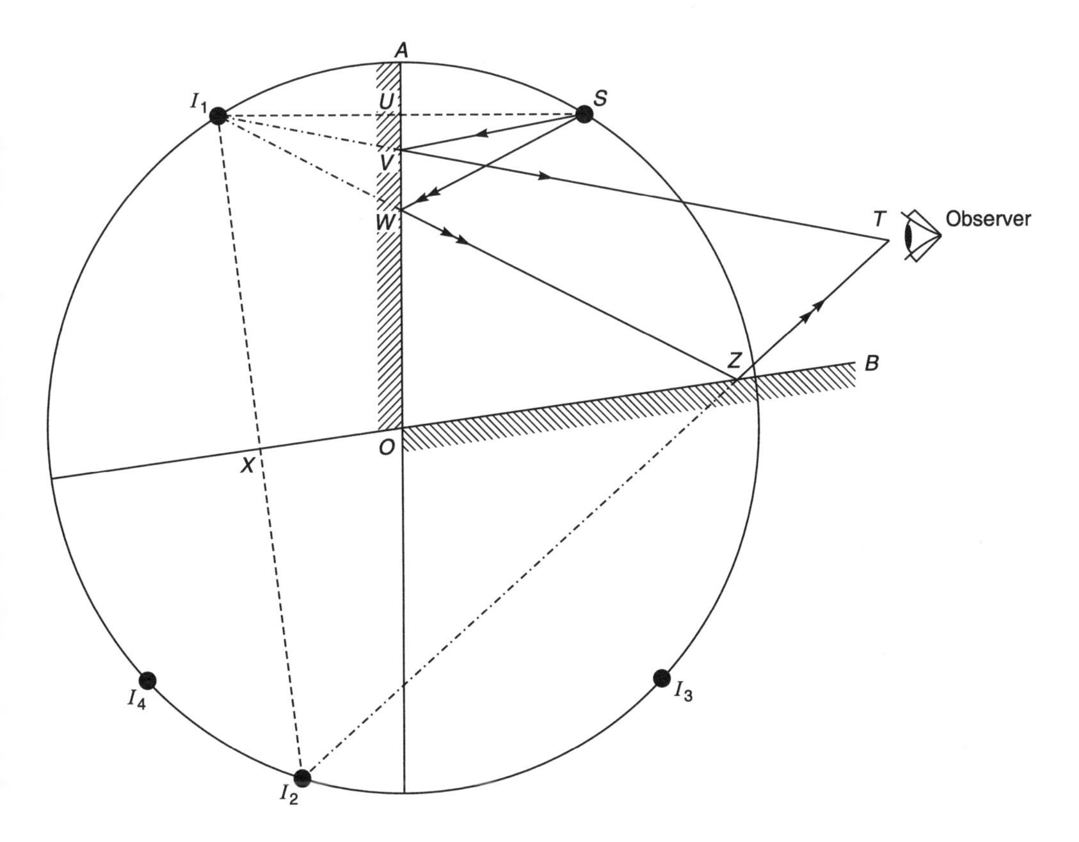

Fig. 6.8 Formation of images by inclined mirrors

is the centre of curvature of the mirror. In Figure 6.9(a) the light source is placed at *O*. Rays are reflected back through *O*. To minimize obstruction to the emerging rays, the lamp filament or arc tube would, in practice, have to be placed to one side of *O*. This also reduces the heating effect of the light energy on the lamp, which may have a deleterious effect.

More useful is the arrangement in Figure 6.9(b). Here the light source is placed at *F*, the focus, one half the distance between *O* and the back of the reflector. Rays from *F* which do not deviate by more than about 10° from the axis are reflected in a direction substantially parallel to the axis.

The parabola (Section 6.5.5) and ellipse (Section 6.5.10, page 219) to be described next provide a more efficient means of directing the light forwards, in that they control more of the luminous flux from the lamp.

6.5.5 PARABOLIC SECTION MIRRORS

The parabola is useful for producing a near parallel beam of light from a source, from a wide collection angle.

The equation of the parabola is $y^2 = 4ax$. Its shape can be generated by plotting this equation or by cutting a right circular cone in a plane parallel to its surface, which is sometimes used for cutting accurate templates.

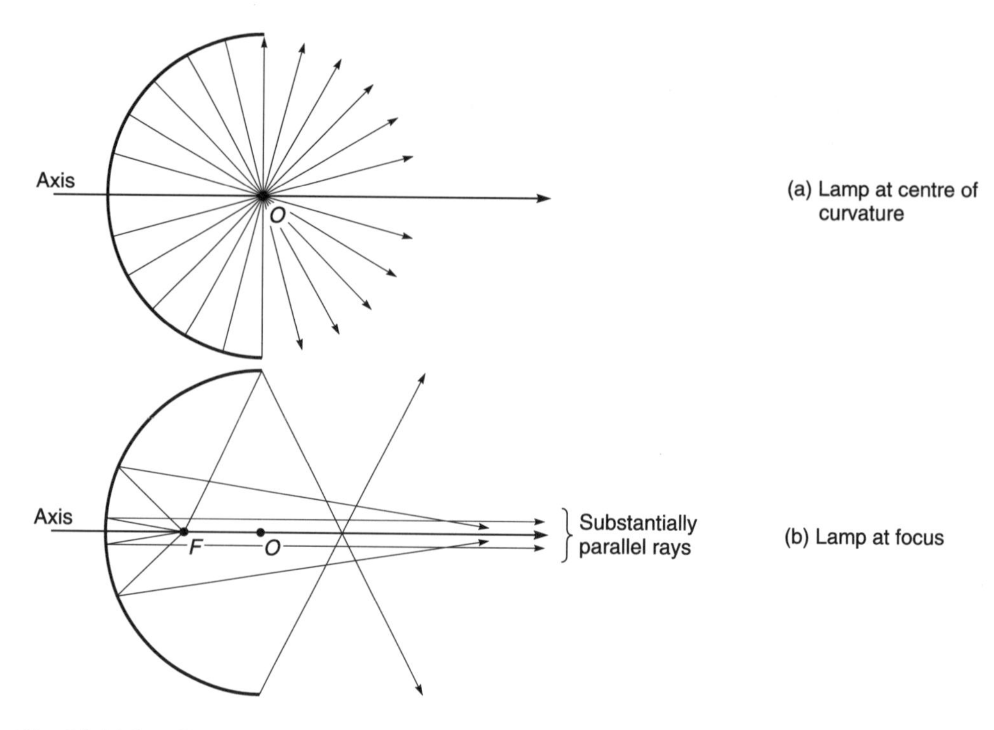

Fig. 6.9 Light reflected from a concave mirror

Figure 6.10 shows the form of the parabola. *O* is the origin, and *F* is the focal point, at which the light source is situated to produce near parallel light. The vertical line at a horizontal distance $2a$ from *F* is called the directrix. The distance $(x + a)$ from *F* to any point *P* on the reflector equals the horizontal distance from *P* to the directrix.

The parabola as a reflector can take a number of forms. It can be rotated about its axis to give a surface of revolution, the paraboloid, or it can be rotated about an inclined axis through the focus, to give a horned quartic, dealt with in Section 6.5.12, or it can be produced as a trough.

6.5.6 *LIGHT DISTRIBUTION FROM A PARABOLIC TROUGH REFLECTOR*

Figure 6.11 shows how the light distribution in the principal plane (perpendicular to the reflector axis) from a parabolic trough reflector can be developed. Figure 6.12 is the light distribution shown schematically. It is assumed that the light source is a cylinder with an even luminance over its surface.

Figure 6.11(a) shows the extent of the flashing of the reflector at 0°, which is complete. As θ, the angle between the centre-line and the direction being considered, is increased, the flashing will remain complete until the situation illustrated in Figure 6.11(b) is reached. Here the direction of view is parallel to the rays of light originating from tangents to the lamp and reflected by the edges of the reflector. This determines θ_1, and between this angle and 0° the luminous intensity will decrease very nearly as cosine θ. The decrease is not exactly proportional to the cosine of the angle because the contribution directly from the lamp stays constant, while the contribution from

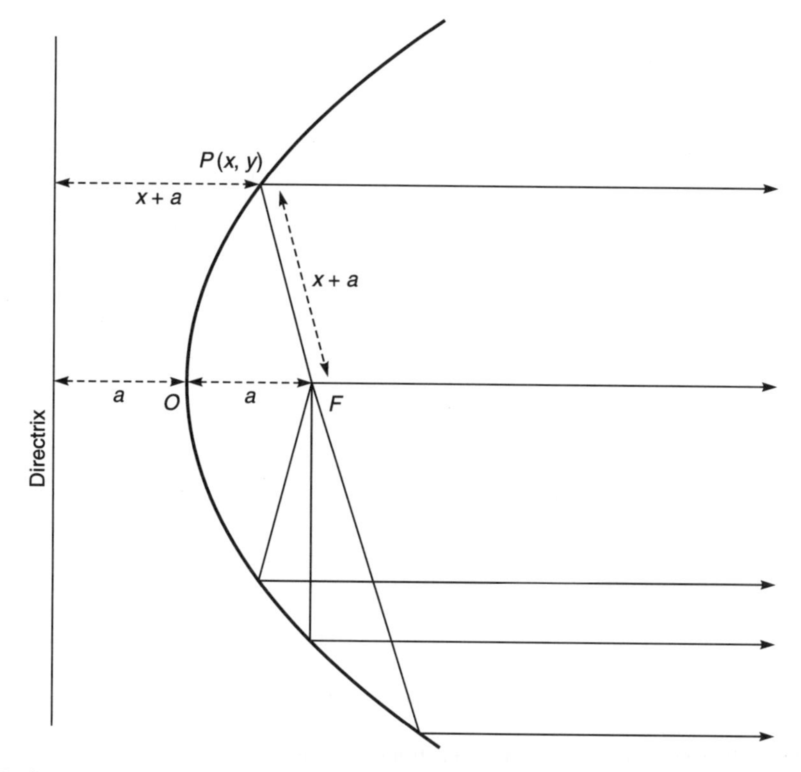

Fig. 6.10 Reflection from a parabola

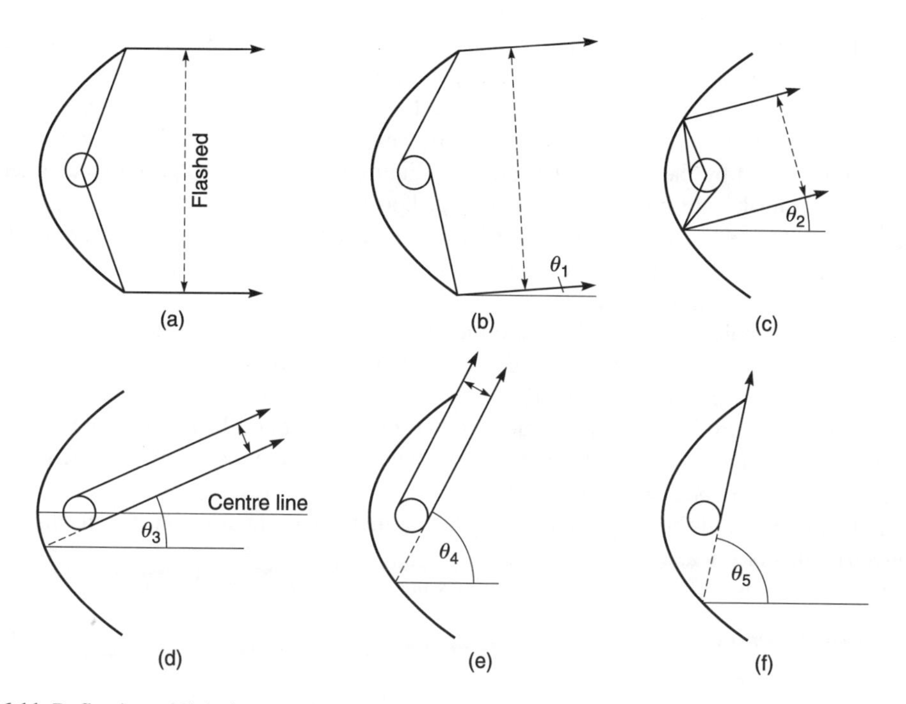

Fig. 6.11 Reflection of light in a parabolic reflector

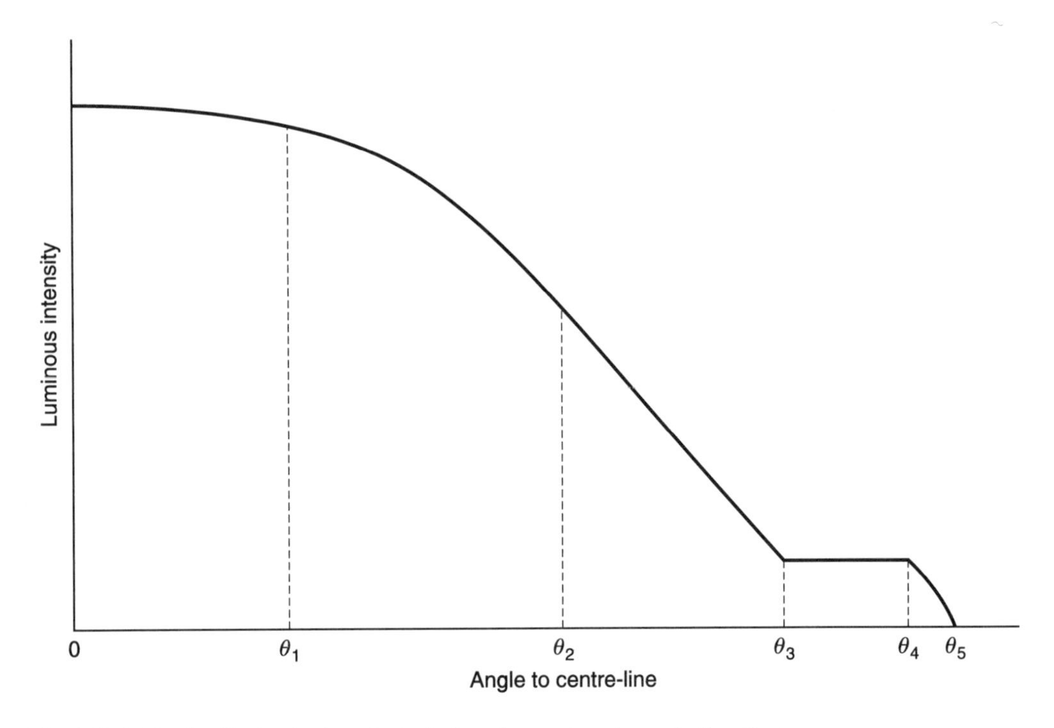

Fig. 6.12 Light distribution in the principal plane from a trough parabolic reflector

the reflector does decrease as the cosine of the angle of view, as it is proportional to the projected area. Past θ_1, the luminous intensity will decrease more rapidly since the reflector will show progressively less flashing (Figure 6.11(c)) until, at Figure 6.11(d), only the lamp appears bright. The luminous intensity will then stay constant and be equal to that of the lamp until the lamp starts to be occluded by the edge of the reflector (Figure 6.11(e)) when the luminous intensity will rapidly diminish until it is completely occluded, as in Figure 6.11(f).

The results obtained in this discussion are based on the assumption that the measurements are taken from a sufficiently great distance that the cones of light from the reflector cross over each other. Figure 6.13 will make this clear.

The cross-over point occurs at *A*, where the tangential ray, *GH*, from the lamp is reflected from the edge of the reflector and crosses over the axis of the parabola, at *A*. It would meet a similar ray from the other side of the reflector at this point, if the reflector is symmetrical about the centre-line. We can now find the distance of *A* from the mouth of the reflector in terms of *W*, the mouth width of the reflector, *r* the radius of the lamp, and *a* the focal length of the parabola.

The origin of the coordinate system is *O*.

First we make use of the property of the parabola that *FH* is equal to *HS*, the distance from the directrix, that is $x + \text{a}$.

Now the triangles *HGF* and *HAK* are similar since they are right-angled and

$$\begin{aligned} \angle FHG &= \angle THA \\ &= \angle HAK \\ (&= \theta) \end{aligned} \tag{6.6}$$

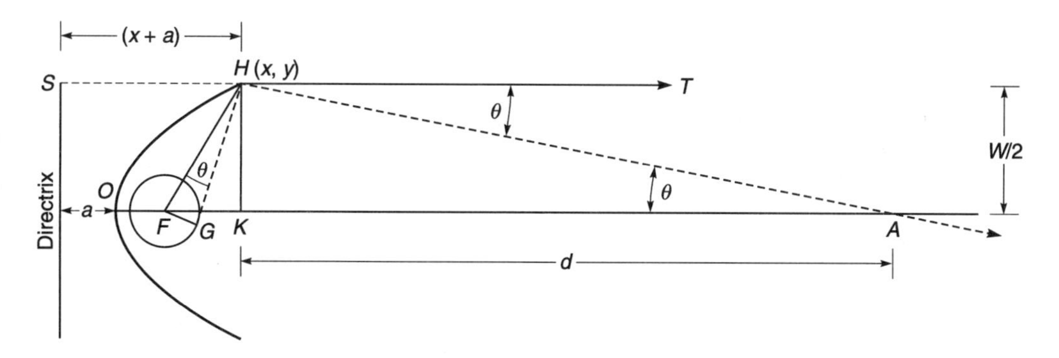

Fig. 6.13 Crossover point for a parabola

hence

$$\frac{FG}{GH} = \frac{HK}{AK} \tag{6.7}$$

so,

$$AK = HK \times \frac{GH}{FG} \tag{6.8}$$

in which

$$\begin{aligned} AK &= d \\ HK &= y \\ FG &= r \\ GH &= \sqrt{(x + a)^2 - r^2} \end{aligned} \tag{6.9}$$

so

$$d = \frac{y}{r} \sqrt{(x - a)^2 - r^2} \tag{6.10}$$

Now we need to express x in terms of a and W, for which we use the equation of the parabola

$$\begin{aligned} x &= \frac{y^2}{4a} \\ &= \frac{(W/2)^2}{4a} \end{aligned} \tag{6.11}$$

This can be substituted in (6.10) to give

$$d = \frac{W}{2r} \sqrt{\left(\frac{W^2}{16a} + a\right)^2 - r^2} \tag{6.12}$$

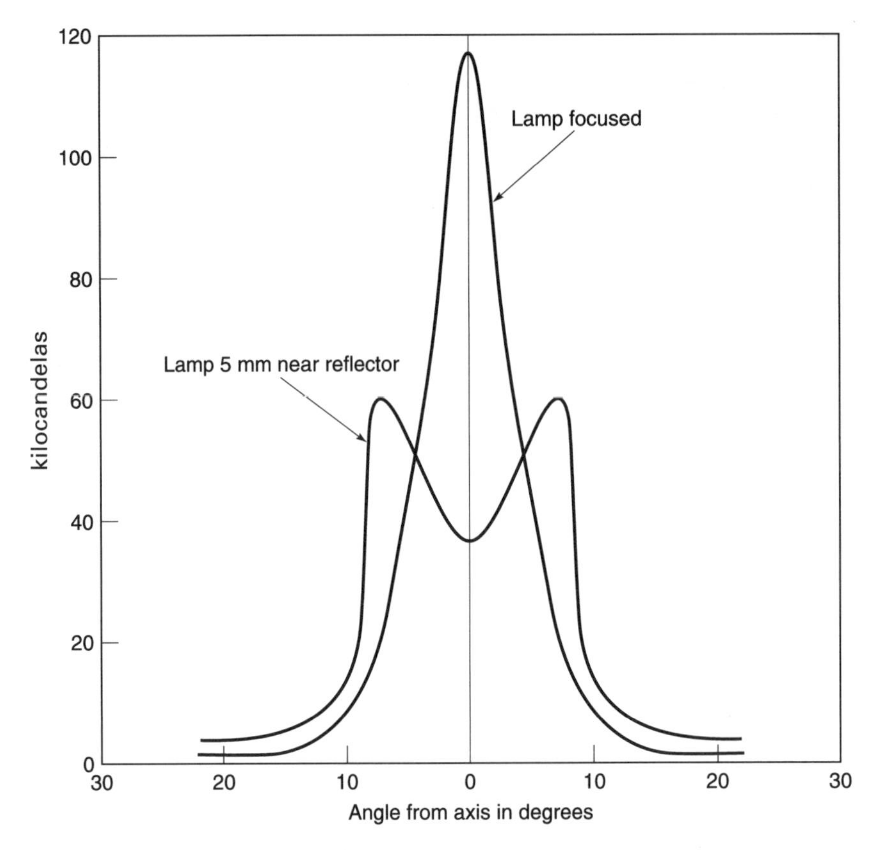

Fig. 6.14 Light distribution curves of 200 W tungsten halogen lamp in parabolic glass trough reflector of focal length 29 mm

6.5.7 PRACTICAL PERFORMANCE

The practical performance to be expected from a parabolic reflector depends on the accuracy of the reflector, the specularity and reflectance of its finish, and the configuration and position of the light source. In addition, the front glass, if present, may modify the beam. In a parabolic trough reflector the side reflectors, if angled correctly, may make an important contribution to the peak luminous intensity.

Very often, a specular finish is used but the reflector is dimpled or faceted to soften the beam. This is preferable to using a matt finish as it is difficult to control the quality of the finish from batch to batch. Defocusing the lamp in an attempt to widen the beam will not produce the desired effect as a trouser leg distribution results, as is shown in Figure 6.14.

Often, the uncontrolled direct light from the lamp is undesirable because it causes glare and light pollution, especially in floodlighting where parabolic trough reflectors are used. This can be overcome to a large extent by positioning a baffle to mask the lamp in the directions in which light emission is not wanted. The baffle itself will mask part of the main reflector so reducing the luminous intensity in some directions. However, it is usually specular so it can reflect light back onto the reflector. If the reflector is fully flashed in the important directions this will not increase the luminous intensity (because the luminance of the reflector cannot be increased) but it will increase the beam width. In particular it can be used to increase the luminous intensity at or near the nadir where, in many cases, the luminous intensity is insufficient for floodlighting purposes.

6.5.8 TESTING PARABOLIC REFLECTORS

It is often difficult to establish the reasons for the performance of a parabolic reflector not coming up to expectations.

Provided that the source is positioned accurately in the focus, the finish of the reflector is as specified, the side reflectors, if present, are accurately angled, and the front glass, if any, is as required, any lack of performance may be due to inaccuracies in the contour. The parts in error can be seen by viewing the lighted mirror on axis from a distance and noting where it is not flashed. It is quite useful to photograph the flashing of the mirror at various angles. However, if the photograph is taken, as it should be, from a great distance, the image may be too small to be of any use unless a powerful telephoto lens is used.

Using a template to fit into the contour is of little value since small local high spots may hold it off the main contour. A plaster or fibre-glass cast can be made and sections taken. Care has to be taken with this method that shrinking does not take place as the material solidifies. A number of thin layers as opposed to one thick layer may obviate this problem.

Determination of the inaccuracy of manufacture in terms of light measurements has to be done on a photometric bench. The set-up for the test on a bench is shown in Figure 6.15. *O* is a small source on the axis of the mirror *MM*, the rays from which are brought to a focus found on a movable screen *S*. Interposed between *O* and *MM* is a mask that allows only a ring of light to reach *MM* in the case of a paraboloid, and two lines of light in the case of a trough reflector. The mask can be made by blacking out appropriate areas on a transparent medium such as glass. By using a series of masks the mirror is divided up into concentric rings or parallel lines, which are

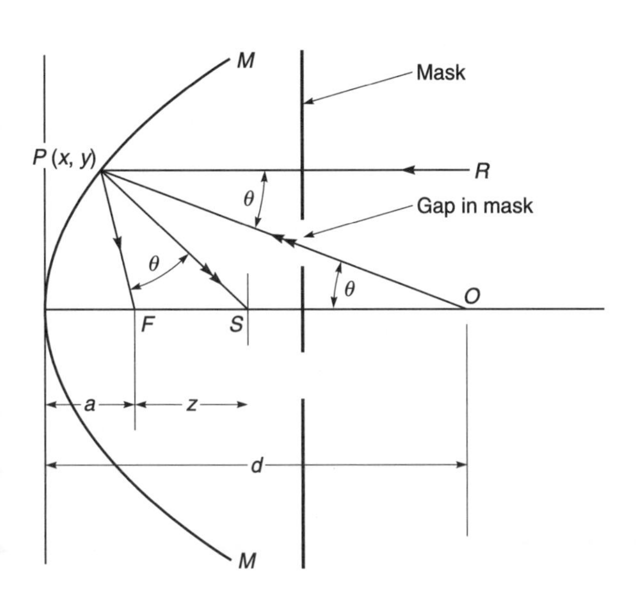

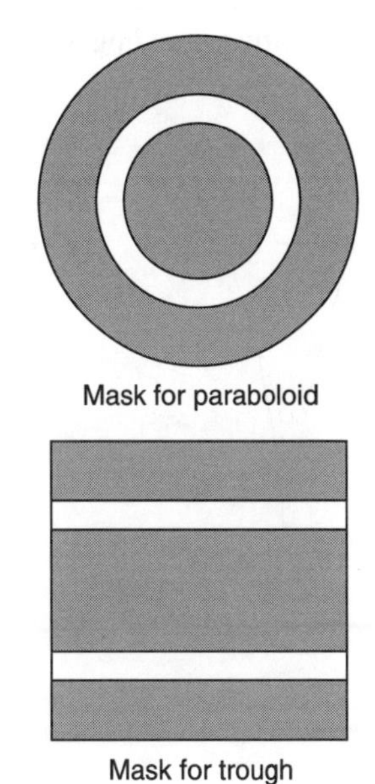

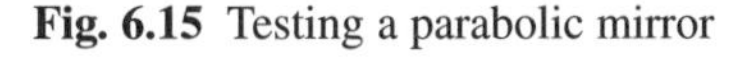
Fig. 6.15 Testing a parabolic mirror

tested one at a time. If the mirror is truly parabolic then all the rings will have their foci in the same position. Allowance is made for the fact that O is not infinitely far away by using a correction derived in the following manner.

In Figure 6.15 the ray of light OP is reflected off a point on the mirror having coordinates (x, y). PR represents a ray of light, parallel to the axis, passing through F, the focal point.

Now

$$\begin{aligned} \angle FPS &= \angle QPR \\ &= \angle POS \end{aligned} \tag{6.13}$$

Triangles FPS and POF are similar since $\angle FPS = \angle POS$, and $\angle FPS$ is common.

Therefore,

$$\frac{PE}{FO} = \frac{FS}{PF}$$
$$\text{so,} \quad FS = \frac{PF^2}{FO} \tag{6.14}$$

By reference to Figure 6.10, page 213, we find that $PF = x + a$ so that

$$z = \frac{(x+a)^2}{d-a} \tag{6.15}$$

where z is the correction FS, and d is the distance of the source from the mirror on its axis.

6.5.9 APPLICATION IN LOUVRES

Figure 6.16(a) shows how specular louvres with a parabolic cross-section can be used to obtain a specified shielding angle. A is the focal point of the parabolic surface of the adjacent wedge so

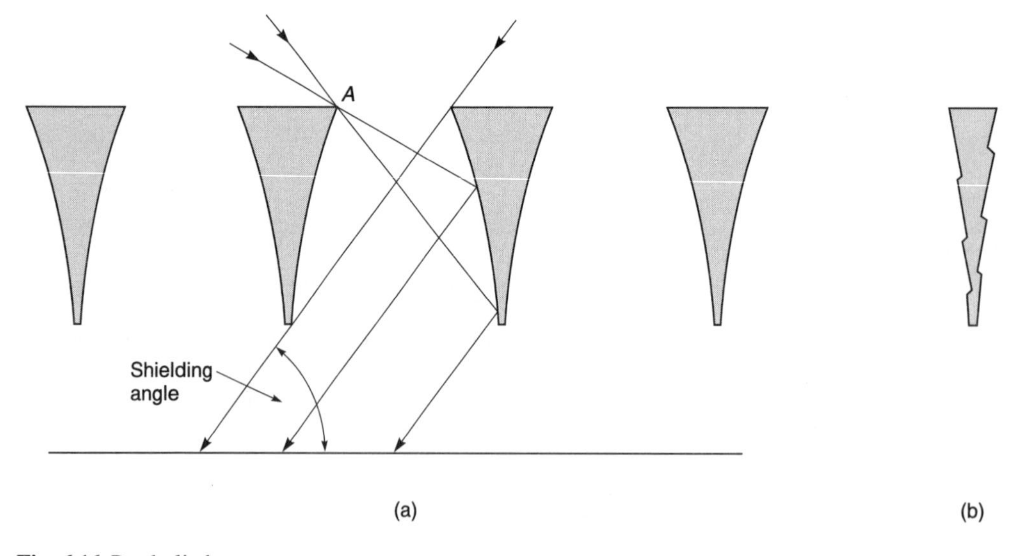

Fig. 6.16 Parabolic louvres

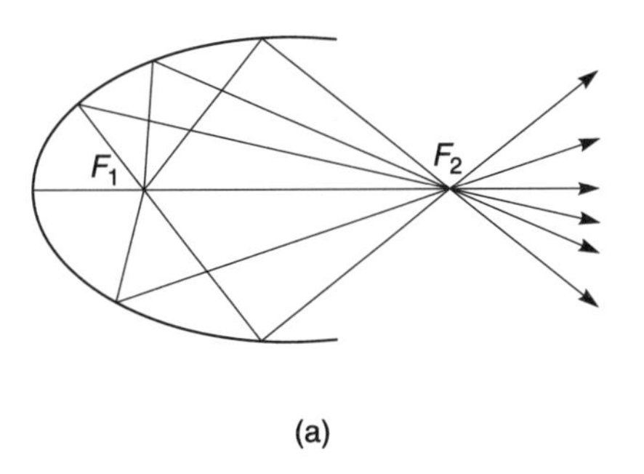

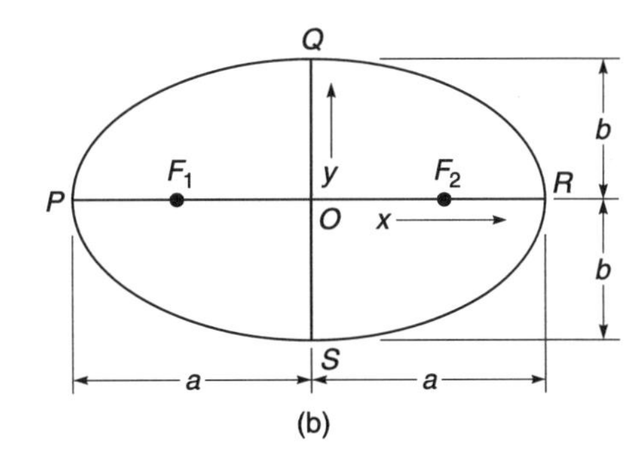

Fig. 6.17 The ellipse

there is no interreflection of the light between the wedges. This undoubtedly improves the light output ratio but the gain is more than offset by the obstruction caused by the top of the wedge-shaped cross-section. To overcome this the curve can be stepped as in Figure 6.16(b).

6.5.10 THE ELLIPSE

The ellipse (Figure 6.17(a)) has two foci, F_1 and F_2, such that if a point source of light is placed at one focus all the light rays pass through the other focus. The parabola may be regarded as a special case of the ellipse in which one focus is at infinity. As with the parabola, the mirror may take the form of a surface of revolution, the ellipsoid, or of a trough.

The equation of the ellipse is (Figure 6.17(b))

$$\frac{x^2}{a^2} + \frac{y^2}{b^2} = 1 \tag{6.16}$$

where a and b are the half lengths of the major (PR) and minor (QS) axes. The distances on the abscissa of the foci from the centre of the mirror, O, are $\pm \sqrt{a^2 - b^2}$.

The ellipse has the property that the sum of F_1P and F_2P is a constant as P is moved around the contour. Whilst this can be used for drawing an ellipse, in practice it does not give a very accurate result. The most accurate way of obtaining an ellipse is by cutting a cylinder or a cone at a suitable angle.

This curve is useful where light has to be directed on to a narrow slit, or in a floodlight where a specific beam divergence is required.

Holmes has shown how a reflector that combines ellipsoidal and parabolic curvatures may be used to produce a fan-shaped beam.[2]

6.5.11 THE HYPERBOLA

Like the ellipse the hyperbola (Figure 6.18(a)) has a second focus but this is behind the mirror and the rays diverge from this point. It is sometimes used in place of the ellipse to obtain a specific divergence. The disadvantage of the hyperbola is that some of the rays may strike the enclosing housing with a resultant loss in light.

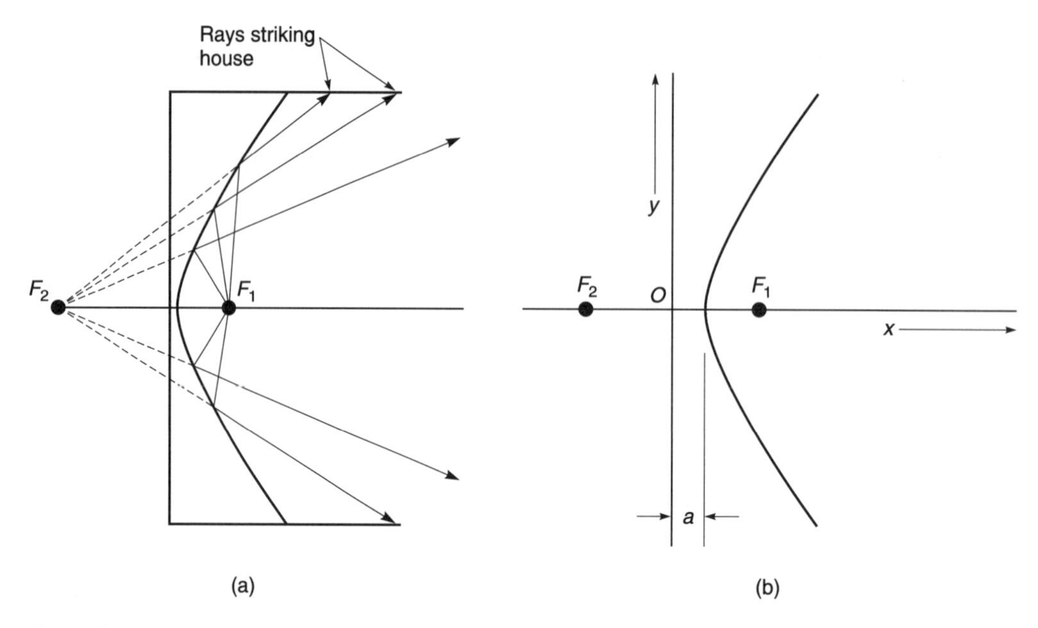

Fig. 6.18 The hyperbola

Like the other curves so far described it can take the form of a surface of revolution about the major axis, the hyperboloid, or of a trough.

The equation of the hyperbola (Figure 6.18(b)) is

$$\frac{x^2}{a^2} - \frac{y^2}{b^2} = 1 \tag{6.17}$$

where a is the distance of the foci from the origin O, and b is a constant. The distance of F_1 from the back of the mirror is $\sqrt{a^2 + b^2}$.

The contour of a hyperbola can be produced by cutting a cone in a plane parallel to its axis.

6.5.12 *HORNED AND PEAKED QUARTICS*

To obtain a beam of greater divergence than that given by a parabola, Spencer[3] developed a family of curves known as horned and peaked quartics. These are obtained by rotating a parabola about a secondary inclined axis passing through the focus.

In Figure 6.19(a) O is the apex of the parabola MOM', OX its axis, and F its focus. The inclined axis about which it is rotated is $O'X'$. The surfaces so generated are shown in Figure 6.19(b). The outer limbs of the parabola form the horned quartic (Figure 6.19(c)), the inner limbs the peaked quartic. Spencer showed that the equation of these surfaces involves the fourth power of the distance along the rotation axis. For this reason and because of the shape of the resultant curves they are called horned and peaked quartics.

The optical properties of these surfaces are shown in Figure 6.20. The horned quartic, Figure 6.20(a), reflects the rays of light away from the optical axis of the reflector and is, in this respect, similar to the hyperbola. The peaked quartic reflects the light so that it crosses over the axis as in Figure 6.20(b). The spread of the rays of light from the direction of the centre of the lamp is

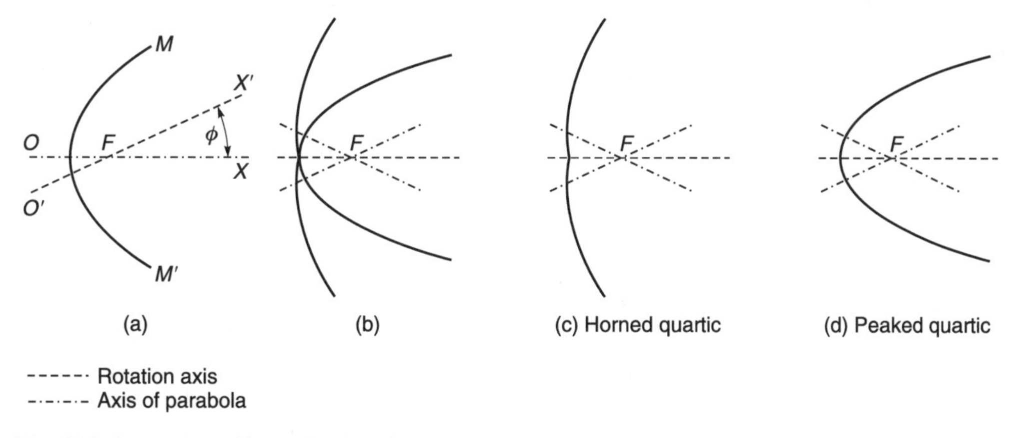

Fig. 6.19 Generation of horned and peaked quartics

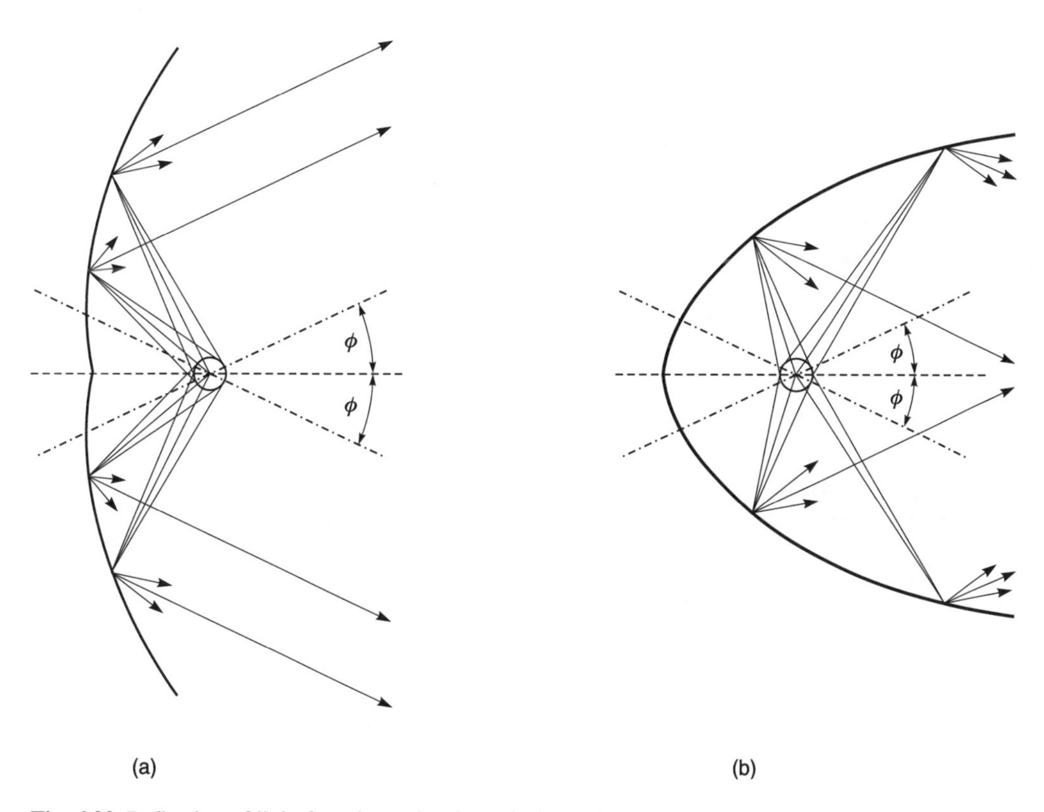

Fig. 6.20 Reflection of light from horned and peacked quartics: (a) horned quartic; (b) peaked quartic

$\pm\phi$ but there is additional spread of the light depending on the size of the light source in relation to its distance from the reflector. ϕ should not be made too great otherwise a dip will be produced in the beam.

Owing to the reflector forming images of the coils of the filament when a tungsten halogen lamp is used, the beam will show striations, a fault not confined to horned and peaked quartics.

Spencer suggested eliminating these by 'modulating the contour'. This amounts to rotating small elements of the contour about vertical axes. The angle must be sufficient to fill in the dips between the striations. This procedure is preferable to matting the reflector for the reasons already stated (Section 6.2, page 201).

Surfaces generated by rotating a parabola about a secondary axis parallel to the true axis have been used in motor-car headlights.[4]

6.5.13 SHARP CUT-OFF REFLECTORS

In some applications, particularly road lighting, it is essential to have a sharp cut-off of the luminous intensity above a certain angle to control glare. If the source can be regarded as point or linear in comparison to the focal length, a parabola can be used with the source itself shielded to stop direct light. A good quality reflector is required otherwise it is necessary to use louvres, which have the drawback of reducing the light output ratio. With tubular lamps, such as the fluorescent type, it is possible to achieve a cut-off by putting a suitable point on the circumference of the lamp at the focus. However, only a part of the parabola can be used without rays being reflected above the cut-off.

To cope with the problem of achieving a sharp cut-off from a tubular source, Stevens developed a family of curves having the property, shown in Figure 6.21(a), that the upper bounding rays of the reflected beams are parallel.[5]

They are defined by the parametric equations ((Figure 6.21(b)):

$$y = \frac{1}{2}\left[x\left(p - \frac{1}{p}\right) - r\left(p + \frac{1}{p}\right)\right] \tag{6.18}$$

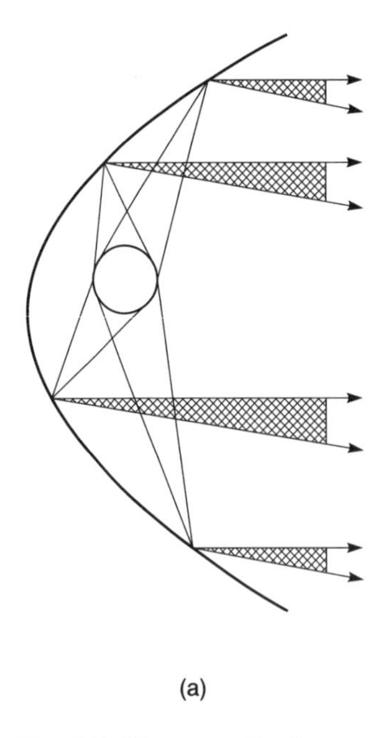

(a)

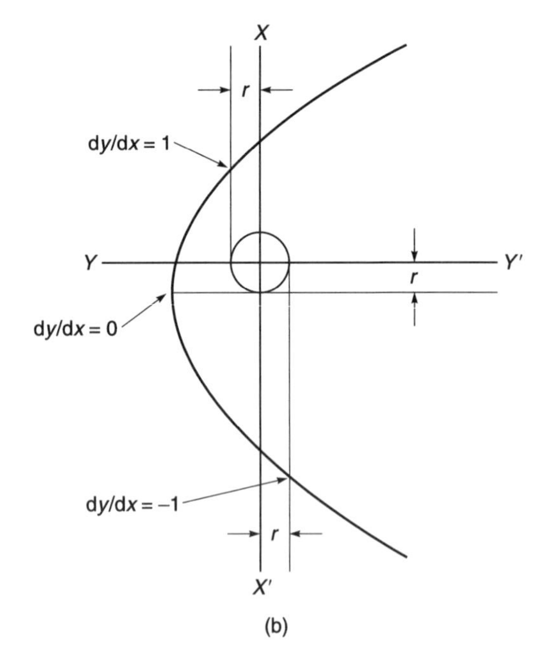

(b)

Fig. 6.21 Sharp cut-off reflector

$$x = p(c - 2r\phi) - r \tag{6.19}$$

where r is the radius of the tube; $p = \tan \phi = dy/dx$; $c = -2y$ when $x = -r$.

As c is a constant for any curve, a family of curves passing further and further from the tube can be obtained by increasing its value.

It is interesting to note that when $r = 0$ the curves become parabolas of focal length $c/2$.

To make a curve go through a specific point (x, y), the values are substituted in the equation

$$p = \frac{y + \sqrt{x^2 + y^2 - a^2}}{x - a} \tag{6.20}$$

ϕ can then be found from $\phi = \tan^{-1} p$, and both p and ϕ substituted in (6.19) to give c. Equations (6.18) and (6.19) can then be used to find values of (x, y).

Whilst the light from the reflector gives the desired sharp cut-off, direct light from the lamp will be uncontrolled. A baffle can be used to block this.

6.5.14 INVOLUTE OF CIRCLE

The involute of the circle (Figure 6.22) has the useful property that no light is reflected back on to the source, when this is circular. This property ensures a high light output ratio for minimum width of reflector.

An involute is produced by the trace of the end of a length of thread as it is unwound from a cylinder. A graphical construction is shown in Figure 6.23.

The circumference of the circle is divided into a number of equal parts, marked P_1, P_2, P_3, etc. Tangents are drawn at these points and distances equal to the circumferential distances of

$$P_0P_1,\ P_0P_2,\ P_0P_3,\ P_0P_4,\ P_0P_5, \text{ etc.} \tag{6.21}$$

are marked off them to give the points R_1, R_2, R_3, etc. These are then joined to give the involute. A more accurate method is to use the parametric equations

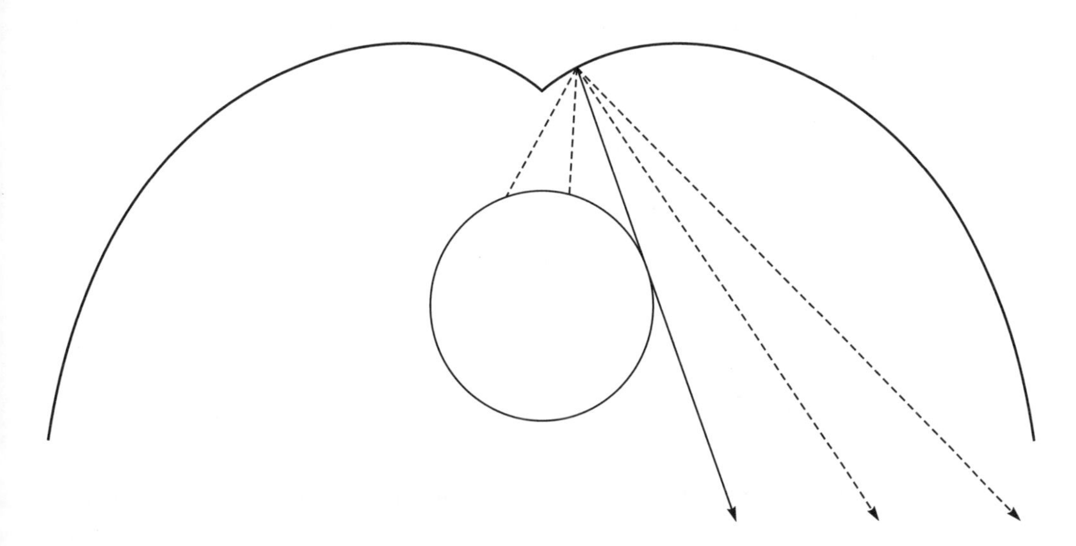

Fig. 6.22 The involute of the circle showing how reflected rays miss the source

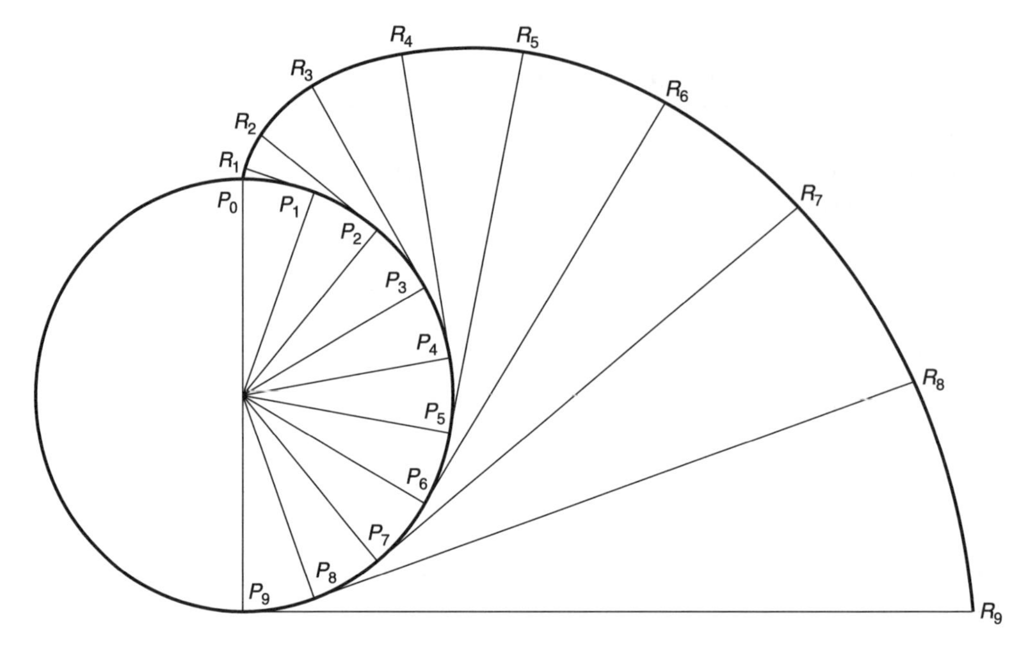

Fig. 6.23 Construction of involute of circle

$$x = r(\cos\theta + \theta\sin\theta)$$
$$y = r(\sin\theta + \theta\cos\theta) \qquad (6.22)$$

where the equation of the circle is $x^2 + y^2 = r^2$.

6.5.15 FACETED REFLECTORS

Faceted reflectors, made up of plane elements, have the advantage over curved reflectors that they give a light distribution that is largely free of striations.[6] This is because the images formed by plane elements are virtual. However, if the elements are small in relation to the size of the source, the reflector as a whole may be considered as approximating to a curve and images of the source may be formed.

Faceted reflectors are not easy to form accurately in manufacturing processes when they are large, as is the case for floodlights. Extrusion can be used but care has to be taken that any polishing processes applied do not reduce the definition of the facets.

The construction for one type of reflector for obtaining a flat topped beam from a linear source, such as a tungsten halogen lamp is shown in Figure 6.24. The method of developing the shape can be followed by means of an example. Suppose we require a beam spread of 20°.

O is the source and the centre line *OX* is drawn. The reflector will be symmetrical about this line so for clarity, only the upper limb of the reflector is shown in the figure. Radial lines emanating from *O* are drawn. The first one is at 10° ($\frac{1}{2} \times 20°$) to *OX* and the remaining ones are at 20° to each other. It should be noted that angles below the horizontal are marked negative.

At *A*, a line is drawn at a convenient distance from *O*, perpendicular to *OX*, to meet the first radial line in *B*. This gives the position of the half-facet *AB*. AI_1 is made equal to *OA* and marks the position of the image of *O* in *AB*. I_1Q is drawn perpendicular to *OX*. I_1B is the position of the most widely diverging ray from facet *AB*. I_2B is then drawn at –10° to the normal to I_1Q to locate

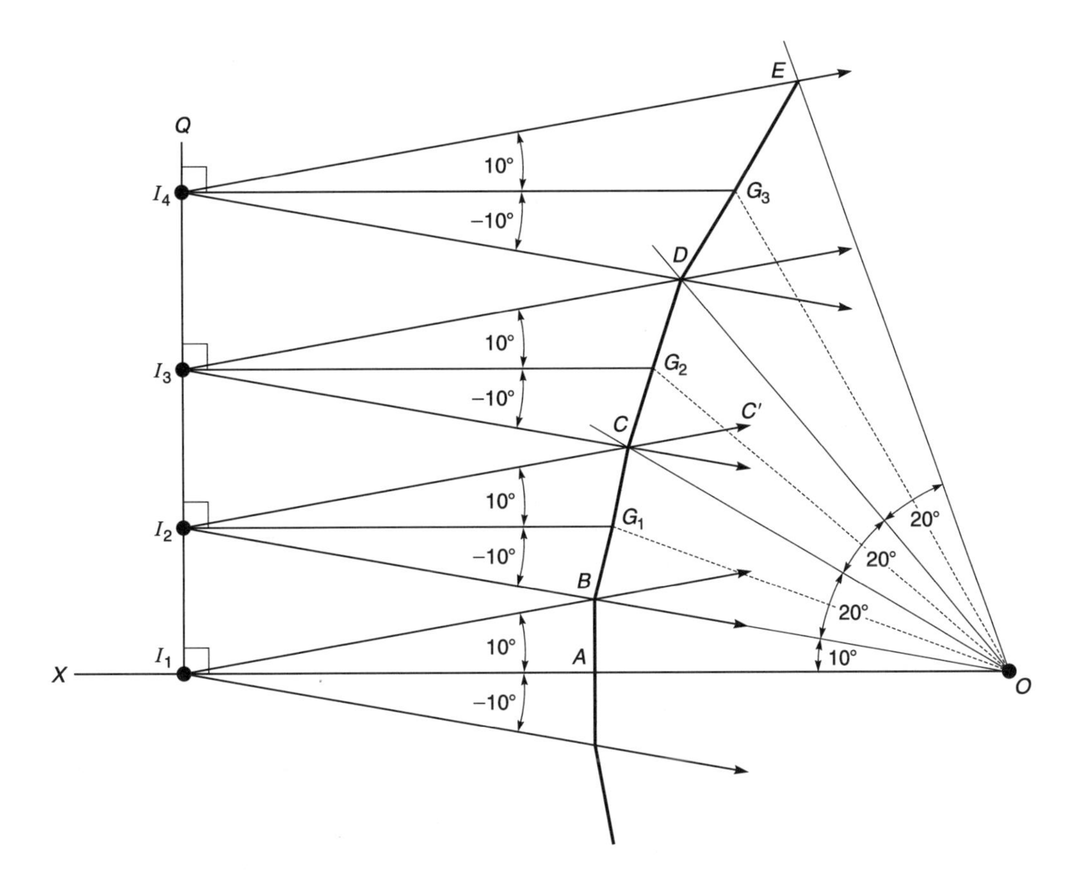

Fig. 6.24 Construction of faceted reflector

the second image I_2. (OBI_2 is a straight line, but this property does not hold for the subsequent facets.) I_2C' is drawn at 10° to the normal to I_1Q to meet the second radial line in C to give the position of the second facet BA. This procedure is then continued to give the positions of the other facets, CD and DE.

In this treatment of faceted reflectors we have considered the source as a line. Providing the reflector is accurately made and the material has no diffuse element this should give a flat-topped beam distribution. In practice, the source will be of finite size and there will be some rounding of the light distribution as indicated in Figure 6.25. This effect can be calculated by regarding each facet as a plane mirror and using the method given in Section 6.5.2.

It is worth noting some of the geometrical properties of this type of faceted reflector. Each facet is tilted at $\frac{1}{2} \times 20° = 10°$ with respect to the adjacent facets. This is because, at the junction of two facets, the upper facet has to turn the light ray from O through 20° more than the lower facet, which is accomplished by turning the facet through half this angle as indicated in Figure 6.5, page 207.

With this information, we can show that triangles such as I_3G_2C and OCG_2 are congruent. This is because the angle G_2CI_3 equals the angle OCG_2 (100°), the angle I_3G_2C equals the angle CG_2O (60°), and the side G_2C is common. Hence OG_2 equals G_2I_3. Similarly it can be shown

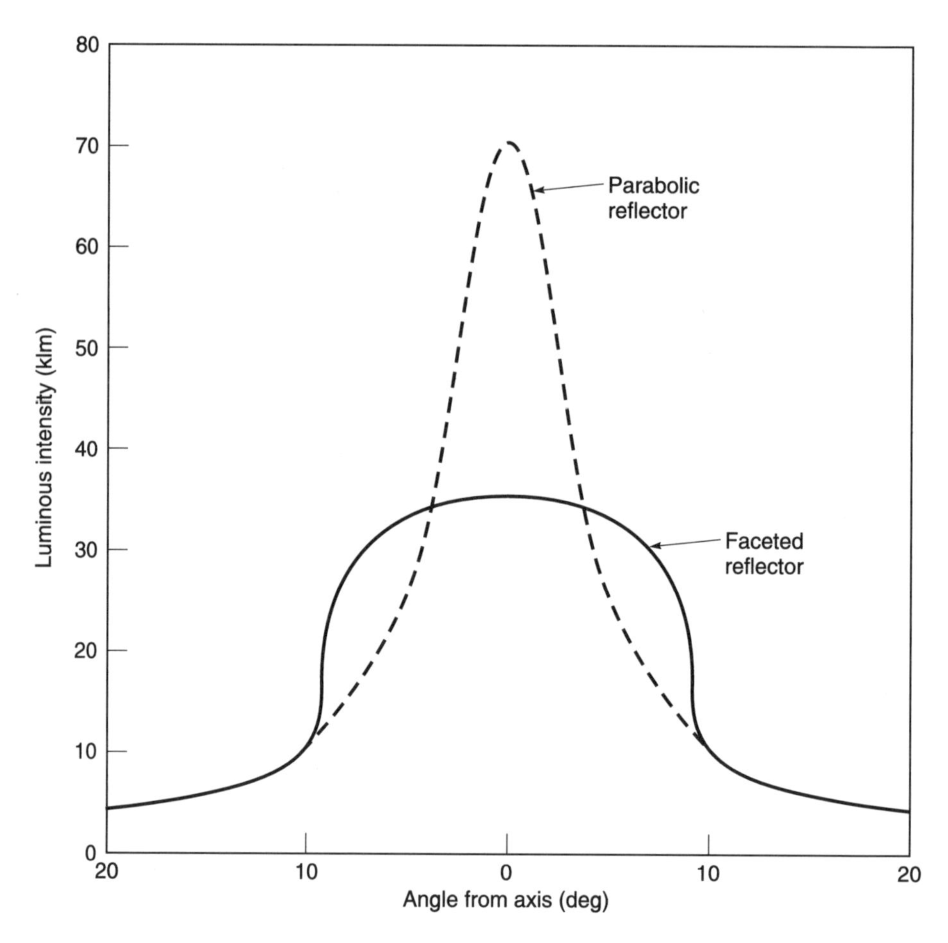

Fig. 6.25 Tranverse distribution curves from parabolic and faceted reflectors using a 1500 W tungsten halogen lamp

that

$$\begin{aligned} OG_1 &= G_1I_2 \\ OG_3 &= G_1I_4 \end{aligned} \tag{6.23}$$

as well as which

$$OA = AI_1 \tag{6.24}$$

Figure 6.10, page 213, indicates that we can regard O as the focus of a parabola, having I_1Q as its directrix, with the points A, G_1, G_2, G_3 and G_4 lying on its surface. The facets are tangential to the parabola at these points so that light rays from O reflected at these points are parallel to the axis.

6.5.16 NON-MATHEMATICALLY DEFINED CONTOURS

When the required light distribution cannot be obtained by one of the contours already described, the following method in which the contour is tailored to the requirements can be used. It only applies to sources of light that may be regarded as being small in relation to the size of the contour. When the sources are large, good accuracy cannot be expected.

The reflector will be a surface of revolution, that is, it will be symmetrical about the main axis, which is a further requirement for the method's use. Similarly, for linear sources the reflector will be trough shaped.

In outline, the method consists of determining the luminous flux available in convenient (for calculation) angular zones from the lamp. Similarly, the zonal luminous flux required in each angular zone is calculated from the required luminous intensity distribution. The luminous flux in each zone from the lamp is then directed into the zones where there is a deficit between the lamp luminous flux and the required luminous flux. The procedure can be split into four parts.

(1) *Selection of reflector system*

Jolley *et al.*[7] have described four possible reflector systems from which the most suitable can be selected (Figure 6.26). In (a) and (d) the reflected light rays diverge whereas in (b) and (c) they crossover. In (b) and (d), light rays reflected from the top zones of the reflector are reflected nearest the vertical, and vice versa for (a) and (d). (b) and (c) give the most compact reflectors. (a) and (d) have the disadvantage, besides their large size, that the light rays from the top of the reflector pass through the lamp envelope and may be obstructed by the internal structure of the lamp or by the lamp envelope itself if it is coated. This may also be true of (b) and (c) if the lamp envelope is large in relation to the reflector. (b) and (c) are the most used types: (b) where a fairly concentrated distribution is required and (c) where a flat surface, such as a chalk board, has to be illuminated as evenly as possible from one side by a linear source. For an example of how the reflector is generated we will select reflector (b).

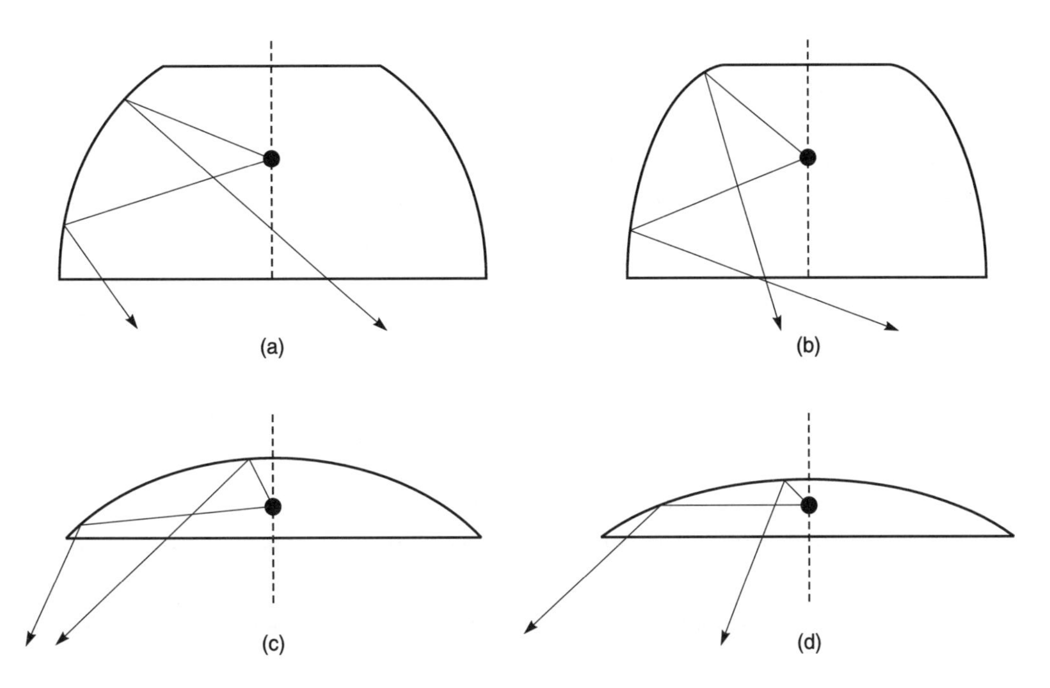

Fig. 6.26 Reflector configurations

Table 6.1 Redistribution of luminous flux from lamp

A	B	C	D	E	F	G	H	I	J	K
Limits of zone (°)	Mid-zonal angle (°)	Bare lamp luminous intensity (cd)	Bare lamp luminous flux (lm)	Lamp luminous flux after losses (lm)	Required luminous intensity (unscaled)	Required luminous flux (unscaled)	Required luminous flux scaled (lm)	Luminous flux reflected (lm)	Cumulative luminous flux reflected (lm)	Cumulative luminous flux from lamp (lm)
0–10	5	720	69	69	34	3.25	245	177	177	
10–20	15	720	204	204	33	9.35	708	503	680	
20–30	25	746	345	345	32	14.81	1120	775	1455	
30–40	35	720	452	452	30	18.85	1425	973	2428	
40–50	45	656	508	508	13	10.07	762	253	2682	
50–60	55	566	508	325						2682
60–70	65	488	484	310						2357
70–80	75	411	435	278						2047
80–90	85	411	448	287						1769
90–100	95	360	393	251						1482
100–110	105	373	395	253						1230
110–120	115	386	383	245						978
120–130	125	386	346	222						733
130–140	135	386	299	191						511
140–150	145	373	234	150						320
150–160	155	360	167	107						170
160–170	165	347	98	63						63
170–180	175	334	32							0
		Totals	5800	4260	–	56.33	4260	2681	–	–

(2) *Redistribution of the reflected luminous flux*

The work can be conveniently laid out in a spreadsheet (see Table 6.1), which is best described column by column.

Column	Description
A	In this example 10° zones are chosen.
B	This is self-explanatory.
C	The bare lamp luminous intensity has to be found by measurement.
D	The zonal luminous flux is most conveniently calculated from the luminous intensity by means of the formula (see Section 1.7.1, page 26)

$$\Omega_{\gamma_1-\gamma_2} = 4\pi I_\gamma \sin\left(\frac{\gamma_2-\gamma_1}{2}\right)\sin\left(\frac{\gamma_2+\gamma_1}{2}\right) \tag{6.25}$$

where

γ_1 is the lower bound of the zone;
γ_2 is the upper bound of the zone;
I_γ is luminous intensity at the mid-zone angle γ;
$\Omega_{\gamma_1-\gamma_2}$ is the zonal luminous flux between γ_1 and γ_2.

E In this example it is assumed that the material of the reflector has a reflectance of 0.8. Also, it is assumed that there is a further fractional loss of 0.2 owing to obstruction by the lamp. This means that the reflected luminous flux has to be multiplied by 0.8 × (1 – 0.2) or 0.64. In addition, it is assumed that all the luminous flux in the 170° to 180° zone is lost because of the provision made for fixing the lamp.

F The required luminous intensity may be arrived at from the illuminance requirements. In this example, the values are unscaled.

G The luminous flux in unscaled units is obtained by the same formula as is used in column D on the figures in column F.

H The luminous flux from the lamp is scaled by multiplying by the total in column E divided by the total in column G, or 4260/56.33 = 75.63. Note there is a 2 lumen rounding error in the total.

I From column H the direct flux in column E, up to 50°, is subtracted.

J Cumulative totals are formed.

K Cumulative totals starting from the bottom of the column are formed from column E. The totals are made from the bottom of the column because luminous flux from the highest angles has to be directed into the lowest angles for the reflector in Figure 6.26(b). It will be noticed that the cumulative total in this column equals the totals in column J and column I disregarding rounding errors.

(3) *Determination of the inclination of the contour elements*

The inclination of the reflector elements to the vertical can now be found by making use of the geometry in Figure 6.27. A is the centre of the light source and XX'. an element of the reflector

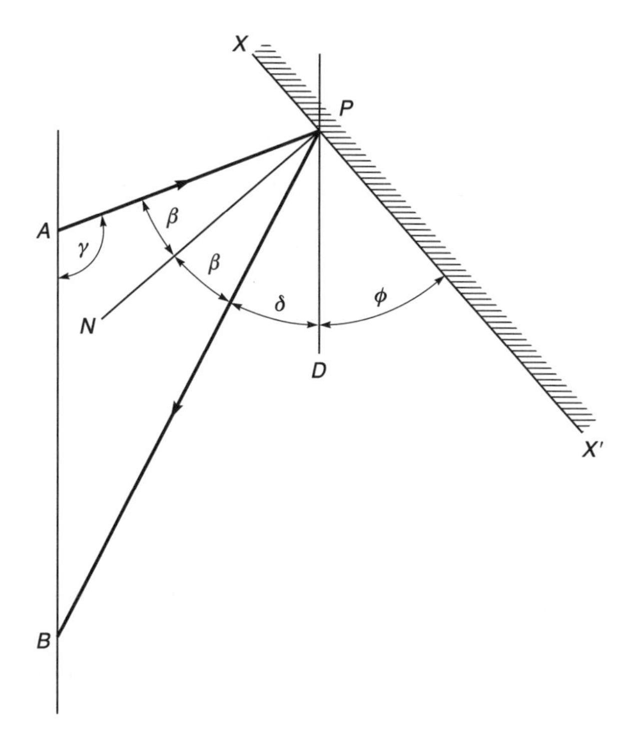

Fig. 6.27 Determination of the inclination of a reflector element

surface. *NP* is the normal to the mirror at *P*, and *AP* and *PB* are the incident and reflected rays. *AB* and *PD* are verticals. The angles are as marked.

Since $\angle NPX' = 90°$,

$$\beta + \delta + \phi = 90° \tag{6.26}$$

In the triangle *APB*,

$$\begin{aligned} \gamma + \delta + 2\beta &= 180° \\ \text{so,}\quad \gamma + \delta + 2\,(90° - \delta - \phi) &= 180° \end{aligned} \tag{6.27}$$

giving

$$\phi = \tfrac{1}{2}(\gamma - \delta) \tag{6.28}$$

In the forms of the reflector where the reflected ray goes to the right of *PD*, δ will be positive in the formula.

To use this formula it is necessary to find γ and δ. This is done by plotting a cumulative flux diagram, as in Figure 6.28. In this, the cumulative luminous flux as calculated in column K of Table 6.1 is plotted against the angle from the downward vertical. As the light from the upper part of the reflector has to be redirected towards the vertical, the angles on the abscissa start with the high angles. By using the cumulative totals in column J, the angular zones into which the luminous flux has to be redirected can be plotted and are shown as dashed lines. The procedure can be clarified by considering the first two zones. From column I it is evident that 177 lumens are to be directed into the 0° to 10° zone. Therefore, the dotted line is drawn on the graph at a cumulative luminous flux of 177 lumens. This is obtained from two zones of the reflector; the

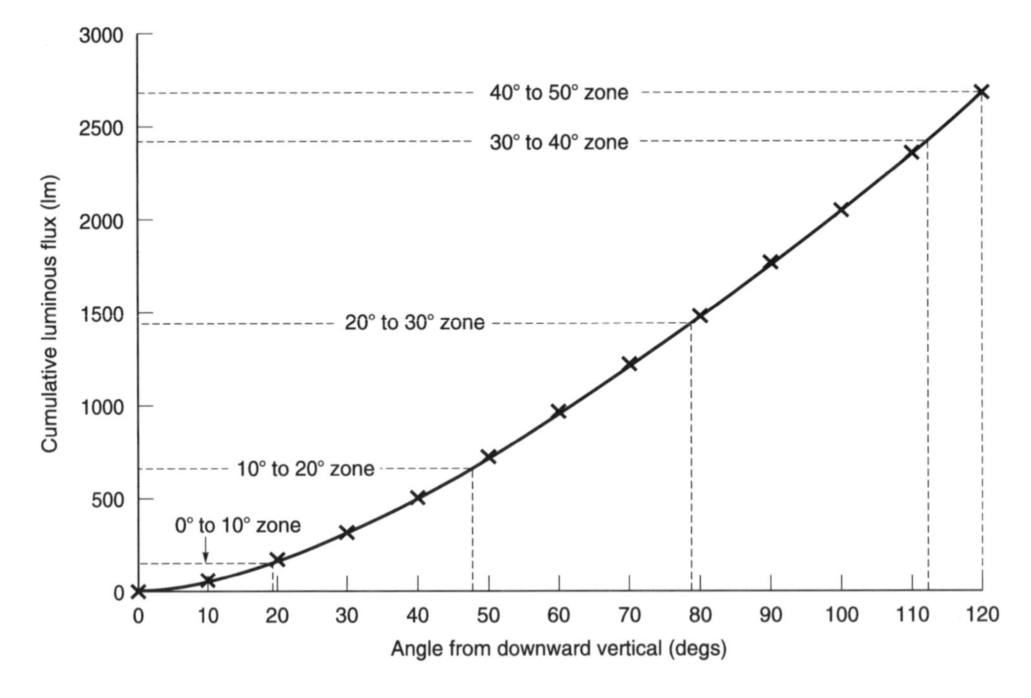

Fig. 6.28 Cumulative luminous flux distribution

Table 6.2 Determination of the inclination of the contour elements

Limits of zone of reflector (deg)	Mid-zone angle γ (deg)	Angle into which light must be redirected δ (deg)	Inclination of element to vertical $\theta = \frac{1}{2}(\gamma - \delta)$ (deg)
50–60	55	45	5
60–70	65	35	15
70–80	75	35	20
80–90	85	35	25
90–100	95	25	35
100–110	105	25	40
110–120	115	25	45
120–130	125	15	55
130–140	135	15	60
140–150	145	15	65
150–160	155	5	75
160–170	165	5	80

170° to 160° zone and the 160° to 150° zone in column K. These values are entered in Table 6.2. Hence δ is 5° and the values for γ are 165° and 155°. The next zone for reflected light is that up to 20°, for which it is evident from column J that 680 lumens are required. The horizontal line in the graph is drawn at this figure and it can be read that cumulative luminous flux from 180° to nearly 120° is required to satisfy this requirement. δ is 15° and γ has three values; 145°, 135° and 125°, which are entered in Table 6.2. This procedure is continued until all the reflected luminous flux is accounted for.

Table 6.2 enables us to generate the reflector contour as we now know the inclination of all the reflector elements to the vertical.

(4) *Generation of the reflector contour*
Starting at 50° to the downward vertical, we draw radial lines at 10° intervals (Figure 6.29). The first element is drawn at 5° to the vertical between the 50° and 60° radial lines at a convenient distance from the source. The other elements are then put in as indicated. A curved contour can be drawn tangentially to all the elements. Finally, the contour can be scaled to give the reflector a convenient size in relation to the lamp.

Adequate ventilation and easy relamping must be allowed for.

Linear sources

Where a trough reflector is to be designed for a linear light source, the procedure is similar to that for the axially symmetric reflector. The important difference is that the zones become sectors (Section 4.4, page 134), and the luminous flux in a sector is proportional to the luminous intensity in the vertical plane perpendicular to the axis of the reflector.

Additional comments on the method

This method gives a systematic approach to achieving a given luminous intensity distribution. It is a powerful method and it can be more useful than many optical ray tracing programs, which simply enable the performance of an existing reflector to be checked.

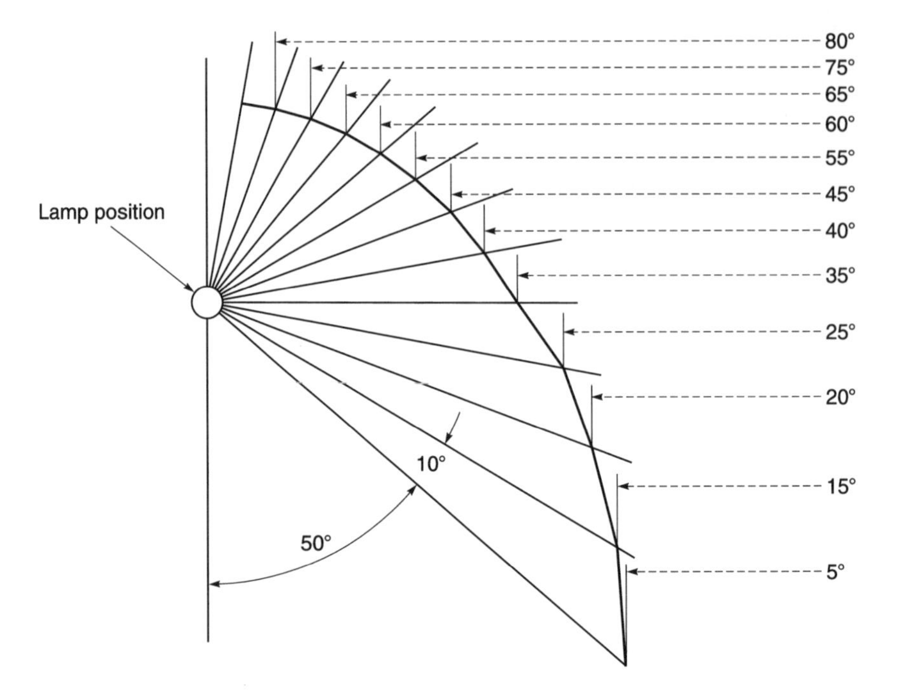

Fig. 6.29 Generation of the reflector contour

Its greatest drawback is that the lamp is regarded as a point or linear source of light. This assumption can lead to serious errors where the assumption is not justified. The effect of the lamp, not only in obstructing light but in redirecting it, can also lead to serious errors.

In the example reflector described above, it was assumed that the light distribution required was to be symmetrical about the vertical axis. Where this is not the case, the method can be extended by dividing each zone into a number of sections by means of vertical planes. Once a computer program or spreadsheet is set up to do this, the process becomes very quick.

If required, smaller angular zones than the 10° used in the example can be taken.

6.6 Metallic light guides using specular reflection

Light guides are used for transporting and distributing light over distances.[8] In this section, we will describe light guides that depend for their main functioning on specular reflection from metallic surfaces; however later (Section 6.8.12, page 265) we will describe light guides where the main light control is by refractive elements. The light losses in metallic light guides are greater than in the refractive types and their main application is in transporting daylight, where the losses are not so critical.

Basically, metallic light guides consist of hollow pipes with specular interior reflecting surfaces, with a collector at one end for directing the light into the pipe and a distributor at the other end. Where light from the sun is to be collected, what is called a dynamic collector (as opposed to a passive collector) is needed to follow the sun to obtain the most effective means of collection. Such a mechanism is expensive and some thought has been given to using other means of collecting the light. One example is illustrated in Figure 6.30 where a prismatic panel is used to increase the angle of incidence of the light on the pipe, so reducing the number of

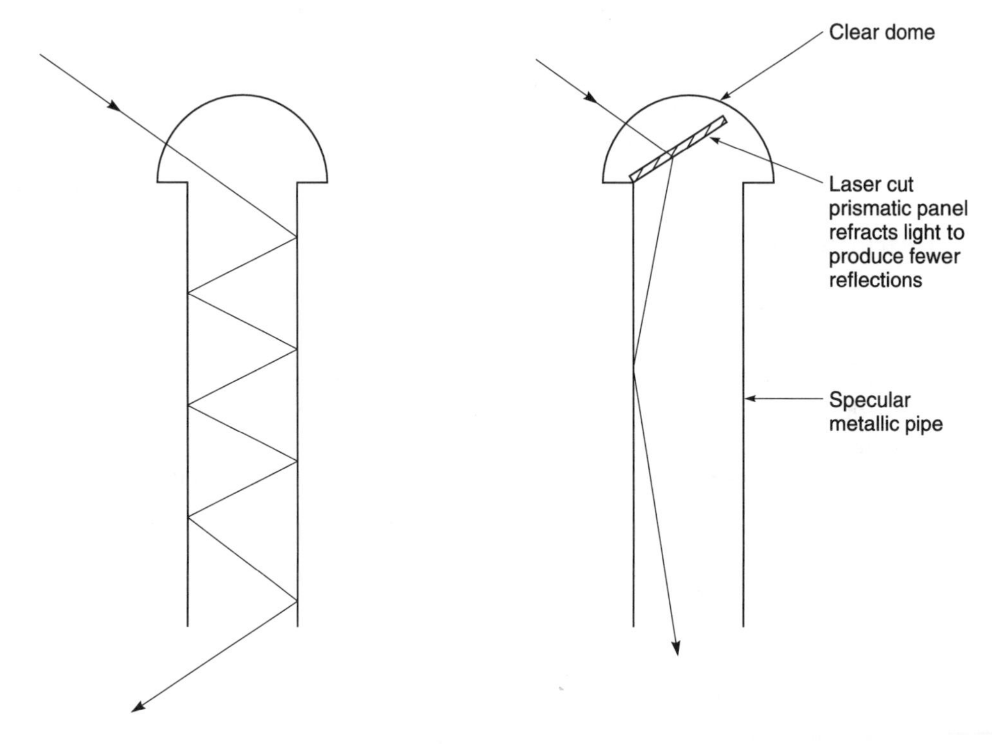

Fig. 6.30 Use of a prismatic panel to reduce the number of reflections in a light pipe (after Edmonds *et al.*)

reflections needed for it to reach the exit.[9] At the other end of the guide relatively conventional means, such as reflectors and refractors, are used to distribute the light.

6.7 Diffuse reflection and transmission

Diffuse reflection and diffuse transmission are achieved by means of white paint, white enamel finishes, or opal plastics. The distribution of light from such surfaces approximately obeys the cosine law for uniform diffusers, with a small specular component superimposed on this. The magnitude of this component will depend on how glossy the surface is. The specular component usually has a negligible effect on the light distribution from luminaires; the purpose of the glossy finish is to make it more durable and to make cleaning easier rather than to control the light distribution. It is worth noting that even the most matt finish will reflect light specularly at grazing angles of incidence.

Except for opal diffusers, the reflectance of the finish should be as high as possible to achieve a high light output ratio. Opal diffusers should have a sufficiently high reflectance to hide the lamp (if this is a requirement).

The light distribution of luminaires with diffusing finishes can be predicted by adopting the following procedure. The light output ratio is found by using the interreflection theory in Chapter 5, page 169. The shape of the polar curve due to reflected light, and transmitted light in the case of opal diffusers, is found by determining the projected areas of the flashed surfaces at appropriate angles of elevation. This curve is scaled so that it represents the luminous flux output of the luminaire less the bare lamp luminous flux emerging directly from the luminaire. The polar curve representing the directly emerging luminous flux is added to the polar curve of the reflected and transmitted light to give the final polar curve. The opposite process of deducing the reflector or diffuser shape to give a required light distribution is not usually undertaken as a mathematical procedure. Reflectors are usually designed to give a certain shielding angle from the lamp, and the size made large enough to give a good light output ratio and to obviate overheating.

6.8 Refractor systems

6.8.1 THE LAWS OF REFRACTION

When a ray of light passes through a boundary between two media of different optical density, such as air and glass, it changes its direction of travel, or is refracted. We can make use of this refraction of the light in designing optical systems for controlling the distribution of luminous flux from a light source.

The following two laws enable the new direction of the light to be found (Figure 6.31).

(a) The refracted ray lies in the same plane as the normal to the boundary and the incident ray, and is on the opposite side of the normal to the incident ray.
(b) The sine of the angle of incidence (i) bears a constant ratio to the sine of the angle of refraction (r) for light of a given wavelength (Figure 6.31(a)), that is,

$$\sin i = \mu_{1,2} \sin r \tag{6.29}$$

where $\mu_{1,2}$ is a constant of proportionality known as the *refractive index* of medium 1 with respect to medium 2.

This is known as Snell's law. For the applications described in this book medium 1 is invariably air. Since the refractive index of a medium measured in air differs by only 0.03% from the *absolute* refractive measured in a vacuum we can use the absolute refractive index without incurring any significant error. For convenience, we can then write the equation as

$$\sin i = \mu \sin r \tag{6.30}$$

where

i is the angle the ray travelling in air makes with the normal;
μ, without any subscripts, is the refractive index of the medium in air. Refractive indices of some common transparent materials are shown in Table 6.3;
r is the angle the ray travelling in the optical medium makes with the normal.

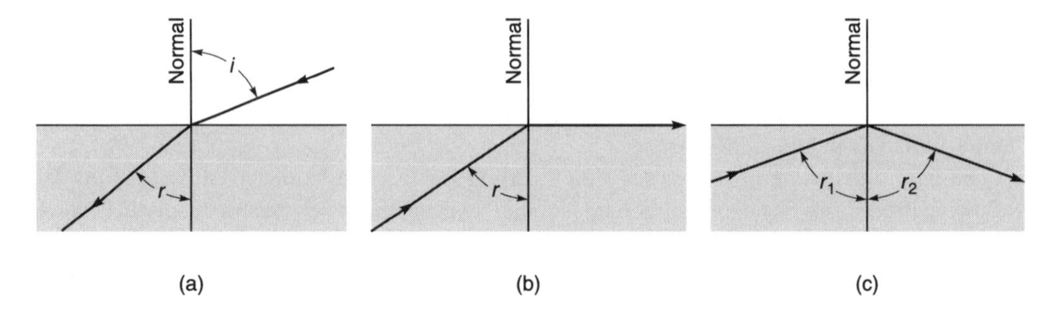

Fig. 6.31 Refraction and total internal reflection of a ray at a boundary

Table 6.3 Refractive indices and critical angles of some transparent media

Optical medium	Refractive index	Critical angle (deg)
Water	1.33	49
Acrylic	1.49	42
Soda glass	1.52	41
Polystyrene	1.59	39
Flint glass	1.62	38

For refraction at curved surfaces, i and r are measured from the normal to the surface at the point of entry of the ray at the boundary, as in Figure 6.53, page 263.

If the direction of a ray is reversed it will follow the same path as before. In other words, in Figure 6.31, the direction of the arrows can be reversed. This is known as the *principle of reversibility*, which can be used to simplify some problems.

Internal reflection and the critical angle

A consequence of Snell's law is that a ray of light is bent towards the normal when passing into an optically dense medium, but when emerging it is bent away from the normal. If the angle of refraction is increased until the emergent ray is parallel to the boundary, the angle of refraction becomes the *critical angle* (Figure 6.31(b)). In this case the angle of emergence (i) is 90°, so that

$$\text{critical angle} = \sin^{-1}\frac{1}{\mu} \tag{6.31}$$

Values of the critical angle for some common optical media are tabulated in Table 6.3.

When the angle of refraction is increased to values greater than the critical angle, *total internal reflection* takes place (Figure 6.31(c)) and the normal laws of reflection are obeyed, that is

$$r_1 = r_2 \tag{6.32}$$

where r_1 and r_2 are angles of reflection.

Reflection at boundaries

We have stated that reflection occurs when the critical angle is exceeded, but in fact a certain proportion of the light is always reflected when passing through a boundary between two media. The proportion depends on the plane of polarization of the light (that is, the plane at right angles to the plane in which the light is vibrating) with regard to the plane of incidence (the plane of the normal and the incident ray), and on the angles of reflection and refraction.

For the plane of polarization at right-angles to the plane of incidence, Fresnel was able to show that the reflectance ρ_1 is given by

$$\rho_1 = \tan^2 (i - r)/\tan^2 (i + r) \tag{6.33}$$

He was also able to show that when the plane of polarization is parallel to the plane of incidence, the reflectance ρ_2 is given by

$$\rho_2 = \sin^2 (i - r)/\sin^2 (i + r) \tag{6.34}$$

Table 6.4 Reflectance at boundaries

Angle to normal (deg)	Refractive index = 1.5		Refractive index = 1.6	
	Reflectance for external angles	Reflectance for internal angles	Reflectance for external angles	Reflectance for internal angles
0	0.000	0.000	0.000	0.000
5	0.040	0.040	0.053	0.053
10	0.040	0.040	0.053	0.053
15	0.040	0.040	0.053	0.054
20	0.040	0.042	0.054	0.056
25	0.041	0.045	0.054	0.062
30	0.042	0.055	0.055	0.080
35	0.043	0.086	0.057	0.151
40	0.046	0.245	0.060	(1.0 at 38°)
45	0.050	(1.0 at 42°)	0.064	1.000
50	0.058	1.000	0.072	1.000
55	0.070	1.000	0.085	1.000
60	0.089	1.000	0.105	1.000
65	0.121	1.000	0.137	1.000
70	0.171	1.000	0.188	1.000
75	0.253	1.000	0.270	1.000
80	0.388	1.000	0.402	1.000
85	0.613	1.000	0.622	1.000
90	1.000	1.000	1.000	1.000

When the light is unpolarized the light vibrates equally in all planes so the reflectance is given by

$$\rho = \tfrac{1}{2}(\rho_1 + \rho_2) \tag{6.35}$$

Table 6.4 shows how reflectance (expressed as a fraction of unity) varies according to the angle the ray makes with the normal.

The reflection of light at boundaries is important to the designer, who must be careful not to make the angles of incidence so great that the losses and the scattering of light into directions where it is not wanted become significant.

Worked example A source of luminance 500 cd/m^2 is viewed through a parallel sided sheet of glass having a refractive index of 1.5. If the angle of view to the normal is 30° and the absorption in the glass can be ignored, find the luminance of the source through the glass.

Solution Since the light paths are reversible we may take the angle of incidence to be 30°. Hence

$$r = \sin^{-1}\left(\frac{1}{\mu}\sin i\right)$$

$$= \sin^{-1}\left(\frac{1}{1.5} \times \sin 30\right)$$

$$= 19.5°$$

From Table 6.4 we find that, for a medium of refractive index 1.5, the reflectance is approximately 0.042 when the external angle is 30°. For an internal angle of 19.5° the reflectance is also approximately 0.042. Hence, the combined losses at the boundaries, expressed as decimal fractions are given by

$$\begin{aligned}\text{fractional light loss} &= \rho_e + \rho_i(1 - \rho_e)\tau \\ &= 0.042 + 0.042(1 - 0.42) \\ &= 0.082\end{aligned}$$

where

ρ_e is the reflectance of the optical material for external surfaces;
ρ_i is the reflectance of the optical material for internal surfaces;
τ is the transmittance of the optical material.

The luminance L as seen through the sheet of glass is, therefore,

$$\begin{aligned}L &= \text{transmittance of glass sheet} \times \text{luminance of source} \\ &= (1 - 0.082) \times 500 \\ &= 459 \text{ cd/m}^2\end{aligned}$$

Comments Very often glass has a green tint due to the presence of iron salts. This coloration can be seen when the glass is viewed through its edge. This will give a further reduction in the light transmitted, the amount depending on the depth of its coloration and the spectral power distribution of the lamp. It is possible to obtain glass that is 'white' but, in the United Kingdom, this is expensive. Plastics free of tinting are readily available.

Some of the reflected light may travel back into the optical control compartment of a system and have a second chance of being emitted, but this is not usually in useful directions.

Polarization by reflection and transmission

As has already been indicated in quoting Fresnel's formulae, reflectance and polarization are linked. If light is incident at an angle, known as Brewster's angle, such that

$$i + r = 90° \tag{6.36}$$

then

$$\sin i = \mu \sin (90° - i) \tag{6.37}$$

so that

$$\tan i = \mu \tag{6.38}$$

and by inserting $i + r = 90°$ in (6.33), page 235, we also obtain

$$\rho_1 = 0 \tag{6.39}$$

This means that all the light that is reflected is completely polarized in the plane parallel to the plane of incidence when the angle of incidence is at Brewster's angle, which is approximately 56° for a refractive index of 1.5. The refracted light is partially polarized in a plane perpendicular to the plane of incidence, and the degree of polarization can be increased by using a stack of plates. Diffusers for luminaires which produce a significant amount of polarized light at Brewster's angle, have been made by embedding glass flakes in a resin. These polarizing materials are known as multilayer polarizers.[10] They are claimed to reduce significantly veiling glare caused by light reflected from non-electrically conducting materials (for example paper and plastics but not metals) and so improve visibility. Light emitted from fluorescent tubes in axial planes is polarized as a result of travelling through the glass envelope at and near Brewster's angle, and the same effect may be increased by a clear enclosure.

Luminance of an image formed by a refractor system

The luminance of an image formed by a refractor system is

$$L = \tau L_o \tag{6.40}$$

where L is the luminance of the image; τ is the transmittance of the refractor system; L_o is the luminance of thc object.

The transmittance is defined as the ratio of the luminous flux transmitted by a material to the incident luminous flux.

We will now show that (6.40) is true for a lens.

In Figure 6.32 the object is at a distance u from the lens. An image is formed at a distance v from the lens and viewed from beyond it, that is from the right-hand side in the diagram.

Let the area of the lens be A. If this is small in relation to u then the solid angle subtended by a point on the object is A/u^2. If a is the area of the object, a small area of it, δa, will have a luminous intensity of $L_o\delta a$, since luminance is measured in luminous intensity per unit area.

The luminous flux reaching the lens is given by the luminous intensity multiplied by a solid angle, that is

$$L_o\delta a \times \frac{A}{u^2} \tag{6.41}$$

It should be noted that this could also have been arrived at by using the inverse square law of illuminance.

If τ is the transmittance of the lens then the luminous flux emerging from the lens will be

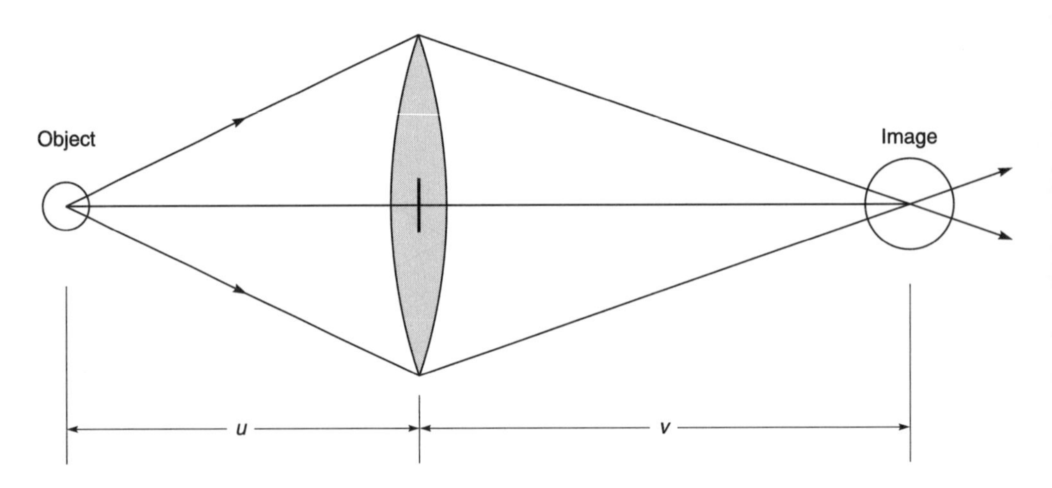

Fig. 6.32 Formation of the image by lens

$$L_o \frac{A}{u^2} \tau\delta a \tag{6.42}$$

From the diagram it is apparent that this is concentrated into a solid angle equal to A/v^2. Hence the luminous intensity of the image is

$$L_o \frac{A}{u^2} \tau\delta a \div \frac{A}{v^2} = L_o \frac{v^2}{u^2} \tau\delta a \tag{6.43}$$

Now the area of the image of δa is

$$\frac{v^2}{u^2} \delta a \tag{6.44}$$

So the luminance L of the aerial image (luminous intensity ÷ area of image) is given by

$$L = L_o \frac{v^2}{u^2} \tau\delta a \div \frac{v^2}{u^2} \delta a$$

$$= \tau L_o \tag{6.45}$$

which proves the proposition.

This is true for an aerial image but does not apply to the illuminance of an image formed on a screen, as will be shown next.

Equation (6.42) gives us the flux falling on the image and, to find the illuminance of the image, we have to divide this by the area of the image, which is given in (6.44). Hence the illuminance E_i of the image is given by

$$E_i = L_o \frac{A}{u^2} \tau\delta a \div \frac{v^2}{u^2} \delta a$$

$$= L_o \frac{A}{v^2} \tau \tag{6.46}$$

In a camera the image is usually formed near the focal plane, so that v is approximately equal to f, the focal length, and the formula becomes $L_o(A/f^2)\tau$. The *f-number* of the aperture of the lens is the diameter of the aperture divided by the focal length, that is, it is proportional to the reciprocal of $\sqrt{(A/f^2)}$.

From (6.46) it is obvious that the illuminance of the image can be increased by increasing the lens aperture. However, in the case of the aerial image the effect of the larger aperture is simply to increase the divergence of the cones of light leaving the image. The additional luminous flux collected would not enter the eye providing the original cone of light was sufficiently large to cover the pupil. Hence, there is no increase of image luminance with lens size.

Deviation of a ray at a boundary

In the calculation of prism angles, the deviation $(i - r)$ at one of the faces is often known and the angle of incidence (i) has to be found. A formula for this purpose can be derived as follows:

$$\sin i = \mu \sin r$$
$$= \mu \sin [i - (i - r)]$$
$$= \mu[\sin i \cos (i - r) - \sin (i - r) \cos i]$$
$$\text{so,} \quad \tan i = \mu[\tan i \cos (i - r) - \sin (i - r)] \tag{6.47}$$

and by rearranging the terms we obtain

$$\tan i = \frac{\mu \sin (i - r)}{\mu \cos (i - r) - 1} \tag{6.48}$$

Parallel sided block Figure 6.33(a) shows that a ray travelling through a parallel sided block is not deviated in angle. This is because the angles of refraction, r and r', are alternate and are equal. However, the ray is displaced sideways and we can find the amount of displacement from Figure 6.33(b). The ray emerging from the block at C is extended back to meet in B the perpendicular at A on the incident ray. Then AB is the displacement.

Now $AB = AC \sin (i - r)$, but $AC = d/\cos r$, where d is the thickness of the block, therefore,

$$AB = \frac{d \sin (i - r)}{\cos r} \tag{6.49}$$

Linear displacement combined with angular deviation can give rise to sharp changes in luminous intensity with angle when the bend in a bowl of a luminaire is aligned with a relatively small source such as the arc tube of a high pressure sodium lamp. This unwanted effect becomes more pronounced as the thickness of the material is increased and is present even if the bowl is of uniform thickness. It can be reduced by angling the axis of the arc tube with respect to the bend and by making the radius of the bend as large as possible. It can be eliminated by placing the arc tube at the centre of curvature, but this is seldom possible as it results in a substantially semicircular bowl, which may not be acceptable for aesthetic reasons.

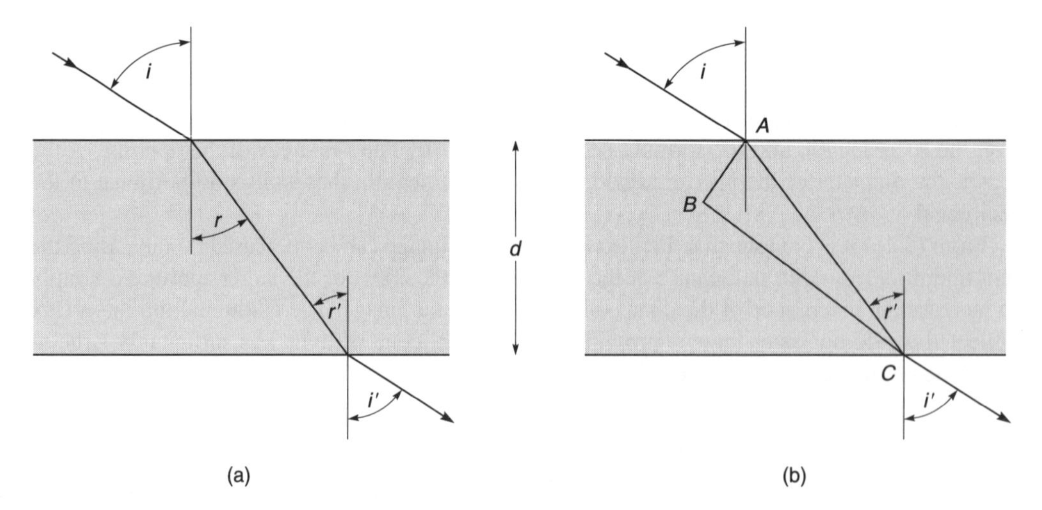

Fig. 6.33 Displacement of a ray in a parallel sided block

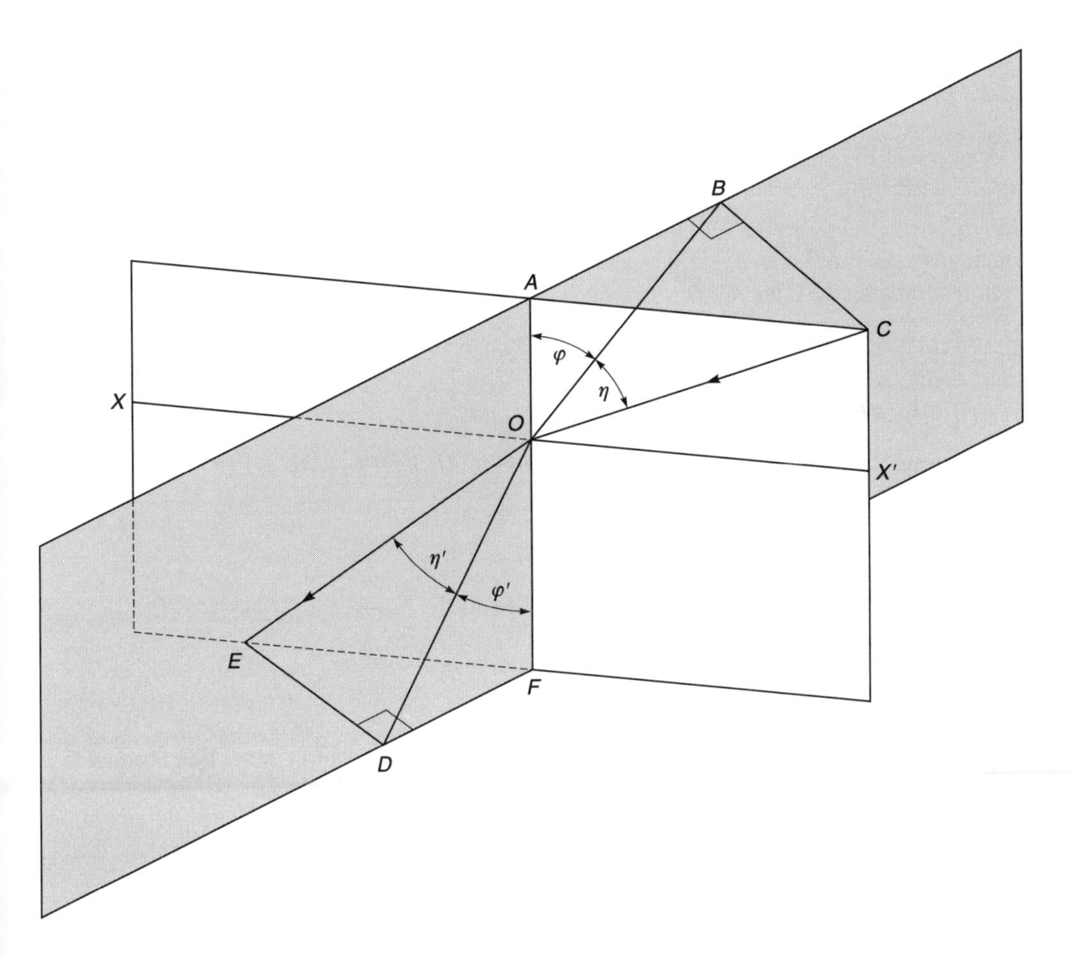

Fig. 6.34 Projection of a ray on an inclined plane

Projection of ray on an inclined plane through the normal

If a trace of a ray through a block is projected on an inclined plane through the normal, a result is obtained that is useful in tracing rays through prisms in planes other than the principal plane – the plane at right angles to the surfaces of the prism. Figure 6.34 will make clear how this is done.

The ray COE is travelling in the plane $ACX'FEX$. XX' represents the boundary between the air and lower optical medium of refractive index μ. AOF is a normal to the boundary. The inclined plane $ABFD$ passes through AOF. BOD is the projection of the ray on the inclined plane $ABFD$, so that angles ABC and EDF are right-angles.

Let $\angle BOC = \eta$, $\angle EOD = \eta'$, $\angle BOA = \phi$ and $\angle DOF = \phi'$.

By construction, make

$$\mu CO = OE \tag{6.50}$$

so $\sin \angle COA = \mu \sin \angle EOF$.

Therefore, using (6.50) we find

$$\frac{AC}{CO} = \mu \frac{EF}{OE}$$

$$= \frac{EF}{CO} \tag{6.51}$$

which gives $AC = EF$.

In the triangles ABC and EFD,

$$\begin{aligned} \angle CAB &= \angle EFD \\ \angle ABC &= \angle EDG \; (= 90°) \\ AC &= EF \end{aligned} \tag{6.52}$$

Therefore, the triangles are equal in all respects, so $AB = DF$ and $BC = ED$:

$$\sin \eta = \frac{BC}{CO}$$

$$= \mu \frac{ED}{OE}$$

$$= \mu \sin \eta' \tag{6.53}$$

Hence, the sine of the angle between the ray and its projection before refraction is equal to μ times that after refraction.

Also,

$$\sin \phi \cos \eta = \frac{AB}{BO} \times \frac{BO}{CO}$$

$$= \mu \frac{AB}{OE} \quad \text{(as } \mu CO = OE\text{)}$$

$$= \mu \frac{DF}{OD} \times \frac{OD}{OE} \quad \text{(as } AB = DF\text{)}$$

$$= \mu \sin \phi' \cos \eta' \tag{6.54}$$

The angle BAC between the two planes in Figure 6.34 can be derived as follows:

$$\tan \angle BAC = \frac{BC}{AB}$$

$$= \frac{BC}{OB} \times \frac{OB}{AB}$$

$$= \frac{\tan \eta}{\sin \phi} \tag{6.55}$$

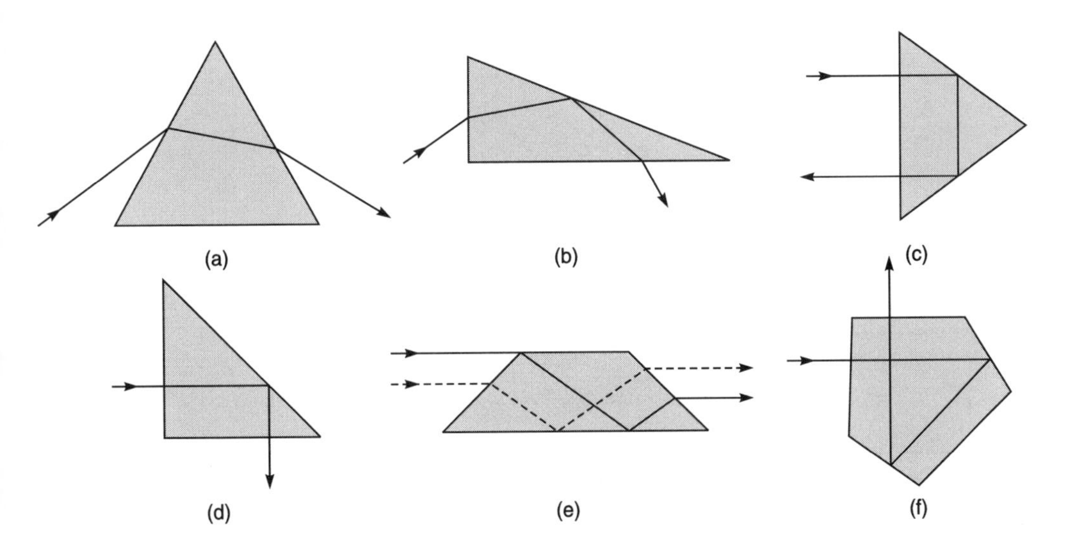

Fig. 6.35 Some common prism forms

6.8.2 PRISM SYSTEMS

Prisms are classed as refracting or reflecting according to whether they make use solely of refraction or make use of total internal reflection as well as refraction.

Figure 6.35 shows some common forms but the angles employed have to be calculated to suit the application. Figure 6.35(a) is a refracting prism and it is this which is commonly used in prism banks in luminaires. Figure 6.35(b) is a reflecting prism, which is also used in luminaire prism banks, and is capable of deviating the light through a greater angle than (a). Figure 6.35(c) is a right-angled reflecting prism that has its principal use in optical instruments, as does Figure 6.35(d). Figure 6.35(e) inverts an image. Finally, Figure 6.35(f) is a pentaprism which turns the light rays through three right-angles and is capable of producing an erect image in cameras. In this the reflecting surfaces may have to be silvered.

Refracting prisms

Figure 6.36 shows the path of a ray through a refracting prism. XX' and the Y ordinates are put in for reference. The ray of light is shown leaving the source at an angle P to Y_1Y_1. The most commonly occurring problem is to find the prism angle A to achieve a specific deviation, D. Usually either the angle B or the angle C is fixed so that the prism can follow the contour of a luminaire. The slope of the base in relation XX' does not affect the deviation of the ray, but it does affect how much of the prism is flashed. This consideration is dealt with in Section 6.8.6, page 252.

For tracing the path of a ray through a refracting prism the following equations may be used:

$$\begin{aligned}
&(1)\ D = (i - r) + (i' - r') \\
&(2)\ D = P - Q \\
&(3)\ A = B + C \\
&(4)\ A = r + r' \\
&(5)\ i' = 90^\circ + C - Q \\
&(6)\ i = B + P - 90^\circ
\end{aligned} \qquad (6.56)$$

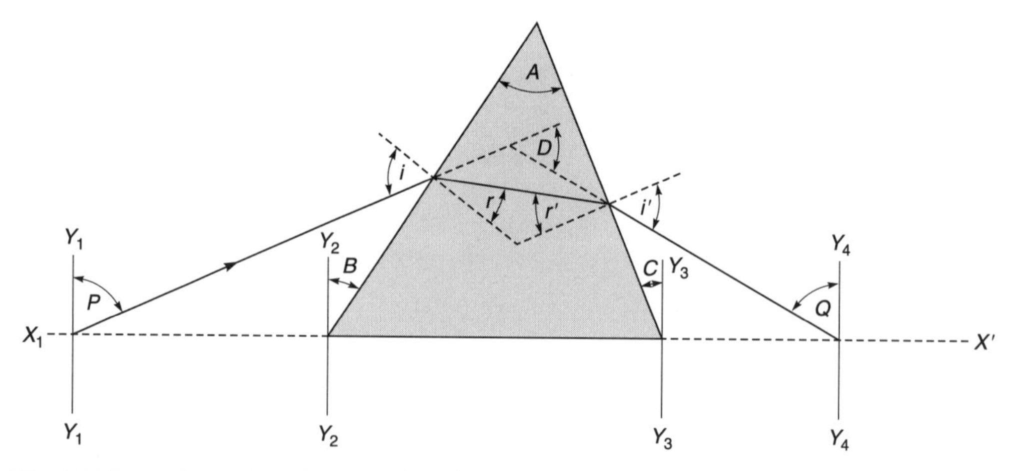

Fig. 6.36 Trace of a ray through a refracting prism

If the angle of incidence, and therefore the angle of refraction, lies on the other side of the normal to that shown on the diagram, then it and the angle of refraction are taken to be negative. D can never be negative since light is always deviated towards the base of the prism.

Maximum deviation

The maximum deviation D is an important consideration when a prism bank is being designed. This is obtained when i and i' each have their maximum practical value, which – by reference to Table 6.4 – can be seen to be approximately 70° before the reflection losses become too great. From this

$$\begin{aligned} D &= (i - r) + (i' - r') \\ &= \left[70° - \sin^{-1}\left(\frac{\sin 70°}{1.5} \right) \right] \times 2 \qquad (6.57) \\ &= 62° \end{aligned}$$

Reflecting prisms

We now turn to the calculation of reflecting prisms. As has already been mentioned these have the advantage over refracting prisms that they can deviate the light through a greater angle. They suffer from the drawback that they are liable to give inaccurate control since their functioning depends on the accuracy of all three faces, and not just two and, moreover, since one face operates by reflection, any angular inaccuracy in this face is doubled (see Figure 6.5, page 207).

Figure 6.37 shows the trace of a ray through a reflecting prism UVW. XX' and the Y ordinates are marked for reference in a similar way to that used for the refracting prism. The ray $GLSMH$ leaves the light source at G at an angle P to the downward vertical Y_1Y_1 and enters the face UV at the point L. It is reflected from the face UW at the point S at an angle θ to the normal and leaves the prism through the face WV at the point M, making an angle Q with the downward vertical Y_5Y_5 and passes through the point H.

Making use of the fact that the angles LSU and WSM are equal, we can find the following relations between the various marked angles:

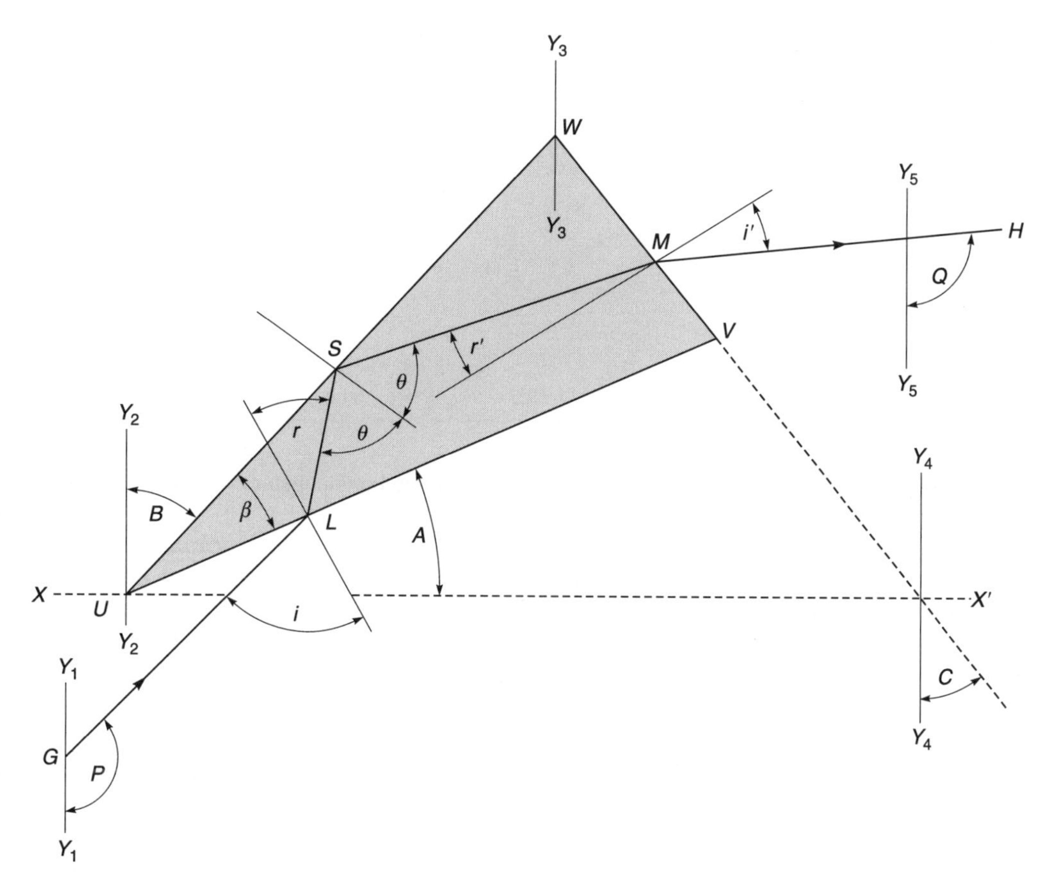

Fig. 6.37 Trace of a ray through a reflecting prism

$$
\begin{aligned}
&(1)\ i = A - P + 180^\circ \\
&(2)\ i' = C - Q + 90^\circ \\
&(3)\ B = \tfrac{1}{2}(90^\circ - C - A + r + r') \\
&(4)\ D = P - Q \\
&(5)\ \beta = 90^\circ - A - B \\
&(6)\ \theta = 90^\circ - B - A + r
\end{aligned}
\qquad (6.58)
$$

where D is the angle of deviation produced by the prism and is equal to the angle between GL and MH.

Relation (3) above follows from the fact that $\angle SUL + \angle SLU = \angle SWM + \angle SMW$, acute angle $SWY_3 = B$, and acute angle $VWY_3 = C$.

These relations enable B to be found once C and A have been set.

Usually the face UV is sloped below XX' to avoid under-cutting the prisms and to make manufacture easier. The formulae given still apply but A is negative.

Worked example Given in Figure 6.37 that P = 140.0°, i = 20.0°, and C = 10.0°, find the remaining prism angles and θ.

Answer The work can be set out in a spreadsheet as in Table 6.5.

Table 6.5 Spreadsheet for calculating the angles of a reflecting prism

Refractive Index $\mu = 1.50$

P	Q	i	r $\sin^{-1}(\sin i/\mu)$	C	i' $C - Q$	r' $\sin^{-1}(\sin i/\mu)$	A $i + P - 180°$	B $\frac{1}{2}(90° - C - A + r + r')$	θ $90° - B - A + r$	β $90° - A - B$	D $P - Q$
140.0	70.0	20.0	13.2	10.0	30.0	19.5	–20.0	66.4	56.8	43.6	70.0
140.0	70.0	30.5	19.8	10.0	30.0	19.5	–9.5	64.4	54.9	35.1	70.0

The refractive index is put in a separate cell so that it can be changed easily, without having to change the formulae in the body of the table.

The first line of calculation gives the values of the required angles and these have been drawn in Figure 6.38. To make sure that the ray does not pass through the face *UV*, it is essential to check that θ is greater than the critical angle, which is true in this case.

A ray entering at the apex *U* of the prism will travel parallel to *SM* and leave the prism at *D*. This means that the part *DV* will be unflashed. To overcome this we can alter the slope of the face *UV* so that it is parallel to *SM* or it converges towards *SM*.

The condition for *UV* to be parallel to *SM* can be found as follows:

$\angle MSW = 90° - B - A$, since *UV* is to be made parallel to *SM*.

In the triangle *SMW*,

$$\angle MSW + \angle SMW + \angle SWM = 180° \tag{6.59}$$

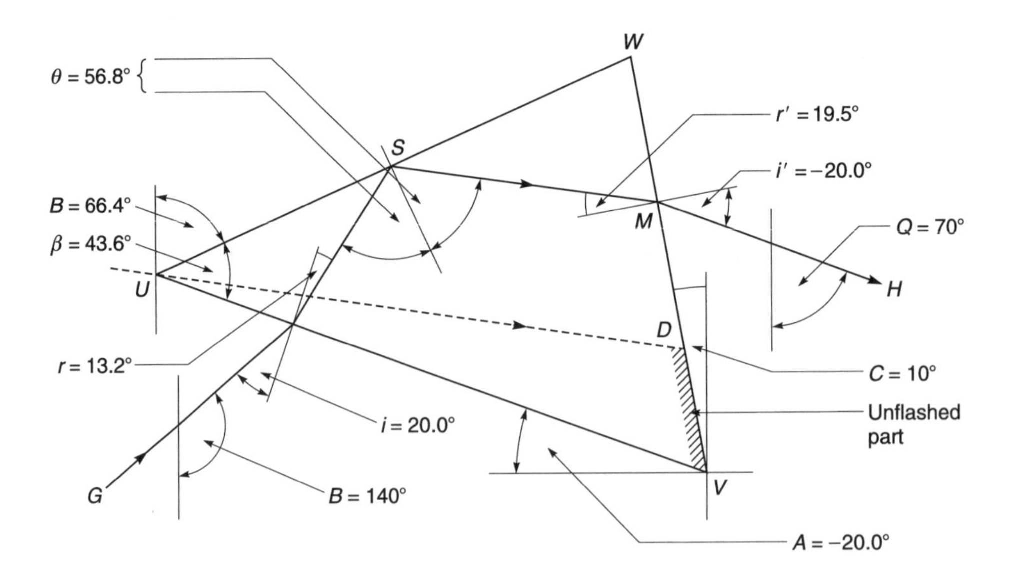

Fig. 6.38 Trace of a ray through a reflecting prism

Therefore,

$$(90° - A - B) + (90° - r') + (B + C) = 180° \quad (6.60)$$

giving

$$(7)\ A = C - r' \quad (6.61)$$

For our problem

$$\begin{aligned} A &= C - r' \\ &= 10° - 19.5° \\ &= -9.5° \end{aligned} \quad (6.62)$$

Now from Relation (1)

$$\begin{aligned} i &= A - P + 180.0° \\ &= -9.5° - 140.0° + 180.0° \\ &= 30.5° \end{aligned} \quad (6.63)$$

This new value of *i* has been inserted in the second line of calculation in Table 6.5 and the prism angles reworked. Figure 6.39 shows the paths of the rays through the prism.

It will be seen that *SM* in the ray *GLSMH* is now parallel to the base of the prism. The figure also shows the trace of a ray *glsmh*, which just misses the apex of a lower prism of the same dimensions as triangle *UVW*. The masking effect of this lower prism is such that *mW* is not flashed. Moreover, it should be noticed that there is the possibility that rays entering the prism

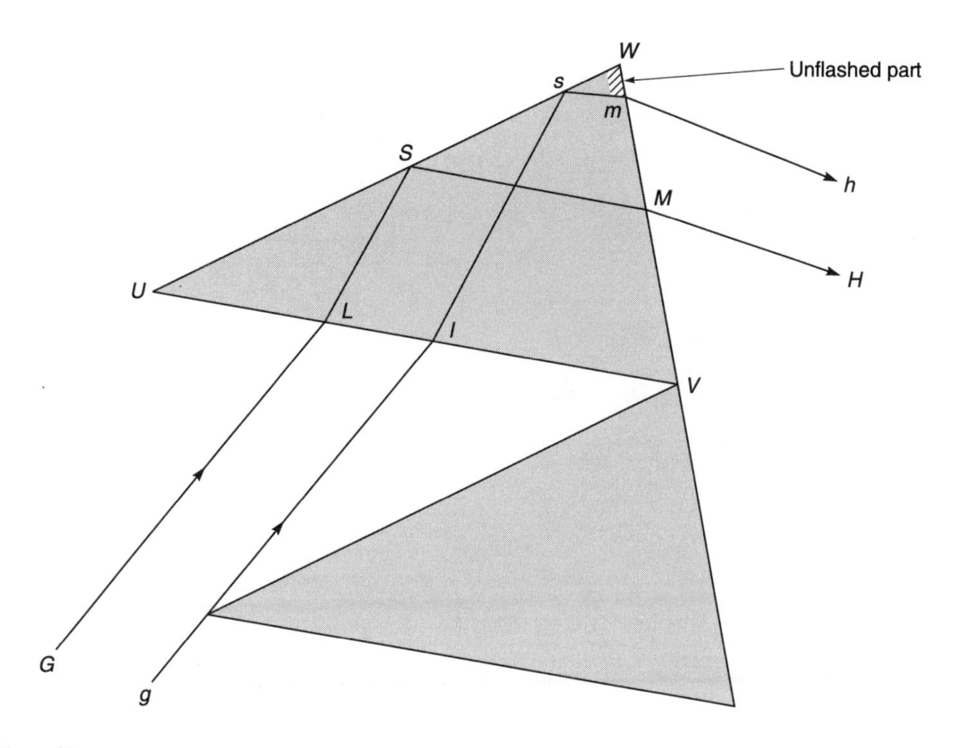

Fig. 6.39 Paths of rays through reworked prisms

at a less steep angle than shown can be refracted upwards, since the prism then acts as a refracting prism. This can have undesirable consequences, especially in street lighting, where the upwardly directed light could cause light pollution.

6.8.3 TRACING RAYS THROUGH A PRISM IN PLANES OTHER THAN THE PRINCIPAL PLANE

So far we have traced rays in a plane that is at right angles to the surfaces of the prism, which we will designate the *principal plane*. We will now consider the path of rays in other planes.

Figure 6.40(a) shows a ray of light *ABDC* entering and emerging from the prism. η and η' are the angles the entering ray *AB* makes with its projection on the principal plane (refer to Figure 6.34, page ??? for further clarification of these symbols) and ϕ is the angle the projection makes with the normal *BN* to the top face of the prism. Similarly, ξ, ξ', and ψ are the corresponding angles for the emerging ray.

η' and ξ' are equal since these are the angles that the ray in the prism makes with the principal plane. Hence, from (6.53), η and ξ are equal. Thus, the incident and emergent rays make the same angle with the principal plane. The equations are:

$$\begin{aligned} \sin\eta &= \mu \sin\eta' \\ \sin\xi &= \mu \sin\eta \\ \eta' &= \xi' \text{ and } \eta = \xi \end{aligned} \tag{6.64}$$

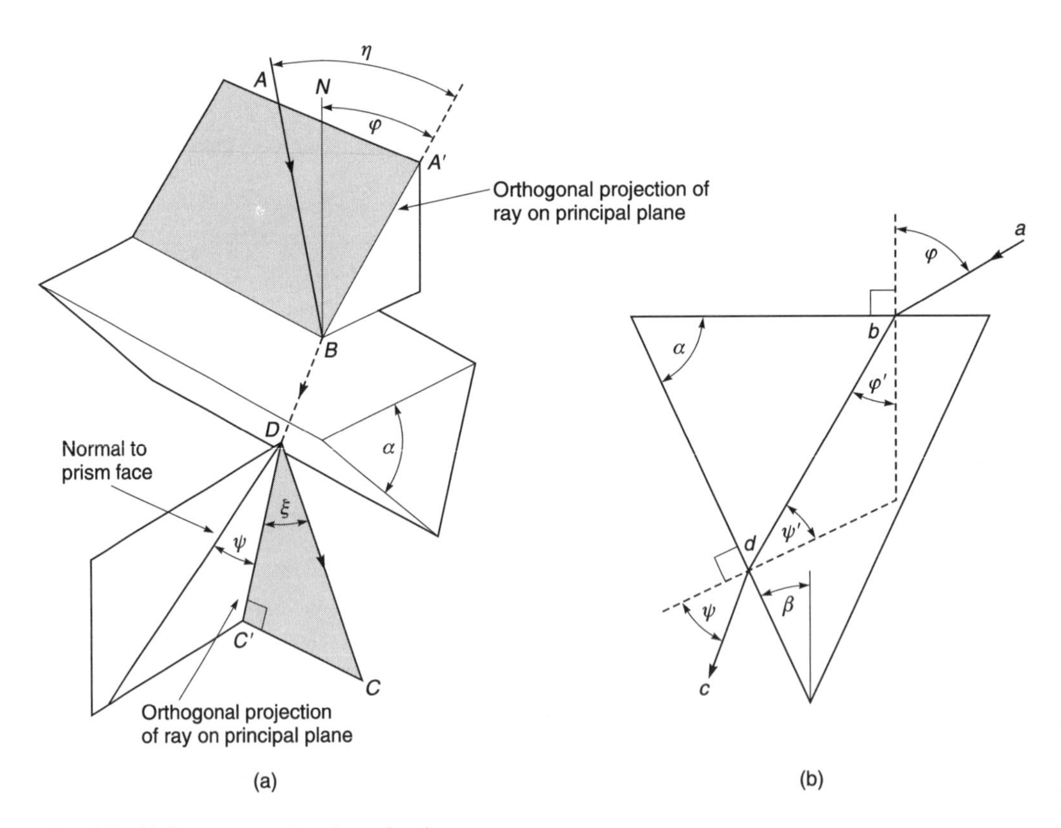

Fig. 6.40 Oblique ray passing through prism

Figure 6.40(b) shows the rays of light projected onto a principal plane. The points in A, B, D and C in Figure 6.40(a) project as points a, b, d and c in Figure 6.40(b). ϕ and ϕ' are the angles the projections make with the normal before and after refraction. In addition, ψ and ψ' are the angles the projections of the emergent ray make with the normal before and after refraction.

Then, from (6.54),

$$\sin \phi \cos \eta = \mu \sin \phi' \cos \xi' \tag{6.65}$$

and

$$\sin \psi \cos \eta = \mu \sin \psi' \cos \xi' \tag{6.66}$$

Also, from Figure 6.40(b),

$$\alpha = \phi' + \psi \tag{6.67}$$

These equations enable a ray to be traced through a prism, and we can use them to investigate how, in a luminaire equipped with a linear light source, such as a fluorescent lamp, linear prisms running parallel to the lamp can control the light in the vertical axial plane.

6.8.4 CONTROL IN THE AXIAL PLANE BY LINEAR PRISMS

By an axial plane we mean a vertical plane that is at right angles to the transverse plane. In Figure 6.40(a), DCC' lies in an axial plane when it is vertical and parallel to the long edges of the prism. In this figure, as before, AB is the incident ray. It makes an angle η with its projection $A'B$ on the principal plane.

CD is the emergent ray, which is in the axial plane, so that the triangle DCC' is vertical and DC' is directly downwards. CD makes an angle ξ with its projection, $C'D$, on the principal plane. $C'D$ makes an angle ψ with the normal.

η will be at its maximum when AB is parallel to the top of the prism. ϕ will then be a right-angle.

In Figure 6.40(b) (which shows the rays projected on to the principal plane) cd will be directly downwards in this instance, since CD is in the vertical axial plane.

Therefore

$$\psi = 90° - \beta \tag{6.68}$$

and

$$90° - \beta = \phi' + \psi' \tag{6.69}$$

Now

$$\sin \phi \cos \eta = \mu \sin \phi' \cos \eta' \tag{6.70}$$

but

$$\phi = 90° \tag{6.71}$$

Therefore,

$$\cos \eta = \mu \sin \phi' \cos \eta' \tag{6.72}$$

which by squaring each side, dividing through by $\mu^2 \sin^2 \phi'$, and subtracting each side from unity gives

$$1 - \frac{\cos^2 \eta}{\mu^2 \sin^2 \phi'} = 1 - \cos^2 \eta'$$
$$= \sin^2 \eta' \qquad (6.73)$$

But

$$\sin \eta = \mu \sin \eta' \qquad (6.74)$$

so,

$$1 - \frac{\cos^2 \eta}{\mu^2 \sin^2 \phi'} = \frac{\sin^2 \eta}{\mu^2} \qquad (6.75)$$

Therefore, by multiplying by $\mu^2 \sin^2 \phi'$ and rearranging the terms so those containing $\cos \eta$ are on the left-hand side, we obtain

$$\cos^2 \eta = (\mu^2 - 1) \tan^2 \phi' \qquad (6.76)$$

We now relate ϕ' to β. Equation (6.72) states

$$\cos \eta = \mu \sin \phi' \cos \eta'$$

moreover

$$\sin \psi \cos \eta = \mu \sin \psi' \cos \eta' \qquad (6.77)$$

Therefore, by division and rearrangement of the terms, we find

$$\sin \psi \sin \phi' = \sin \psi' \qquad (6.78)$$

Now, by substituting (6.68) and (6.69) in (6.78), we obtain

$$\cos \beta \sin \phi' = \cos (\beta + \phi')$$
$$= \cos \beta \cos \phi' - \sin \beta \sin \phi' \qquad (6.79)$$

hence by dividing by $\cos \phi'$ and rearrangement of the terms we obtain

$$\tan \phi' = \frac{\cos \beta}{\cos \beta + \sin \beta}$$
$$= \frac{1}{1 + \tan \beta} \qquad (6.80)$$

which by substitution in (6.76) gives

$$\cos^2 \eta = (\mu^2 - 1) \left(\frac{1}{1 + \tan \beta} \right)^2 \qquad (6.81)$$

so

$$\cos \eta = \sqrt{(\mu^2 - 1)} \left(\frac{1}{1 + \tan \beta} \right) \qquad (6.82)$$

We will now investigate the control of light in the transverse plane.

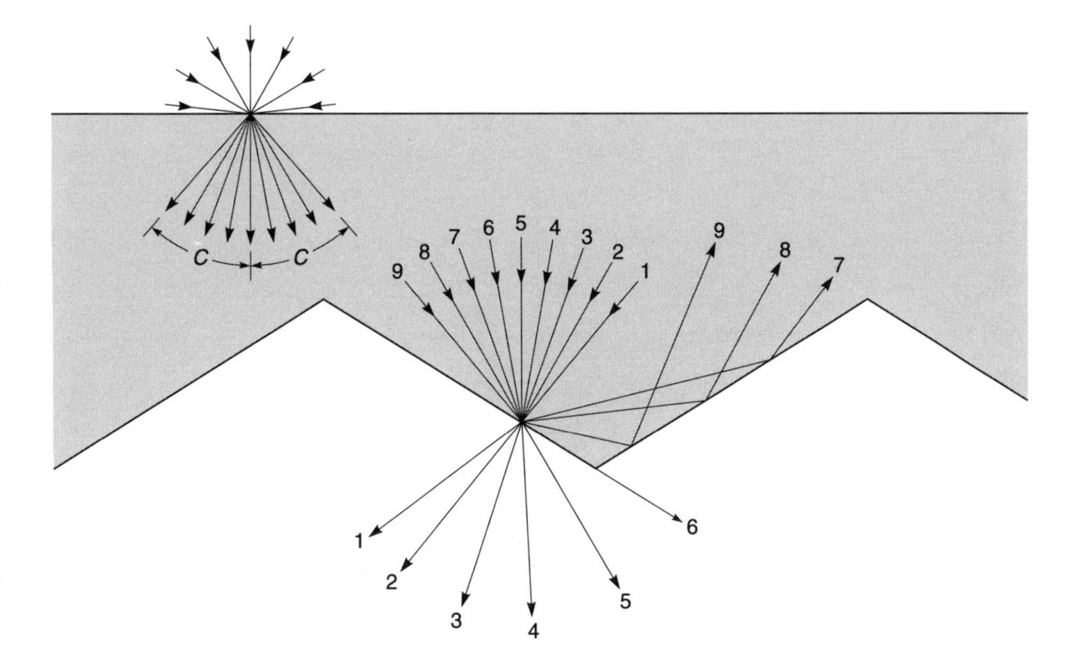

Fig. 6.41 Transverse section of base prism

6.8.5 CONTROL IN THE TRANSVERSE PLANE BY LINEAR PRISMS

In the previous section we have shown how control of the light distribution in the axial plane may be obtained by linear prisms. In this section we will show how to obtain control in the transverse plane.

Figure 6.41 shows the action of a base prism. Rays entering from all directions are concentrated into a cone having an angle at the apex equal to twice the critical angle. The rays are numbered for reference purposes. Ray 9 travels in a direction parallel to the critical angle. The angle of the prism is made such that this ray is reflected off the second face at the critical angle. If the apical angle were made any larger it would pass through this face. If it were made any smaller, ray 1 would leave the first face at a higher angle than is necessary.

Figure 6.42 shows the trace of ray 9, *PQRS*. We are required to find *x*, the half apical angle of the prism, whose faces are *LM* and *MN*. The ray is reflected at *Q* and *R* at the critical angle *c*. *TM* is a vertical construction line.

$$\begin{aligned} \angle UQP &= c \\ \text{now } \angle UQL &= \angle TMQ \\ &= x \\ \therefore \angle RQM &= \angle PQL \\ &= x - c \end{aligned} \tag{6.83}$$

In the triangle *RQM*,

$$\angle RQM + \angle QMR + \angle QRM = 180° \tag{6.84}$$

so,

$$(x - c) + (2x) + (90° - c) = 180° \tag{6.85}$$

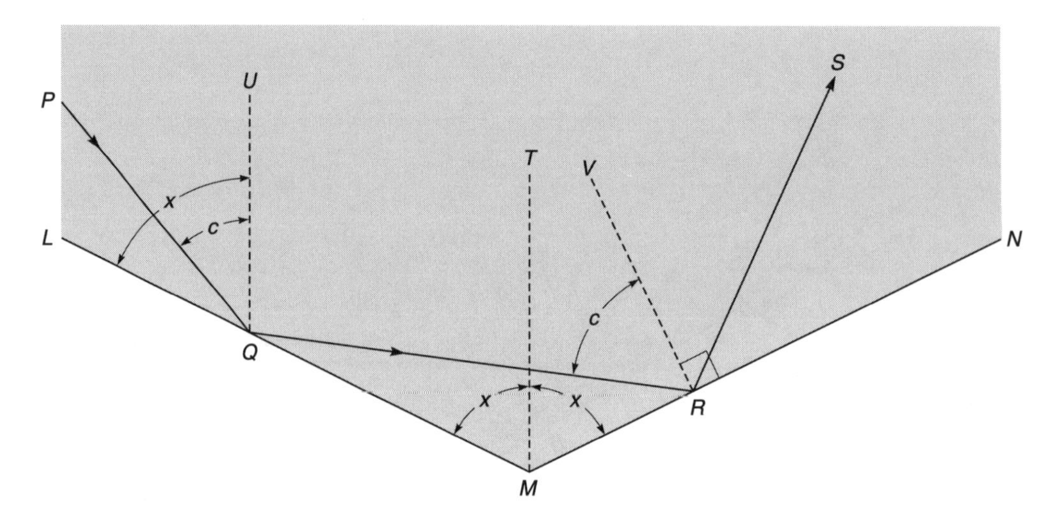

Fig. 6.42 Trace of ray 9 through prism

Hence

$$x = 30° + \frac{2}{3}c \tag{6.86}$$

If μ is 1.5, then c is approximately 42° and x is 58°. Thus, the longitudinal prisms give a cut-off of 58° in the transverse plane. To find the cut-off in the axial plane we use (6.82).

$$\begin{aligned}\cos\eta &= \sqrt{(\mu^2 - 1)}\left(\frac{1}{1 + \tan\beta}\right)\\ &= \sqrt{(1.5^2 - 1)}\left(\frac{1}{1 + \tan 58°}\right)\\ &= 0.43\end{aligned} \tag{6.87}$$

From this we obtain

$$\eta = 65^{\circ} \tag{6.88}$$

So we obtain a cut-off of 65° in the transverse plane and 58° in the vertical axial plane. At these angles the luminous intensity will be zero but at angles just smaller than these there will not be an abrupt increase of luminous intensity, for two reasons. First, because of the high percentage of the incident light which is reflected at large angles of incidence, as is apparent from Table 6.4. Second, because of the limited size of the lamp or lamps and their spacing from the top surface of the prism panel the cone angle of the incident light will be less than 180°.

6.8.6 DESIGN OF LINEAR PRISM BANKS

Linear prism banks are used on the sides of indoor and road lighting luminaires. In the former case they are used to obscure a direct view of the lamps by directing light upwards or downwards, thereby controlling discomfort glare. For road lighting luminaires, prism banks are used to direct a light beam on to the road at an angle that provides a long tail to the T-shaped light patch (see Section 9.5, page 350).

For a linear prism bank to be effective, the light source itself also needs to be linear, usually fluorescent or low pressure sodium.

The design process can be split into a number of steps which we will consider separately.

(1) *Choice of lamp and its attitude*
The choice of lamp will mainly depend on its luminous flux in relation to the luminous flux required and the number of luminaires to be used. Other considerations will be its size and possibly the heat generated.

For the great majority of applications the lamp is used horizontally and the lines of the prisms will also run in this direction.

(2) *Width of luminaire*
Normally, the objective is to make the luminaire as narrow as possible. The controlling factor is that if the prism bank is too close to the lamp the acceptance angle of the light reaching each prism will be too great for the prism to provide effective control.

In the design of a road lighting luminaire, the angle of the peak luminous intensity is the controlling factor. This can be understood from Figure 6.43.

Suppose that we want the peak luminous intensity to occur at 65°, which is the angle commonly used for road lighting luminaires. At the same time we want the lamp to be obscured by the prism bank at 80° to give good glare control. *BC* is drawn tangentially to the lamp at 80° to meet *AC*, which is drawn at 65° to the vertical and tangentially to the lamp. The prism bank is designed on the line *CD* so that it provides a peak luminous intensity at 65° and obscures the lamp at 80°. The bare lamp will be visible below 65°.

(3) *Attitude of prisms*
We now have to decide whether to form the prisms so that their apices face inwards, towards the lamp, or outwards.

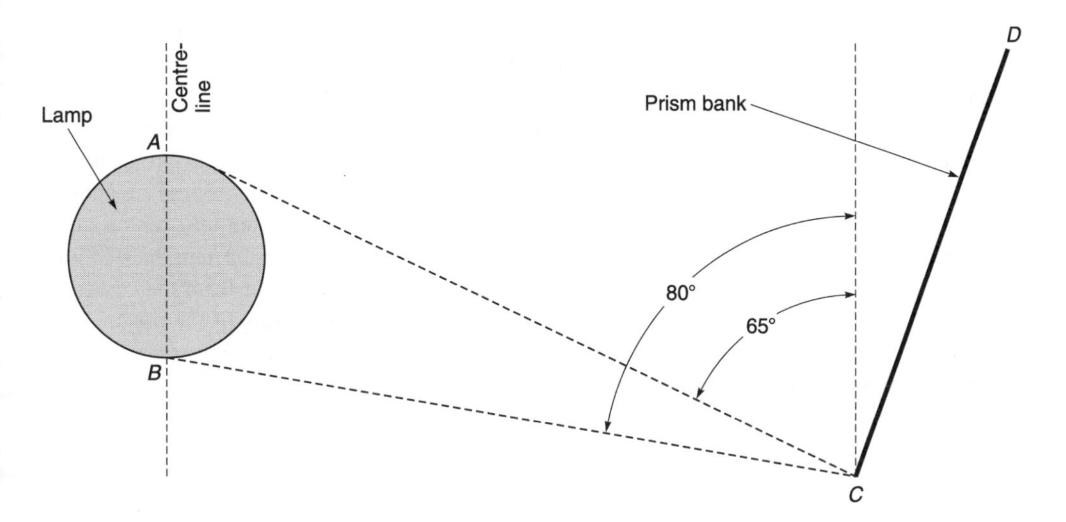

Fig. 6.43 Determining the width of a luminaire

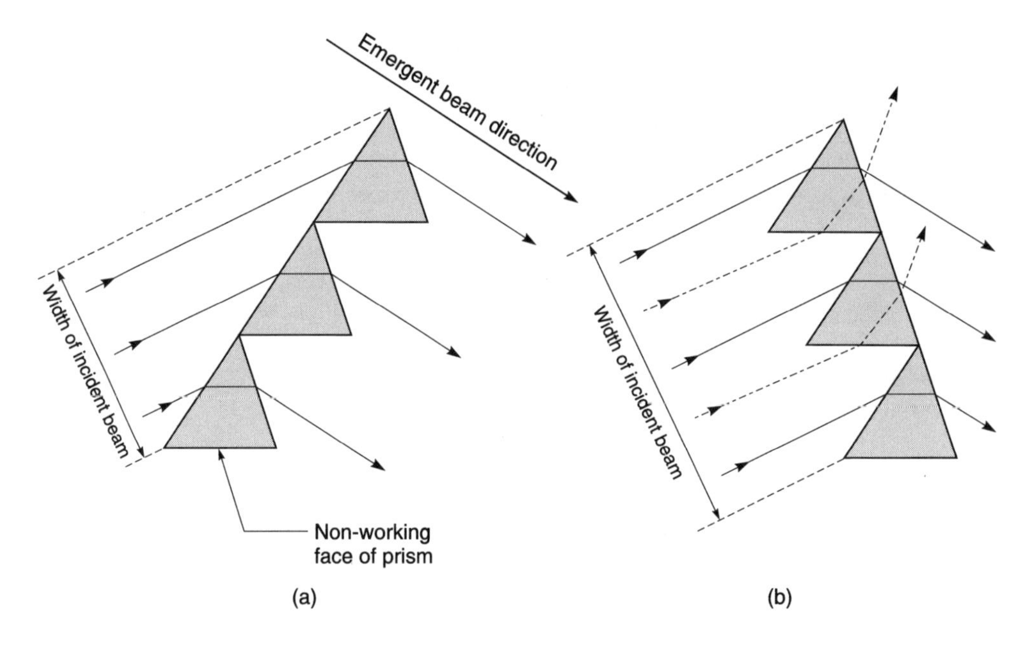

Fig. 6.44 Dark areas caused by the non-working prism faces

The outward facing prisms have two disadvantages. First, they are likely to get dirtier quicker, especially in outdoor luminaires. Second, it is impossible to flash the whole of the prism surface. This can be seen from Figure 6.44, which compares the action of outwardly and inwardly facing prisms.

In Figure 6.44(a) and (b) the flashed area in the direction of the beam is the same, but in (a) this area is broken up by the dark areas of the non-working faces. To achieve flashing of all three prisms in (b) a wider incident beam of light is required than in (a), moreover the incident light striking the non-working faces is directed upwards, which for outdoor applications is usually undesirable as it causes light pollution (see Section 9.1, page 346).

With regard to the overall shape of the prismatic diffuser, (a) will tend to narrow towards the bottom, whereas the opposite is true for (b). However, it is possible to counteract these trends by suitable prism design.

For road lighting applications, prism panels are sometimes designed as separate items for cementing on to the inside of clear bowls. The prisms are outwardly facing so are sealed to prevent the deposition of dirt in the prism angles. This construction has the additional advantages that the bowl narrows towards the bottom, which makes its manufacture easier as it can be easily withdrawn from the forming tool, and the panels can be used for different lengths of lamp. Sealing the panel to the bowl may be a problem because the solvent vapour from the cement is trapped in the space between the panel and the bowl, and may cause crazing of the prisms. This problem can be overcome in manufacture by passing a stream of air through the space until the cement is dry. However, this is labour intensive.

In the example to be calculated we will consider outwardly facing prisms.

(4) *Calculation of prism angles of working faces*

The work is done by making use of (6.56) page 243, (6.58) page 245 and (6.48) page 240, and setting out the work in a spreadsheet.

The angles of each prism in a bank of prisms can be calculated separately so that they all differ from each other. However, to save tool costs they can be repeated so as to form groups each comprising prisms having the same angles.

The effect of the first approach is to form repeated images of one part of the lamp in the direction of the peak luminous intensity. Brighter flashing results if this part of the lamp is of higher luminance than the rest of the lamp, as it often is with high intensity discharge lamps. The second method produces an image of the source or part of the source, depending on the width of the prism bank. For the purposes of illustration, we will use this method for a fluorescent lamp, the luminance of which across a diameter is substantially even.

Table 6.6 shows how the work may be set out on the spreadsheet. As in Table 6.5, the refractive index of the optical medium is entered in a separate cell so that it can be referred to in the formulae where the angle of refraction is calculated from the angle of incidence or vice versa. This is better than entering the value directly in the formulae as only one figure needs to be changed when the refractive index is changed. P, Q and B (Figure 6.36, page 244) are entered in the appropriate cells and the values in the remaining cells in the row are automatically calculated. i' is calculated from $i' - r'$ by means of (6.48), page 240.

The first step is to enter values for P, Q and B. From Figure 6.43, page 253, it is evident that P is 80°. B is the slope of the side of the bowl, which we will make 15° from aesthetic considerations, and because the greater the slope the greater will be the refraction at the surface facing the lamp. This leaves less work for the other prism face to do, thereby making a greater deviation D possible. Q is 70°. These values give a prism angle A of 19.2°. Angles i and r' must be examined in relation to Table 6.4, page 236, to find if the reflection losses are acceptable. At 15.8°, r' will give a loss of about 0.05, which is acceptable, as is the loss of about 0.04 for i at 5.0°. Later we will consider the angle of the non-working face, which forms the base of the prism.

We now have to decide how far upwards Prism Group 1 can be extended upwards before it directs light into the 80° to 90° zone. For this we use Table 6.7, which determines the value of P for a ray to be emitted at 80°.

P works out to be 89.6°. In Figure 6.45 we draw a line E_1C_1 at this angle to the vertical and tangentially to the lamp to meet the prism bank in C_1. C_1C_0 is the length of Prism Group 1 which, by measurement or calculation, is found to be less than the diameter of the lamp. Therefore, the whole of the Prism Group 1 will be flashed at the beam angle and no adjustment need be made to the angle of E_1C_1, which would be the case if C_0C_1 were greater than the diameter of the lamp.

For Prism Group 2, we enter Table 6.6, again, with a value of P equal to 89.6°. A is now calculated as 35.1°. Examination of the losses at the two refracting surfaces shows they are

Table 6.6 Determination of prism angles A and C. Angles in degrees

Refractive index of optical medium (μ) = 1.5

Prism Group	P	Q	B	i $B+P-90°$	r	$i-r$	D $P-Q$	$i'-r'$	i'	r'	A $r'+r'$	C $A-B$
1	80.0	70.0	15.0	5.0	3.3	1.7	10.0	8.3	24.2	15.8	19.2	4.2
2	89.6	70.0	15.0	14.6	9.7	4.9	19.6	14.7	40.1	25.4	35.1	20.1
3	98.4	70.0	15.0	23.4	15.4	8.0	28.4	20.4	52.1	31.7	47.1	32.1
4	106.2	70.0	15.0	31.2	20.2	11.0	36.2	25.2	60.8	35.6	55.8	40.8

Table 6.7 Determination of extent of prism groups. Angles in degrees

Refractive index of optical medium (μ) = 1.5

Prism Group	Q	A	C	i' $90°+C-Q$	r'	$i'-r'$	r $A-r'$	i	$i-r$	D $(i-r)+(i'-r')$	P $D+C$
1	80.0	19.2	4.2	14.2	9.5	4.7	9.7	14.6	4.9	9.6	89.6
2	80.0	35.1	20.1	30.1	19.7	10.4	15.4	23.4	7.9	18.4	98.4
3	80.0	47.1	32.1	42.1	26.7	15.4	20.4	31.2	10.9	26.2	106.2

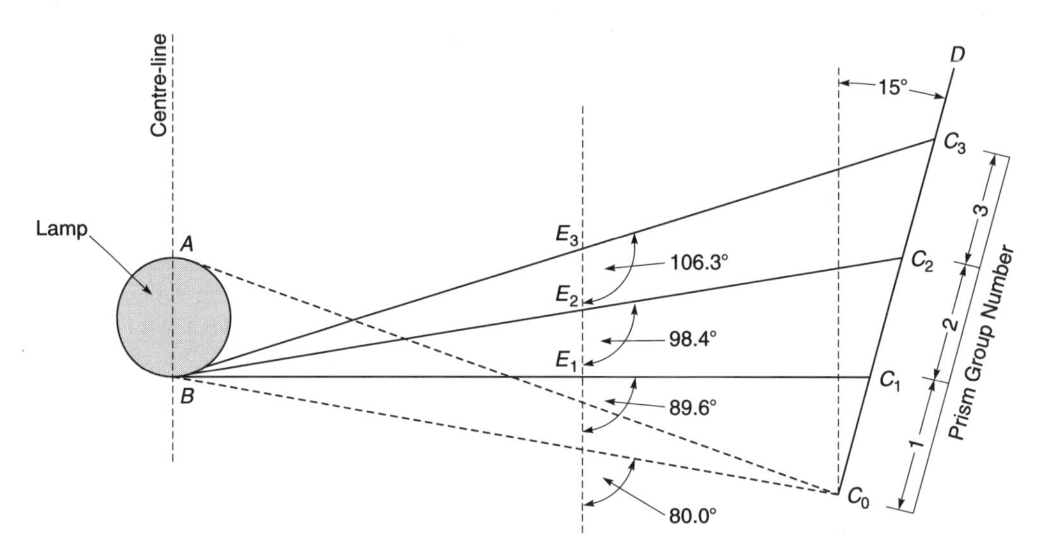

Fig. 6.45 Positioning of the prism groups

acceptable. In Table 6.7 we enter 35.1° for A and 20.1° for C, which give 98.4° for A. Adopting the same procedure as for Prism Group 1 we draw E_2C_2 at 98.4° to the vertical (Figure 6.45) to find the extent of Prism Group 2. C_1C_2 is less than the diameter of the lamp so no adjustment need be made to the angle of E_2C_2.

For Prism Group 3 we return to Table 6.6 with a value of A of 98.4°. After checking that the angles of incidence and refraction do not give unduly high values of reflection, we enter the new values of 47.1° for A and 32.1° for C in Table 6.7.

For Prism Group 4 we return to Table 6.6, this time with a value of A of 106.2°. r' is 35.6°, which gives a reflectance greater than 0.086 and is on a part of the reflectance table where the reflectance is rising rapidly with angle. We decide to reject this, although it may be worth making a prism bank with the angles of Prism Group 4 and assessing it experimentally.

(9) *Angle of non-working faces*

We now have to decide the angle of the non-working face or base of the prism. Ideally, this should be parallel to the ray of light that forms the beam to achieve the maximum projected area of flashing in the beam direction.

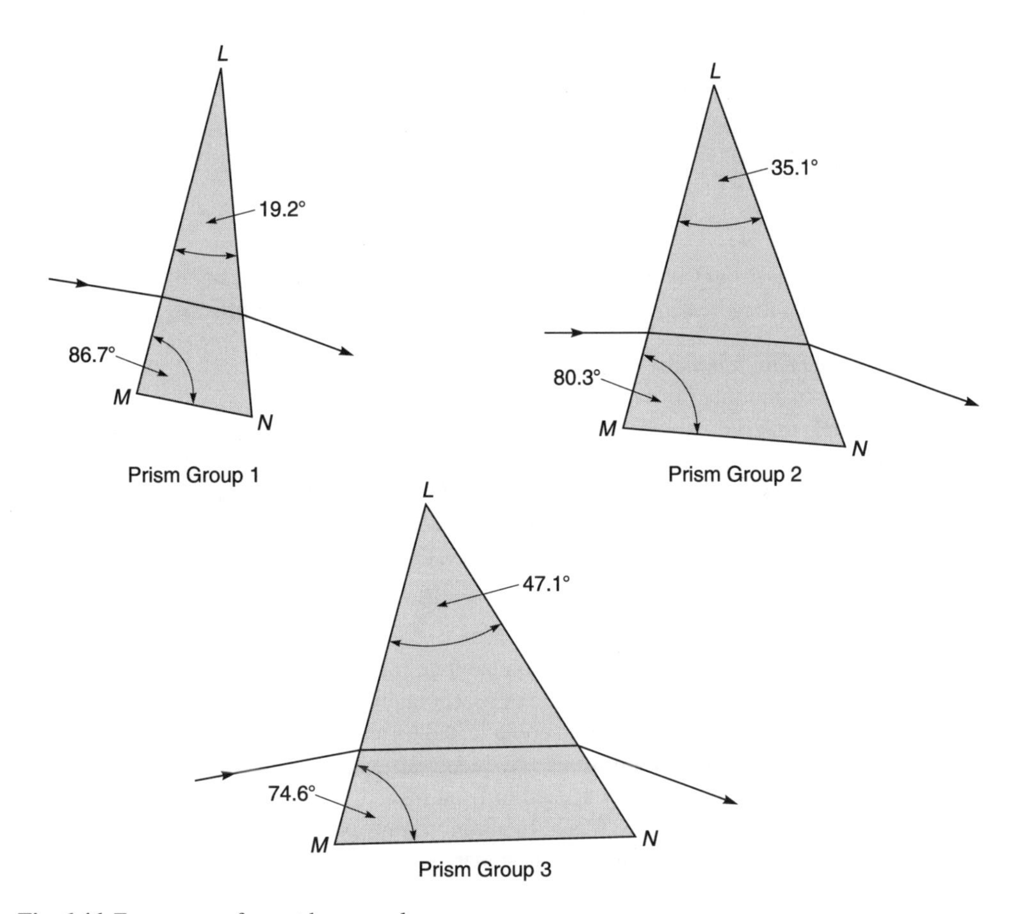

Fig. 6.46 Emergence of ray at beam angle

Figure 6.46 shows a prism from each of the prism groups. The ray shown emerges at the beam angle, 70° to the downward vertical. The face *MN* is made parallel to the path of this ray as it travels in the prism by using the equation

$$\angle LMN = 90^\circ - r \tag{6.89}$$

At this point, consideration has to be given as to whether it is possible to withdraw the manufacturing tool. This is usually possible providing the $\angle LMN$ is one or two degrees less than a right-angle, which is the case in this illustration.

(10) *Luminous intensity of the prism bank in the direction of beam*
The ratio of the orthogonally projected height of the prism bank in the direction of the beam to the diameter of the lamp can be found from Figure 6.45 to be 3.1. This means that its luminous intensity will be 3.1 times the luminous intensity of the lamp (in the direction of the prism bank) not corrected for the reflection losses and absorption losses in the optical material, which are taken to be zero. On the surface facing the lamp, the reflection losses are 0.04 approximately, and on the internal surface they vary from 0.04 to 0.09 approximately, for which we can take an average of about 0.06, to be conservative. There will also be losses of about 0.04 at each of the bowl's surfaces. This means the transmission will be

$$(1 - 0.04)(1 - 0.06)(1 - 0.04)(1 - 0.04) = 0.85$$

So the effective luminous intensity of the prism bank will be 0.85×3.1 which is 2.6. To calculate the total luminous intensity of the beam, account has to be taken of the fact that an image of the lamp will be visible below the prism bank. The luminous intensity of this will be reduced by the reflection losses of the bowl, about 0.08.

(7) *Upward light*
If the control of upward light is an important consideration, then an investigation needs to be made of the light that may be reflected off the non-working faces of the prisms.

(8) *Increasing the beam luminous intensity*
This may be done in a number of ways.

The prism bank may be extended upward by using reflecting prisms. A specular reflector above the lamp can be used to reflect light above the prism bank. A clear window can be left beneath the prism bank so that the whole of the lamp is visible through it, and beneath this, further prisms can be added to refract light into the beam. This may produce a dip in the polar curve and it may be necessary to leave spaces between the prisms to make this less pronounced.

6.8.7 LENSES

We will not give proofs of the various lens formulae, but, for reference, give a summary of them as they apply to thin lenses. These formulae are only accurately applicable to rays that do not deviate more than about 10° from the axis. More complicated formulae have to be used in other cases.

Figure 6.47 shows how images are formed by biconvex and biconcave lenses. Each lens has two foci, F_1 and F_2, which are equally spaced in front of and behind the lens. Rays parallel to the longitudinal axis pass through the focus. In addition, rays passing through the centre of the lens are not deviated. The use of these two rules enables the image to be located. There are two sorts of image. A real image, which can be focused on a screen, and a virtual image, which cannot be formed on a screen, because the rays are diverging from the lens.

The formulae for locating the image are:

$$\frac{1}{u} + \frac{1}{v} = \frac{1}{f}$$
$$(\mu - 1)\left(\frac{1}{r_1} + \frac{1}{r_2}\right) = \frac{1}{f} \tag{6.90}$$

where

u is the distance from the object, O, to the centre of the lens;
v is the distance of the image, I, to the centre of the lens;
f is the focal length of the lens;
r_1 and r_2 are the radii of curvature of the surfaces of the lens, which are taken as positive for a converging (convex) lens and negative for a diverging lens;
μ is the refractive index of the material of the lens.

To make these formulae applicable to all cases the following sign convention can be used.

All distances to real points are reckoned as positive, and all distances to virtual points as negative.
The focal length of a converging lens is positive, that of a diverging lens negative.

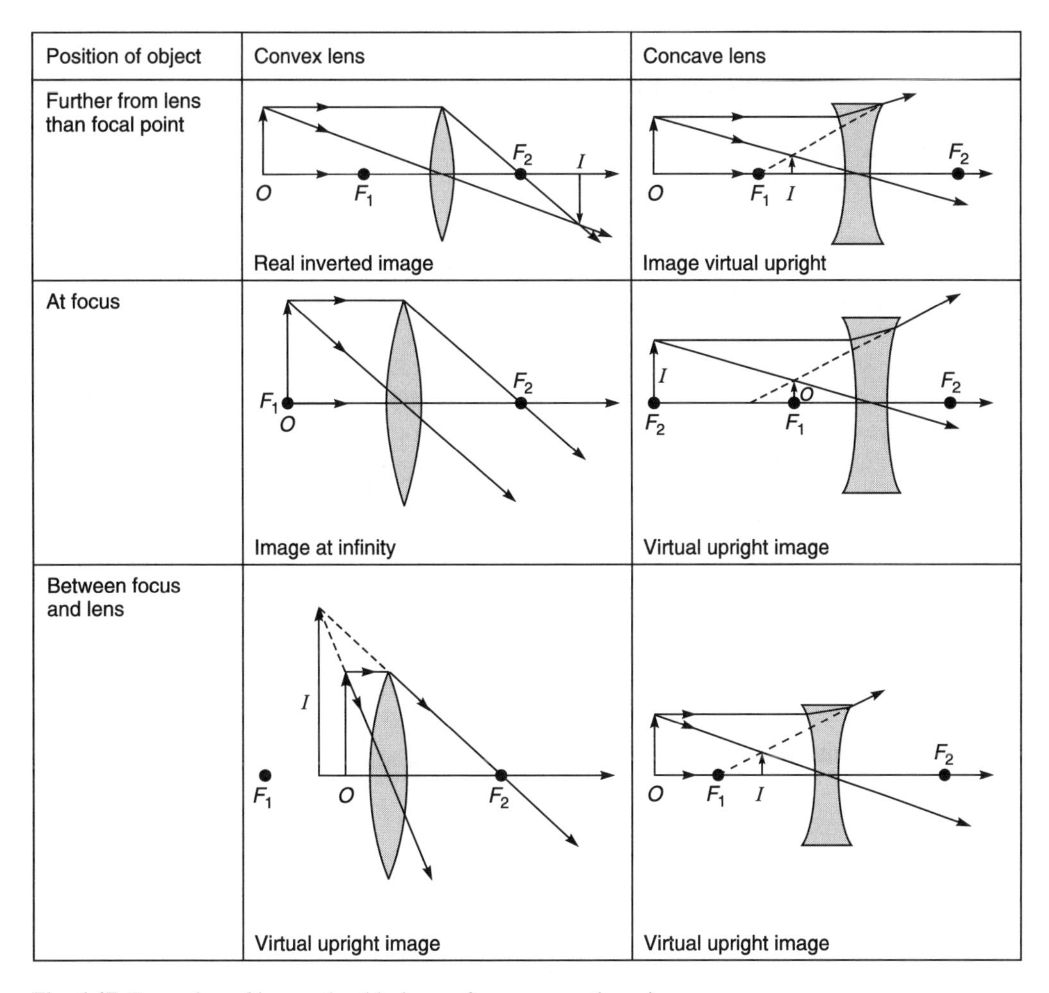

Fig. 6.47 Formation of images by thin lenses for rays near the axis

Where two thin lenses are placed together, their combined focal length f_c is given by

$$\frac{1}{f_c} = \frac{1}{f_1} + \frac{1}{f_2} \tag{6.91}$$

where f_1 and f_2 are the focal lengths of the two lenses. The focal length of a converging lens is taken as positive whilst that of a diverging lens is taken as negative. This formula may be extended to the combination of any number of lenses,

$$\frac{1}{f_c} = \frac{1}{f_1} + \frac{1}{f_2} + \ldots + \frac{1}{f_n} \tag{6.92}$$

where f_n is the focal length of the nth lens.

Linear magnification of the image = v/u.

6.8.8 *FLASHING OF LENSES*

When a lens produces an image of a light source, the area of the lens which appears flashed depends on the distance of the eye from the lens. Therefore, to calculate the luminous intensity provided by the lens, it is necessary to calculate this area. In this section we will describe how this is done for points on the axis of the lens.

(1) *Image of source smaller than lens*

Four cases are considered in Figure 6.48, in which all the images are indicated by *I*. In (a) D_1 is the point on the axis where the rays of light from the rim of the lens cross over each other and the axis. If the eye is placed at this point the whole of the lens, from A_1 to A_2, would appear flashed. In (b) the distance of the eye from the lens is increased by moving it to D_2 so the image subtends a smaller angle at the eye than the lens. The result is that the lens is only flashed from A_3 to A_4. If we now return to (a) and move the eye nearer the lens, the whole of it will appear flashed until D_3 in (c) is reached where the inner rays from the cones of light from the rim (A_5 and A_6) of the lens cross over. When the eye is brought nearer to the lens than this, progressively less and less of it, A_7 to A_8, will appear flashed as in (d).

For points beyond the image, the same effect as described above can be obtained by replacing the image with a diaphragm having an aperture the same size as the image, and regarding the lens as being fully flashed. This aperture is known as the exit pupil.

(2) *Image of source greater than or equal in size to that of the lens*

In Figure 6.49 the inner rays in the cones of light from the rim of the lens cross over at *D*. At any point on the axis beyond *D* the lens will appear fully flashed. Closer to the lens the flashing will diminish in area in a similar way to that shown in Figure 6.48(d).

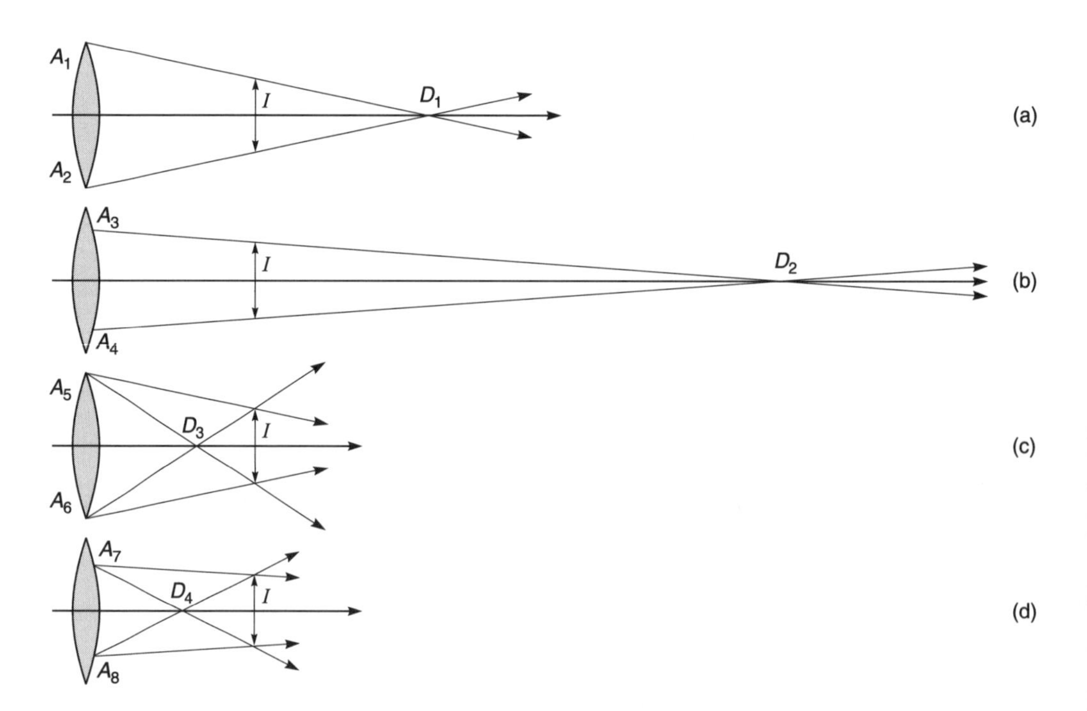

Fig. 6.48 Flashing of lens when the image is smaller than the lens

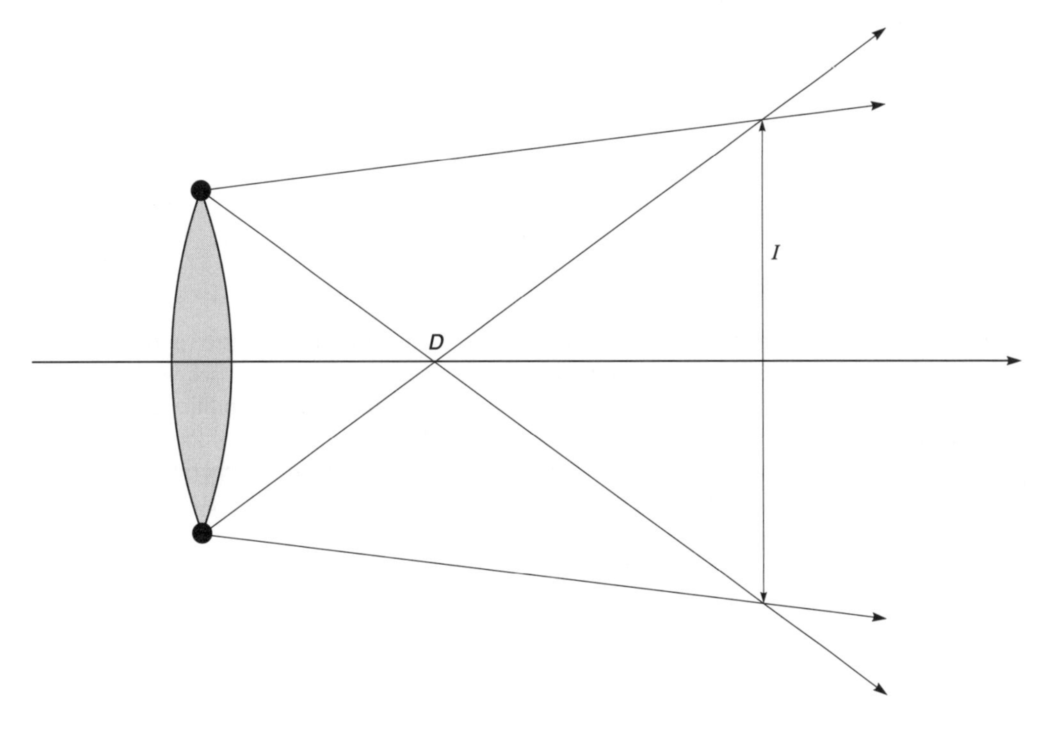

Fig. 6.49 Size of image greater or equal to that of lens

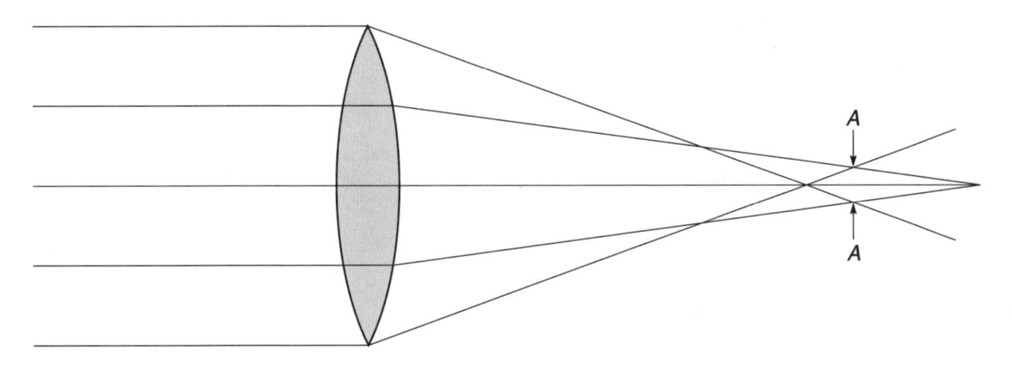

Fig. 6.50 Spherical aberration

6.8.9 SPHERICAL ABERRATION

Parallel rays of light passing through a lens near its rim will meet the axis nearer to the lens than more central rays, as illustrated in Figure 6.50 for rays travelling from left to right. This is known as spherical aberration.

This means that if we consider light travelling from right to left and want the whole of the lens to appear flashed when viewed on axis, we require a light source at least as large as *AA*. This may be impossible to attain for very bright light sources or for lenses with diameters large in comparison to their focal length; that is, for lenses with large *f-numbers*. In these cases it is necessary to reduce the diameter of *AA* by using special lens designs. Photographic lenses reduce spherical aberration to negligible proportions but are normally too expensive for lighting engineering purposes. The two solutions commonly used are lenses with parabolic surfaces and stepped lenses.

6.8.10 LENSES WITH PARABOLIC SURFACES

Figure 6.51 shows two arrangements of lenses employing parabolic surfaces.

6.8.11 STEPPED LENSES

In certain applications, such as signal lights, including lighthouses and theatre spotlights, large diameter lenses (over about 50 mm) are required. Biconvex or parabolic lenses would be very thick. It was suggested by Buffon that they could be considerably reduced in thickness by stepping them in concentric zones, as shown in Figure 6.52. However, because of their large diameter in comparison to their focal length they suffer considerably from spherical aberration. Fresnel overcame this by calculating the curvature on each of the separate steps to make the

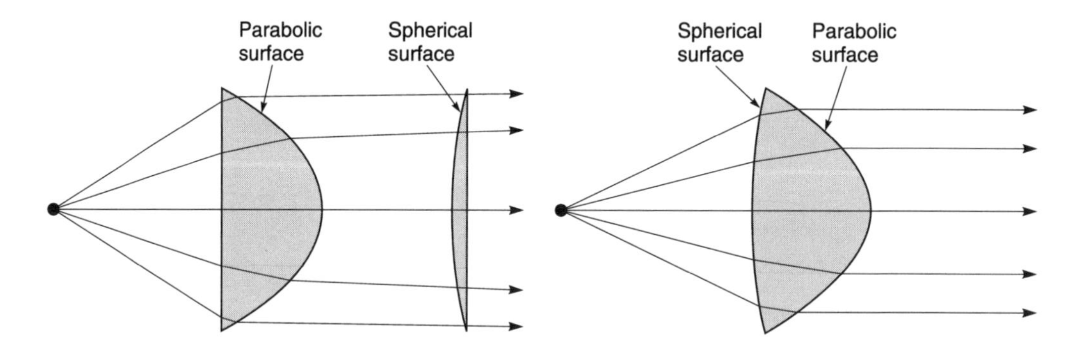

Fig. 6.51 Lenses with a parabolic surface

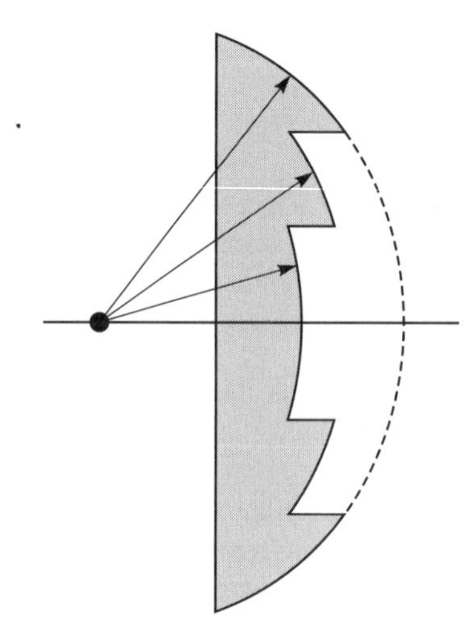

Fig. 6.52 Buffon lens

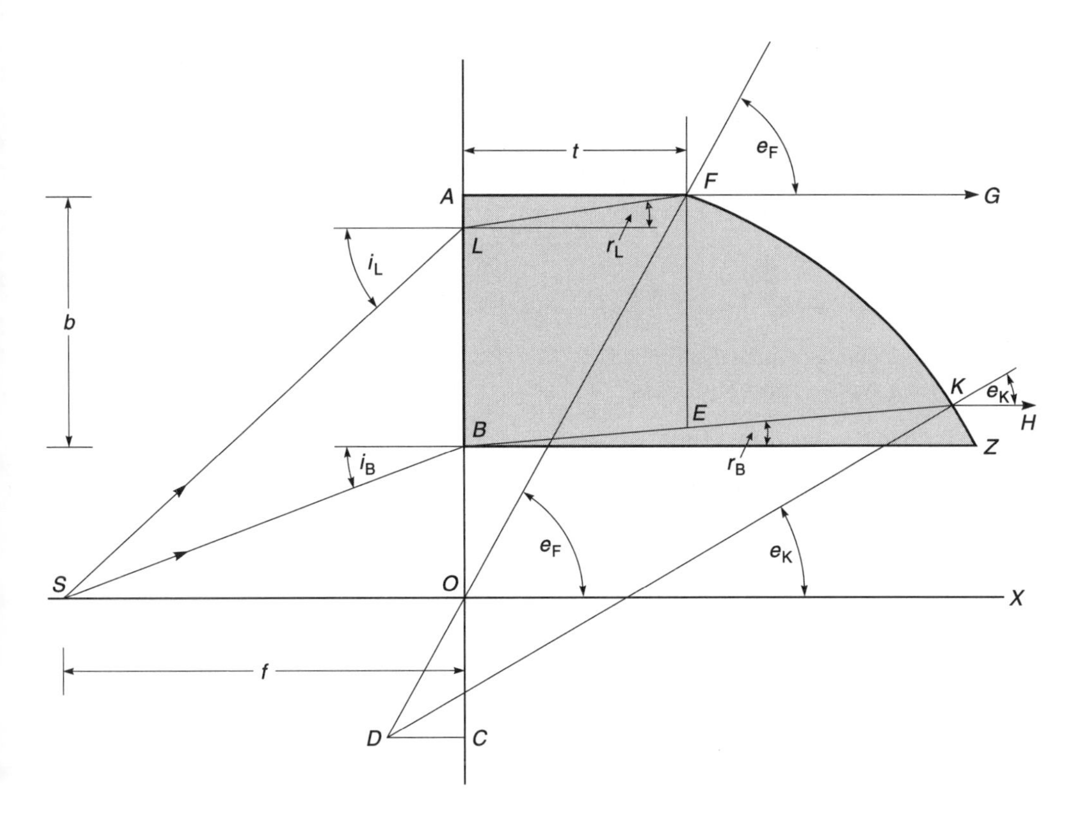

Fig. 6.53 Fresnel lens

rays emerging from the top and the bottom of the step face parallel to the lens axis. Modern developments include sheets of fine stepped lenses for use in overhead projectors.

The method by which the curvature of the steps in a Fresnel lens is derived can be followed by reference to Figure 6.53.

In the diagram, the stepped element is *AFKZB*. The source is at *S*. We will consider two rays *SLFG* and *SBKH* from the source. The radius (*DF* and *DK*) and centre of curvature (*D*) of the face *FZ* have to be such that the extreme rays *FG* and *KH* emerge parallel to *SX*.

Let *DF* and *DK* make angles of e_F and e_K with *SX*.

In addition, let the angles of incidence and refraction be respectively i_B and r_B at *B*, and i_L and r_L at *L*.

Therefore,

$$\sin i_L = \mu \sin r_L \tag{6.93}$$

and

$$\sin i_B = \mu \sin r_B \tag{6.94}$$

Let the angles of emergence at *F* and *K* be e_F and e_K respectively, so that

$$\sin e_F = \mu \sin (e_F - r_L) \tag{6.95}$$

and

$$\sin e_K = \mu \sin (e_K - r_B) \tag{6.96}$$

Therefore, from (6.48), (page 240)

$$\tan e_F = \frac{\mu \sin r_L}{\mu \cos r_L - 1} \tag{6.97}$$

and

$$\tan e_K = \frac{\mu \sin r_B}{\mu \cos r_B - 1} \tag{6.98}$$

from which e_F and e_K can be determined.

Let $AB = b$, $SO = f$ and $AF = t$, then

$$b = f(\tan i_A - \tan i_B) + t \tan r_L \tag{6.99}$$

Draw FE parallel to AO to meet BK in E.

Consider the triangle FEK:

$$\begin{aligned} \angle FEK &= 90° - r_B \\ \angle FKE &= \angle FKD - \angle BKD \\ &= 90° - \tfrac{1}{2}(e_F - e_K) - (e_K - r_B) \\ &= 90° - \tfrac{1}{2}(e_F + e_K) + r_B \end{aligned} \tag{6.100}$$

$$\frac{\sin \angle FKE}{FE} = \frac{\sin \angle FEK}{\text{chord } FK}$$

by the sine rule for triangles.

Therefore,

$$\begin{aligned} \text{chord } FK &= \frac{FE \cos r_B}{\cos\left[\frac{1}{2}(e_F + e_K) - r_B\right]} \\ &= \frac{(b - t \tan r_B)\cos r_B}{\cos\left[\frac{1}{2}(e_F + e_K) - r_B\right]} \end{aligned} \tag{6.101}$$

If r is the radius of curvature of FK then

$$\begin{aligned} r &= DF \\ &= DK \end{aligned} \tag{6.102}$$

and

$$FK = 2r \sin \tfrac{1}{2}(e_F - e_K) \tag{6.103}$$

In addition, the coordinates of D with respect to O are

$$\begin{aligned} DC &= r \cos e_F - t \\ OC &= r \sin e_F - f \tan i_L - t \tan r_B \end{aligned} \tag{6.104}$$

This completes the determination of the radius and centre of curvature of FKL.

The Fresnel lens may be generated in three ways. The profile may be rotated about the axis SX. This produces a beam of light in one direction when the source is located at S. Second, it may be rotated about an axis through S and parallel to AC, which produces a beam in all directions of azimuth, providing the source emits light in all the relevant directions. This is useful for beacons. Third, it can be generated as a linear prism panel in a plane through *AC* and perpendicular to *SX*, which is useful for controlling the light from linear sources.

6.8.12 LIGHT GUIDES USING REFRACTION

One fascinating phenomenon occurring in optics is the way that light can be guided or 'piped' along a rod, block or sheet of transparent material by making use of total internal reflection. This is a very much more efficient method of conduction or transport than is provided by conventional mirror surfaces (see Section 6.6, page 232). At best, these have a reflectance of about 95% so that there is a considerable loss of light after comparatively few reflections. On the other hand, total internal reflection gives a reflectance approaching 100% and, provided the material of the light guide has a good transparency, light loss is considerably less.

Figure 6.54 shows how light is transported in a sheet or rod *RSTU* of optically clear material. Light enters the face *RU* from all directions and, at each point, is concentrated into a cone having a total apical angle equal to twice the critical angle *c*. An extreme ray *VW* is shown reaching the face *RS* at *W*. It makes an angle 90° – *c* with the normal *NW*. If the material has a refractive index of 1.5, *c* will be 43°, so that the angle of incidence at *W* will be 90° – 43° or 47°, which is greater than the critical angle for the material. The ray will then be reflected and will zigzag down the sheet and leave parallel to *ST*. Rays incident on *RU* at less extreme angles will either be reflected in a similar manner or travel straight through.

The material can be bent (Figure 6.55) provided that the bend is not so sharp that the rays are incident at less than the critical angle. To facilitate bending, fibres, typically 50 μm to 150 μm in diameter, are used bundled together in a 'cable' about 10 mm in diameter. These are encased in a protective sheath. Each fibre is coated by a material with a lower refractive index than the fibre itself to prevent the leakage of light due to surface imperfections and contact with other fibres. Low voltage tungsten halogen lamps are the sources commonly used but high intensity discharge lamps may be used for large applications. The emitting end of the fibre-optic cable may be shaped or have an optical device to distribute the light in the required fashion. Light attenuation may vary from 2% to 12% per metre. As well as for producing decorative effects, fibre optics have found applications in situations where having a remote light source is beneficial. Instances include the lighting of the contents of museum cabinets, which can be kept sealed for lamp changing, the lighting of hazardous areas where there is a danger of an explosive vapour being ignited by a spark and, in rare instances, general lighting.[11]

Light guides in the form of sheets have been used in a number of applications. A decorative pattern can be engraved on the surface of the sheet and the sheet edge lit. The engraving will

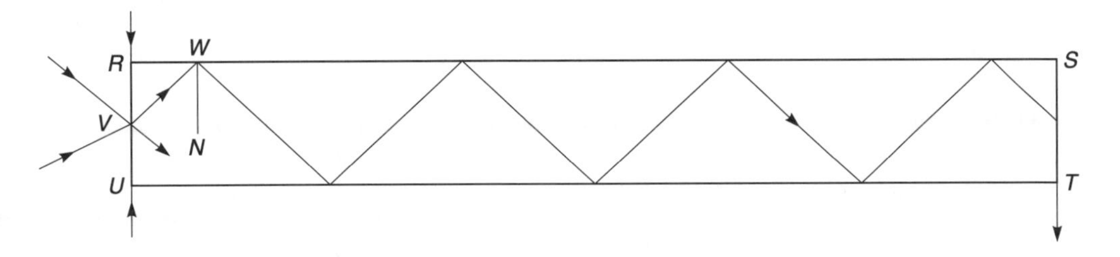

Fig. 6.54 Light transport by total internal refraction

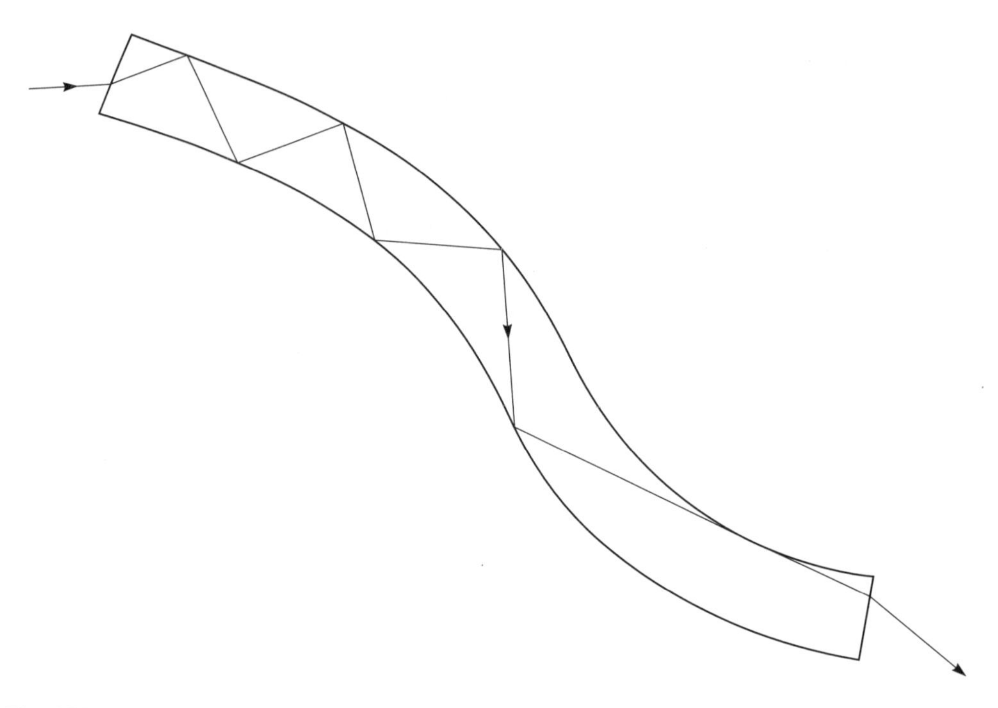

Fig. 6.55 Ray travelling through curved sheet or rod

allow light to escape, so producing a bright image. Alternatively, if the sheet is written on with a wax crayon light escapes through the wax, once again producing a bright image. In both these instances a black background enhances the effect.

Sheet light guides have also been used in aircraft instrument panels.[12] Figure 6.56(a) shows a section through such a panel. Light is provided by the midget lamp and its light is transmitted through the clear plastic after passing through the red filter. It is overlaid with a layer of white plastic, and finished with a layer of black plastic. This final layer is machined away to give the desired legend, which is seen as red by transmitted light and white by reflected light. Figure 6.56(b) shows how the edge of the sheet can be finished at an angle to direct light onto an instrument face.

Hollow light guides consisting of prismatic surfaces are used to conduct light and provide self-luminous surfaces. Typically these consist of right-angle prisms facing outwards (Figure 6.57) and running longitudinally. In the original design by Whitehead[13] the prisms were made in 6.35 mm thick plastic sheet, but to save material and weight the prisms have subsequently been made in very thin sheet by using precision methods of manufacture.[14] The result is a film only 0.5 mm in thickness with a prism depth of 0.18 mm. Since the film is so thin, it can be rolled to form a hollow cylinder if required.

There is some light leakage due to irregularities in the prisms and it is this that makes the hollow light pipe glow. Alternatively, an opal diffuser can be inserted into the pipe or holes can be cut into it to allow the light to escape in a controlled fashion. High intensity discharge sources in parabolic reflectors at one end provide the light, the other end having either a plane mirror or another light source, which can be of a different colour to provide directional guidance. For correct functioning, these reflectors should concentrate the light within a

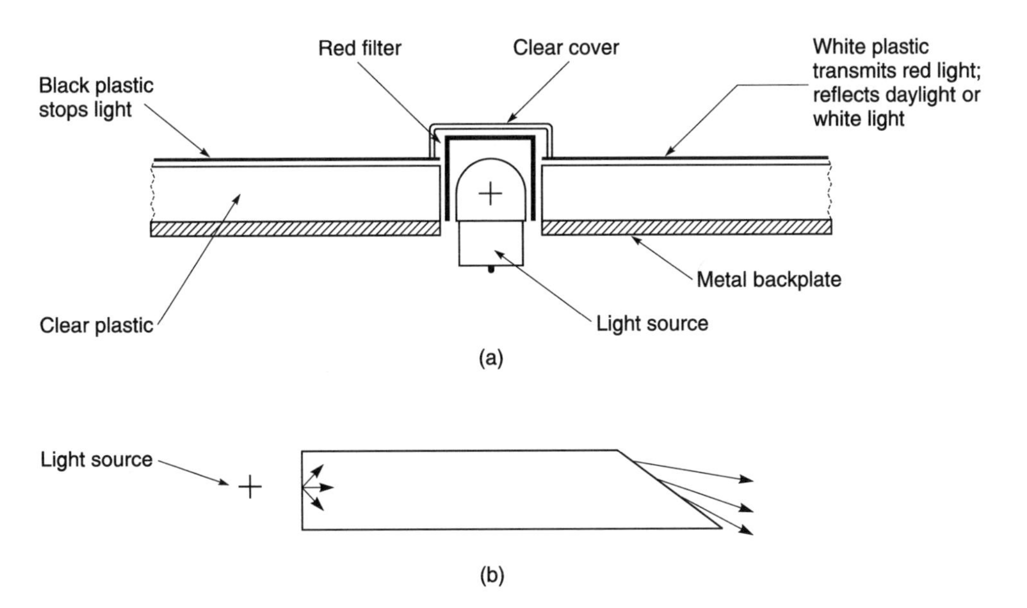

Fig. 6.56 Sheet light guides for instrument panels

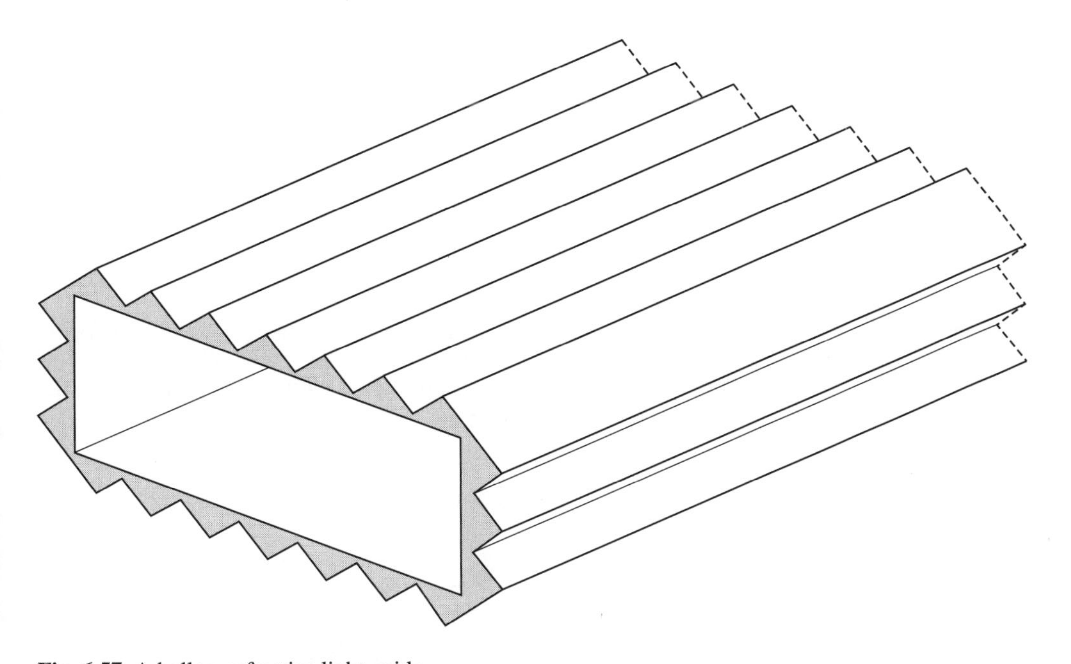

Fig. 6.57 A hollow refractive light guide

cone angle of 55.2° (that is, 2 × 27.6°), the axis of the cone being parallel to that of the light guide.

These pipes can give an appearance of even brightness over several metres. Suggested applications are mainly in road lighting; for lighting tunnels, for providing low mounting height guidance in the form of lines of light, and lighting overhead signs.

Problems

1. Light is incident on one face of a refracting prism at an angle of 35°. What prism angle will produce an angle of deviation of 20° if the refractive index of the optical material is 1.5?

   Answer: [36.8°]

2. If in problem 1 a ray is incident at 20° (in Fig. 6.36, page 244) what will be the deviation?

   Answer: [20.1°]

3. A ray passes through a parallel-sided block of thickness 20 mm and of refractive index 1.5. There is no absorption in the optical material of the block. What is its linear displacement and the approximate transmission factor when the angle of incidence is (1) 0°, (2) 45°, (3) 60°?

   Answers: [(1) 0, 0; (2) 6.6 mm, 0.90; (3) 10.3 mm, 0.83]

4. A specular paraboloid reflector of reflectance ρ has a radius of r (m). What, approximately, is the apparent peak luminous intensity in candelas if a source of luminance L (cd/m^2) is placed at the focus?

   Answer: [$\pi\rho r^2$ (cd)]

5. An object is placed between two vertical mirrors inclined towards each other at angles of (a) 90° and (b) 75°. How many images are visible when the mirrors are viewed in the plane at right angles to their line of intersection?

   Answers: [(a) 3, (b) 4]

6. A parabolic trough reflector has a width in cross-section of 200 mm and a focal length of 25 mm. Find the distance of the cross-over point on the axis of the reflector from the plane of the mouth of the reflector if the centre of a cylindrical source of diameter 3 mm is placed at its focus.

   Answer: [8333 mm]

7. A ray at 150° to the downward vertical is incident on one refracting face of a reflecting prism at 25° and emerges from the prism at 35° to the normal to the other refracting face, which is tilted at 15° to the downward vertical (angle C in Figure 6.37). Find the prism angles, the deviation, and the angle of incidence of the ray on the reflecting face if the refractive index of the optical material is 1.5.

   Answers: [Prism angles: 35°, 75° and 70°. Deviation: 80°. Angle of incidence: 52°]

Bibliography

Coaton, J.R. and Marsden, A.M. (1997) *Lamps and Lighting*, 4th edition (Arnold).
Elmer, W.B. (1980) *The Optical Design of Reflectors* (New York: Wiley).

References

1. Harding, G. (1999) Material Differences. *Light and Lighting*. February, 22.
2. Holmes, J.G. (1979) Para-ellipsoid and fan shaped beams. *Lighting Research and Technology*, **11**.
3. Spencer, D.E. (1962) *Illum Engng (New York)*, **57**, 166.

4. Anon. (1965) *The Engineer*, **219**, 318.
5. Stevens, R.W. (1953) *Trans IES (London),* **18**, 243.
6. Keitz, H.A.E. (1955) *Light Calculations and Measurements,* Philips Technical Library.
7. Jolley, L.B.W., Waldram, J.M. and Wilson, G.H. (1930) *The Theory and Design of Illuminating Engineering Equipment* (Chapman & Hall).
8. Ayers, M.J. and Carter, D.J. (1995) Remote source electric lighting systems: a review. *Lighting Research and Technology*, **27**(1), 1–15.
9. Edmonds, I.R., Moore, G.I., Smith, G.B. and Swift, P.D. (1995) Daylighting enhancement with light pipes coupled to laser-cut light-deflecting panels. *Lighting Research and Technology*, **27**, 27–35.
10. Marks, A.M. (1959) *Illum Engng (New York)*, **54**, 123.
11. McGowan, T.K., Davenport, J.M. and Hansler, R.L. (1993) High performance fiber optic illuminator systems for general lighting applications. CIBSE. *Proc Lux Europa*. **1**, 419–416.
12. Strange, J.W. and Stevens, B. (1958) *Trans IES (London),* **23**, 65.
13. Whitehead, L.A. (1981) *Prism light guide having surfaces which are in octature.* US Patent 4,260,220, 7 April.
14. Kneipp, K.G. (1994) Use of prismatic films to control light distribution. *3M Company, Traffic Control Materials Division, 3M Center, Building 260–5N–14, Saint Paul, MN 5514–1000, USA.*

7
Colour

7.1 Introduction

In the foregoing chapters, the calculations have been devoted, in the main, to problems associated with luminous flux distribution. The nature of that luminous flux has not been addressed other than to describe the light field as a region of space filled with electro-magnetic radiation within the visible range.

The feature that determines whether this electro-magnetic radiation is within the visible range (that is, being capable of stimulating the eye to vision), is the frequency of the radiation or, alternatively, the wavelength; since $C = f\lambda$, where C is the velocity of light in a vacuum, f the frequency and λ the wavelength.

The visible range, in terms of wavelength, lies between about 380 nm and 780 nm (1 nm = 10^{-9} m). The relative stimulation of vision varies with the wavelength and, for daytime or photopic vision, the maximum value is at about 555 nm. Night-time or scotopic vision has a similar response curve, but the maximum is at about 500 nm. Scotopic and photopic vision are a consequence of the eye containing two basic receptor systems; the rods, which are very sensitive to light, but give no sensation of colour (scotopic vision) and cones, which are less sensitive to light but which give the sensation of colour (photopic vision). The initial mechanism of vision is photochemical and at high levels of illuminance the rods become saturated and play little part in daytime vision. Under night-time or low light conditions the lack of sensitivity of the cones means that they play much less of a role in the seeing process and the rods then become the main active visual receptors. As the light level falls the balance between the two sets of receptors changes and a series of intermediate responses occur in what is termed the mesopic region (Figure 7.1).

In this chapter, we will assume that the cone receptors are fully operative and are producing colour vision.

The frequency (or wavelength) response of the eye in this condition not only determines the magnitude of the sensation of vision, but also the sensation of colour produced. A continuous change of colour with wavelength occurs as the wavelength progresses from 380 nm through to 780 nm; with the colour stimulated by the different frequencies changing from violet/blue, through greens to yellows, then orange and reds, as classically seen in the rainbow where white light has been split up by refraction.

Experimentally, a remarkable discovery was made. This was that, within certain limitations, particular colours of light can be exactly imitated by combining three suitable primary colours or sets of radiations. It was also found that additivity held in this colour mixture over a wide range of visual conditions.

This means that two colours can appear visually identical while a spectral analysis would

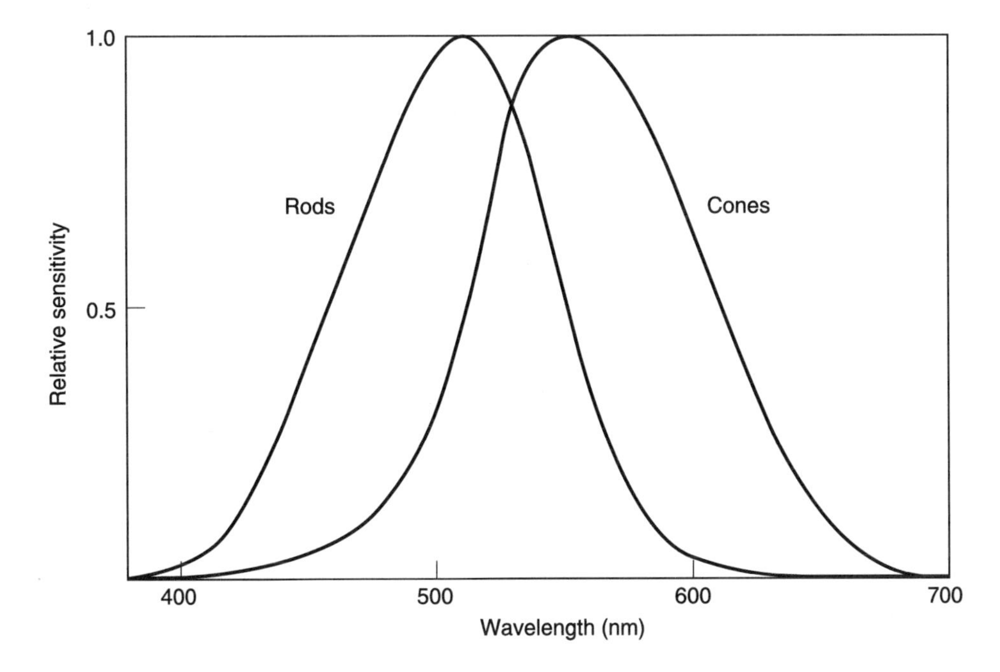

Fig. 7.1 Relative sensitivity curves for the eye

reveal that they are not identical in spectral distribution. This also means that a system of colour measurement, based on the amounts of the three primary colours – or radiations – that are required to match the test colour, is possible. This metamerism is the basis of *trichromatic colorimetry*.

This property of the eye of accepting matched colours as identical when their spectral distributions are different has led to the hypothesis that the cone receptors have three different forms or sensitivities; some responding mainly to red light, some to green and some to blue. In fact, there is evidence that there are three different types of cone receptors but that their spectral responses, in terms of maximum response, are yellow-orange, green and blue-violet. The spectral sensitivity curves for these cones, given symbols ρ, γ and β respectively, are shown in Figure 7.2.

The number of cones is not divided equally between these three responses and one estimate suggests ratios of 40 to 20 to 1 for ρ, γ and β respectively.

An important feature of the eye is that almost all the cones are concentrated on the optical axis of the eye at one small spot on the retina (the light sensitive layer at the back of the eye). The rods are distributed over the rest of the retina. There are about 6 million cones and 120 million rods in the eye. The fovea, or yellow spot, where the cones are situated is about 0.25 mm in diameter. This means that the angular extent of the measuring system becomes a factor in the specification. Thus, the CIE standard data produced in 1931 is based on the angle being between 1° and 4°, usually taken as 2°, but in 1964 data for a 10° field was also adopted by the CIE for applications that exceed the 4° field size.

It is necessary to make certain assumptions for a measurement system to have a wide application. We must assume that all eyes have the same colour response and also that any match will hold for a wide range of luminance and observer adaptation.

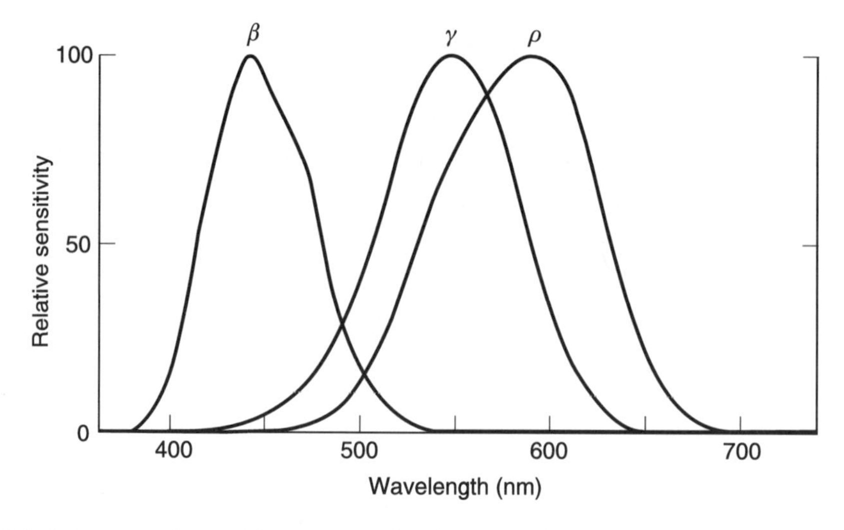

Fig. 7.2 Relative spectral sensitivity curves for the three types of cones

The psycho-physical aspects of colour are both fascinating and complex and are beyond the scope of this chapter which, of necessity, concerns itself mainly with calculations. It is sufficient for our purposes to note that a colour specification merely expresses equivalence between two sets of physical stimuli and their capacity to produce a particular colour sensation. The primaries commonly chosen are red, green and blue.

7.2 The *R, G, B* system

The basic concept is expressed in the following way. Let a colour to be matched be denoted by (C) and the reference stimuli by $[R]$, $[G]$ and $[B]$. Then after the match is achieved

$$c(C) = R[R] + G[G] + B[B] \tag{7.1}$$

where R, G and B are the amounts of $[R]$, $[G]$ and $[B]$ required for the match. The (C), $[R]$, $[G]$ and $[B]$ terms are, in effect, 'labels' and C, R, G and B the quantities.

So,

$$c = R + G + B \tag{7.2}$$

R, G and B are called the *tristimulus values.*

Dividing through equation (7.1) by equation (7.2) gives:

$$1.0(C) \equiv r[R] + g[G] + b[B]$$

where

$$r = \frac{R}{R+G+B} \quad g = \frac{G}{R+G+B} \quad b = \frac{B}{R+G+B}$$

and

$$r + g + b = 1.0$$

r, g and b are the *chromaticity coordinates* of (C).

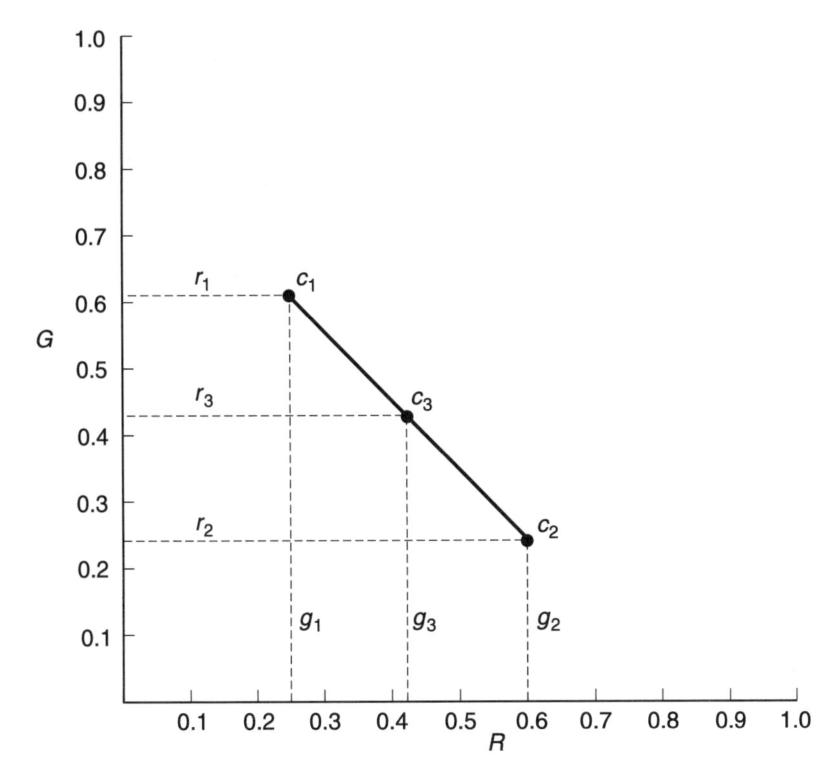

Fig. 7.3 The simple *R*, *G* chromaticity diagram

This equation is called a *unit trichromatic equation* and the amount of (C) represented is one trichromatic unit.

The advantage of this unit equation is that once two of the coordinates are known, the other coordinate is also fixed. Thus, a chromaticity diagram could be produced as shown in Figure 7.3.

As a simple example, let us consider two colours C_1 and C_2 combined to produce a third colour C_3. Let us assume that each colour contributes one trichromatic unit to the mixture. Then

$$\begin{aligned} c_3(C_3) &= (C_1) + (C_2) \\ &= (r_1 + r_2)[R] + (g_1 + g_2)[G] + (b_1 + b_2)[B] \end{aligned}$$

c_3 represents, in this case, the two trichromatic units present in the mixture. To obtain the unit equation values for r, g and b we must divide the sum of the coordinates by 2, that is:

$$(C_3) \equiv r_3[R] + g_3[G] + b_3[B]$$

where

$$r_3 = \tfrac{1}{2}(r_1 + r_2)$$

$$g_3 = \tfrac{1}{2}(g_1 + g_2)$$

$$b_3 = \tfrac{1}{2}(b_1 + b_2)$$

The equivalent sign (≡) used above can be used to indicate that a match exists between (*C*) and the [*R*], [*G*], [*B*] mixture in respect of colour appearance only and not of spectral distribution.

Experimentally, it is found that the pure spectral colours cannot be matched by adding other spectral colours in this way. This is because they are fully saturated colours and adding spectral colours together desaturates them; that is, it moves them towards the white position on the diagram.

However, it is possible to obtain a match that will enable the coordinates of the spectral colours to be obtained if the spectral colour to be matched is itself desaturated by adding an amount of the appropriate primary to it and matching this mixture against the mixture of the other two primaries. Algebraically, this is expressed as:

$$r + c = g + b$$

So,

$$c = g + b - r$$

Here we encounter the concept of a negative amount of colour. Thus, a new chromaticity diagram can be produced which shows the locus of all the spectral colours, as long as we allow for negative values (see Figure 7.4).

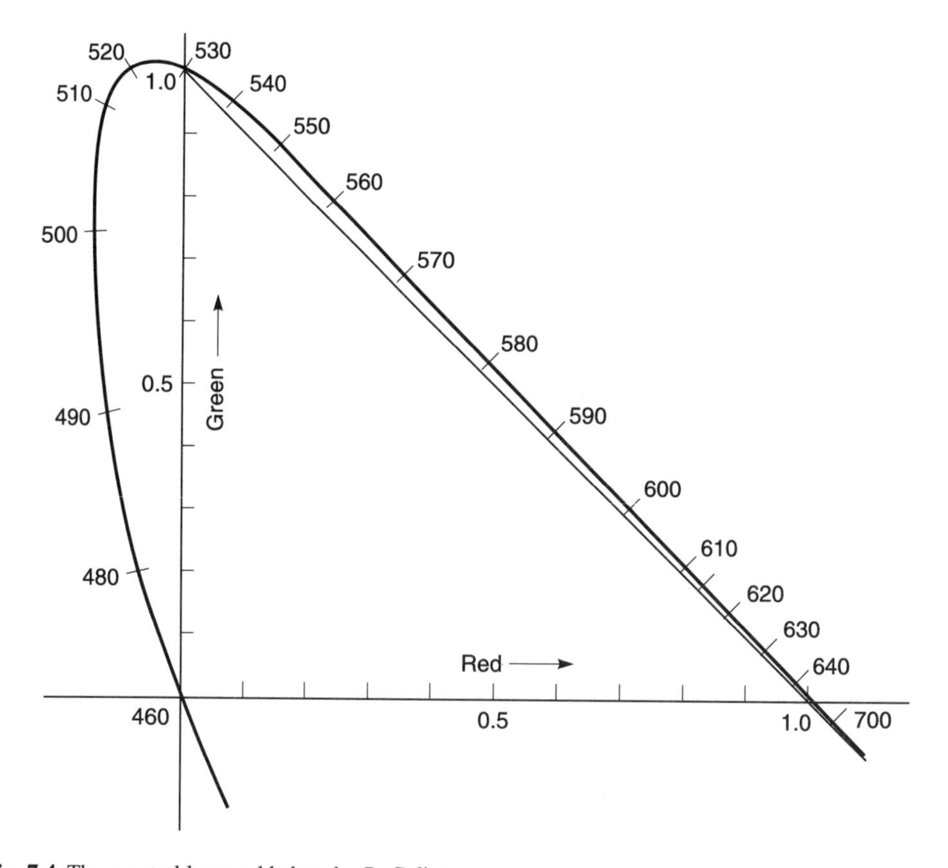

Fig. 7.4 The spectral locus added to the *R*, *G* diagram

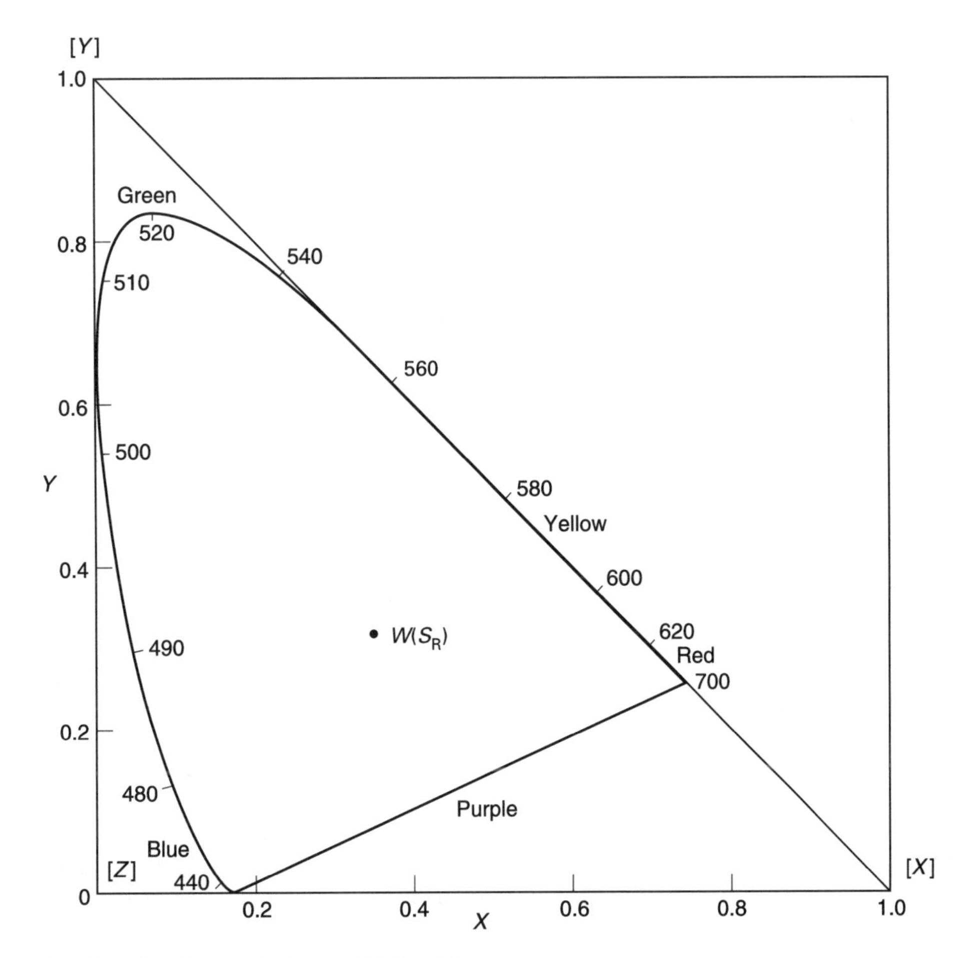

Fig. 7.5 The CIE diagram in terms of *X*, *Y* and *Z*

7.3 The CIE system

In 1931, the CIE adopted a colorimetric system that was designed for practical use in colour specification and measurement. It took the system described above and modified it to make it more convenient to use in practice.

The main changes were

1. A new set of theoretical primaries were adopted (*X*, *Y* and *Z*) that eliminated the need for negative quantities in the chromaticity calculations.
2. By a theoretical device, the *Y* primary was made the measure of the luminance and, consequently, the *X* and *Z* primaries have zero luminance.
3. Units for *X*, *Y* and *Z* were chosen that placed equal energy white at the centre of the [*XYZ*] triangle. This theoretical illuminant has a distribution of constant power per unit wavelength interval throughout the visible spectrum.
4. *X*, *Y* and *Z* values were selected that gave the most convenient shape for the spectral locus on the [*XYZ*] triangle.

The resulting CIE 1931 (*x*, *y*) chromaticity diagram is shown in Figure 7.5.

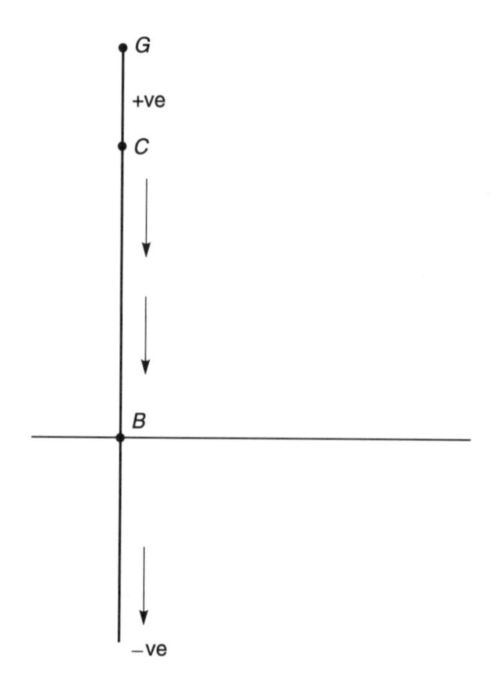

Fig. 7.6 A simplified illustration of locating a zero luminance point

It is perhaps easy to see that, in terms of a diagram, moving the reference points to positions outside the spectral locus means that all measurements to real colours must be positive. It is certainly not immediately obvious how two of these new primaries can have zero luminance. A very simple illustration as to how it is possible to arrange this will now be attempted. Consider Figure 7.6.

A colour *C* lies on the *GB* axis of the *RGB* chromaticity diagram and so consists of a mixture of *G* and *B* primaries. If the amount of the *G* primary is steadily reduced, *C* will eventually be located at *B* and contain no *G* primary. If point *C* is moved even further down the diagram the amount of the *G* primary in the mixture becomes theoretically negative. Of necessity, each colour carries with it not simply the attribute of colour but also luminance (although the diagram is not designed to indicate this directly).

Once the contribution of the *G* primary to the mixture becomes negative so, theoretically, does its luminance. In terms of luminance it is possible to see that a position will be reached where the negative luminance of the *G* primary component exactly equals the positive luminance of the *B* primary component. Thus, in algebraic terms, a point of zero luminance has been reached, relative to the *G* and *B* primaries.

This means that, on a diagram, and algebraically, zero luminance points can be specified (although this theoretical mixture cannot be realized in practice) and related to the real primaries *G* and *B* for calculation purposes.

It is by similar considerations to those of the simplified example given above that both the *X* and *Z* reference stimuli points are placed at positions of zero luminance relative to the real primaries used to set up the system.

Results obtained by using real *RGB* primaries can, therefore, be transformed algebraically into terms of the CIE *XYZ* primaries.

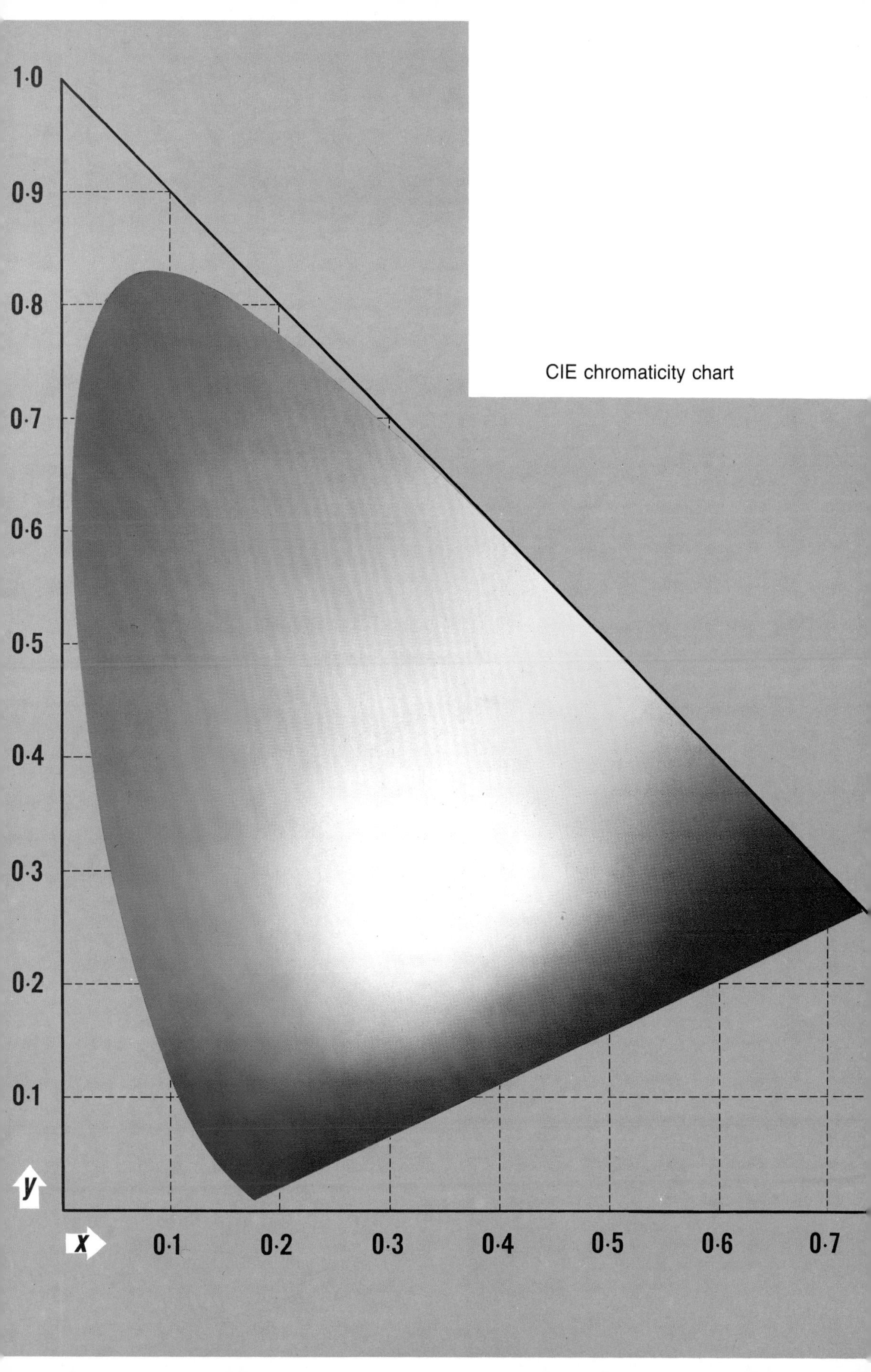

CIE chromaticity chart

There is a very real advantage in employing two reference stimuli that have zero luminance, since the amount of light represented by a colour equation can then be computed directly from the third tristimulus value, in this case *Y*.

The locus of spectral colours was determined experimentally using a number of observers in terms of *RGB* and then translated into *XYZ* coordinates, as shown in Figures 7.4 and 7.5.

The assumption was made that equal quantities of the *XYZ* primaries would produce an equal energy white. This is a theoretical illuminant that has a distribution of constant power per unit wavelength interval throughout the visible spectrum.

Since the *Y* value is an exclusive measure of luminance, the CIE spectral tristimulus curve (colour matching function) for *Y* is identical with the $V(\lambda)$ response curve for the standard observer (assumed to represent the average observer's visual response), see Figure 7.1.

Thus, the colour matching function denoted

$$\bar{y}(\lambda) = V(\lambda)$$

The $\bar{x}$ and $\bar{z}$ functions for the spectral locus are adjusted accordingly. If the $\bar{x}$, $\bar{y}$ and $\bar{z}$ values are separately totalled they will be found to give three equal totals (see Table 7.1).

This will be seen to be necessary once the assumption was made of equal energy white giving equal tristimulus values of *X*, *Y* and *Z*.

Since,

$$X = k \int \psi(\lambda)\bar{x}(\lambda)\, d\lambda \quad \text{and} \quad \psi = \rho(\lambda)S(\lambda)$$
$$Y = k \int \psi(\lambda)\bar{y}(\lambda)\, d\lambda \quad \text{or} \quad = \beta(\lambda)S(\lambda)$$
$$Z = k \int \psi(\lambda)\bar{z}(\lambda)\, d\lambda \quad \text{or} \quad = T(\lambda)S(\lambda)$$

where

$$k = \frac{100}{\int S(\lambda)\bar{y}(\lambda)\, d\lambda}$$

Table 7.1 The Tristimulus values of spectral colours

Wavelength (nm)	$\bar{x}$	$\bar{y}$	$\bar{z}$	Wavelength (nm)	$\bar{x}$	$\bar{y}$	$\bar{z}$
400	0.0143	0.0004	0.0679	560	0.5945	0.9950	0.0039
410	0.0435	0.0012	0.2074	570	0.7621	0.9520	0.0021
420	0.1344	0.0040	0.6456	580	0.9163	0.8700	0.0017
430	0.2839	0.0116	1.3856	590	1.0263	0.7570	0.0011
440	0.3483	0.0230	1.7471	600	1.0622	0.6310	0.0008
450	0.3362	0.0380	1.7721	610	1.0026	0.5030	0.0003
460	0.2908	0.0600	1.6692	620	0.8544	0.3810	0.0002
470	0.1954	0.0910	1.2876	630	0.6424	0.2650	
480	0.0956	0.1390	0.8130	640	0.4479	0.1750	
490	0.0320	0.2080	0.4562	650	0.2835	0.1070	
500	0.0049	0.3230	02720	660	0.1649	0.0610	
510	0.0093	0.5030	0.1582	670	0.0874	0.0320	
520	0.0633	0.7100	0.0782	680	0.0468	0.0170	
530	0.1655	0.8620	0.0422	690	0.0227	0.0082	
540	0.2904	0.9540	0.0203	700	0.0114	0.0041	
550	0.4334	0.9950	0.0087	710	0.0058	0.0021	

$\rho(\lambda)$ is the spectral reflectance
$\beta(\lambda)$ is the spectral luminance factor
$T(\lambda)$ is the spectral transmittance
$S(\lambda)$ is the relative spectral power distribution

k is a normalizing factor that cancels when the chromaticity coordinates are calculated. k is chosen such that $Y = 100$ for the perfect diffuser with a reflectance factor $\rho(\lambda)$ equal to 1.0 for all wavelengths. If k is put equal to 683 and $S(\lambda)$ is replaced by $P(\lambda)$ in watts per steradian per square metre, then Y is the luminance in candelas per square metre (cd/m^2). The chromaticity coordinates are the ratio of each of the tristimulus values to their sum.

So,

$$x = \frac{X}{X+Y+Z}, \quad y = \frac{Y}{X+Y+Z} \quad \text{and} \quad z = \frac{Z}{X+Y+Z}$$

Example 1 Below, in spreadsheet form, Table 7.2 is the calculation of the *X, Y* and *Z* values for a particular light source and, hence, the chromaticity coordinates. In practice, the integration is replaced by a summation. To avoid lengthy repetition the wavelength bands are chosen as 20 nm; in practice 5 nm and 10 nm bands are usual. The range has also been limited to 400 to 700 nm:

Table 7.2 Spreadsheet calculation of the *XYZ* values for Example 1

λ	$\bar{x}(\lambda)$	$\bar{y}(\lambda)$	$\bar{z}(\lambda)$	$S(\lambda)$	$S(\lambda) \times \bar{y}(\lambda)$	$\rho(\lambda)$	X/k	Y/k	Z/k
400	0.0143	0.0004	0.0679	63.3	0.02532	0.10	0.1	0	0.43
420	0.1344	0.0040	0.6456	98.1	0.3924	0.10	1.32	0.04	6.333
440	0.3483	0.0230	1.7471	122	2.7945	0.10	4.23	0.279	21.23
460	0.2908	0.0600	1.6692	123	7.386	0.10	3.58	0.739	20.55
480	0.0956	0.1390	0.8130	124	17.2221	0.10	1.18	1.722	10.07
500	0.0049	0.3230	0.2720	112	36.2083	0.10	0.1	3.621	3.049
520	0.0633	0.7100	0.0782	96.9	68.799	0.10	0.61	6.88	0.758
540	0.2904	0.9540	0.0203	102	97.4034	0.10	2.97	9.74	0.207
560	0.5945	0.9950	0.0039	105	104.7735	0.10	6.26	10.48	0.04
580	0.9163	0.8700	0.0017	97.8	85.086	0.10	8.96	8.509	0.02
600	1.0622	0.6310	0.0008	89.7	56.6007	0.10	9.53	5.66	0
620	0.8544	0.3810	0.0002	88.1	33.5661	0.10	7.53	3.357	0
640	0.4479	0.1750	0.0000	87.8	15.365	0.10	3.93	1.537	0
660	0.1649	0.0610	0.0000	87.9	5.3619	0.10	1.45	0.536	0
680	0.0468	0.0170	0.0000	84	1.428	0.10	0.39	0.143	0
700	0.0114	0.0041	0.0000	76.3	0.31283	0.10	0.1	0.03	0
			Totals		532.72505		52.2	53.27	62.69

$$k = \frac{100}{532.7}$$

$X = 9.79439244$
$Y = 10$
$Z = 11.7681911$

$$X = k \sum_{400}^{700} S(x)\bar{x}(\lambda)$$

$$Y = k \sum_{400}^{700} S(\lambda)\bar{y}(\lambda)$$

$$Z = k \sum_{400}^{700} S(\lambda)\bar{z}(\lambda)$$

$$x = \frac{X}{X+Y+Z} = \frac{97.94}{97.94 + 100 + 117.7} = 0.310$$

$$y = \frac{Y}{X+Y+Z} = \frac{100}{315.6} = 0.316$$

$z = 1 - x - y = 1 - 0.310 - 0316 = 0.374$ (see Figure 7.7(a))

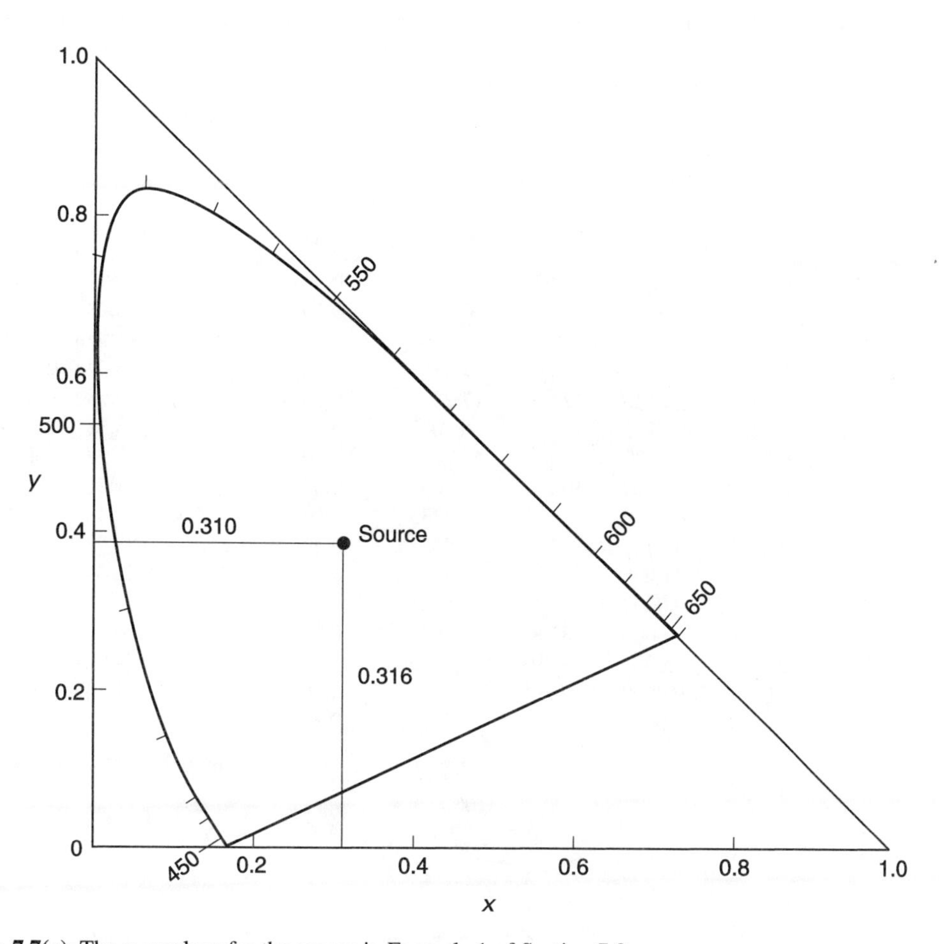

Fig. 7.7(a) The x, y values for the source in Example 1 of Section 7.3

Example 2 Below, in spreadsheet form, Table 7.3 is the calculation for the *X*, *Y* and *Z* values for the light reflected from a test sample. The light source is the same as that used in Example 1 and the reflectance values $\rho(\lambda)$ are given in the table.

$$X = k \sum_{400}^{700} S(\lambda)\rho(\lambda)\bar{x}(\lambda)$$

$$Y = k \sum_{400}^{700} S(\lambda)\rho(\lambda)\bar{y}(\lambda)$$

$$Z = k \sum_{400}^{700} S(\lambda)\rho(\lambda)\bar{z}(\lambda)$$

$$x = \frac{X}{X+Y+Z} = \frac{44.73}{44.73 + 39.2 + 84.4} = 0.266$$

$$y = \frac{Y}{X+Y+Z} = \frac{39.2}{168.3} = 0.233$$

$z = 1 - x - y = 0.501$ (see Figure 7.7(b))

Table 7.3 Spreadsheet calculation of the *XYZ* values for Example 2

λ	$\bar{x}(\lambda)$	$\bar{y}(\lambda)$	$\bar{z}(\lambda)$	$S(\lambda)$	$S(\lambda) \times \bar{y}(\lambda)$	$\rho(\lambda)$	X/k	Y/k	Z/k
400	0.0143	0.0004	0.0679	63.3	0.02532	0.40	0.36	0.01	1.719
420	0.1344	0.0040	0.6456	98.1	0.3924	0.60	7.91	0.235	38
440	0.3483	0.0230	1.7471	122	2.7945	0.70	29.6	1.956	148.6
460	0.2908	0.0600	1.6692	123	7.386	0.75	26.8	5.54	154.1
480	0.0956	0.1390	0.8130	124	17.2221	0.80	9.48	13.78	80.58
500	0.0049	0.3230	0.2720	112	36.2083	0.70	0.38	25.35	21.34
520	0.0633	0.7100	0.0782	96.9	68.799	0.60	3.68	41.28	4.547
540	0.2904	0.9540	0.0203	102	97.4034	0.30	8.9	29.22	0.622
560	0.5945	0.9950	0.0039	105	104.7735	0.10	6.26	10.48	0.04
580	0.9163	0.8700	0.0017	97.8	85.086	0.30	26.9	25.53	0.05
600	1.0622	0.6310	0.0008	89.7	56.6007	0.40	38.1	22.64	0.03
620	0.8544	0.3810	0.0002	88.1	33.5661	0.50	37.6	16.78	0
640	0.4479	0.1750	0.0000	87.8	15.365	0.70	27.5	10.76	0
660	0.1649	0.0610	0.0000	87.9	5.3619	0.75	10.9	4.021	0
680	0.0468	0.0170	0.0000	84	1.428	0.80	3.15	1.142	0
700	0.0114	0.0041	0.0000	76.3	0.31283	0.80	0.7	0.25	0
			Totals		532.72505		238.2	209	449.6

$$k = \frac{100}{532.7}$$

$X = 44.73$
$Y = 39.23$
$Z = 84.40$

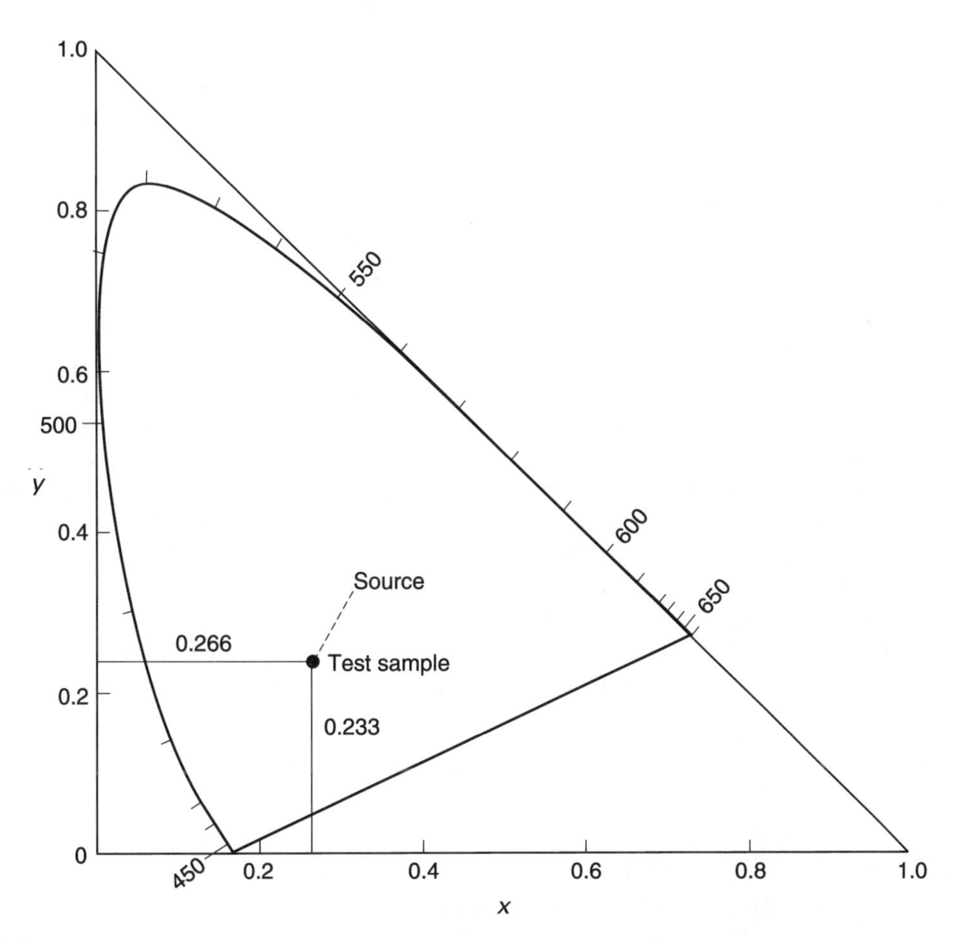

Fig. 7.7(b) The x, y values for the reflected light in Example 2 of Section 7.3

Example 3 Below, in spreadsheet form, Table 7.4 is the calculation for X, Y and Z values for the light reflected from a test sample. The test sample is the same as that in Example 2, but the light source has been changed.

$$X = k \sum_{400}^{700} S(\lambda)\rho(\lambda)\bar{x}(\lambda)$$

$$Y = k \sum_{400}^{700} S(\lambda)\rho(\lambda)\bar{y}(\lambda)$$

$$Z = k \sum_{400}^{700} S(\lambda)\rho(\lambda)\bar{z}(\lambda)$$

$$x = \frac{X}{X + Y + Z} = \frac{49.0}{49.0 + 38.29 + 25.6} = 0.434$$

$$y = \frac{Y}{X + Y + Z} = \frac{38.29}{112.9} = 0.339$$

$$z = 1 - x - y = 1 - 0.434 - 0.339 = 0.227$$

It will be noted how the change of light source in this example compared with Example 2 has resulted in the colour moving on the CIE chromaticity diagram towards the red-yellow area of the diagram (see Figure 7.7(c)).

7.3.1 CHROMATIC ADAPTATION

The objective assessment of colour shifts due to changes in illuminant is complicated by chromatic adaptation. When the eye is exposed to a saturated colour, the colour receptors most highly stimulated become fatigued and the visual sensation moves towards the complementary colour to the stimulating colour. In addition, the eye/brain combination also appears to adjust when different light sources are experienced successively, so that what are seen as large changes when seen simultaneously are readily accepted when viewed successively. This is called the *colour constancy effect.*

Von Kries (1902) suggested that a correction could be made for chromatic adaptation, if the assumption was made that the shape of the sensitivity curves for the three types of receptor did not change due to adaptation, but the relative magnitude of the responses of the receptors to

Table 7.4 Spreadsheet calculation of the *XYZ* values for Example 3

λ	$\bar{x}(\lambda)$	$\bar{y}(\lambda)$	$\bar{z}(\lambda)$	$S(\lambda)$	$S(\lambda) \times \bar{y}(\lambda)$	$\rho(\lambda)$	X/k	Y/k	Z/k
400	0.0143	0.0004	0.0679	15	0.006	0.40	0.1	0	0.407
420	0.1344	0.0040	0.6456	21	0.084	0.60	1.69	0.05	8.135
440	0.3483	0.0230	1.7471	29	0.667	0.70	7.07	0.467	35.47
460	0.2908	0.0600	1.6692	38	2.28	0.75	8.29	1.71	47.57
480	0.0956	0.1390	0.8130	48	6.672	0.80	3.67	5.338	31.22
500	0.0049	0.3230	0.2720	60	19.38	0.70	0.21	13.57	11.42
520	0.0633	0.7100	0.0782	72	51.12	0.60	2.73	30.67	3.378
540	0.2904	0.9540	0.0203	86	82.044	0.30	7.49	24.61	0.524
560	0.5945	0.9950	0.0039	100	99.5	0.10	5.95	9.95	0.04
580	0.9163	0.8700	0.0017	114	99.18	0.30	31.3	29.75	0.06
600	1.0622	0.6310	0.0008	129	81.399	0.40	54.8	32.56	0.04
620	0.8544	0.3810	0.0002	144	54.864	0.50	61.5	27.43	0.01
640	0.4479	0.1750	0.0000	158	27.65	0.70	49.5	19.36	0
660	0.1649	0.0610	0.0000	172	10.492	0.75	21.3	7.869	0
680	0.0468	0.0170	0.0000	185	3.145	0.80	6.93	2.516	0
700	0.0114	0.0041	0.0000	198	0.8118	0.80	1.81	0.649	0
			Totals		539.2948		264	206.5	138.3

$$k = \frac{100}{539.7}$$

$X = 49.02$
$Y = 38.29$
$Z = 25.64$

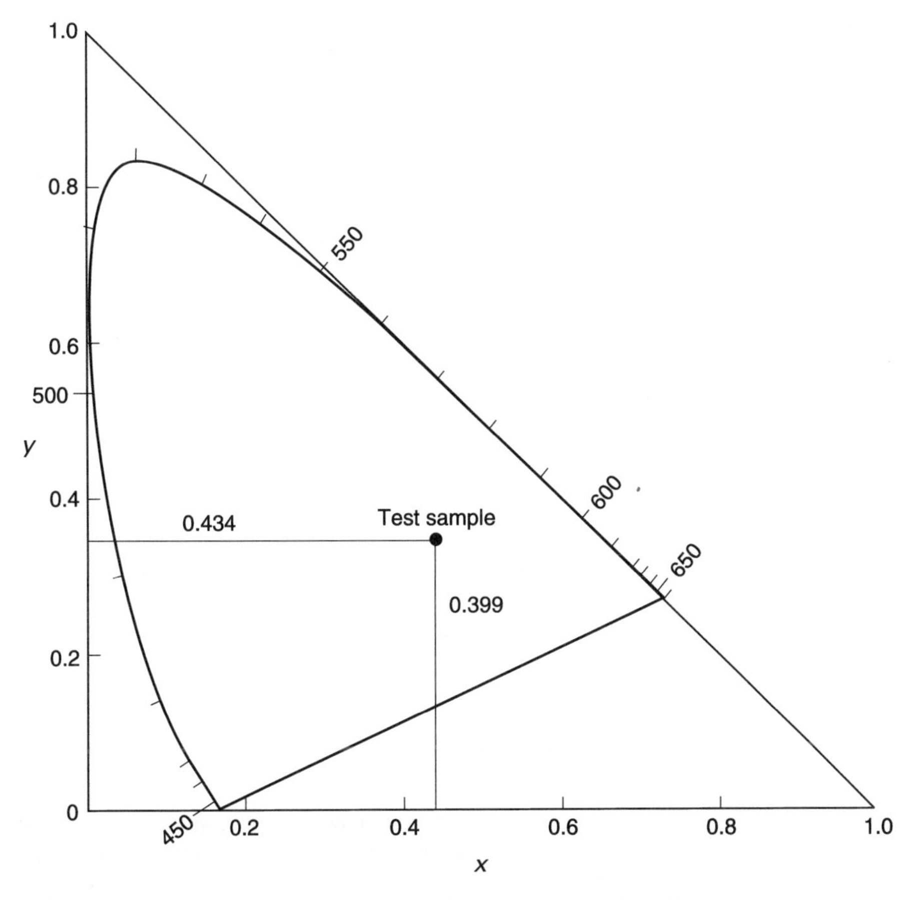

Fig. 7.7(c) The *x*, *y* values for the reflected light in Example 3 of Section 7.3

each other did change. The Von Kries method of correction still finds use in the CIE method of calculating the General Colour Rendering Index of light sources (see Section 7.9.2), even though experimental evidence suggests that predictions using this type of correction do not give very satisfactory results. Work is still continuing on the production of a more satisfactory method.

7.3.2 COLOUR MIXTURE ON THE CHROMATICITY DIAGRAM

In the chromaticity diagram of Figure 7.8, three colours are indicated, C_1, C_2 and C_3. C_3 is produced by a mixture of C_1 and C_2 and is, therefore, on the line joining them.

Let us choose an amount of C_1 such that $X = x_1$, $Y = y_1$ and $Z = z_1$. The luminance of this amount of C_1 will be proportional to Y. Let this luminance be $L_y Y$ (where L_y is a constant) and from above $Y = y_1$ so it is equal to L_y, y_1.

Thus, for 1 luminance unit of C_1 we may write

$$X = \frac{x_1}{L_y y_1}, \quad Y = \frac{y_1}{L_y y_1}, \quad Z = \frac{z_1}{L_y y_1}$$

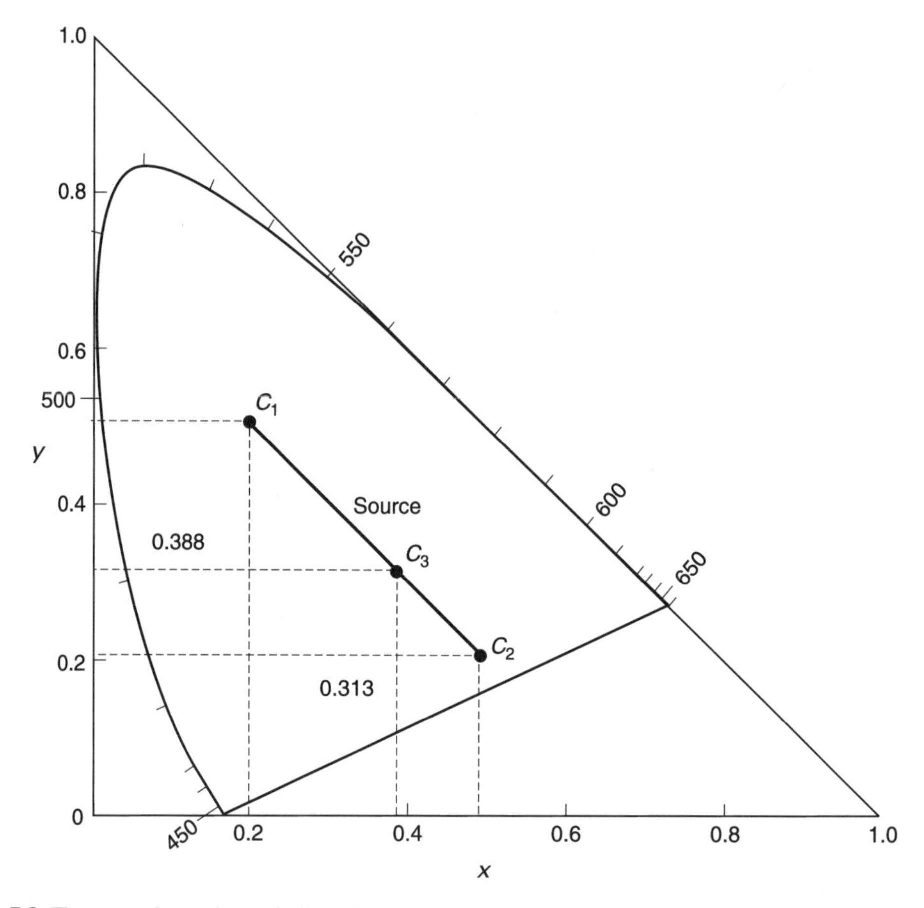

Fig. 7.8 The two colour mixture indicated on the CIE chromaticity diagram

and for m_1 luminance units of C_1 we may write

$$X = \frac{m_1 x_1}{L_y y_1}, \quad Y = \frac{m_1 y_1}{L_y y_1}, \quad Z = \frac{m_1 z_1}{L_y y_1}$$

Similarly for C_2,

$$X = \frac{m_2 x_2}{L_y y_2}, \quad Y = \frac{m_2 y_2}{L_y y_2}, \quad Z = \frac{m_2 z_2}{L_y y_2}$$

Let the amounts of these two colours be additively mixed together:

$$X = \frac{m_1 x_1}{L_y y_1} + \frac{m_2 x_2}{L_y y_2}, \quad Y = \frac{m_1 y_1}{L_y y_1} + \frac{m_2 y_2}{L_y y_2}, \quad Z = \frac{m_1 z_1}{L_y y_1} + \frac{m_2 z_2}{L_y y_2}$$

and so

$$X + Y + Z = \frac{m_1 x_1}{L_y y_1} + \frac{m_2 x_2}{L_y y_2} + \frac{m_1 y_1}{L_y y_1} + \frac{m_2 y_2}{L_y y_2} + \frac{m_1 z_1}{L_y y_1} + \frac{m_2 z_2}{L_y y_2}$$

$$= \frac{m_1(x_1 + y_1 + z_1)}{L_y y_1} + \frac{m_2(x_2 + y_2 + z_2)}{L_y y_2}$$

$$= \frac{m_1}{y_1} + \frac{m_2}{y_2} \text{ since } (x + y + z) = 1.0.$$

This gives

$$x_3 = \frac{\dfrac{m_1 x_1}{y_1} + \dfrac{m_2 x_2}{y_2}}{\dfrac{m_1}{y_1} + \dfrac{m_2}{y_2}}$$

$$y_3 = \frac{m_1 + m_2}{\dfrac{m_1}{y_1} + \dfrac{m_2}{y_2}}$$

and

$$z_3 = 1 - x_3 - y_3$$

Example Let three luminance units of a colour of chromaticity coordinates $x = 0.2$, $y = 0.5$ be mixed with two luminance units of a colour of chromaticity coordinates $x = 0.50$, $y = 0.20$.

The new chromaticity coordinates are given by

$$x = \frac{\dfrac{3 \times 0.2}{0.5} + \dfrac{2 \times 0.5}{0.2}}{\dfrac{3}{0.5} + \dfrac{2}{0.2}}$$

$$= 0.388$$

$$y = \frac{3 + 2}{\dfrac{3}{0.5} + \dfrac{2}{0.2}}$$

$$= 0.313$$

$$z = 1 - 0.388 - 0.313$$

$$= 0.299$$

Note To complete the specification of a surface colour under a particular illuminant, it is necessary to quote both the chromaticity coordinates and the luminance factor Y. The luminance factor is defined as follows:

> The ratio of the luminance of the body to that of a perfectly reflecting or transmitting diffuser identically illuminated, symbol β.

7.3.3 DOMINANT WAVELENGTH AND PURITY

If a line is drawn from a chosen white point on the chromaticity diagram, such that it passes through the colour to be specified and intersects the spectral locus (Figure 7.9), the wavelength at the point of interception is called the *dominant wavelength*. This is because the colour can be matched by adding appropriate amounts of this dominant wavelength and the white. When a colour is on the purple side of the white, where there is a gap in the spectrum locus, the convention is to take the complementary wavelength as the dominant wavelength.

The purity is the measure of the saturation of the colour and is determined by the proportions of the monochromatic source and the white that are needed to match the test colour additively.

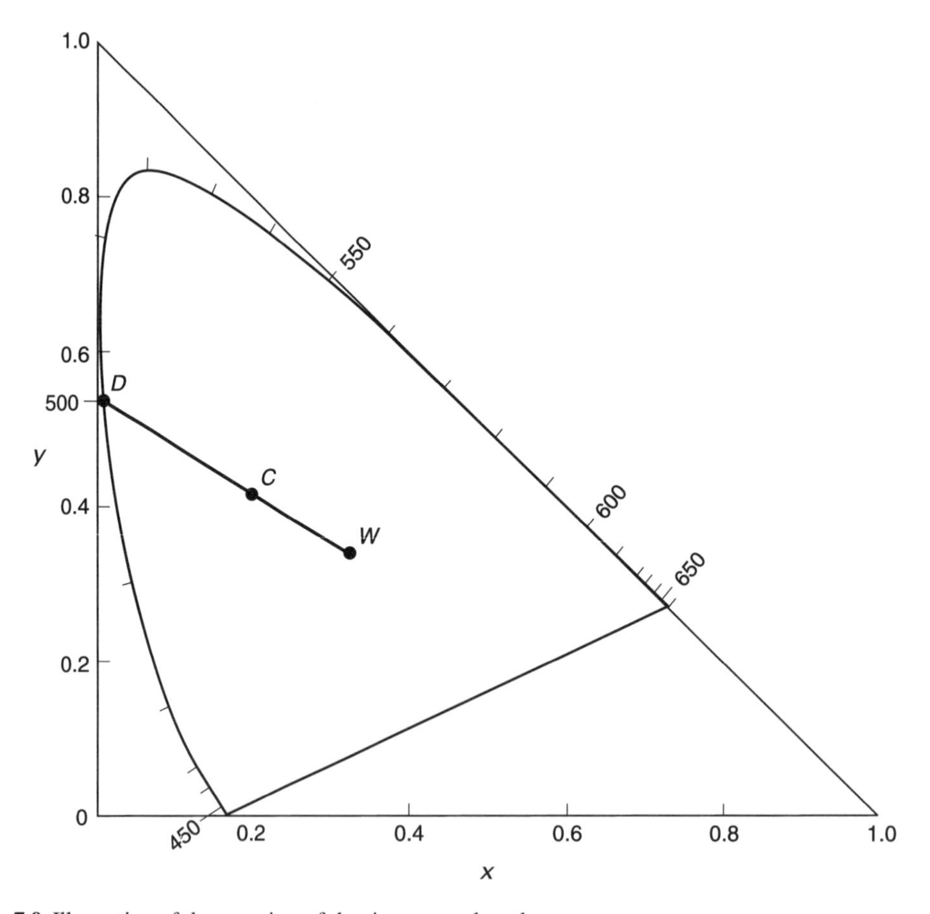

Fig. 7.9 Illustration of the meaning of dominant wavelength

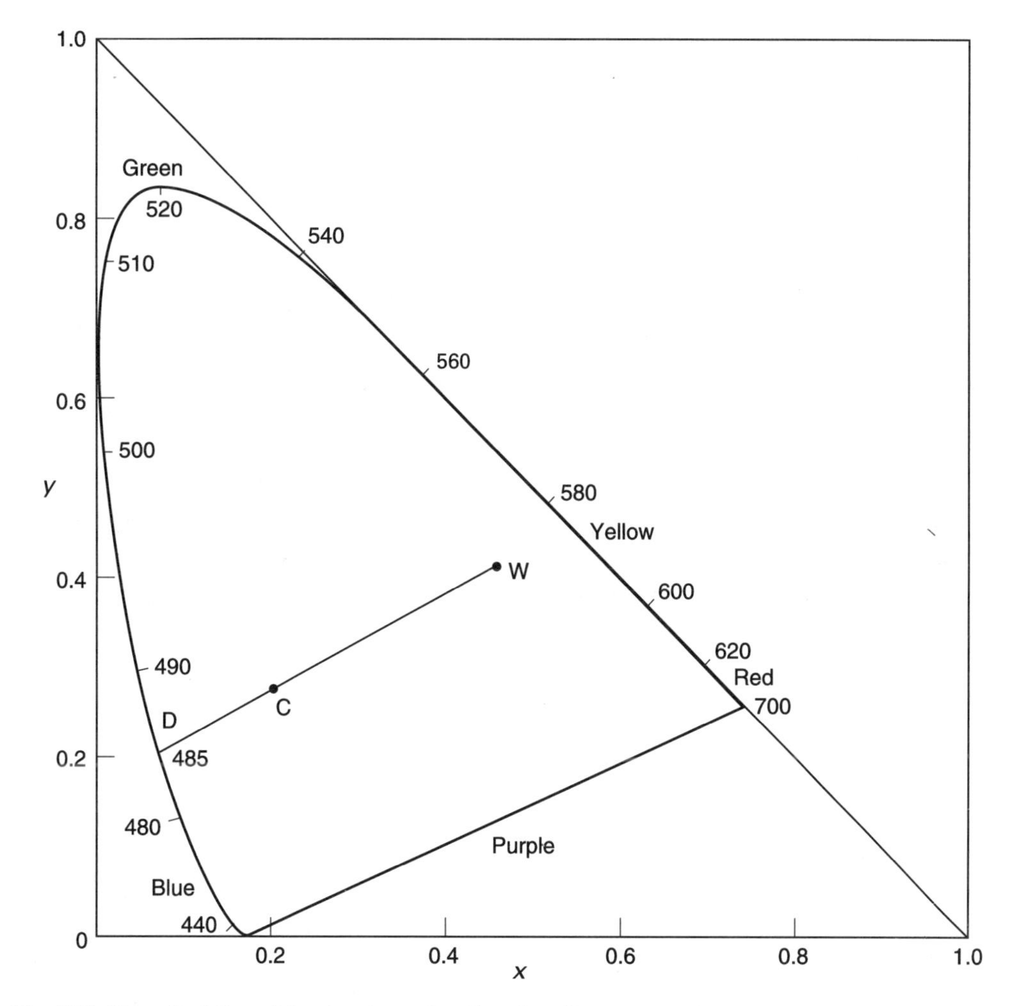

Fig. 7.10 The calculation of dominant wavelength and purity

The excitation purity (P_e) can be calculated from the ratio of the distance from the chosen white point to the colour, to the distance from the white point to the dominant wavelength *WC*/*WD* in Figure 7.9.

There are a number of ways that the purity can be calculated, and so the appropriate CIE document should be consulted for full information (CIE 1986a: *Colorimetry*, 2nd edition, Publication No. 15.2).

A simple example is given below of the calculation of this means of specification.

Example Given that the chromaticity coordinates of a test colour are $x = 0.2$, $y = 0.273$ and assuming a reference white with chromaticity coordinates of $x = 0.48$, $y = 0.408$, find the dominant wavelength and the excitation purity.

From Figure 7.10 it will be seen that a line from the reference white point through the test colour intersects with the spectral locus at $x = 0.069$, $y = 0.201$ which are the coordinates for $\lambda = 485$ nm.

The dominant wavelength is therefore 485 nm.
The excitation purity P_e is given by $P_e = WC/WD$.
So,

$$P_e = \frac{WC}{WD} = \frac{x_w - x_c}{x_w - x_d} + \frac{0.448 - 0.2}{0.448 - 0.069}$$

$$= 0.65$$

Obviously, the closer to 1.0 the value of P_e, the more saturated the colour.

7.4 Non-uniformity of the CIE (1931) diagram

In the 1940s, researchers discovered serious weaknesses in the 1931 CIE diagram, in that the 'noticeable difference' in colour terms of x and y varied across the diagram. Various attempts have been made to produce a more uniform chromaticity scale diagram. The one that we shall use in the next section is the CIE 1960 uniform chromaticity scale (UCS) diagram. This diagram exchanges x and y for u and v. This scale gives improved uniformity in the centre of the diagram near the full radiator locus – see Figures 7.11 and 7.12. The MacAdam ellipses shown in these diagrams relate to the minimum perceptible colour differences across the diagrams.

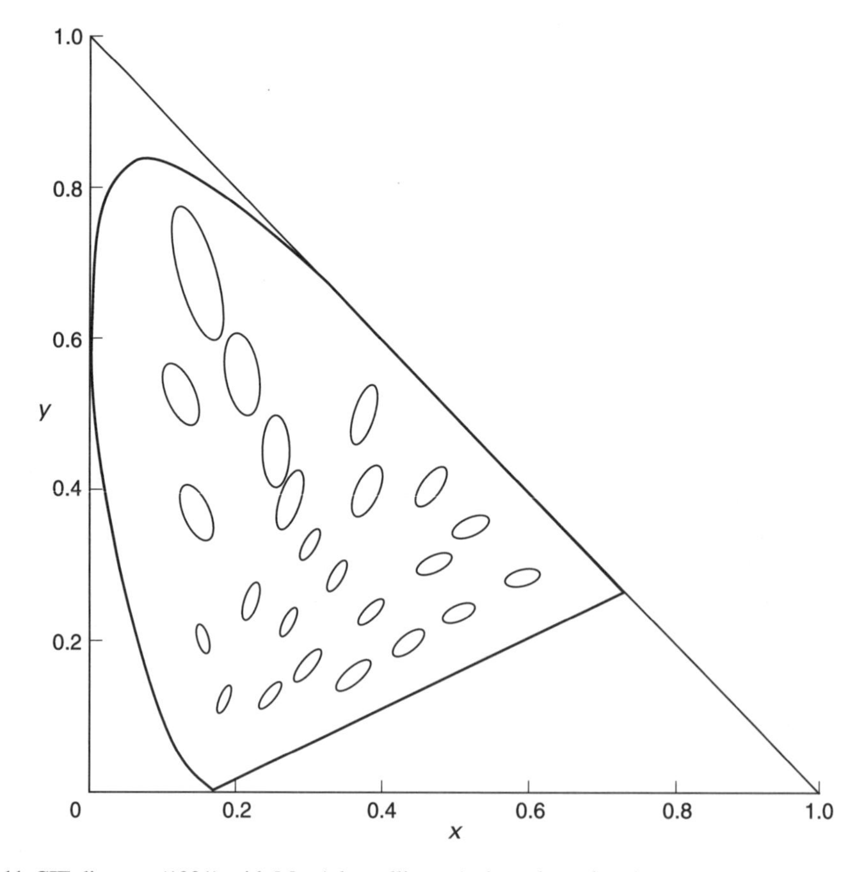

Fig. 7.11 CIE diagram (1931) with MacAdam ellipses (enlarged ten times)

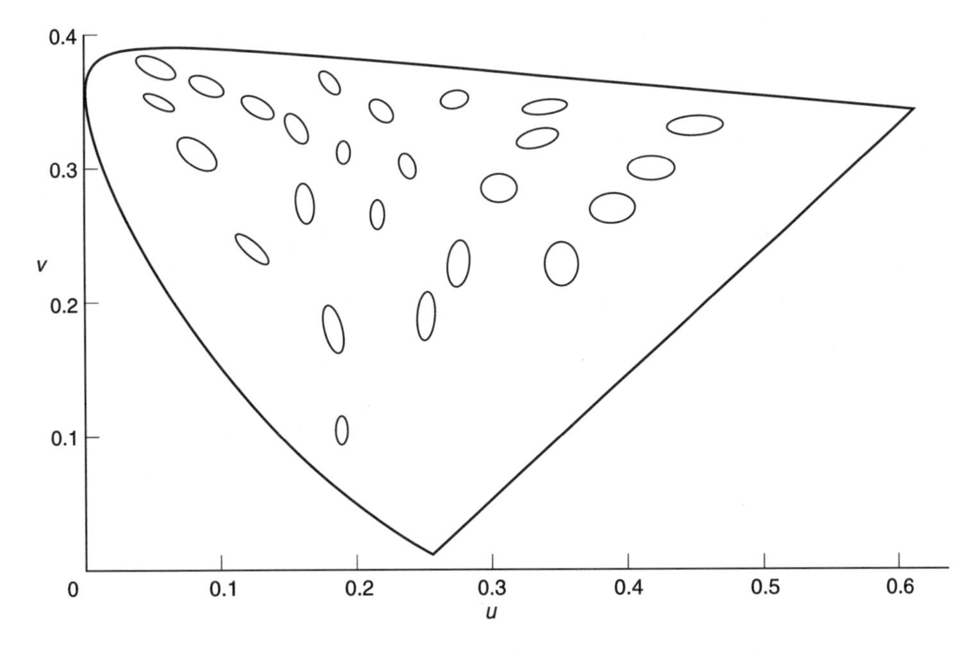

Fig. 7.12 CIE UCS diagram (1960) with MacAdam ellipses (enlarged ten times)

The transformation from the 1931 system to the 1960 diagram is achieved as follows:

$$u = \frac{4X}{X + 15Y + 3Z} = \frac{4x}{-2x + 12y + 3}$$

$$v = \frac{6Y}{X + 15Y + 3Z} = \frac{6y}{-2x + 12y + 3}$$

At the time of writing, attempts are still being made to construct a fully acceptable uniform colour diagram. A footnote to CIE document 13–3 (1995) states:

> The 1960 UCS diagram and 1964 Uniform Space are declared obsolete recommendation in CIE 15·2 (1986), but have been retained for the time being for calculating colour rendering indices and correlated colour temperature.

7.5 Correlated colour temperature

This is a method of relating the colour appearance of a light source to that of the Planckian (Full) Radiator. The Planckian Radiator is a thermal radiator with a spectral distribution defined uniquely by the temperature.

The defining equation is:

$$M_{e\lambda}^{th} = \frac{C_1}{\lambda^5} (e^{(C_2/\lambda T)} - 1)^{-1} \times 10^{-9}\ (\mathrm{Wm^{-2}\ nm^{-1}})$$

where

$C_1 = 3.7418 \times 10^{-16}$ (Wm^{-2})
$C_2 = 1.4388 \times 10^{-2}$ (mK)
λ is in metres

The result is expressed in Wm^{-2} per nm band to bring it in line with the general use of nm for spectral distributions (hence the 10^{-9} multiplier).

From the calculated spectral distribution the x and y coordinates at each temperature may be obtained and this gives rise to the Full Radiator locus shown in Figure 7.13. For the purpose of calculating the correlated colour temperature, the 1931 CIE (x, y) chromaticity coordinates are converted into the 1960 CIE UCS (u, v) coordinates. The appropriate part of the chromaticity diagram is shown in Figure 7.14.

If the test source has u, v values that lie on the Planckian locus then its colour temperature is that of the Full Radiator at that temperature. However, such exact coincidence is unlikely and so if the u, v values lie close to the Full Radiator locus it can be referred to by its correlated colour temperature.

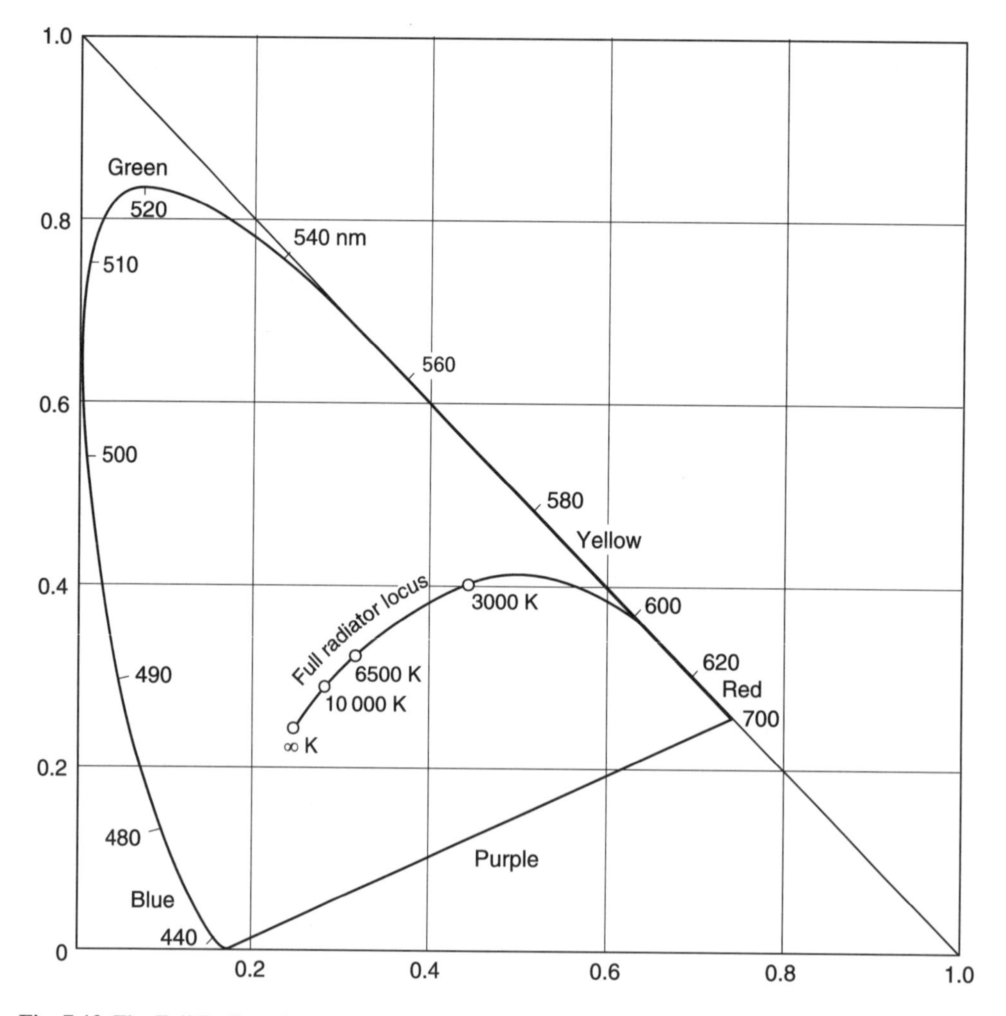

Fig. 7.13 The Full Radiator locus (1931 diagram)

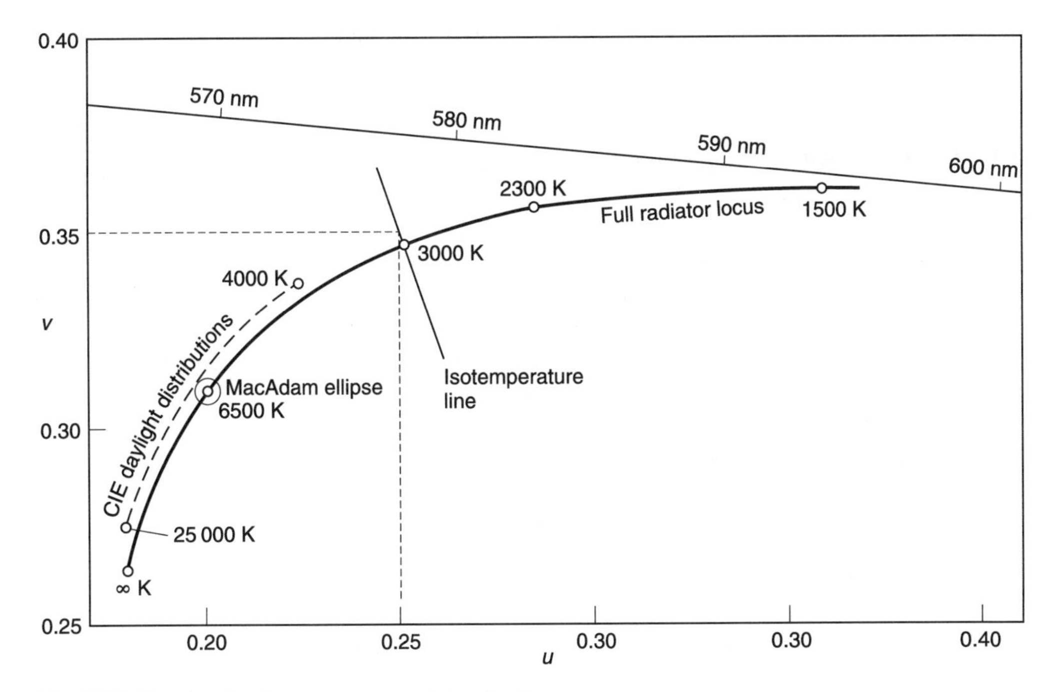

Fig. 7.14 Correlated colour temperature determination

The correlated colour temperature is found by drawing a normal from the Full Radiator locus that passes through the u, v plot for the test source. (This is called an isotemperature line.) Obviously, correlated colour temperatures have most significance when they lie close to the Full Radiator locus.

Example A lamp has chromaticity coordinates in the CIE (1931) system of $x = 0.439$ and $y = 0.412$. Determine its correlated colour temperature.

First the x, y values must be converted into the u, v values of the CIE (1960) UCS system.

$$u = \frac{4x}{-2x + 12y + 3}$$

$$= \frac{4 \times 0.439}{-2 \times 0.439 + 12 \times 0.412 + 3}$$

$$= 0.249$$

$$v = \frac{6y}{-2x + 12y + 3}$$

$$= \frac{6 \times 0.412}{-2 \times 0.439 + 12 \times 0.412 + 3}$$

$$= 0.350$$

When these u, v values are plotted on the chromaticity diagram in Figure 7.14 it gives a correlated colour temperature of 3000 K.

7.6 Colour sample systems

7.6.1 THE MUNSELL SYSTEM

The *Munsell System* is a well-known logical means of ordering colour samples according to three attributes; namely, hue, value and chroma.

Hue Hue is defined as an attribute of visual sensation that gives rise to colour names, such as blue, green, yellow, red, purple, etc.

In the Munsell System, hues are indicated by initials, so that BG is a hue lying between B and G, where B represents blue and G green. The system has 10 basic hues, each of which have 10 steps, as the colour balance changes between one colour and the next.

The 10 basic hues of the Munsell System are:

B, BG, G, GY, Y, YR, R, RP, P and PB

Value This is a specific Munsell term related to luminance factor and so indicates the increasing lightness on a scale from 0 to 10; 0 being black and 10 white; the greys within this range having values from 1 to 9. This attribute of lightness or value is extended to the non-achromatic colours.

Chroma In the Munsell System, this represents numerically the increasing intensity or vividness of the colour.

The arrangement of the Munsell samples The arrangement of the colour samples in this system is indicated in Figures 7.15(a) and (b). The numerical scale of hue changes is clockwise and, for each of the basic hue segments, there are 10 steps as the colour changes towards the next basic hue and then beyond it. When the number reaches 5 the centre of the hue segment is reached and that hue is a maximum. As the number increases above 5 the influence of that hue diminishes. The colour samples for each hue are placed in rows according to their value (lightness); 0 being at the bottom and 10 at the top. (In practice, 0 and 10 are omitted being simply black and white. The samples would then be in value rows 1 to 9.) The intensity or chroma of the colour increases from left to right from 0 to 12.

The Munsell order code is sometimes termed the *HVC system*, because it gives a specification in the order Hue, Value and Chroma. Thus, a colour is specified in terms of three numbers such as 10R5/8. In this case the specification calls for the saturated red hue 10R of a lightness half-way between black and white and the / indicates that the following number represents chroma which, in this case, is becoming quite strong at 8. If a sample was given the specification 2.5GY, it would mean that it was intended to lie mid-way between the samples of Munsell hues 2GY and 3GY.

The Munsell System is based on an attempt to make the changes in steps that are perceptually equal (obviously, such perceptual equality is not necessarily perfect). The Munsell System has the advantage that it is widely known and used.

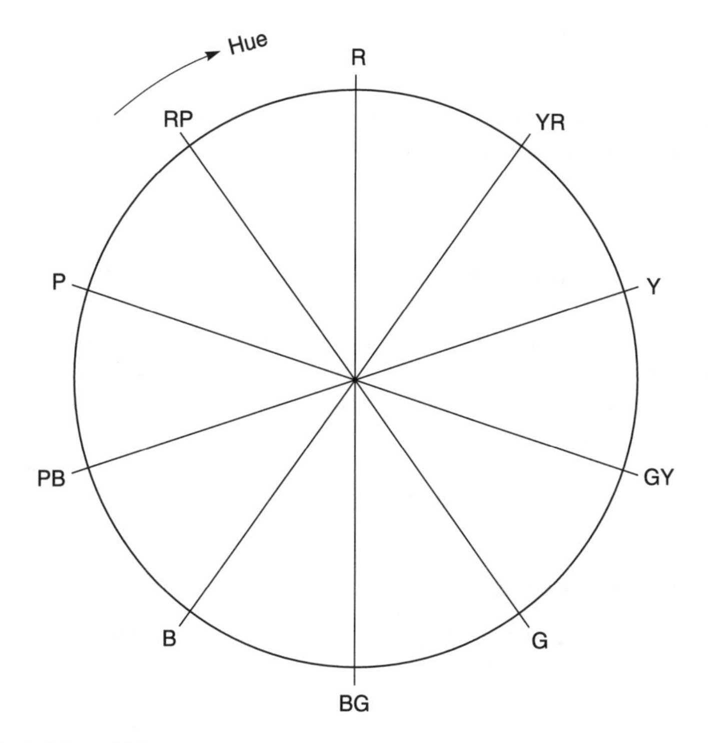

Fig. 7.15(a) Basic Munsell Hues

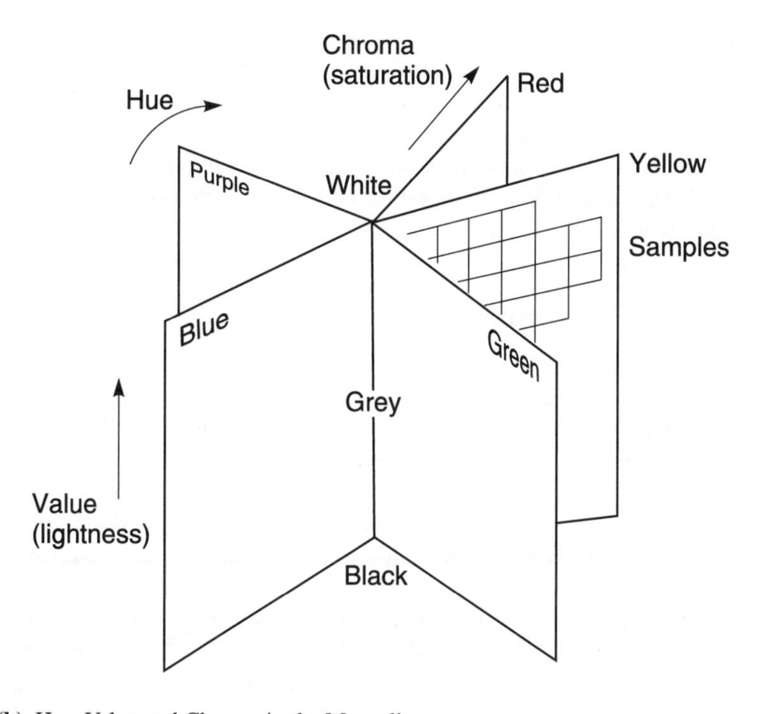

Fig. 7.15(b) Hue, Value and Chroma in the Munsell arrangement

7.6.2 THE NCS SYSTEM (NATURAL COLOUR SYSTEM)

This system is based on six basic chromatic colours. Four of these are red, yellow, green and blue. They are chosen because no trace of any other chromatic colour can be seen in them. Black and white, which contain no trace of the chromatic colours, are then included on a constant hue triangle to indicate the degree of desaturation of the hue.

In atlas form, the hues represent the pages and the constant hue triangle the arrangement of the samples on the pages. Figures 7.16(a) and (b) show the form of the hue circle and the constant hue triangle. (Note that on the triangle black is represented by *S*.)

The hue specification R3OB indicates a hue where there is a 70% resemblance to red and a 30% resemblance to blue; the sum of the resemblances always adding to 100.

On the constant hue triangle, S + W + C = 100. Arrays of constant blackness and whiteness lie parallel to the WC and SC sides of the triangle, while arrays of constant chromaticness lie parallel to the WS side of the triangle.

This system differs from the Munsell System in that the spacings do not coincide with visual equality. However, the system is being used in practice.

7.7 Standard illuminants

The colour appearance of a reflective sample or filter will depend, in part, on the spectral power distribution of the light source used. In 1931, the CIE laid down spectral distributions for three standard illuminants designated S_A, S_B and S_C. In fact, the spectral distribution used for Examples 1 and 3 in Section 7.3 was an abbreviated version of S_C; while that for Example 2 was S_A. The use of S_B is no longer recommended.

S_A is produced by a non-halogen gas-filled tungsten lamp operating at a colour temperature of 2856 K, which is produced by running the lamp at about 2790 K. The difference between the colour temperature and the actual temperature is a consequence of the practical lamp having a spectral distribution close to the theoretical Planckian Radiator distribution at this lower temperature. The S_C distribution is obtained by using the source S_A in conjunction with a liquid filter. The colour temperature of S_C is 6774 K.

In 1967, the CIE introduced a recommendation for a standard illuminant to represent daylight at 6500 K, designated D_{65}. In addition, a method for obtaining spectral power distribution in the range 4000 to 25 000 K was given (Figure 7.14).

7.8 Subtractive colour mixture

This is the name given to changing colours by removing wavelengths from the original spectral distribution. Consider a simple example.

If an equal energy white spectrum is passed through a yellow filter that absorbs mainly blue wavelengths then the transmitted light will contain mainly green, yellow and red wavelengths and, when reflected from a white surface, the light will appear a weak yellow.

If this light is then reflected from a surface that reflects mainly blue-green wavelengths, so that under a complete white spectrum it would appear as cyan, it actually appears a weak green; this is because the blue wavelengths have already been effectively removed from the white spectrum by the yellow filter and the red wavelengths have mainly been removed by absorption at reflection.

Tables 7.5 and 7.6 indicate the process; Table 7.5 shows the result of combining the light source with the yellow filter and Table 7.6 the result of also reflecting this light from the blue-green surface.

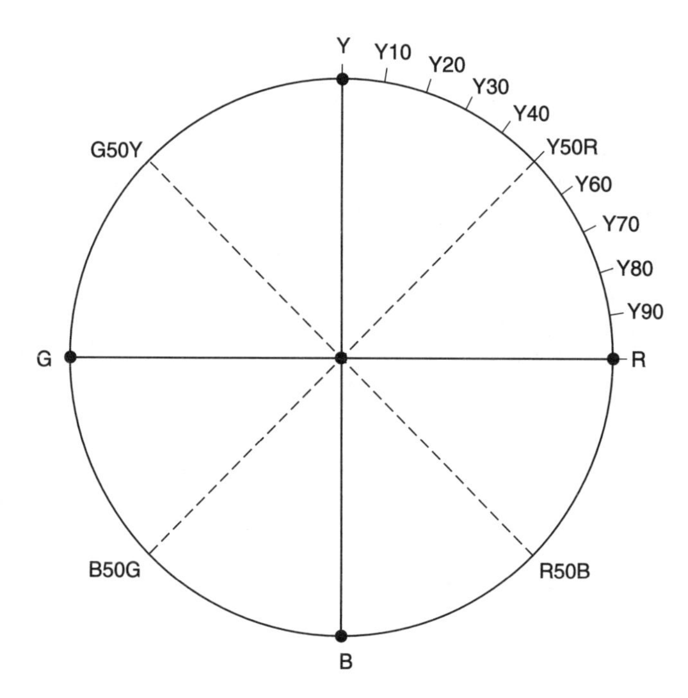

Fig. 7.16(a) NCS Hue Circle

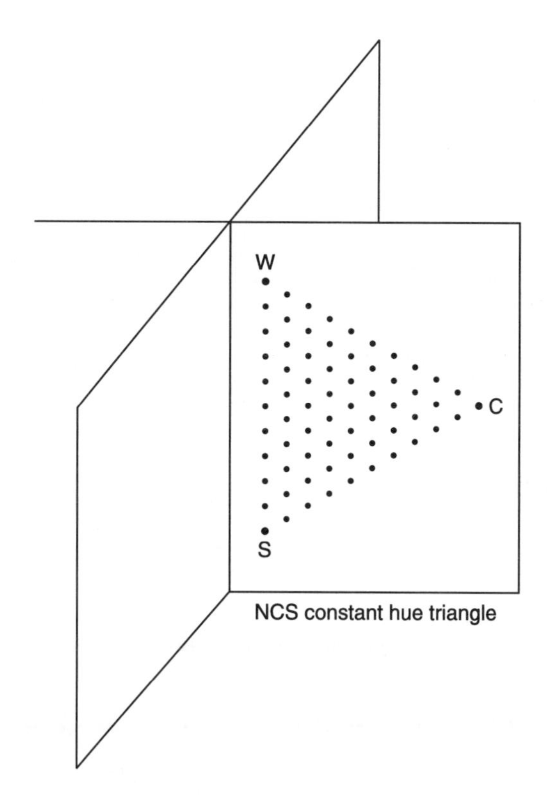

Fig. 7.16(b) NCS Constant Hue Triangle

Table 7.5 Spread calculation of the *XYZ* values for the result of passing an equal energy white spectrum through a yellow filter

λ	$\bar{x}(\lambda)$	$\bar{y}(\lambda)$	$\bar{z}(\lambda)$	$S(\lambda)$	$S(\lambda) \times \bar{y}(\lambda)$	$\tau(\lambda)$	X/k	Y/k	Z/k
400	0.0143	0.0004	0.0679	1	0.0004	0.40	0	0	0.03
420	0.1344	0.0040	0.6456	1	0.004	0.30	0	0	0.194
440	0.3483	0.0230	1.7471	1	0.023	0.40	0.14	0	0.699
460	0.2908	0.0600	1.6692	1	0.06	0.50	0.15	0.03	0.835
480	0.0956	0.1390	0.8130	1	0.139	0.70	0.1	0.1	0.569
500	0.0049	0.3230	0.2720	1	0.323	0.80	0	0.258	0.218
520	0.0633	0.7100	0.0782	1	0.71	0.80	0.1	0.568	0.06
540	0.2904	0.9540	0.0203	1	0.954	0.80	0.23	0.763	0.02
560	0.5945	0.9950	0.0039	1	0.995	0.80	0.48	0.796	0
580	0.9163	0.8700	0.0017	1	0.87	0.80	0.73	0.696	0
600	1.0622	0.6310	0.0008	1	0.631	0.80	0.85	0.505	0
620	0.8544	0.3810	0.0002	1	0.381	0.80	0.68	0.305	0
640	0.4479	0.1750	0.0000	1	0.175	0.80	0.36	0.14	0
660	0.1649	0.0610	0.0000	1	5.3619	0.80	0.13	0.05	0
680	0.0468	0.0170	0.0000	1	0.017	0.80	0	0.01	0
700	0.0114	0.0041	0.0000	1	0.0041	0.80	0	0	0
			Totals		10.6484		3.96	4.235	2.625

$$k = \frac{100}{10.65}$$
$$X = 37.22$$
$$Y = 39.77$$
$$Z = 24.65$$

From Table 7.5

$$x = \frac{X}{X+Y+Z} = \frac{37.2}{37.2 + 39.8 + 24.6} = \frac{37.2}{101.6} = 0.366$$

$$y = \frac{Y}{X+Y+Z} = \frac{39.8}{101.6} = 0.392$$

$$z = 1 - 0.366 - 0.392 = 0.242$$

From Table 7.6

$$x = \frac{X}{X+Y+Z} = \frac{10.28}{10.28 + 15.54 + 11.48} = \frac{10.28}{37.3} = 0.276$$

$$y = \frac{Y}{X+Y+Z} = \frac{15.54}{37.3} = 0.417$$

$$z = 1 - x - y = 1 - 0.276 - 0.417 = 0.307$$

The colour shifts from the equal energy white, to the white plus yellow filter, and to the colour after reflection as indicated in Figure 7.17.

Table 7.6 Spreadsheet calculation of the *XYZ* values obtained when the equal energy white spectrum is first passed through the yellow filter and then reflected from a blue-green surface

λ	$\bar{x}(\lambda)$	$\bar{y}(\lambda)$	$\bar{z}(\lambda)$	$S(\lambda)$	$S(\lambda) \times \bar{y}(\lambda)$	$\tau(\lambda)$	$\rho(\lambda)$	X/k	Y/k	Z/k
400	0.0143	0.0004	0.0679	1	0.0004	0.40	0.20	0	0	0.01
420	0.1344	0.0040	0.6456	1	0.004	0.30	0.30	0	0	0.058
440	0.3483	0.0230	1.7471	1	0.023	0.40	0.40	0	0	0.2795
460	0.2908	0.0600	1.6692	1	0.06	0.50	0.50	0.1	0.02	0.4173
480	0.0956	0.1390	0.8130	1	0.139	0.70	0.50	0	0.05	0.2846
500	0.0049	0.3230	0.2720	1	0.323	0.80	0.60	0	0.155	0.1306
520	0.0633	0.7100	0.0782	1	0.71	0.80	0.60	0	0.341	0.038
540	0.2904	0.9540	0.0203	1	0.954	0.80	0.50	0.12	0.382	0.01
560	0.5945	0.9950	0.0039	1	0.995	0.80	0.40	0.19	0.318	0
580	0.9163	0.8700	0.0017	1	0.87	0.80	0.30	0.22	0.209	0
600	1.0622	0.6310	0.0008	1	0.631	0.80	0.20	0.17	0.101	0
620	0.8544	0.3810	0.0002	1	0.381	0.80	0.20	0.14	0.06	0
640	0.4479	0.1750	0.0000	1	0.175	0.80	0.10	0	0.01	0
660	0.1649	0.0610	0.0000	1	5.3619	0.80	0.10	0	0	0
680	0.0468	0.0170	0.0000	1	0.017	0.80	0.10	0	0	0
700	0.0114	0.0041	0.0000	1	0.0041	0.80	0.10	0	0	0
			Totals		10.6484			1.09	1.655	1.223

$$k = \frac{100}{10.65}$$

$X = 10.28$

$Y = 15.54$

$Z = 11.48$

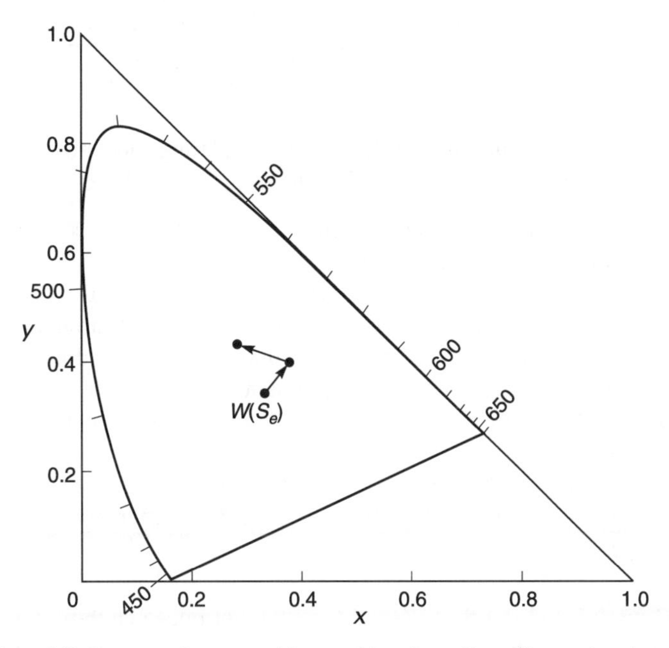

Fig. 7.17 The colour shift from equal energy white to white plus yellow filter and to the colour after reflection (Section 7.7)

A similar effect occurs with the mixture of paints or dyes where the different colours mixed together progressively absorb more wavelengths until, if the process is continued, eventually the light will all be absorbed and the perceived colour will be black.

It will be obvious from the above that, since the process is one of absorption, the luminance of the colour is reduced at each stage.

From this it will be seen that the perceived colour of an object or surface depends both on the spectral reflectance of the object or the surface and also the spectral emission from the source or the source-filter combination.

7.9 Colour rendering and the CIE colour rendering index

7.9.1 COLOUR RENDERING

In the previous section it was pointed out that the colour appearance of an object depends both upon its spectral reflectance and the spectral distribution of the light source. It is common to think of daylight as giving the 'correct' appearance of a colour or the correct rendering of the object's natural colour. The distortion of a colour that can be produced by other light sources is, therefore, a matter of importance for lighting engineers and architects, among others.

In Section 7.3, Examples 2 and 3 taken together illustrated the change in colour that different light source spectral distributions could produce in terms of x and y chromaticity values.

7.9.2 THE CIE COLOUR RENDERING INDEX

The CIE have produced a method for evaluating the colour rendering properties of light sources that enables a single colour Rendering Index (R_a) to be calculated for a particular lamp.

This General Colour Rendering Index is based on combining the results of calculations of the Special Colour Rendering Index (R_i) for a set of sample surface colours. The CIE General Colour Rendering Index is the mean of the Special Colour Rendering indices for a specified set of eight samples.

The procedure for determining the Special Colour Rendering Index for each sample is given in detail in the CIE document, CIE 13.3 (1995). The steps involved are:

Step 1. Determine the CIE 1931 tristimulus values for each of the test colours for the source to be tested and for the reference source.
Step 2. Transform these tristimulus values into coordinates of the 1960 UCS diagram.
Step 3. Account for adaptive colour shift due to the different state of chromatic adaptation under the test lamp compared with the reference illuminant. This is done by means of a von Kries type adjustment for chromatic adaptation (see Section 7.3.1). The somewhat complex formula for this adjustment of the u, v values is given in the CIE document 13.3 (1995) Section 5.7.
Step 4. Transform this modified colorimetric data into 1964 Uniform Space coordinates.
Step 5. Determine the resultant colour shift using the colour difference equation for the 1964 colour space.

Note The 1964 modification of the 1960 UCS system introduced a third dimension related to luminance called lightness. The u, v 1960's values are converted into the 1964 system using the following formulae, which include the luminance factor Y.

$W^* = 25Y^{1/3} - 17$ where Y lies between 1 and 100
$U^* = 13W^*(u - u_n)$
$V^* = 13W^*(v - v_n)$

where u_n and v_n are usually those of the illuminant.

The colour difference equation is

$$\Delta E = [(\Delta U^*)^2 + (\Delta V^*)^2 + (\Delta W^*)^2]^{1/2}$$

Step 6. Calculate the special Colour Rendering Index R_i from

$$R_i = 100 - 4.6\Delta E$$

Step 7. Repeat for the eight CIE samples and calculate the general Colour Rendering Index from

$$R_a = \frac{1}{8}\sum_{i=1}^{8} R_i$$

The CIE have produced a program on disk for the calculation of Colour Rendering Indices.

It will be seen from the equation for R_i in Step 6 above that R_i or R_a of 100 represents no colour shift between the test source and the reference illuminant. An R_a value above 80 means good to excellent colour rendering; a value between 60 and 80 moderate colour rendering; while with a value below 60 the colour rendering becomes progressively poor and below 40 there is marked colour distortion.

7.10 Visualization and colour

In Section 8.8 there is a brief discussion of the problems of visualization, which is the generation of an image on a computer display screen of an interior where changes in the lighting can be simulated. That discussion does not dwell on colour matters. It is, of course, perfectly possible to divide up the light output of lamps into colour bands such as *R*, *G* and *B* bands. If this is done and the surface reflectances are also specified in the same bands, then the radiosity method of Chapter 5 can be applied separately for the *R* band, the *G* band and the *B* band. The result would be for each surface a set of *R, G, B* values from which the final colours could be determined. However, when it is necessary to display these colours by means of the *RGB* primaries of the CRT screen, the bands covered by these primaries have to be accommodated in the calculations and hence the computer program.

For the purposes of colour display a very simple model of colour space will usually suffice, based on the primaries that the CRT monitor can produce (see Figure 7.18).

In this colour space, black is the coordinate origin (0, 0, 0) and white is (1, 1, 1); the primaries being given by *R*(1, 0 ,0), *G*(0, 1, 0) and *B*(0, 0, 1). Clearly, all the colours within the scope of the three primaries mixture can be represented on this scale.

The developed programs vary in their sophistication in dealing with colour. A common simplification is to display the colour produced by the first reflection of the light flux (that is, ignore interreflection). A full radiosity program would be capable of taking interreflection into account.

Another simplification, where the interest is in the room rather than the lighting as such, is to use a particular white light source specification rather than to analyse the actual light source. It has been found that the eye/brain combination is not very discriminating where

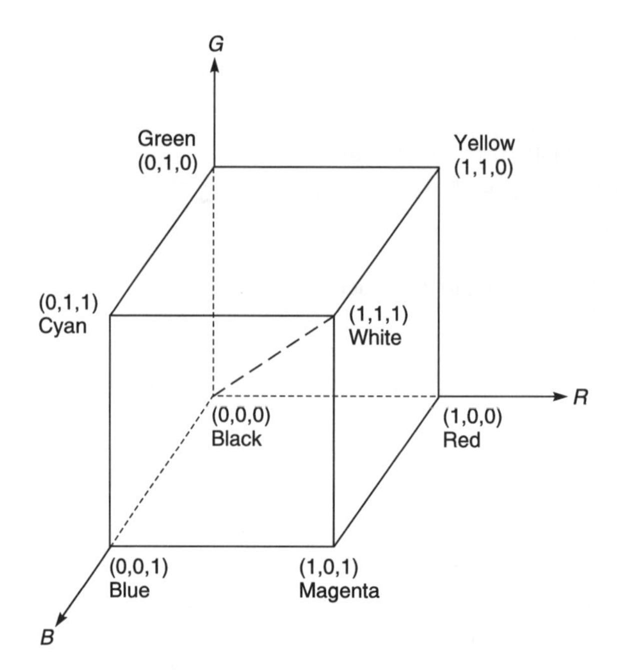

Fig. 7.18 A simple model of colour space for visualization

scenes are viewed successively, whilst being very discriminating when scenes are viewed simultaneously.

Bibliography

Coaton, J.R. and Marsden, A.M. (1997) *Lamps and Lighting* (Arnold) pp. 46–69.
Hunt, R.W.G. (1991) *Measuring Colour*, 2nd Edition (Chichester: Ellis Horwood).

8 Interior Lighting

8.1 General

Interior lighting has a number of aspects:

(1) the lighting required for the tasks that have to be performed in the space,
(2) the visual comfort of the occupants of the space,
(3) the satisfaction of the occupants with appearance of surfaces and the other people occupying the space.

8.1.1 TASK LIGHTING

The lighting level required to perform a particular task depends upon the visual capability of the person seeking to perform the task. Obviously, this varies from individual to individual both with age and the condition of their eyes. The CIBSE Lighting Code gives detailed guidance for the specification of lighting levels for particular tasks, including adjustments for some of the above factors.

The level of lighting specified in the CIBSE Schedule is the maintained illuminance level, which is the level at which the maintenance procedure is carried out to ensure that the illuminance does not fall below this level.

The *Design Maintained illuminance level* is the level specified for designing the installation. It takes into account any variation from the assumed task conditions, such as the task details being more difficult than normal in this type of interior or whether errors pose a serious risk.

The design calculations then include allowances for reduction in lamp output with burning hours, lamp failure with burning hours, dust and dirt collected on the luminaires with the passage of time and dust and dirt collected on the room surfaces with the passage of time. Guidance on all these factors is given in the CIBSE Code for Interior Lighting.[1]

8.1.2 VISUAL COMFORT

Although the requirement is to ensure visual comfort, it is usual to approach this problem by eliminating visual discomfort. The main method used is to establish a Glare Index or Glare Rating for an installation which, if not exceeded, will make it unlikely that serious glare will be experienced. For example, the Glare Index recommended in the CIBSE Code for general offices is 19 and that for a warehouse 25.

These values are related to the Glare Index system developed in the UK and described in the CIBSE Technical Memorandum TM 10.

The basic research upon which this is based was carried out by Petherbridge and Hopkinson of the, then, Building Research Station (BRS), now the Building Research Establishment (BRE).[2] The formula for obtaining the Glare Index for a lighting installation is:

$$\text{Glare Index} = 10 \log_{10} 0.45 \sum \frac{(L_s)^{1.6} (\omega_s)^{0.8}}{L_b P^{1.6}}$$

where L_s and ω_s are the luminance and solid angular size of each luminaire, L_b is the background luminance and P relates to the position of each source relative to the observer's direction of view.

The CIBSE is likely to adopt the CIE unified glare rating to replace the UK Glare Index, the formula for which is:

$$\text{Unified Glare Rating, UGR} = 8 \log \frac{0.25}{L_b} \sum \frac{L_s^2 \omega}{P^2}$$

This gives Glare Rating numbers very close to the UK Glare Index numbers.

8.1.3 VISUAL SATISFACTION

For a long time, it has been appreciated that it is not just the surfaces within a room that need illuminating, but also the space enclosed within the room envelope. It is this space that is occupied by the people using the room, and it is the quality and quantity of the light density in this space that determines the appearance of those people to each other. This is a very important factor relative to that of the satisfaction of the occupants with their visual environment.

Vector scalar ratio

A recent pioneer in this field is Lynes who took the mathematical work of Gershun[3] and related it to practical lighting design. The quantity that Gershun called 'Space Illumination' Lynes rechristened Scalar Illuminance, and Gershun's 'Light Vector' became the Illumination Vector (or Vector Illumination).

These are two elegant concepts that are mathematically attractive and have met with some success in promoting consideration of the quality of the lighting of the space within the room envelope. They are discussed in detail in the paper 'The flow of light into buildings'[4] and in the book *Principles of Natural Lighting*.[5]

The calculation of scalar illuminance has already been considered in Chapters 3, 4 and 5 and the concept of illuminance as a vector quantity has also been introduced in Chapter 3, and it will be considered further in this chapter.

The Scalar Illuminance (E_s) at a point is defined as the average illuminance on the surface of an infinitesimally small sphere at that point, due to light reaching the point from all directions.

The Illumination Vector (**E**) at a point can be defined as the maximum difference of the illuminance across diameters of an infinitesimally small sphere at that point.

There will be illuminance differences in other directions across the sphere, which are vector quantities resulting from considering the individual illuminances on either side of each diameter as vectors and then combining them, but the Illumination Vector is the resultant of *all* these individual vectors, which are therefore components of it.

The value of this approach is that the vector/scalar ratio can be related to the appearance of people's faces within the room envelope. For example, when the vector/scalar ratio is 0.5, the modelling due to illuminance variation is weak and the appearance can be dull, while if the value is about 1.5 the modelling is usually considered to be pleasant. If the vector/scalar ratio approaches 4 then the modelling will be harsh and unacceptable.

The ratio of cylindrical to horizontal illuminance

The cylindrical illuminance is the mean vertical illuminance at a point, taking into account all azimuth directions (see Chapters 3, 4 and 5).

It has been found that the ratio of cylindrical to horizontal illuminance can also be used to indicate the modelling properties of a general lighting installation, where the lighting is mounted on or suspended from the ceiling.

When $E_{cyl}/E_h = 0.1$ the modelling is very harsh; when $E_{cyl}/E_h = 0.6$ the modelling is good; when $E_{cyl}/E_h = 0.9$ the modelling is very weak.

Although in, say, an office, the height for considering facial features is between 1 m and 1.4 m above the floor, the vector/scalar and cylindrical/horizontal calculations are often done at 0.8 m, since the horizontal illuminance is routinely calculated at this height. This is taken as the working plane height (for example, desk height).

Detailed calculations covering each of the three aspects of interior lighting described above will be carried out for a simple example to illustrate the procedure.

8.2 Example

Design the lighting for a general office given the following information.

Size 4 m long by 4 m wide, floor to ceiling height 2.8 m. Reflectances: ceiling 0.7, walls 0.5, floor cavity 0.3 (working plane).

The CIBSE Code specifies 500 lux for a general office and a Glare Index of 19. The vector/scalar ratio and the cylindrical horizontal ratio will also be calculated to check on the degree of modelling obtained in the installation.

The first step is to select the luminaire, because without knowing the luminous intensity distribution of the luminaire the calculations cannot be carried out.

The luminaire chosen is a recessed ceiling luminaire with a prismatic panel fitted flush with the ceiling. The luminaire measures 600 mm by 600 mm. The intensity distribution is symmetrical about a vertical axis and is given in Table 8.1 and the polar curve is illustrated in Figure 8.1.

The second step is to calculate the direct flux to the room surfaces. This cannot be done until we specify the number of luminaires and their spacing. The manufacturer has specified that, for acceptable uniformity, the luminaires should not be spaced apart further than 1.33 times the mounting height above the working plane.

The working plane is generally assumed to be at 0.8 m above the floor (desk height) and so the mounting height for spacing purposes is given by $h_m = 2.8 - 0.8 = 2$ m.

So, $2 \times 1.33 = 2.66$ m is the maximum distance between luminaires for acceptable uniformity. The office is 4 m square and so two rows of luminaires are required in each direction; that is, 4 luminaires. It is usual to locate the luminaires a distance from the wall that is half the distance between adjacent luminaires; consequently the spacing is chosen as 2 m and this is shown in Figure 8.2. The spacing is measured from the centre of the luminaire.

Now that the position of the luminaires within the room is fixed and the intensity distribution per 1000 lumens is also known, we can calculate the flux to each of the room surfaces (treating the working plane as a surface).

Because there are four luminaires placed symmetrically within the room, only one luminaire needs to be considered, as the result will apply to each of the luminaires.

Table 8.1

Angle (deg)	Mean luminaire intensity in vertical plane cd/1000 lumens
0	218
5	218
10	215
15	211
20	205
25	197
30	186
35	167
40	139
45	107
50	82
55	70
60	44
65	31
70	23
75	17
80	13
85	8
90	0

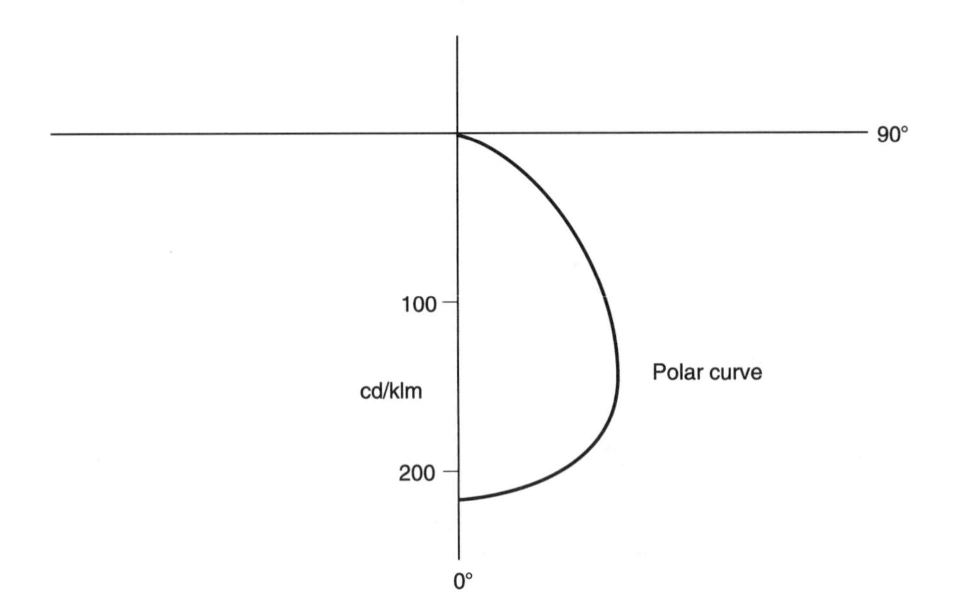

Fig. 8.1 Polar curve for the intensity data given in Table 8.1

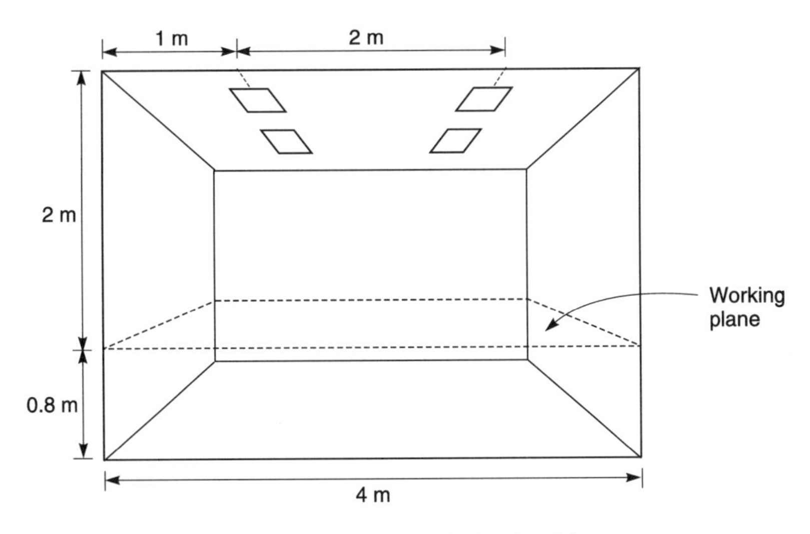

Fig. 8.2 The general office to be lit in the example given in Section 8.2

The direct flux could be calculated by approximating the practical distribution to a cosine power distribution as shown in Chapter 4, or calculated directly using the inverse-square law. In this case we will use the inverse-square law. The working plane area will be divided into 16 areas and the illuminance at the centre of each area calculated. These values will be summed and divided by 16 to give an average value. Multiplying this average value by the working plane area will give the direct flux to the working plane per 1000 lumens of installation lamp flux. Thus, to obtain the total direct flux to the working plane it will be necessary to multiply by 4 times the lamp flux of one luminaire in kilolumens. Figure 8.3(a) shows the working plane divided into 16 areas and Figure 8.3(b) shows the dimensions for the inverse-square law calculations.

In Figure 8.3(a) the position of the centre of the luminaire is shown as a large dot. Because of the relative positions of the target squares and the luminaire, only six values of illuminance need to be calculated to determine the 16 point values.

Consider square 3, the coordinates of its centre point (x_c, y_c) are given by (2.5, 2.5) and the luminaire is at (x_l, y_l) = (1.0, 1.0). So the horizontal distance on the working plane from beneath the source of the working plane to the centre of square 3 is given by

$$\sqrt{(x_c - x_1)^2 + (y_c - y_1)^2} = \sqrt{(2.5-1)^2 + (2.5-1)^2} = 2.12 \text{ m}$$

$$\theta = \tan^{-1} \frac{2.12}{2} = 46.7°$$

From Table 8.1, by linear interpolation (see Chapter 2, Section 2.3.1):

$$I_\theta = 98.5$$
$$\cos^3 \theta = 0.3228$$

$$E_3 = \frac{I_\theta \cos^3 \theta}{h_m^{\ 2}} = \frac{98.5 \times 0.3228}{2^2} = 7.95 \text{ lux}$$

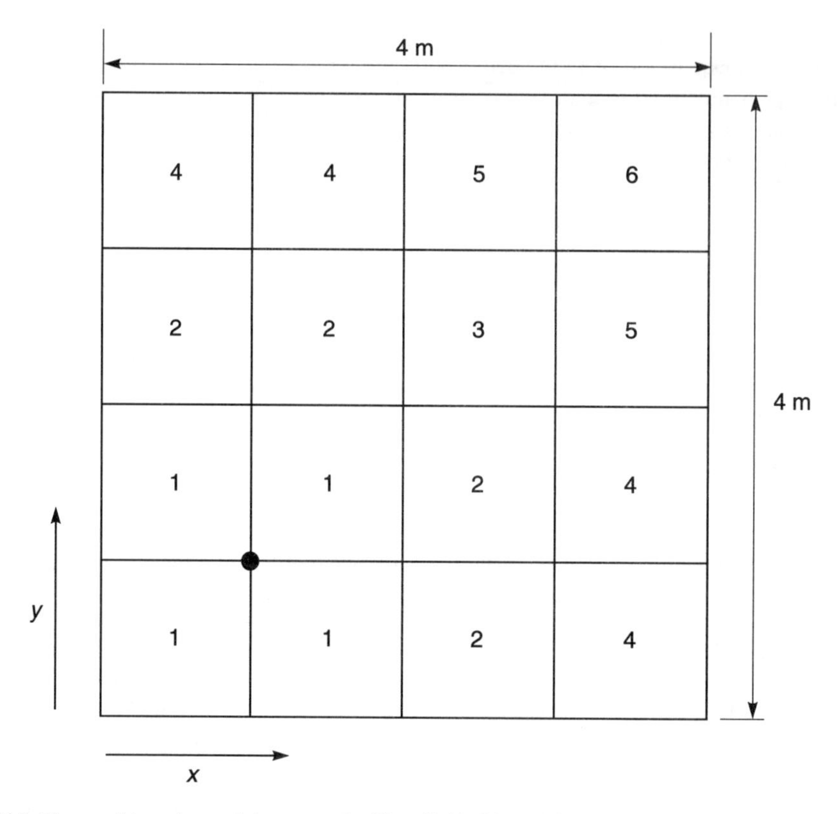

Fig. 8.3(a) The working plane of the general office divided into 16 areas for the calculation

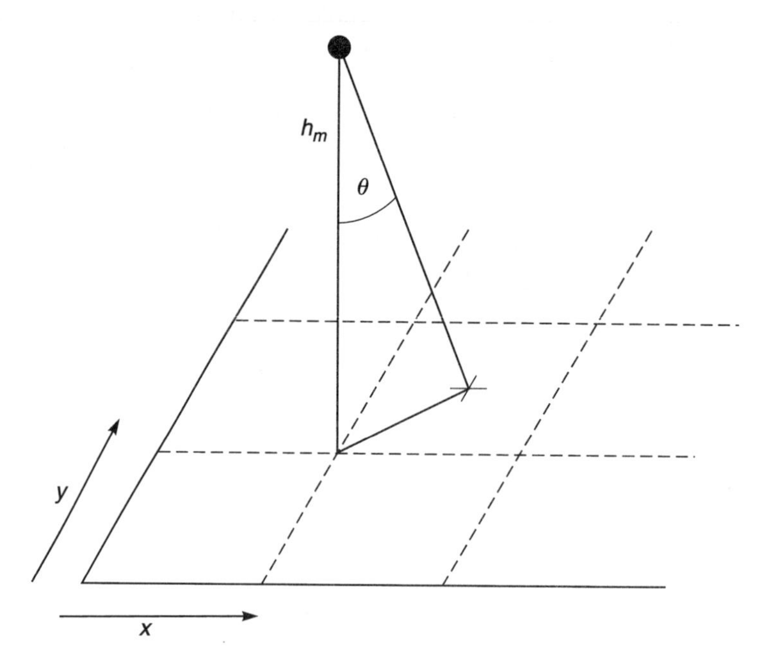

Fig. 8.3(b) The dimensions required for the inverse-square law calculations

Similar calculations for the other squares give

4.58	4.56	2.72	1.28
18.97	18.97	7.95	2.72
43.14	43.14	18.97	4.56
43.14	43.14	18.97	4.56

$\Sigma E_h/klm = 109.83 + 109.81 + 48.61 + 13.12 = 281.4$

Since we have chosen 16 squares, dividing by 16 to obtain the average and multiplying by 16 to obtain the flux cancels, and so the answer is 281.4 lumens (per 1000 lamps lumens).

The fraction of lamp flux received by the working plane is therefore 281.4/1000 = 0.281.

0.281 is the distribution factor (*DF*(*F*)) required to calculate the utilization factor (*UF*(*F*)) for the working plane (see Section 5.4.4).

The distribution factor for the ceiling (*DF*(*c*)) is zero, since there is no upward light from the luminaires.

The distribution factor for the walls (*DF*(*W*)) is given by

$$DF(W) = DLOR - DF(F)$$

where *DLOR* is the downward light output ratio of the luminaire (see Chapter 16) and is given by the manufacturer for this type of luminaire as 0.47.

So, $DF(W) = 0.47 - 0.281 = 0.189$.

The next step is to calculate the utilization factors for the working plane, ceiling and walls (see Section 5.4.4).

The expression for utilization factors is of the form

$$UF(F) = TF(C, F)DF(C) + TF(F, F)DF(F) + TF(W, F)DF(W)$$

where *TF*(*C*,*F*) is the transfer factor from ceiling to floor, as in this case, or floor to floor etc., which takes into account interreflection (see Section 5.4.5).

Transfer factors for a room index of 1.0 and reflectances of $\rho_c = 0.7$, $\rho_w = 0.5$ and $\rho_F = 0.3$ have been calculated in the example given in Section 5.4.4. These are:

$$TF(C, C) = 1.156, \quad TF(F, C) = 0.187, \quad TF(W, C) = 0.248$$
$$TF(C, F) = 0.437, \quad TF(F, F) = 1.105, \quad TF(W, F) = 0.285$$
$$TF(C, W) = 0.694, \quad TF(F, W) = 0.342, \quad TF(W, W) = 1.453$$

(The notation used in Chapter 5 for developing the basic interreflection theory employed numbers, but here *C* for ceiling, *F* for floor (cavity) and *W* for walls are used to identify more directly the room surfaces.)

Using these transfer factors the utilization factors are calculated as follows:

$$\begin{aligned} UF(C) &= TF(C, C) \times DF(C) + TF(F, C) \times DF(F) + TF(W, C) \times DF(W) \\ &= (1.156 \times 0) + (0.187 \times 0.281) + (0.248 \times 0.189) \\ &= 0.099 \end{aligned}$$

$$\begin{aligned} UF(F) &= TF(C, F) \times DF(C) + TF(F, F) \times DF(F) + TF(W, F) \times DF(W) \\ &= (0.437 \times 0) + (1.105 \times 0.281) + (0.285 \times 0.189) \\ &= 0.364 \end{aligned}$$

$$\begin{aligned} UF(W) &= TF(C, W) \times DF(C) + TF(F, W) \times DF(F) + TF(W, W) \times DF(W) \\ &= (0.649 \times 0) + (0.342 \times 0.281) + (1.453 \times 0.189) \\ &= 0.371 \end{aligned}$$

The average surface illuminances are calculated from the following equations

$$E_C = \frac{N \times F_L \times UF(C) \times MF}{A_C}$$

$$E_F = \frac{N \times F_L \times UF(F) \times MF}{A_F}$$

$$E_W = \frac{N \times F_L \times UF(W) \times MF}{A_W}$$

where N is the number of luminaires
F_L is the total lamp flux of a luminaire
MF is the maintenance factor
A_C, A_F and A_W are the areas of the ceiling, working plane and walls respectively.

The lamp flux is taken initially as 1000 lumens and later adjusted to obtain the required working plane illuminance.

Only the maintenance factor (MF) is unknown on the right-hand side of the equation.

$$MF = LLMF \times LSF \times LMF \times RSMF$$

where $LLMF$ is the lamp lumen maintenance factor
LSF is the lamp survival factor
LMF is the luminaire maintenance factor
$RSMF$ is the room surface maintenance factor

Each of these factors may be defined as below.

(x_n) the proportion of the initial light output remaining after a specified period of time, taking into account (y) but not including other losses.
And where $x_1 = LLMF$ and y_1 = fall in lamp lumens output with time
$x_2 = LSF$ and y_2 = lamp survival after a specified time
$x_3 = LMF$ and y_3 = accumulation of dust and dirt on the luminaires after a specified time
$x_4 = RSMF$ and y_4 = accumulation of dust and dirt on the room surfaces after a specified time.

The data required to determine these factors can be obtained from the manufacturer's data and from data given in the CIBSE Code. For our example, we will obtain data from the CIBSE Code.

Let us assume a two-year period for cleaning and relamping.

$LLMF = 0.91$
$LSF = 1.0$
$LMF = 0.77$
$RSMF = 0.82$

where $MF = LLMF \times LSF \times LMF \times RSMF$.

This gives a value for $MF = 0.91 \times 1.00 \times 0.77 \times 0.82 = 0.57$.

The room surface illuminances can now be calculated (per 1000 lumens):

$$E_C/\text{klm} = \frac{N \times F_L \times UF(C) \times MF}{A_C}$$

$$= \frac{4 \times 1000 \times 0.099 \times 0.57}{4 \times 4}$$

$$= 14.1 \text{ lux/klm}$$

$$E_F/\text{klm} = \frac{N \times F_L \times UF(F) \times MF}{A_F}$$

$$= \frac{4 \times 1000 \times 0.364 \times 0.57}{4 \times 4}$$

$$= 51.9 \text{ lux/klm}$$

$$E_W/\text{klm} = \frac{N \times F_L \times UF(W) \times MF}{A_W}$$

$$= \frac{4 \times 1000 \times 0.371 \times 0.57}{2 \times 4 \times 4}$$

$$= 26.4 \text{ lux/klm}$$

The specified illuminance for the working plane is 500 lux so the lamp flux in kilolumens required from each luminaire is:

$$\frac{500}{51.6} = 9.63$$

Two 55 W warm white compact fluorescent lamps would give 9.60 kilolumens and so this solution is chosen.

The design illuminances are therefore:

$$\begin{aligned} E_C &= E_C/\text{klm} \times \text{klm} \\ &= 14.1 \times 9.6 \\ &= 135.4 \text{ lux} \end{aligned}$$

$$\begin{aligned} E_F &= E_F/\text{klm} \times \text{klm} \\ &= 51.9 \times 9.6 \\ &= 498.2 \text{ lux} \end{aligned}$$

$$\begin{aligned} E_W &= E_W/\text{klm} \times \text{klm} \\ &= 26.4 \times 9.6 \\ &= 253.4 \text{ lux} \end{aligned}$$

At this stage it is convenient to calculate the values of scalar illuminance and cylindrical illuminance on the working plane.

For each of these we require (1) the average direct illuminance value and (2) the contribution made by interreflection from ceiling, walls and working plane (floor cavity).

In Section 3.8.1 the formula for scalar illuminance from a point source is given, namely,

$$E_S = \frac{E_{max}}{4}$$

where E_{max} is the maximum illuminance produced by the light source at the specified point

$$= \frac{E_h}{4 \cos \theta}$$

where E_h is the horizontal plane illuminance and where θ is the angle between the normal to the plane containing the point and the line joining the light source to the point.

In addition, in Section 3.8.2, the formula for cylindrical illuminance from a point source is given,

$$E_{cyl} = E_{max} \frac{\sin \theta}{\pi}$$

$$= E_h \frac{\tan \theta}{\pi}$$

The values for E_h at 16 points have already been calculated and so by multiplying by $1/(4 \cos \theta)$ or by $(\tan \theta)/\pi$ we can obtain the values of scalar illuminance and cylindrical illuminance respectively.

First, for scalar illuminance, this procedure gives,

1.86	1.86	1.20	0.76
6.04	6.04	2.89	1.20
11.44	11.44	6.04	1.86
11.44	11.44	6.04	1.86

$$\Sigma E_S/\text{klm} = 30.78 + 30.78 + 16.17 + 5.68 = 83.41$$

Thus, the average direct scalar illuminance per 1000 lumens for one luminaire

$$= \frac{83.41}{16} = 5.21 \text{ lux/klm}$$

Each luminaire gives 9.6 kilolumens and there are four luminaires, so,

$E_S = E_S/\text{klm} \times \text{klm}$
$= 5.21 \times 9.6 \times 4$
$= 200$ lux (direct scalar illuminance)

Secondly, for cylindrical illuminance, we get,

1.86	1.86	1.26	0.72
4.77	4.77	2.68	1.26
4.86	4.86	4.77	1.86
4.86	4.86	4.77	1.86

$\Sigma\, E_{cyl}/\text{klm} = 16.35 + 16.35 + 13.48 + 5.7 = 51.88$

Thus, the average direct cylindrical illuminance for 1000 lumens for one luminaire is

$$= \frac{51.88}{16} = 3.2 \text{ lux/klm}$$

So,

$$\begin{aligned} E_{cyl} &= E_{cyl}/\text{klm} \times \text{klm} \\ &= 3.2 \times 9.6 \times 4 \\ &= 122.8 \text{ lux} \\ &\text{(direct cylindrical illuminance)} \end{aligned}$$

The interreflected components of scalar illuminance and cylindrical illuminance are obtained by using the inter-illuminance factors (*Ilf*) introduced in Section 5.4.9, together with the results of the scalar and cylindrical zonal multiplier calculations obtained in the example given in the same section.

From that example,

for $RI = 1.0$ $D_{S\ (scalar)} = 0.234$
$D_{S\ (cylindrical)} = 0.103$

$$E_S = (Ilf_{S(c)} \times M_C) + (Ilf_{S(f)} \times M_F) + (Ilf_{S(W)} \times M_W)$$

where M_C etc. are the luminous exitances of the ceiling, floor and walls respectively.

$$\begin{aligned} Ilf_{S(C)} &= D_{S(scalar)} \\ Ilf_{S(f)} &= 0.5 \\ Ilf_{S(W)} &= 0.5 - D_{S(scalar)} \end{aligned}$$

So, the reflected component of E_S is given by,

$$\begin{aligned} E_S &= 0.24 \times M_C + 0.5 \times M_F + (0.5 - 0.234)M_W \\ &= (0.234 \times 135.4 \times 0.7) + (0.5 \times 498.2 \times 0.3) + (0.266 \times 253.4 \times 0.5) \\ &= 130.6 \text{ lux (interreflected component)} \end{aligned}$$

The total value of E_S is given by adding to this the direct scalar illuminance already calculated, so,

$$E_S = 200 + 130.6 = 330.6 \text{ lux}$$

Similarly, the reflected component of

$$\begin{aligned} E_{cyl} &= (0.103 \times 135.4 \times 0.7) + (0.5 \times 498.2 \times 0.3) + (0.397 \times 253.4 \times 0.5) \\ &= 9.76 + 74.3 + 50.6 \\ &= 134.8 \text{ lux} \end{aligned}$$

So, $E_{cyl} = 122.8 + 134.8 = 257.6$ lux.

To calculate the vector/scalar ratio, the value of the illumination vector (E) for the installation is required.

Since the predominant direction of the light is downward, a good estimate of the average vector can be obtained by calculating the difference of the working plane illuminance and the upward illuminance produced by reflection from the floor cavity (see Chapter 3, Section 3.1).

$$\begin{aligned} E &= E_F - \rho_F E_F \\ &= E_F(1 - \rho_F) \\ &= 498.2(1 - 0.3) \\ &= 348.7 \text{ lux} \end{aligned}$$

$$\frac{E}{E_S} = \frac{348.7}{330.6} = 1.05$$

The CIBSE Code suggests that this ratio should lie between 1.2 and 1.8 ('where perception of faces is important').

If the working plane cavity reflectance had been 0.2 instead of 0.3 the vector/scalar ratio would have been above 1.2.

The ratio of cylindrical to horizontal plane illuminance is more directly obtained

$$\frac{E_{cyl}}{E_F} = \frac{257.6}{498.2} = 0.52$$

The work of Hewitt, Bridgers and Simons[6] shows that this value of the ratio indicates acceptable modelling, but modelling that is subtle rather than strong.

It should be noted that this ratio gives less prominence to the effect of the floor cavity than is the case in the vector/scalar ratio.

It is of particular interest to note that our calculations give a value for mean wall illuminance of

$$E_W = 253.4 \text{ lux}$$

and for cylindrical illuminance of

$$E_{cyl} = 257.6 \text{ lux}$$

The values are very close. In practice, it has been found that in general lighting installations this is a common outcome. This means that for a practical check on the cylindrical to horizontal illuminance ratio, that value of mean wall illuminance can be used as a good estimate of E_{cyl}.

To complete the example, it is now necessary to calculate the Glare Index. The procedure is given in CIBSE Technical Memoranda TM10 (The Calculation of Glare Indices 1985). (The CIE has proposed a unified glare rating formula (UGR) which, in practice, gives values very close to those of the UK system.)

The first step is to calculate the glare constant g for each of the four luminaires see Figures 8.4(a) and (b). The position of the observer is at the centre of the wall as shown in Figure 8.4(a).

From considerations of symmetry it will be seen that only two calculations are required.

The glare constant

$$g = 0.9 \frac{L_S^{1.6} \omega^{0.8}}{L_b P^{1.6}}$$

L_S and L_b are source luminance and background luminance respectively (cd/m^2). ω is the solid angle subtended at the eye by the source (steradians). P is the position index (see later).

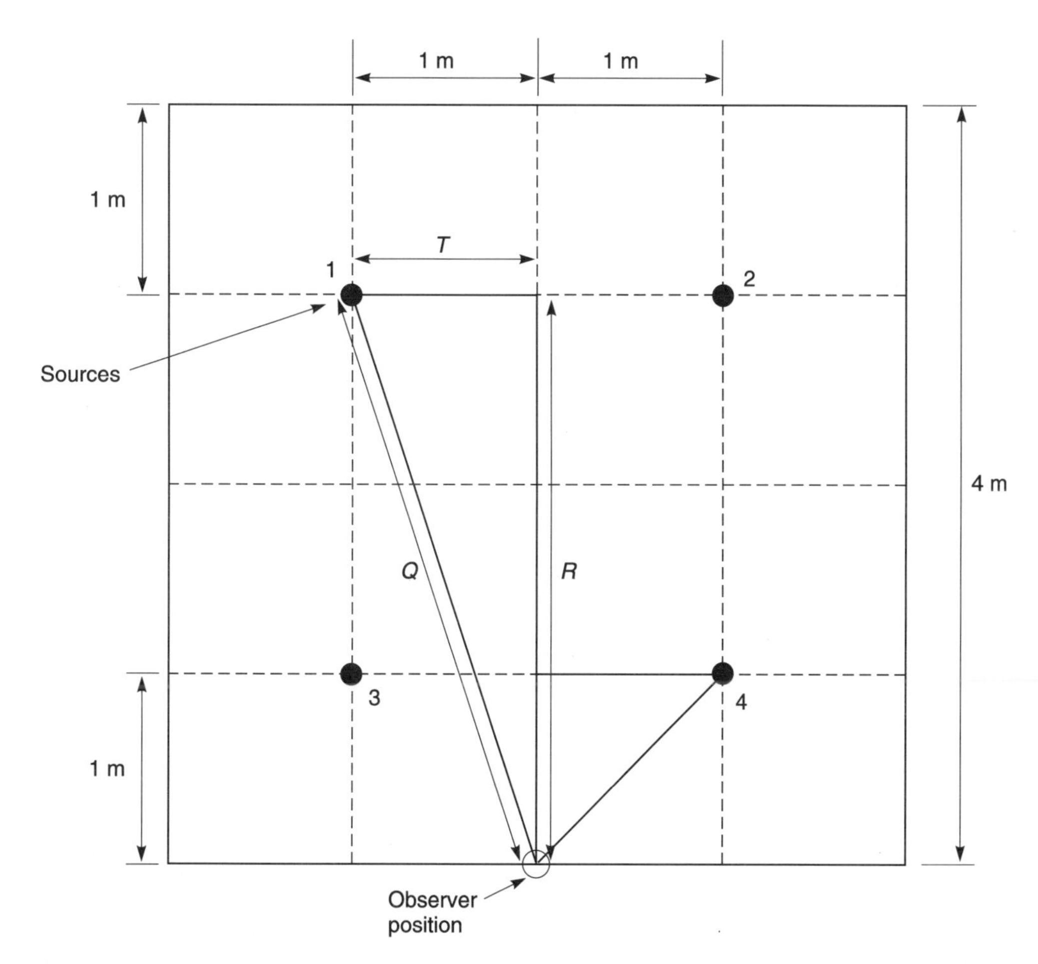

Fig. 8.4(a) Horizontal plane parameters required to enter the glare table

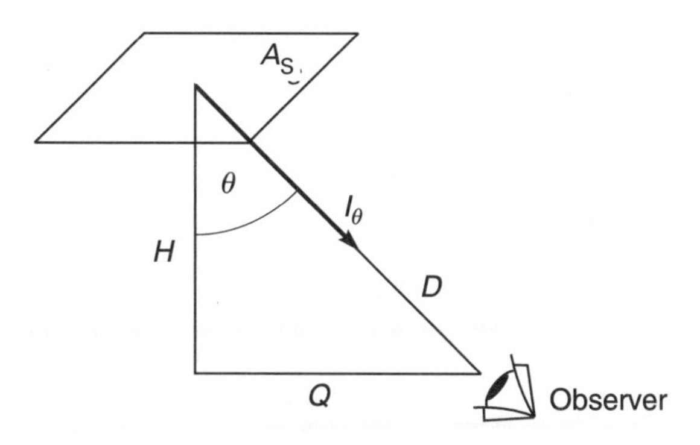

Fig. 8.4(b) Vertical plane parameters required to enter the glare table

$$L_S = \frac{I_\theta}{A_S \cos\theta} \quad \text{where } \theta \tan^{-1} \frac{Q}{H}$$

First, Sources (1) and (2),

$$Q = \sqrt{1^2 + 3^2}$$
$$= 3.16$$

$$\theta = \tan^{-1} \frac{3.16}{1.6}$$
$$= 63.15°$$

Note: $H = 1.6$ since TM10 specifies 1.2 m as the observer's eye level. So, in this case 2.8 – 1.2 = 1.6 m.

As

$$\cos\theta = 0.6 \times 0.6 \times 0.452$$
$$= 0.163$$

by interpolation, I_θ at 63.15° is 35.4 cd/1000 lumens (see Chapter 2).

$$L_S = \frac{34.5 \times 9.6}{0.163}$$

$$= 2087 \text{ cd/m}^2$$

The solid angle:

$$\omega = \frac{A_S \cos\theta}{D^2} \text{ (TM10 approximation):}$$

$$D = \sqrt{Q^2 + H^2}$$

$$D^2 = 3.16^2 + 1.6^2$$
$$= 12.55$$

$$\omega = \frac{A_S \cos\theta}{D^2}$$

$$= \frac{0.163}{12.55} = 0.013 \text{ steradians}$$

TM10 defines L_b as the reflected component of wall illuminance at the observer position divided by π. This is because the reflected component of wall illuminance is proportional to the average luminance of the room surfaces that it faces and, when this illuminance value is divided by π (assuming uniform diffusion), it is equal to it, that is L_b (see Chapter 3).

So,

$$L_b = \frac{E_W - \text{direct wall illuminance}}{\pi}$$

where $E_W = 253.4$ lux.

Direct wall illuminance $E_{W(D)} = DF_{(W)} \times F_L \times MF \div A_W$

$$= \frac{0.189 \times 9600 \times 4 \times 0.57}{2 \times 4 \times 4}$$

$$= 129.3 \text{ lux}$$

giving

$$L_b = \frac{E_W - E_{W(D)}}{\pi}$$

$$= \frac{253.4 - 129.3}{\pi}$$

$$= 39.5 \text{ cd/m}^2$$

The position index P is the ratio of the actual source luminance to the luminance of a source on the line of sight of the same angular size and having the same degree of discomfort glare. It is obtained from a table in TM10 in terms of H/R and T/R.

The values required for our example are as given in Table 8.2.

From Figure 8.4(a), $R = 3.0$ and $T = 1.0$:

$$\frac{T}{R} = \frac{1.0}{3.0} = 0.33 \quad \text{and} \quad \frac{H}{R} = \frac{1.6}{3.0} = 0.53$$

By interpolation $P = 2.88$.

Thus, for luminaires (1) and (2),

$$g_1 = \frac{0.9 \times L_S^{1.6} \omega^{0.8}}{L_b P^{1.6}}$$

$$= \frac{0.9 \times (2087)^{1.6} \times (0.013)^{0.8}}{39.5 \times (2.88)^{1.6}}$$

$$= \frac{0.9 \times 204{,}756 \times 0.031}{39.5 \times 5.43}$$

$$= 26.6$$

Table 8.2

T/R	H/R			
	0.000	0.50	0.60	1.60
0.00				
0.30		2.70	3.25	
0.40		2.80	3.30	
1.00				12.95

The two other luminaires (3) and (4) are closer to the observer and give a different value of glare constant g.

For this new position

$$Q = \sqrt{1^2 + 1^3}$$
$$= 1.41$$

So,

$$\tan^{-1} \frac{Q}{H} = \frac{1.41}{1.6}$$

and

$$\theta = 41.4° \quad \cos \theta = 0.75$$
$$A_S \cos \theta = 0.6 \times 0.6 \times 0.75$$
$$= 0.27$$

$$\omega = \frac{0.27}{1.6^2 + 1.41^2} = \frac{0.27}{4.56} = 0.05936$$

By interpolation

$$I_\theta = 130 \text{ cd/klm}$$

So,

$$L_S = \frac{130 \times 9.6}{0.27}$$

$$= 4622 \text{ cd/m}^2$$

Position index P is again obtained from Table 8.2:

$$\frac{T}{R} = \frac{1.0}{1.0} = 1.0, \quad \frac{H}{R} = \frac{1.6}{1.0} = 1.6$$

The value of P is obtained directly as 12.95.

The second value for g is:

$$g_2 = \frac{0.9 L_S^{1.6} \omega^{0.8}}{L_b P^{1.6}}$$

$$= \frac{0.9 \times (4622)^{1.6} \times (0.05936)^{0.8}}{39.6 \times (12.95)^{1.6}}$$

$$= 28.7$$

Then,

$$\begin{aligned}\text{Glare Index} &= 10\log_{10}[0.5(2g_1 + 2g_2)]\\ &= 10\log_{10}[0.5(2\times 26.6 + 2\times 28.7)]\\ &= 10\log_{10}[0.5(53.2 + 57.4)]\\ &= 17.4\end{aligned}$$

This is below the CIBSE Code limit of 19 and is therefore satisfactory.

8.3 Designed appearance lighting

At this point it is convenient to discuss the concept of apparent brightness. In a given room, the illuminances and the reflectances of the individual surfaces will result in a luminance pattern. This overall luminance pattern will determine the adaptation level of the observer's eye and this adaptation level will dictate the apparent brightness of each surface. Without taking the adaptation level into account, the effects on the observer of the individual luminances cannot be judged in advance (see Figure 8.5(a)). If the adaptation level is known then the relative effects of the different surface luminances can be estimated. In other words, the apparent brightnesses of the various surfaces can be estimated in relative terms. However, in a well lighted interior with a fairly uniform distribution of light and particularly where the light sources can be seen, the eye–brain combination can easily differentiate between two surfaces that have the same apparent brightness but different illuminances and reflectances.

Thus, apparent brightness as a lighting design tool has serious limitations but, nevertheless, it can be used with success provided that these limitations are kept in mind.

The relationship between apparent brightness and luminance has to be taken into account in visualization computer programs and this is discussed further later in this chapter (Section 8.8).

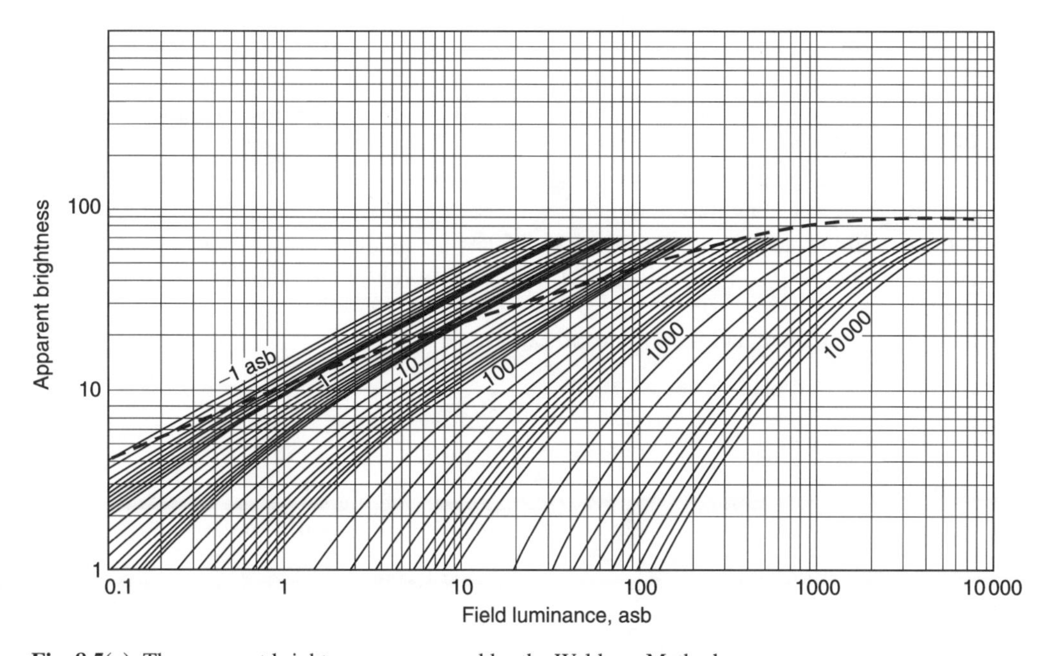

Fig. 8.5(a) The apparent brightness curves used by the Waldram Method

A prominent UK lighting engineer, J.M. Waldram[7] proposed a more logical approach to lighting design which began with a specification of the preconceived appearance of the interior.

The idea is to state the apparent brightness that is required for each room surface and, working from this specification, work back to the lighting equipment and its disposition needed to achieve the desired effect.

The most important step in this procedure relies on having (a) some idea of a scale of apparent brightness that can be used to specify the designer's requirements and (b) some link between this apparent brightness scale and the surface luminances needed to produce these values. Once the surface luminances are known the required illuminances can be calculated.

The next step is to separate from the total illuminance on each surface the direct illuminance required. Once the required direct illuminance is known the problem is to select the lighting equipment and its layout. This is quite difficult, since the direct illuminances depend both on the luminaire distributions and their positions within the lighted space.

This is the reason that Waldram applied the method mainly to the lighting of buildings such as cathedrals, where the positions of the luminaires could be individually chosen and a variety of luminaires could be used – most of them in concealed positions. Using concealed luminaires also eliminated the problem of the effect of luminaire brightness on observer adaptation, which is ignored in the basic method.

To illustrate the steps required in using this design method, the detailed example already introduced will be used. The design is for a general office 2.8 m high, 4 m square and with reflectances of: ceiling 0.7, walls 0.5, workplane 0.3.

To guide the choice of apparent brightness values we will use a scale provided by Waldram[7] based upon experience (Table 8.3).

Choosing the values to relate to both room decoration and the importance of each surface to the room function, we choose 60 units for the working plane, 55 units for the walls and 45 units for the ceiling. Taking a simple average to obtain the adaptation level gives 51.6 units. To convert the apparent brightness units into luminance values we use the curves given in Figure 8.5(a). We determine which curve to use for the conversion by first tracing the apparent brightness value for adaptation across the diagram until it intersects with the dotted line. This intersection occurs at the average value of luminance in the field of view that would result in this adaptation level.

The curve corresponding to this intersection is the one to be used for translating the apparent brightness values into luminance values. Commensurate with the accuracy of the diagram we will use an apparent brightness of 50 as the adaptation level. Tracing across the diagram gives a value of about 100 asb for the adaptation luminance and so we use the 100 asb curve.

Note: 'asb' is an abbreviation for 'apostilb' where

$$1 \text{ asb} = \pi \times 1 \text{ cd/m}^2$$

Table 8.3

10	20	30	40	50	60	70	80	90	100
Non critical work					Critical work				
Shadows		General level			High lights			Lighting equipment	

Reading from that curve we have:

Apparent brightness	45 units
Corresponding luminance	85 asb
Apparent brightness	55 units
Corresponding luminance	120 asb
Apparent brightness	60 units
Corresponding luminance	150 asb

We now have the surface luminance values which are:

$$\text{Ceiling} \quad 85 \text{ asb or } \frac{85}{\pi} = 27 \text{ cd/m}^2$$

$$\text{Walls} \quad 120 \text{ asb or } \frac{120}{\pi} = 38 \text{ cd/m}^2$$

$$\text{Workplane} \quad 150 \text{ asb or } \frac{150}{\pi} = 48 \text{ cd/m}^2$$

The required total illuminances are:

$$E_C = \frac{85}{0.7} = 121 \text{ lux}$$

$$E_W = \frac{120}{0.5} = 240 \text{ lux}$$

$$E_F = \frac{150}{0.3} = 500 \text{ lux} \qquad \text{(see Figure 8.5(b))}$$

Each of these surfaces can now be treated as area sources and the contribution to the light reflected onto the other surfaces calculated. Subtracting the reflected flux from the total surface flux gives the required direct flux.

For these calculations we require the surface distribution factors (or form factors) used in the calculation of utilization factors in the example in Section 5.4.4.

These are:

$$f_{cf} = f_{fc} = 0.415 \text{ (from Table 5.1)}$$

$$f_{cw} = f_{fw} = (1 - 0.415) = 0.585$$

$$f_{wf} = f_{wc} = f_{cw}\left[\frac{RI}{2}\right] = 0.585 \times 0.5 = 0.293$$

$$f_{ww} = (1 - (RI)f_{cw}) = (1 - 0.585) = 0.415$$

$$\text{Flux reflected from the ceiling} = E_c \times A_c \times \rho_c = 121 \times 4 \times 4 \times 0.7 = 1355.2 \text{ lumens}$$

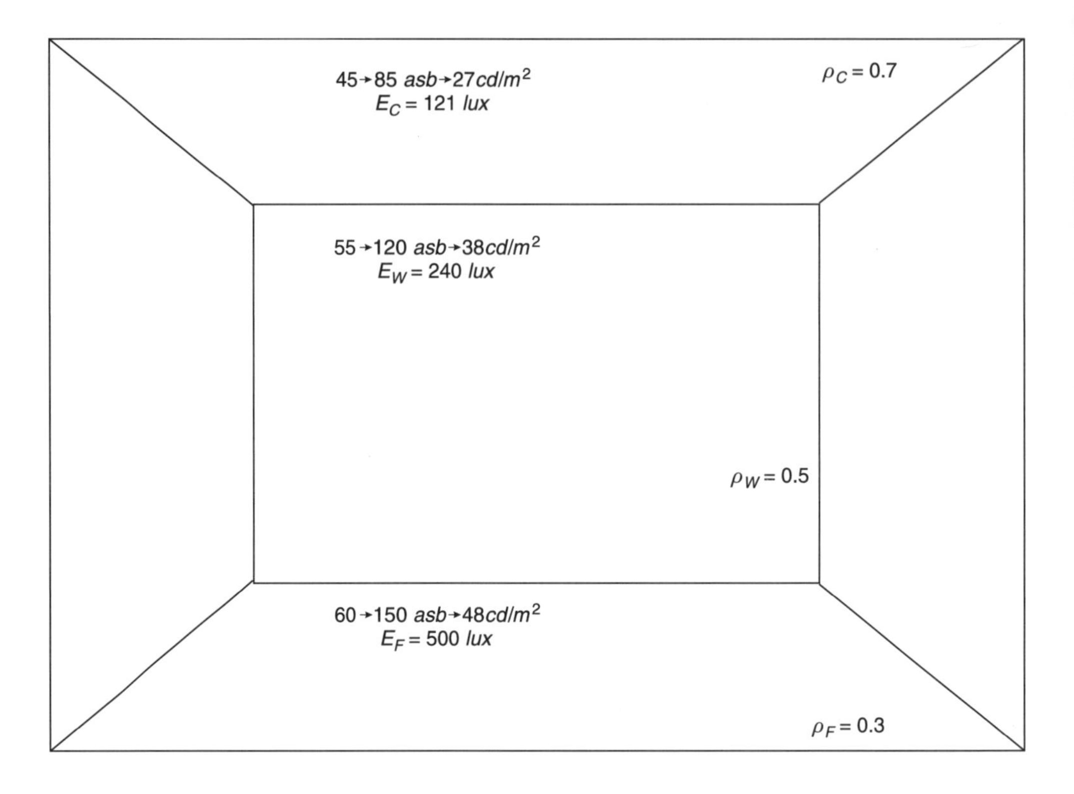

Fig. 8.5(b) The calculational steps for the Waldram Method

Received by the floor cavity from the ceiling = 1355.2 × 0.415 = 562.4 lumens

Received by the walls from the ceiling = 1355.2 × 0.585 = 792.8 lumens

Flux reflected from the floor cavity = $E_F \times A_F \times \rho_F$
$= 500 \times 4 \times 4 \times 0.3 = 2400$ lumens

Received by the ceiling from the floor cavity = 2400 × 0.415 = 996 lumens

Received by the walls from the floor cavity = 2400 × 0.585 = 1404 lumens

Flux reflected from the walls = $E_W \times A_W \times \rho_W$
$= 240 \times 2 \times 4 \times 4 \times 0.5 = 3840$ lumens

Received by the ceiling from the walls = 3840 × 0.293 = 1125 lumens

Received by the floor cavity from the walls = 3840 × 0.293 = 1125 lumens
(since with uniformly diffuse reflection the same amount of flux is reflected downwards as upwards)

Received by the walls from the walls = 3840 × 0.415 = 1593.6 lumens

Subtracting the reflected components on a surface from the total flux incident on the surface gives the required direct flux

Total flux received by ceiling = $E_C \times A_C$
$= 121 \times 16 = 1936$ lumens

Reflected flux received by ceiling = reflected flux from the floor cavity + reflected flux from the walls
= 996 + 1125 = 2121 lumens

Direct flux required on ceiling = 1936 – 2121 = –185 lumens or – 12 lux
(this result will be commented on later)

Total flux received by floor cavity = $E_F \times A_F$
= 500 × 16 = 8000 lumens

Reflected flux received by floor cavity = reflected flux from the ceiling + reflected flux from the walls
= 562.4 + 1125 = 1687 lumens

Direct flux required on floor cavity = 8000 – 1687 = 6313 lumens

Total flux received by the walls = 240 × 32 = 7680 lumens

Reflected flux received by the walls = reflected flux from the ceiling + reflected flux from the floor cavity + reflected flux from walls = 792.8 + 1404 + 1593.6 = 3790 lumens

Note: When treated as one surface the walls 'see' each other

Direct flux required on the walls = 7680 – 3790 = 3890 lumens

Before proceeding any further, it is necessary to comment on the negative value obtained for the required direct flux for the ceiling. This means that we have chosen apparent brightness values that cannot exist together in this three-surface enclosure; that is, ceiling, walls and floor with these reflectance values. The apparent brightness values chosen for the walls and floor cavity (working plane) are such that the reflected light from these surfaces to the ceiling exceeds that required by our specified apparent brightness value for the ceiling.

Since we cannot provide negative illuminance on the ceiling and the difference is not large for a non-working surface, in practice we would assume that we required no direct ceiling illuminance. The rather strange outcome is that the ceiling luminance would be higher than specified and this would slightly increase the reflected light onto both the walls and the floor cavity.

If we ignore these slight increases we can calculate the increase in ceiling luminance obtained in practice and see how it affects the value for apparent brightness of the ceiling.

The flux received by the ceiling from the walls and the floor is 2121 lumens and that required by the original apparent brightness value was 1936 lumens.

The increase in percentage luminous flux and hence luminance is

$$\frac{2121 - 1936}{1936} = 9.6\%$$

If we increase the original luminance of the ceiling by this percentage we can determine the apparent brightness that would actually be achieved:

$$85 \times 1.096 = 93 \text{ asb}$$

From Figure 8.5(a) we obtain a new value of apparent brightness of 50 units.

This compares with the 45 units originally specified, but since we chose this value from Waldram's table, which was based on experience, the difference is unlikely to be significant, particularly for a non-working surface such as the ceiling. In fact, the ceiling will be slightly brighter than originally planned, but still less bright than the walls.

Thus, apparent brightness values depend upon surface reflectances as well as the direct fluxes received, and so sometimes what is required is not a change in direct flux but in surface reflectance. This is one of the weaknesses of this method when there is significant reflected light.

Returning to the example, we now know the direct luminous flux required on each surface and we must choose a luminaire and a layout that will achieve this result. Here again we have a problem, since it is unlikely that a luminaire exists that will exactly fit our requirements. In fact, this example was chosen so that we would know of a luminaire and an arrangement that would be acceptable; that is, the luminaire used in the previous example. Let us compare our requirements with the performance of the luminaire used for that example.

The luminaire had a distribution factor (DF_F) for the working plane (floor cavity) of 0.279 and a distribution factor (DF_W) for the walls of 0.191. The ceiling distribution factor (DF_C) was zero.

Our requirement is

$$DF(F) = \frac{\text{direct flux}}{\text{lamp flux} \times MF}$$

$$= \frac{6313}{9600 \times 4 \times 0.57} = 0.288$$

$$DF(C) = 0$$

$$DF(W) = \frac{3890}{9600 \times 4 \times 0.57} = 0.164$$

Comparing this with the values for the luminaire used in the previous example, we find

DF(*F*) our requirement 0.288
DF(*F*) the previous value 0.281

DF(*W*) our requirement 0.164
DF(*W*) the previous value 0.189

DF(*C*) our requirement 0
DF(*C*) the previous value 0

Light output ratio

(*LOR*) our requirement 0.288 + 0.164 = 0.45
(*LOR*) the previous value = 0.47

In practice, this luminaire would be acceptable as meeting the requirements, and the wall illuminance of 255 lux instead of 240 lux would be accepted.

The argument for accepting the lack of precision in the Waldram method is that, because the visual system operates on a logarithmic or power scale, such changes are acceptable and of little visual consequence.

8.4 Accuracy in calculations

This brings us to a very important issue with regard to lighting calculations. This is the argument that accuracy is of little importance because the eye is tolerant of quite large differences, say 10–20%, and so calculations need not be precise. There are two main points here; the first being that cumulative errors rapidly become serious errors. An acceptable error at each stage can become an unacceptable one, say 50%, even in visual terms.

An important point is that someone has to pay for the lighting installation. If the calculations are in error so that 20% more luminaires and lighting points are required and 20% more electricity is consumed, both the initial and the running costs will be increased by a similar order.

Perhaps even more important is the comparison of competitive lighting tenders, where the cheaper tender may be cheaper because the scheme will provide less light, even though it claims to give the same lighting level. In the absence of accurate calculations, no satisfactory judgement can be made.

Let the final words go to the writer of the introduction to IES Technical Report No.2 which introduced more accurate methods for calculating coefficients of utilization (utilization factors) in 1961.

> The opinion is often expressed that there are so many other sources of error in practice . . . that errors in the basic data are of little account. This is akin to arguing that because the aim of a rifle is affected by wind, there is no point in making the barrel straight and true.

8.5 Cubic illuminance

8.5.1 CALCULATION OF CUBIC ILLUMINANCE

It has been suggested by Cuttle[8] that a useful way to describe the illuminance conditions at a point is in terms of 'cubic illuminance'. The cubic illuminance specifies six illuminance values that would be present on the six faces of a small cube centred at that point. The methods needed to calculate these illuminances have been developed in earlier chapters.

Certainly, a lighting engineer or designer would find such information very useful in analysing the performance of a lighting installation. In a rectangular room, the sides of the cube would most conveniently be arranged to be parallel with the adjacent walls, the ceiling and the floor. However, in its simplest form, cubic illuminance could be related to average values.

If this was so then the illuminance on the top of the cube would be the average horizontal illuminance; the illuminance on the bottom face of the cube would be the average horizontal illuminance multiplied by the floor cavity reflectance, and the illuminance on each of the vertical faces would be equal to the value of cylindrical illuminance. This assumes that the cube is not only moved to every possible position on the working plane, but also orientated in every possible azimuth direction.

If we use the values obtained in the example of Section 8.2 the cubic illuminances would be:

On the top of the cube	498.2 lux
On the bottom of the cube	149.5 lux (that is 498.2×0.3)
On each of the four sides	257.6 lux

Perhaps the most important use of cubic illuminance is in studying the distribution of illuminance throughout the lighted space. This means identifying points of interest in the room envelope and calculating the cubic illuminance at those points.

Clearly, a three-dimensional system of specification is needed. The x, y, z coordinate system in common use for three-dimensional mathematical problems is an obvious choice and was the one adopted in the original paper.

The previous example of Section 8.2 will now be further developed to illustrate the calculation of cubic illuminance. Consider Figure 8.6(a).

Here, we have assumed that no planes or surfaces beneath the working plane need be considered. Accordingly, our room is now considered to be 2 m high and the working plane is treated as the 'floor'. The right-handed coordinate system used in Section 3.1 is used here, but instead

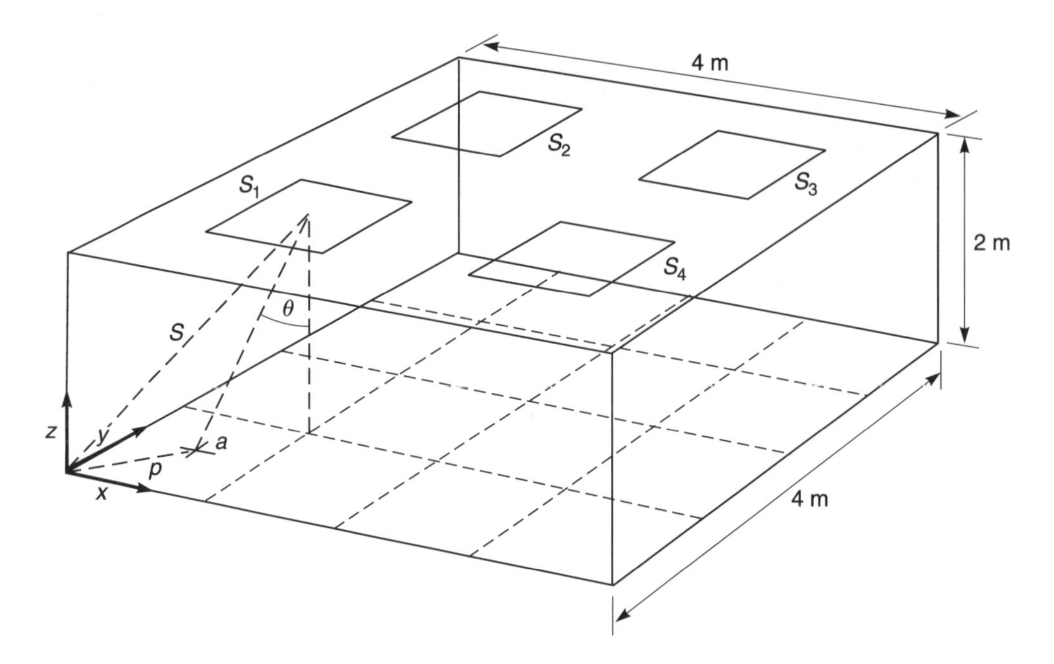

Fig. 8.6(a) The position of point ***a*** relative to the coordinate system and the luminaires

of making the origin the illuminated point, the corner of the room is now the origin, as shown, so that all the values measured from the origin are positive. Vector notation will be used in this section.

The intention is to calculate the cubic illuminance at point **a** from all four luminaires.

Let the position vector of the point **a** be $\boldsymbol{P}_a$ and that of source S be $\boldsymbol{S}_1$, then the distance from S_1 to P_a is $|\boldsymbol{S}_1 - \boldsymbol{P}_a|$ and the direction of S_1 at P_a is defined by the unit vector $q(x, y, z)$.

$$\text{Distance } S_1P_a = |\boldsymbol{S}_1 - \boldsymbol{P}_a| = [(S_{1(x)} - P_{a(x)})^2 + (S_{1(y)} - P_{a(y)})^2 + (S_{1(z)} - P_{a(z)})^2]^{0.5}$$

$$= [(1.0 - 0.5)^2 + (1.0 - 0.5)^2 + (2.0 - 0)^2]^{0.5}$$

$$= 2.121 \text{ m}$$

See Figure 8.6(b):

$$\boldsymbol{q}_{(x)} = \left(\frac{0.5}{2.121}, 0, 0\right) \quad \boldsymbol{q}_{(y)} = \left(0, \frac{0.5}{2.121}, 0\right) \quad \boldsymbol{q}_{(z)} = \left(0, 0, \frac{2}{2.121}\right)$$

$$= (0.2357, 0, 0) \qquad = (0, 0.2357, 0) \qquad = (0, 0, 0.94295)$$

$$\text{So, } \boldsymbol{q} = \boldsymbol{q}_x + \boldsymbol{q}_y + \boldsymbol{q}_z$$

$$= (0.2357, 0.2357, 0.94295)$$

Now $\boldsymbol{q}\cdot\boldsymbol{n} = \boldsymbol{q}_x\cdot\boldsymbol{n}_x + \boldsymbol{q}_y\cdot\boldsymbol{n}_y + \boldsymbol{q}_z\cdot\boldsymbol{n}_z$. Consider first the horizontal plane illuminance $E_{(-z)}$. $\boldsymbol{n}$ is the vector direction of the normal to the illuminated point, that is $\boldsymbol{P}_a$, and in this case is the normal to the horizontal plane.

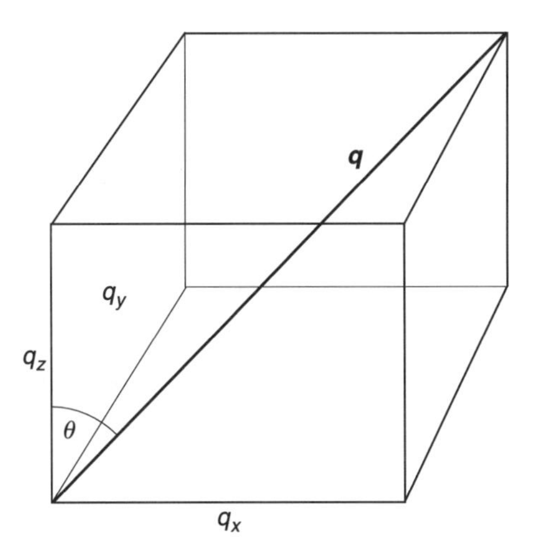

Fig. 8.6(b) The unit vector ***q***

So, $|\boldsymbol{n}_x| = 0$, $|\boldsymbol{n}_y| = 0$, $|\boldsymbol{n}_z| = 1.0$ or in vector notation $\boldsymbol{n} = (0, 0, 1)$.

Thus,

$$\boldsymbol{q}\cdot\boldsymbol{n} = 0.2357 \times 0 + 0.2357 \times 0 + 0.94295 \times 1.0 = 0.94295 = \cos\theta$$

and $\theta = 19.45°$

It is convenient for cubic illuminance calculations to convert the inverse square law equation

$$E = \frac{I_\theta \cos\theta}{d^2}$$

to

$$E = \boldsymbol{q}\cdot\boldsymbol{n}\,\frac{I_\theta}{|S_1 - P_a|^2}$$

From the intensity distribution table (Table 8.1) and using linear interpolation, the intensity at 19.45° is found to be 205.8 cd.

Giving

$$E_{(-z)} = 0.94295 \times \frac{205.8}{(2.121)^2}$$
$$= 0.94295 \times (45.746)$$
$$= 43.1 \text{ lux}$$

(this value is per 1000 lumens).

Also, for the vertical plane illuminance,

$E_{(-x)}$ the normal $\boldsymbol{n} = (1, 0, 0)$

So,

$$E_{(-x)} = 0.2357 \times (45.745)$$
$$= 10.78 \text{ lux}$$

$E_{(x)}$ = 0 since this face of the cube cannot be illuminated by the source until its x coordinate is greater than that for the source.

$E_{(-y)} = 0.2357 \times (45.745)$
$= 10.78$ lux

Also,

$E_{(y)}$ = zero

and

$E_{(z)}$ = zero

These are the direct illuminances on the cube from source S_1 (per 1000 lumens).

Let us now find the direct illuminances on the cube from source S_2.

$$\text{Distance } S_2P_a = |\boldsymbol{S}_2 - \boldsymbol{P}_a| = [(S_{2(x)} - P_{a(x)})^2 + (S_{2(y)} - P_{a(y)})^2 + (S_{2(z)} - P_{a(z)})^2]^{0.5}$$
$$= [(1.0 - 0.5)^2 + (3.0 - 0.5)^2 + (2.0 - 0)^2]^{0.5}$$
$$= 3.24 \text{ m}$$

and

$$|\boldsymbol{q}_{(x)}| = \frac{0.5}{3.24} \quad |\boldsymbol{q}_{(y)}| = \frac{2.5}{3.24} \quad |\boldsymbol{q}_{(z)}| = \frac{2}{3.24}$$
$$= 0.1543 \quad = 0.7715 \quad = 0.6173$$

For the horizontal plane $\boldsymbol{n} = (0, 0, 1)$.

So,

$\cos\theta = 0.6173 \quad \theta = 51.88°$
$I_\theta = 77.5$ cd

$$E_{(-z)} = 0.6173 \times \left(\frac{77.5}{(3.24)^2}\right)$$

$= 0.6173(7.38)$
$= 4.56$ lux (per 1000 lumens) (direct illuminance source 2)

$E_{(-x)} = \boldsymbol{q}_{(x)}(7.38)$
$= 0.1543 \times 7.38$
$= 1.14$ lux

$E_{(-y)} = \boldsymbol{q}_{(y)}(7.38)$
$= 0.\ 7715 \times 7.38$
$= 5.69$ lux

By symmetry, the same value of illuminance is produced on the cube from source 4, but the x and y values are interchanged.

$E_{(-z)} = 4.56$ lux (per 1000 lumens)
$E_{(-x)} = 5.69$ lux (per 1000 lumens)
$E_{(-y)} = 1.14$ lux (per 1000 lumens)

To complete the direct illuminance total we calculate the values for source 3.

Distance $S_3P_a = |\boldsymbol{S}_3 - \boldsymbol{P}_a| = [(S_{3(x)} - P_{a(x)})^2 + (S_{3(y)} - P_{a(y)})^2 + (S_{3(z)} - P_{a(z)})^2]^{0.5}$

$$= [(3 - 0.5)^2 + (3 - 0.5)^2 + (2.0 - 0)^2]^{0.5}$$

$$= 4.06 \text{ m}$$

$$|\boldsymbol{q}_{(x)}| = \frac{2.5}{4.06} \quad |\boldsymbol{q}_{(y)}| = \frac{2.5}{4.06} \quad |\boldsymbol{q}_{(z)}| = \frac{2}{4.06}$$

$$= 0.616 \qquad = 0.616 \qquad = 0.493$$

$\cos\theta = 0.493 \quad \theta = 60.46°$

$I_\theta = 43$ cd

$$E_{(-z)} = 0.493\left(\frac{43}{4.06^2}\right)$$

$= 0.493 \times 2.60$
$= 1.28$ lux (per 1000 lumens) (direct illuminance source 3)

$E_{(-x)} = 0.616 \times 2.60$
$= 1.60$ lux

$E_{(-y)} = 1.60$ lux
$E_{(x)} = 0$
$E_{(y)} = 0$

Total direct illuminances for point P_a (per 1000 lumens)

$E_{(-z)} = 43.14 + 4.56 + 4.56 + 1.28$
$= 53.54$ lux (per 1000 lumens)

$E_{(-x)} = 10.78 + 1.14 + 5.69 + 1.60$
$= 19.21$ lux (per 1000 lumens)

$E_{(-y)} = 10.78 + 5.69 + 1.14 + 1.60$
$= 19.21$ lux (per 1000 lumens)

To complete the calculation of cubic illuminance it is necessary to add the reflected components on each of the six surfaces of the cube.

Since the lengthy calculations carried out so far to illustrate the methods of calculation would normally be executed using a computer program, the best way of proceeding would be to divide the room surfaces into discrete areas and apply the inverse square law to each area. Each area is treated as a light source having a uniformly diffuse intensity distribution. The intensity is calculated from the known mean surface luminance and the projected area of the surface element being considered. The procedure would be the same as that used above in calculating the direct illuminances.

An alternative method, which will be illustrated here, would be to treat each of the room surfaces as an area source and to use the formulae developed in Chapter 3 for area sources (see Figure 8.7).

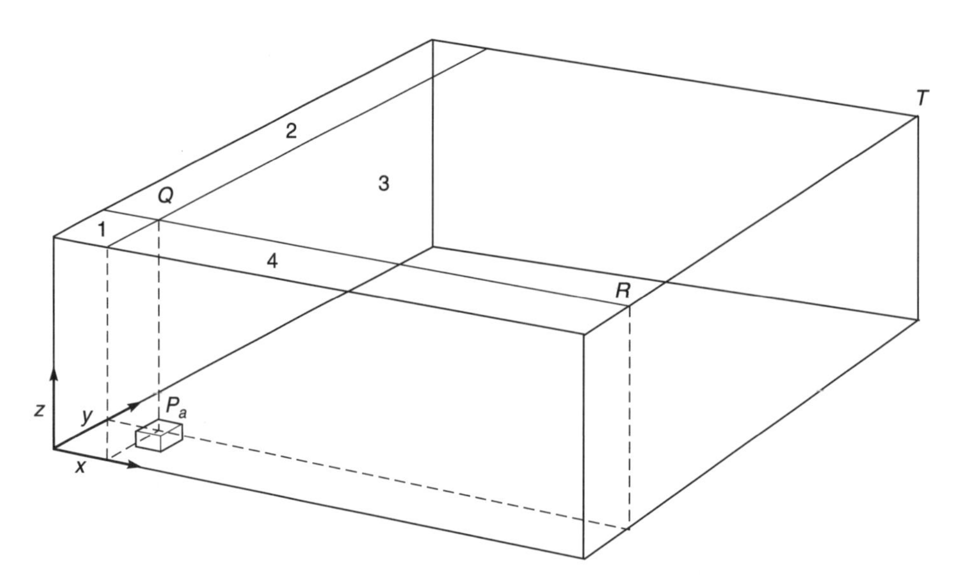

Fig. 8.7 The four ceiling areas providing reflected illuminance at point 'a' (P_a)

The formulae are

$$E = \frac{L}{2}(B_1 \sin A + A_1 \sin B) \text{ for surfaces parallel to the area source and}$$

$$E = \frac{L}{2}(B - B_1 \cos A) \text{ for surfaces at right angles to the area source (see Figures 3.18 and 3.20).}$$

The ceiling is divided into four component rectangular sources. Each of the formulae will now be applied to area 3 in Figure 8.7 to illustrate the procedure. Consider Figures 8.8(a) and 8.8(b).

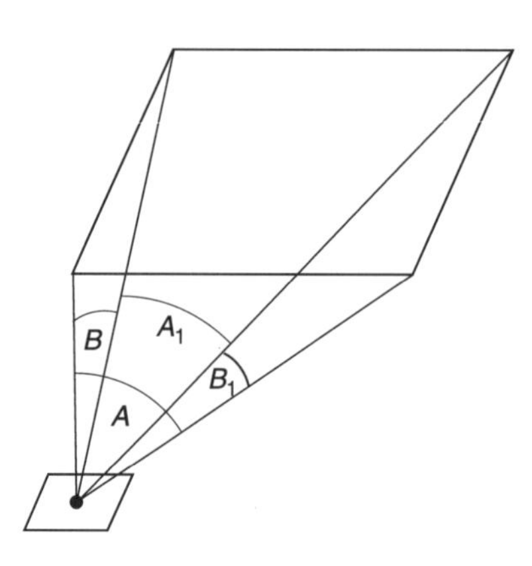

Fig. 8.8(a) The area source configuration for calculating the horizontal component of reflected illuminance at P_a

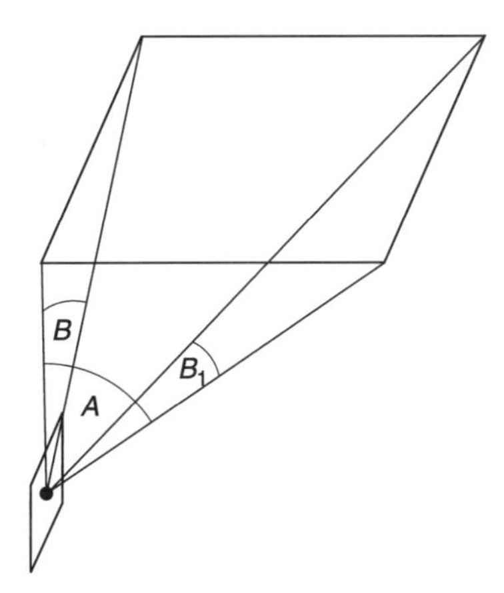

Fig. 8.8(b) The area source configuration for calculating the vertical component of reflected illuminance at P_a

In this case, the illuminance on the parallel surface is $E_{(-z)}$ and is given by the formula

$$E_{(-z)} = \frac{L}{2}(B_1 \sin A + A_1 \sin B)$$

Angle B_1 is the angle between the lines RP and TP.

$$\begin{aligned}\text{Distance } RP = |\boldsymbol{R} - \boldsymbol{P}| &= [(R_{(x)} - P_{(x)})^2 + (R_{(y)} - P_{(y)})^2 + (R_{(z)} - P_{(z)})^2]^{0.5} \\ &= [(0.5 - 0.5)^2 + (4.0 - 0.5)^2 + (2.0 - 0.0)^2]^{0.5} \\ &= 4.03 \text{ m}\end{aligned}$$

$$|\boldsymbol{q}_{(x)}| = 0, \quad |\boldsymbol{q}_{(y)}| = \frac{3.5}{4.03} = 0.868 \quad |\boldsymbol{q}_{(z)}| = \frac{2}{4.03} = 0.496$$

$$\begin{aligned}\text{Distance } TP = |\boldsymbol{T} - \boldsymbol{P}| &= [(T_{(x)} - P_{(x)})^2 + (T_{(y)} - P_{(y)})^2 + (T_{(z)} - P_{(z)})^2]^{0.5} \\ &= [(4.0 - 0.5)^2 + (4.0 - 0.5)^2 + (2.0 - 0)^2]^{0.5} \\ &= 5.34 \text{ m}\end{aligned}$$

$$|\boldsymbol{q}_{(x)}| = \frac{3.5}{5.34} = 0.656 \quad |\boldsymbol{q}_{(y)}| = 0.656 \quad |\boldsymbol{q}_{(z)}| = 0.3745$$

$$\begin{aligned}\cos B_1 = \boldsymbol{q}_{RP} \cdot \boldsymbol{q}_{TP} &= 0 \times 0.656 + 0.868 \times 0.656 + 0.496 \times 0.3745 \\ &= 0.755\end{aligned}$$

$$B_1 = 40.96° = 0.715 \text{ rad}$$

Angle A is the angle between the lines QP and RP. RP has already been evaluated. Distance QP = 2.0 m and since QP is the normal at P

$$\boldsymbol{q}_{QP} \cdot \boldsymbol{n} = (0, 0, 1)$$

So,

$$\cos A = \boldsymbol{q}_{RP} \cdot \boldsymbol{n} = 0.496$$

$$A = 60.26°$$

$$\sin A = 0.868$$

Angle $A_1 = B_1 = 40.96° = 0.715$ rad

Angle $B = A = 60.26°$, $\sin B = 0.868$

So,

$$E_{(-z)} = \frac{L}{2}(0.715 \times 0.868 + 0.715 \times 0.868)$$

$$= \frac{L}{2}(1.241)$$

L is the ceiling luminance in cd/m^2 and applies to all the ceiling areas. It will be inserted after all the ceiling components have been summed.

Area 3 also produces vertical illuminance on the cube in directions $-x$ and $-y$. For these calculations the formula

$$E_{(\ )} = \frac{L}{2}(B - B_1 \cos A) \text{ is used}$$

The required angles have already been evaluated in using the other formula.

$$E_{(-x)} = \frac{L}{2}(B - B_1 \cos A)$$

$$= \frac{L}{2}(1.05 - 0.715 \times 0.496)$$

$$= \frac{L}{2}(0.695)$$

and $E_{(-y)} = \frac{L}{2}(0.695)$

Similar calculations for all the ceiling areas produces the following results for the bracketed terms:

Area	x	y	$-x$	$-y$	$-z$
1	0.014	0.014			0.115
2	0.412			0.183	0.359
3			0.695	0.695	1.241
4		0.412	0.183		0.359
Total for the ceiling	0.434	0.434	0.878	0.878	2.074

The ceiling luminance L is given by:

$$L = \frac{E_c \times \rho_c}{\pi} \quad \text{where } E_c \text{ is obtained from Section 8.1}$$

$$= \frac{135.4 \times 0.7}{\pi}$$

$$= 30.17 \text{ cd/m}^2$$

So,

$$\frac{L}{2} = 15.09$$

Multiplying each result by the ceiling luminance divided by 2 gives the reflected illuminance components on the cube for each direction produced by the whole ceiling.

$E_{(x)} = 6.55$ lux, $E_{(y)} = 6.55$ lux, $E_{(-x)} = 13.25$ lux
$E_{(-y)} = 13.25$ lux, $E_{(-z)} = 31.31$ lux

To these results must be added the illuminance components due to light reflected from the walls. The wall areas are divided in a similar manner to the ceiling so that the two formulae can be applied to calculate these components (see Figure 8.9).

Symmetry greatly reduces the number of calculations required, but care must be exercised to ensure that each component is allocated to its correct surface and also that the correct formula is used.

Consider the calculations for area 12. Each area produces illuminance on three sides of the cube and in the case of area 12 these are directions $+y$, $-x$ and $-z$. The formulae to be used for this particular case are:

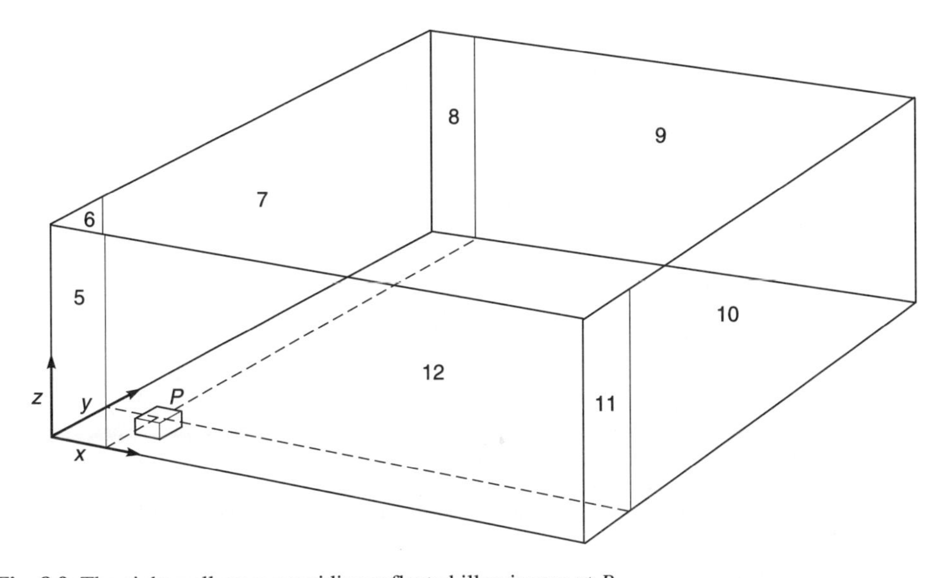

Fig. 8.9 The eight wall areas providing reflected illuminance at P_a

$$E_{(-z)} = \frac{L}{2}(B - B_1 \cos A)$$

$$E_{(y)} = \frac{L}{2}(B_1 \sin A + A_1 \sin B)$$

$$E_{(-x)} = \frac{L}{2}(B - B_1 \cos A)$$

where B, B_1, and A, A_1 are in radians.

It should be noted that, although the formulae for $E_{(-z)}$ and $E_{(-x)}$ are the same, the allocation of angle values are not, since the surfaces are at 90° to each other.

Applying the method already employed for the ceiling, the values of the components on the cube for light reflected from area 12 are:

$$E_{(-z)} = \frac{L}{2}(B - B_1 \cos A)$$

$B = 81.87°$ or 1.43 rad

$B_1 = 59.5°$ or 1.038 rad

$A = 75.96°$ or $\cos A = 0.2425$

$$E_{(-z)} = \frac{L}{2}(1.43 - 1.04 \times 0.2426)$$

$$= \frac{L}{2}(1.178) \text{ lux}$$

$$E_{(-x)} = \frac{L}{2}(1.253) \text{ lux}$$

$$E_{(y)} = \frac{L}{2}(1.52) \text{ lux}$$

If we apply these methods to all the vertical surfaces numbered 5 to 12 in Figure 8.9, we achieve the following results for the bracketed terms:

Area	x	y	$-x$	$-y$	$-z$
5	0.456	1.100			0.727
6	1.100	0.456			0.727
7	1.520			1.253	1.178
8	0.009			0.133	0.035
9			0.247	0.680	0.164
10			0.680	0.247	0.164
11			0.133	0.009	0.035
12		1.520	1.253		1.178
Total for the walls	3.085	3.085	1.185	1.185	4.208

The wall luminance L is given by,

$$L = \frac{E_w \times \rho_w}{\pi}$$

$$= \frac{255.4 \times 0.5}{\pi}$$

$$= 40.65 \text{ cd/m}^2$$

So,

$$\frac{L}{2} = \frac{40.65}{2} = 20.3$$

Multiplying each result by the wall luminance divided by 2 gives the reflected illuminance components on the cube produced by the walls for each direction.

$$E_{(x)} = 62.6 \text{ lux}, \quad E_{(y)} = 62.6 \text{ lux}, \quad E_{(-x)} = 24.1 \text{ lux}$$
$$E_{(-y)} = 24.1 \text{ lux}, \quad E_{(-z)} = 85.4 \text{ lux}$$

This completes the calculation of the direct and reflected components on the cube from the cavity above the working plane. The components from below the working plane now have to be considered.

In a practical office situation, the cavity below the working plane will be complex and liable to change because of the disposition of desks, etc. This is why it is usual to specify an assumed working plane reflectance (although it could be calculated).

In this example, a relatively high value of 0.3 has been assumed. A more common value is 0.2 when the furnishing is unknown. We have assumed light finishes for the floor, furniture and lower walls.

The light reflected from the working plane will produce illuminance on the bottom of the cube $E_{(z)}$ and on the sides of the cube $E_{(x)}$, $E_{(y)}$, $E_{(-x)}$ and $E_{(-y)}$.

If the cube is assumed to be very small then the local value of reflected illuminance can be considered as coming from an infinite plane (as far as the cube is concerned). The illuminance on the underside of the cube would be equal to the value of the local value of horizontal illuminance multiplied by the working plane reflectance, and the illuminance on the vertical sides of the cube would be half that value (that is, the infinite plane illuminance, when only half the plane can illuminate the surface).

This gives,

$$E_{(z)} = E_{(-z)} \times \rho_F$$
$$= E_{(-z)} \times 0.3$$

and

$$E_{(x)} = E_{(y)} = E_{(-x)} = E_{(-y)} = 0.5 \times E_{(-z)} \times 0.3$$

The values of these components cannot be calculated until a final value for $E_{(-z)}$ has been determined.

Final value $E_{(-z)}$ = direct component + reflected component from the ceiling + reflected component from the walls.

The direct component is given by direct illuminances from light sources 1 to 4 scaled up to 9.6 kilolumens and multiplied by the maintenance factor 0.57. The reflected illuminances have been calculated from the average illuminance values of the ceiling and wall illuminances, obtained in Section 8.1, and have already been reduced by the maintenance factor.

So,

$$E_{(-z)} = 53.54 \times 9.6 \times 0.57 + 31.3 + 85.4$$
$$= 409.7 \text{ lux}$$

and

$$E_{(z)} = 409.7 \times 0.3$$
$$= 122.9 \text{ lux}$$
$$E_{(x)} = E_{(y)} = E_{(-x)} = E_{(-y)} = 0.5 \times 122.9$$
$$= 61.5 \text{ lux}$$

The total illuminance on each of the six surfaces of the cube can now be determined.

$E_{(x)}$ = Direct illuminance + illuminance produced by reflections from the ceiling + illuminance produced by reflections from the walls + illuminance produced by reflections from the working plane

$$= 0 + 6.55 + 62.6 + 61.5$$
$$= 130.6 \text{ lux}$$

Similarly,

$$E_{(y)} = 0 + 6.55 + 62.6 + 61.5$$
$$= 130.6 \text{ lux}$$

$$E_{(z)} = 122.9 \text{ lux}$$

$$E_{(-x)} = 19.21 \times 9.6 \times 0.57 + 13.25 + 61.5$$
$$= 105.1 + 13.25 + 61.5$$
$$= 179.9 \text{ lux}$$

$$E_{(-y)} = 105.1 + 13.25 + 61.5$$
$$= 179.9 \text{ lux}$$

$$E_{(-z)} = 409.7 \text{ lux}$$

For practical purposes these values are:

$E_{(x)} = 131$ lux, $E_{(y)} = 131$ lux, $E_{(z)} = 123$ lux
$E_{(-x)} = 180$ lux, $E_{(-y)} = 180$ lux, $E_{(-z)} = 410$ lux

It is of interest to see if the uniformity criteria for horizontal illuminance ($E_{(-z)}$) has been met. The CIBSE Code specifies that the uniformity should be such that the ratio of minimum illuminance to the average illuminance should not fall below 0.8 for the main areas, excluding areas closer than 0.5 m to the walls.

In our example the value of this ratio at our point of calculation is

$$\frac{410}{498} = 0.82, \text{ which is quite satisfactory.}$$

The steps required in obtaining the cubic illuminance for one point in a simple situation have been followed through in detail to illustrate what calculations have to be carried out. Obviously, once these or similar steps have been included in a computer program these calculations can be routinely carried out quickly and accurately.

The most laborious part of the calculations was the determination of the reflected components, and Cuttle has suggested that a useful estimate of the average reflected component could be used instead of these detailed calculations. This estimate method is based on the integrating sphere theory given in Section 5.2.

The indirect component of mean room surface illuminance could be taken as the mean room surface exitance:

$$M = FRF/A(1 - R)$$

where FRF is the first reflected flux
A is the total room surface area
R is the weighted mean of the room reflectances

The total lamp flux is 4×9600 lumens.
The direct flux to the working plane is $38400 \times 0.279 = 10714$ lumens.
The direct flux to the walls is $38400 \times 0.191 = 7334$ lumens.
There is no direct flux to the ceiling:

$$\begin{aligned} FRF &= (10714 \times 0.3 + 7334 \times 0.5 + 0) \times MF \\ &= (3214 + 3667)0.57 \\ &= 3992.17 \end{aligned}$$

Taking a weighted average of room reflectance, we find

$$R = \frac{16(0.7 + 0.3) + 32 \times 0.5}{64} = 0.5$$

$$\begin{aligned} A &= 2 \times 4 \times 4 + 4 \times 2 \times 4 \\ &= 64 \text{ m}^2 \end{aligned}$$

$$M = \frac{3922}{64(1 - 0.5)} = 122 \text{ lumens per square metre}$$

which in terms of incident illuminance on the cube becomes 122 lux.

If we compare this value with the individual values of reflected illuminance for each side of the cube, we can see how appropriate this estimate would be in the case of our example.

Values of reflected illuminance

$E_{(x)}$ = 131 lux (122 lux) – 7%
$E_{(y)}$ = 131 lux (122 lux) – 7%
$E_{(z)}$ = 123 lux (122 lux) – 0.8%
$E_{(-x)}$ = 75 lux (122 lux) + 63%
$E_{(-y)}$ = 75 lux (122 lux) + 63%
$E_{(-z)}$ = 117 lux (122 lux) + 4%

The large errors are in the vertical illuminances where it swings from –7° to +63°. This is a consequence of the cube being close to one corner of the room. If the point was nearer to the centre of the room these errors would be reduced.

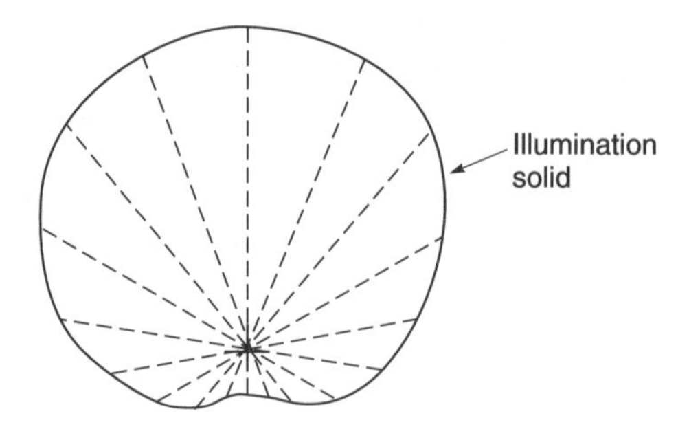

Fig. 8.10 The illumination solid

8.6 The illumination solid

The illumination solid is a polar solid of the illuminances surrounding a point. It is constructed from the planar illuminances normal to each direction from the point. The distances from the illuminated point to the surface of the solid represent these planar illuminances. The solid is, therefore, an indication of the illumination condition at the point in question in a three-dimensional form[3] (see Figure 8.10).

To construct the solid accurately the illuminances in every direction from the point must be calculated or measured. Cuttle uses the values obtained from cubic illuminance to give an approximate illumination solid and develops a series of equations for determining planar illuminance, scalar illuminance, hemispherical illuminance, cylindrical illuminance and semi-cylindrical illuminance.[8]

In the analysis of the illumination solid the following important aspect of the illuminance vector is found to be useful.

Consider two cosine waves of different magnitude, but the same frequencies, displaced by 180° and added together algebraically. This is equivalent to small size light sources illuminating opposite sides of a plane containing the point of interest (Figure 8.11).

As the plane containing the point is rotated, the resultant is $A \cos \theta + B \cos (\theta + \pi) = (A - B) \cos \theta$, since $\cos(\theta + \pi) = -\cos \theta$.

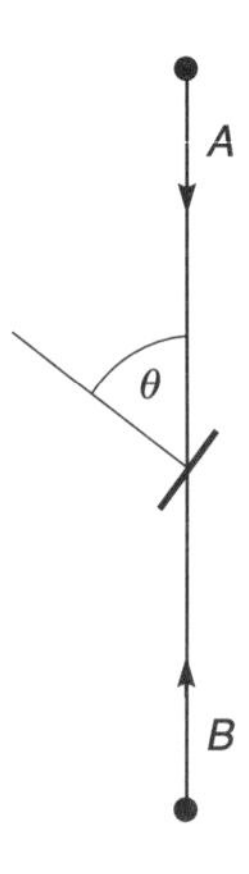

Fig. 8.11 Two small light sources illuminating opposite sides of a plane

This is obviously another cosine wave of the same frequency, but different magnitude; the change in magnitude being the difference of the illuminances on each side of the plane.

It can be shown that this is true for all angles of displacement of the point light sources about the illuminated point.[4] Thus, the illumination vector obtained is the difference between the illuminances on opposite sides of the rotated plane. This is true in general because any large source can be considered to consist of many point sources. The polar curve of the illuminance vector about the point of interest is therefore the tangent sphere associated with the cosine law of illumination. It must be noted that, because treating illuminance as a vector quantity gives the difference of the illuminances when the sources are on different sides of the illuminated plane, care is needed in the calculation. That is why, in using the vector concept in Chapter 3, Sections 3.1 and 3.2, care was taken to keep the light sources on one side of the plane and not allow the illuminated plane to separate the sources or cut through an area source.

The fact that the vector component of the illumination solid is always a tangent sphere to the illuminated point means that this component has the same distribution as that which would be produced by a point source located in the direction of the diameter of the sphere that passes through the point (Figure 8.12). Since this sphere represents the magnitude and direction of the differences in illuminance in each direction at the illuminated point, subtracting this vector component from the illumination solid would leave another solid that would be symmetrical in the sense that all opposing illuminance values would be equal (Figures 8.13(a) and (b)). That is $E_1 = E_2$ and $E_3 = E_4$, etc.

This property of the illumination solid can be used to calculate the scalar illuminance at the point. The vector component can be treated as though it was due to a point source and, in Section 3.8.1, we showed that

$$E_S = \frac{E_{max}}{4} = \frac{|E|}{4}$$

To this must be added the contribution of the symmetrical component. The symmetrical component is equal to

$$\frac{1}{N} \Sigma E_{SM}$$

where N is the number of values of illuminance measured or calculated to produce ΣE_{SM}. This mean value of E_{SM} is added to the result for the vector component to give the total value for E_S.

If the values of illuminance measured or calculated are those of cubic illuminance (that is 6) then the resulting value for E_S will be an approximation.

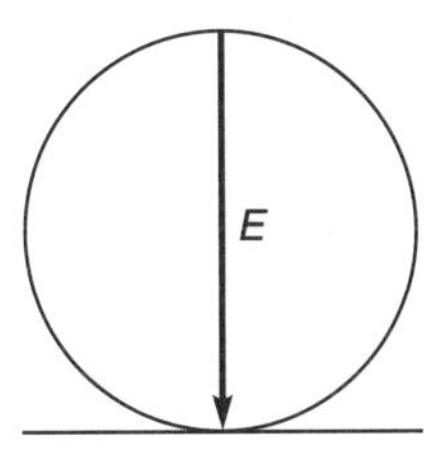

Fig. 8.12 The tangent sphere distribution of the vector component

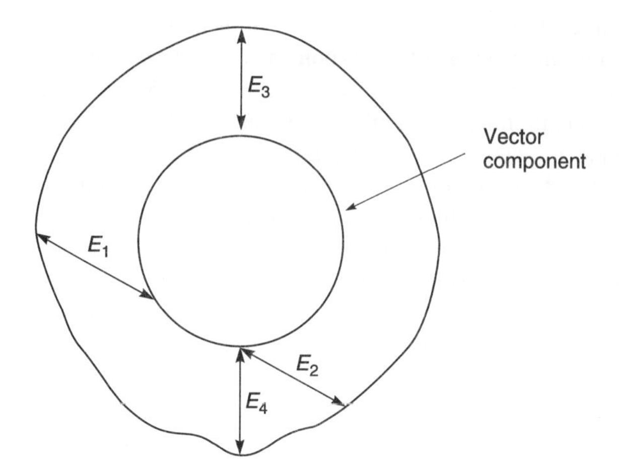

Fig. 8.13(a) The tangent sphere component within the illumination solid

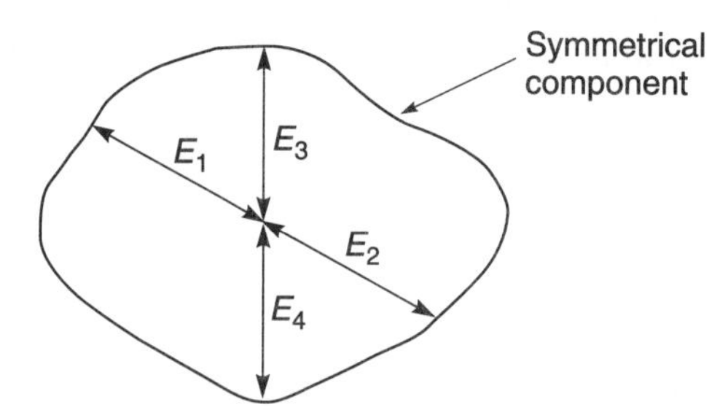

Fig. 8.13(b) The symmetrical component left once the vector component has been removed

Let us calculate the value for average E_S for our previous example and compare it with the more rigorously calculated value obtained using zonal multipliers and inter-illuminance factors (see Section 8.2).

The average illuminance vector would act in a downward direction and would be given by $E_F - \rho_F E_F$.

So,

$$\begin{aligned} |E| &= E_F(1 - \rho_F) \\ &= 498(1 - 0.3) \\ &= 349 \text{ lux} \end{aligned}$$

The vector component of E_S is therefore 349/4 = 87.3 lux.

The symmetrical component is obtained from each of the pairs of values:

$$\begin{matrix} E_{(x)}, & E_{(-x)}, & E_{(y)}, & E_{(-y)}, & E_{(z)}, & E_{(-z)} \\ 258, & 258, & 258, & 258, & 149, & 498 \end{matrix}$$

taking rounded values from Section 8.5.1.

In the cases of the x and y axes, the differences to be subtracted are zero, so that the magnitude of the symmetrical component is 258 lux.

In the case of the z axis, once the difference in the two values has been subtracted the symmetrical component values will be 149 lux.

The contribution to the scalar illuminance of the symmetrical component is therefore the mean of these three values.

$$E_{S(SM)} = \frac{258 + 258 + 149}{3}$$

$$= 222 \text{ lux}$$

The total value for E_S

$$= 87 + 222$$

$$= 309 \text{ lux}$$

The value obtained in our earlier example (Section 8.2) was 330 lux, a difference of 6% in this case.

Cuttle[8] lists five approximate illuminance equations based upon the vector and symmetrical components of the illumination solid derived from cubic illuminance. In these equations $\boldsymbol{n}$ is the Normal unit vector, defining the direction relative to the plane specified by the suffix, and $\boldsymbol{e}$ is the Illumination unit vector, defining the direction of the illumination vector, specified by components on the x, y and z axes.

For planar illuminance

$$E_{pr} = |E|\,\boldsymbol{e}\cdot\boldsymbol{n} + E_{SM(x)}\boldsymbol{n}_{(x)} + E_{SM(y)}\boldsymbol{n}_{(y)} + E_{SM(z)}\boldsymbol{n}_{(z)}$$

where $\boldsymbol{e} = E_{(x,y,z)}/|E|$.

If $e\cdot n<0$, then set $e\cdot n = 0$.

For scalar illuminance

$$E_S = |E|/4 + (E_{SM(x)} + E_{SM(y)} + E_{SM(z)})/3$$

For hemispherical illuminance

$$E_{hem} = |E|(1 + \boldsymbol{e}\cdot\boldsymbol{n})/4 + (E_{SM(x)} + E_{SM(y)} + E_{SM(z)})/3$$

For cylindrical illuminance

$$E_{cyl} = |E|\,\boldsymbol{e}\cdot\boldsymbol{e}_{(x,y)}/\pi + (E_{SM(x)} + E_{SM(y)})/2$$

For semi-cylindrical illuminance

$$E_{semcyl} = |E|\,(\boldsymbol{e}\cdot\boldsymbol{e}_{(x,y)})(1 + e_{(x,y)})/\pi + (E_{SM(x)} + E_{SM(y)})/2$$

8.7 CSP

So far this chapter has been devoted to a simple example of office lighting in order to illustrate thoroughly the steps required to calculate the various physical parameters by which to judge the quality of the lighting.

From these calculations we have determined that the working plane illuminance should be satisfactory; that the glare should be well controlled and that the brightness of the walls and ceiling should be acceptable. In addition, we have established that the modelling of the features of the occupants should be acceptable. However, the question still remains 'Will the users of the office be satisfied?'

An attempt has been made to answer this question in the case of office lighting by developing an empirical formula, based on experimentation, which seeks to combine the physical parameters to produce a simple number index to indicate whether occupants of an office would be satisfied.[9]

The weakness of this work is that it is based on testing 650 people in 44 offices and combining the results in a histogram. This means that, for a small office with only a few occupants, the sample would be small and therefore biased. The value of this index is that it indicates that people using that office should or should not be satisfied. The index is called the CSP Index because it is based on considering Visual Comfort (absence of glare), Visual Satisfaction (relative illuminance of vertical and horizontal surfaces) and Visual Performance (lighting conditions at the task).

Another issue is that combining these three elements of visual quality in a single number suggests that there is a trade-off between them; that more visual comfort can offset less visual performance or satisfaction. However, the index has been designed to restrict the degree of this trade-off. The histogram produced from the experiments suggests that there is, within limits, some such effect.

The CSP formulation has been designed to deal with offices that do not employ VDUs, offices that do and offices where there is only a percentage of VDU use. Most offices today employ VDUs and so only one of the formulae would be needed.

Table 8.4 gives the formulae for the method of assessment and Figure 8.14 gives the curve of the percentage of occupants likely to be satisfied plotted against the CSP Index, which was derived from the histogram obtained from applying the Index to the observer responses. It is interesting to see how the office lighting employed in the example used in this chapter would be rated by this Index.

We will assume 100% VDU use.

(1) The Comfort Index (C) (derived from the Glare Index)

$$\begin{aligned} GI &= 17.4 \text{ (see Section 8.1)} \\ C &= 10 - 0.3(GI - 14) \\ &= 10 - 0.3(17.4 - 14) \\ &= 8.98 \end{aligned}$$

Table 8.4 VDU Lighting

Condition	Formula
$GI>14$:	$C = 10 - 0.3(GI = 14)$
$GI<14$:	$C = 10$
$E_{cyl}/E_h<2/3$:	$S = 15E_{cyl}/E_h$
$E_{cyl}/E_h>2/3$:	$S = 10$
$P = 0.075K_{Evd}K_{DR}\left(1 + \dfrac{K_U + K_R}{60}\right)$	
If $P>10$, then assume $P = 10$	
$E_h<500$:	$K_{Evd} = E_h/50$
$E_h>500$:	$K_{Evd} = 5000/E_h$
DR at $RI = 0.75<0.5$:	$K_{DR} = 20DR$
DR at $RI = 0.75>0.5$:	$K_{DR} = 10$
	$K_U = 10E_{min}/E_{av}$
	where $E_{av} = E_h$
$R_a<90$:	$K_R = 0.111R_a$
$R_a>90$:	$K_R = 10$
$Q_{iv} = 3CSP/(C + S + P)$	

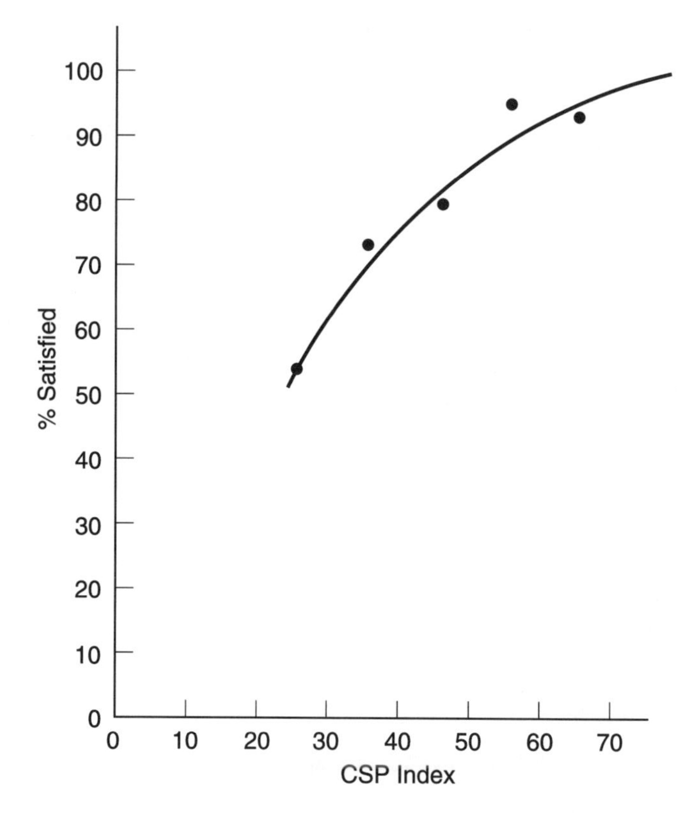

Fig. 8.14 The experimentally obtained curve of satisfaction against CSP Index

(2) *The Satisfaction Index* (S) (derived from the cylindrical to horizontal illuminance ratio)

$$S = 15\frac{E_{\text{cyl}}}{E_{\text{h}}}$$

$$\frac{E_{\text{cyl}}}{E_{\text{h}}} = 0.52$$

$$S = 15 \times 0.52$$
$$= 7.8$$

(3) The Performance Index (P) (derived from the illuminance level, the uniformity, the direct ratio and the colour rendering index)

$$P = 0.075 K_{\text{Evd}} K_{\text{DR}} \left[1 + \frac{K_{\text{u}} + K_{\text{R}}}{60} \right]$$

$$E_{\text{av}} = E_{\text{h}} = 498 \text{ lux}$$

$$\text{Uniformity} = \frac{E_{\text{min}}}{E_{\text{av}}} = 0.82$$

$$DR = \text{Distribution factor divided by } LOR$$
$$= 0.24/0.47 \qquad (DF \text{ from manufacturer})$$
$$= 0.51$$

R_a has not been specified so far. We will choose a moderate value of 70 (see CIBSE Code).

$$K_{Evd} = E_h/50$$

$$= \frac{498}{50} = 9.96$$

$$K_u = 10\frac{E_{min}}{E_{av}}$$

$$= 10 \times \frac{410}{498}$$

$$= 8.2$$

$$K_{DR} = 10$$
$$K_R = 0.111R_a$$
$$= 0.111 \times 70$$
$$= 7.77$$

$$P = 0.075 \times 9.96 \times 10\left(1 + \frac{8.2 + 10}{60}\right)$$

$$= 7.47 \times 1.30$$
$$= 9.7$$

$$\text{CSP Index} = \frac{3 \times C \times S \times P}{C + S + P}$$

$$= \frac{3 \times 8.98 \times 7.8 \times 9.7}{8.98 + 7.8 + 9.7}$$

$$= \frac{2038}{26.48}$$

$$= 77$$

This value is at the top of the practical CSP range and suggests that over 90% of occupants should be satisfied.

It has been suggested that the CSP Index could be used as a 'go', 'no go' gauge. If the calculated CSP Index value is 40 or above, the installation should be rated reliable and if it is below that value it should be rated unreliable since a low value of CSP implies that occupants' satisfaction or dissatisfaction would be governed mainly by their visual capacity.

For example, if the client required the designer to produce an installation that was rated above 40 on the CSP Index scale and with a power consumption no greater than 4 $w/m^2/100$ lux, they would be controlling the visual quality of the installation as well as its energy efficiency, without interfering with its detailed design.

8.8 Visualization

In Section 8.3 we dealt with Waldram's 'designed appearance' method of lighting design and pointed out that the first step was to have some idea of a scale of apparent brightness to enable the designer to specify his or her requirements. This aspect of the method relied on being able to imagine the required brightness relationships within the room and then translate these into numerical values, a very sophisticated task.

Visualization is an attempt to enable both the designer and the client to have a visual image of the final result and to vary that image until the requirements are met. The technique relies on the ability to produce a satisfactory image of the room and its surface brightnesses and then relate the image to a practical lighting installation. Needless to say, visualization had to await the advent of very powerful computers working at high speed to produce such images in real time.

Bodman[10] has pointed out the following differences between a real scene and computer generated images on a TV monitor:

1. The screen picture is much smaller than the real room.
2. A three-dimensional room appears in perspective on a two-dimensional screen.
3. The viewing distances are different.
4. The screen is self-radiating, whereas most real surfaces reflect light.
5. The observer views the screen in a room, whereas in the real room he experiences it.
6. The light levels (luminances) on the screen are generally less than in reality.

Bodman makes the following two important points:

1. The linear transformation of the calculated high luminances on to the lower luminances of the display screen produces a too dark overall impression.
2. Tranformations involving the apparent brightness/luminance relationships derived by such people as Adams/Cobb or Stevens gives a good or very good impression of the real room.

He states these apparent brightness/luminance relationships as follows:

Stevens: $H \sim (L - L_s)^n$

Adams/Cobb: $H \sim \dfrac{L}{L + L_A}$

where H = apparent brightness of the object
L = Luminance of the object
L_s = Threshold luminance
L_A = Adaptation luminance
n = Exponent

Since most people are used to viewing television screens and to interpreting the display in relation to reality, this method of generating lighting designs has much to commend it. It is, of course, of particular value for major design projects where the work involved in producing the image is justified.

The current VDUs cannot display the full range of luminances or colours in the scene, nor do they fill the entire visual field. This has important consequences:

1. Glare from bright sources such as windows and luminaires cannot be adequately represented; a bright patch on a VDU might have a luminance of 120 cd/m^2 compared with a luminaire brightness of 3000 cd/m^2.

2. The luminance range of the scene must be compressed on to the luminance range of the VDU.

Realistic images demand vast numbers of calculations to determine the detailed luminance patterns that would exist in the true installation. Practical visualization programs, therefore, contain many approximations and optimizations to reduce the number of equations to be solved or the geometric calculations needed.

Much of this work depends upon programming skills concerned with substructuring and establishing sophisticated hierarchical algorithms. All this is beyond the scope of this book. The radiosity method described in Chapter 5 and applied to calculating the average illuminance of three surfaces can be extended to any number of surfaces. A room could be divided into hundreds or thousands of elementary areas and solutions obtained. However, the problem would be the size of the matrix, since this increases as the square of the number of individual areas considered.

One simple technique is to use the radiosity method to determine the average luminance of each major surface produced by secondary reflections and to add to this the detailed luminance pattern produced by the first reflected flux.

To do this a calculation is carried out using the average direct illuminance on the surface to calculate an average luminance value. The first reflected element of the average luminance value is then subtracted and replaced by a detailed luminance pattern generated by point by point values of direct illuminance. This, of course, greatly reduces the amount of computation required.

In addition, to give more realism, ray tracing can be undertaken to determine where the highlights occur due to the semispecular nature of many room surfaces.

Because visualization depends not upon the development of lighting theory but upon computer technique, it is part of an enormous effort being made continually in the computer field. Obviously, visualization techniques will be continuously refined and their use in lighting design will become common for the most important designs.

8.9 Detailed requirements for interior lighting

Since 1936, in the United Kingdom, a Lighting Code has been published that contains recommendations for the lighting levels to be used in specific situations. This Code has been developed over the years into a valuable guide to the design of interior lighting for different situations. The 1994 Code[1] gave an outline of the many different aspects to be considered in interior lighting design, both in general and for specific applications and this is a valuable aid to the lighting designer.

This book is mainly devoted to the fundamentals of lighting calculations and to their application. It cannot include all the detailed recommendations for the dozens of applications listed in the Code. That Code, together with such books as *Lamps and Lighting*[11] is an important source of reference for designers. This present book provides the fundamental lighting calculational techniques for meeting the design requirements outlined in the Code and in other design guides.

References

1. CIBSE (London) (1994) *Code for Interior Lighting*.
2. Petherbridge, P. and Hopkinson, R.G. (1950) Discomfort glare and the lighting of buildings. *Trans. Illum. Eng. Soc.* (*London*) **15**, 39–79.

3. Gershun, A. (1939) The light field. *J. Maths and Physics*, **18**, 51–151.
4. Lynes, J.A., Burt, W., Jackson, G.K. and Cuttle, C. (1966) The flow of light into buildings. *Trans. Illum. Eng. Soc. (London)* **31**, 65–91.
5. Lynes, J.A. (1968) *Principles of Natural Lighting* (Elsevier).
6. Hewitt, H., Bridgers, D.J. and Simons, R.H. (1965) Lighting and the environment. Some studies in appraisal and design. *Trans. Illum. Eng. Soc. (London)*, **30**, 91–112.
7. Lynes, J.A. (ed) (1978) *Developments in Lighting*, Chapter 5 (London: Applied Science Publishers).
8. Cuttle, C. (1997) Cubic illumination. *Lighting Research and Technology*, **29**(1), 1–14.
9. Bean, A.R. and Bell, R.T. (1992) The CSP Index: a practical measure of office lighting quality as perceived by the office worker. *Lighting Research and Technology*, **24**(4), 215–225.
10. Bodman, H. (1988) How do our eyes see light levels on the graphic picture screen? *Licht* **6**.
11. Coaton, J.R. and Marsden, A.M. (ed) (1997) *Lamps and Lighting* (Arnold).

9

Main Road and Motorway Lighting

9.1 Introduction

The principal purpose of main road and motorway lighting is to improve visibility for the driver, in contrast to residential road lighting (Chapter 10), which is installed for the benefit of the pedestrian. Main road lighting is installed in the hope that it will reduce accidents. In fact, the economic justification for installing it rests mainly on costing exercises to evaluate the savings from the estimated reduction in accidents. An added benefit is that road lighting makes the task of driving easier. This is particularly the case for the older driver, for whom the glare from oncoming traffic can result in a marked impairment of the capacity to discern objects. Some light from main road lighting will fall onto adjacent footways, which is of help to the pedestrian, and in many instances installation designs are made with the intention of providing sufficient light for pedestrians as well as for drivers.

The lighting can improve the appearance of the environment at night, but care has to be taken to limit severely upward light from the luminaires, which can form a bright aurora. This constitutes 'light pollution'; it is not only unsightly but is the bane of astronomers because it makes observation of objects in the sky difficult or impossible.[1] Light pollution, or perhaps more aptly, *light trespass*, may also be caused when the installation is close to housing where even the useful light from the installation may be regarded as a nuisance by some residents.

9.2 Lighting and accidents

In the UK, the accident rate at night, weighted for distance travelled, is about 80% higher than by day. For fatal accidents the risk is about three times greater at night than that by day. Whilst it is tempting to ascribe this to the greater difficulty of seeing by night, other factors such as tiredness and the greater intake of alcohol at night are at least partly responsible. Because of these uncertainties, many studies have been carried out to find whether road lighting does reduce accidents. Other studies have tried to discover which lighting parameters (average illuminance, average luminance, uniformity of luminance, etc.) are important and how levels are related to accident rate. Accounts of these studies have been collated in CIE 93,[2] *Road Lighting as an Accident Countermeasure*. Generally, these studies suggest that installing some lighting will reduce the ratio of accidents at night to those during the day by 30%. This ratio, the night–day accident ratio, is often used as a measure of the effectiveness of road lighting because any non-lighting change to the installation should affect the accidents during both periods equally, whereas a change to the lighting will affect only the night accident rate.

Three of the studies may be singled out as being on a large scale and providing significant data. In the UK, a study of 100 30 mph roads was carried out.[3] Various lighting parameters were measured. It was found that the parameter that most strongly correlated with the night–day accident ratio of fatal accidents was the average luminance of the road surface. A steady decrease in the night–day accident ratio was produced as the luminance was increased from 0.5 to 2.0 candelas per square metre. However, not all the studies cited in CIE 93 gave results as straightforward as this. Box, in the USA, in two studies, one on 22 freeway sites the other on 329 urban sites, found a U-shaped relation between horizontal illuminance and the night–day accident ratio, the minimum of the U occurring at about 5,5 lux.[4,5] Janoff *et al.* found the night–day accident ratio increased with illuminance but decreased with an index of visibility which took into account disability glare as well as other lighting criteria.[6] On the basis of this they conjectured that the increased disability glare at the higher illuminances was responsible for the increased night–day accident ratio. These findings of the work done in the USA have provided the impetus for work on Small Target Visibility, which will be discussed later (Section 9.13, page 373).

The need to install road lighting is often decided on the basis of cost benefit. The projected savings as a result of reduction in accidents are set against the capital and running costs, including maintenance, of the installation.

9.3 Visibility of objects on the road

In road lighting, the economic upper limit for average illuminance on the road surface is about 20 lux, but for most roads half this value is used. This compares with the CIBSE recommended horizontal illuminance for offices of 500 lux.[7] The question immediately arises; how can road lighting, where decisions affecting life and death are being taken, often very quickly, be effective in revealing objects on the road when it operates, as it does, at low illuminances?

The answer is provided by Waldram's principle of silhouette vision, enunciated in the early 1930s. Waldram was the first to realize that there was not sufficient light available to illuminate effectively the objects of interest, as is done with most artificial lighting, but use had to be made of the reflection properties of the road to provide a bright background against which objects would be seen in silhouette. This implies that the illuminance on vertical surfaces should be as low as possible to produce a silhouette and achieve the maximum contrast between the object and the road surface. It is, however, not possible to produce this condition for all objects wherever they may be positioned on the road. The luminance of objects on the road will vary according to their reflectances as well as the illuminance falling on them. It is, therefore, possible for an object to be brighter than the road surface (reversed silhouette), and there will be positions where the luminance of the object will match that of the background, and therefore become very difficult or impossible to see. Direct light from the road lighting luminaires may cause disability glare, which – as the name implies – decreases the ability to discern objects. This effect can be calculated, but there is no satisfactory method of calculating discomfort glare. It is possible that for short spells of driving, discomfort glare is not of much importance, but over longer periods it has been argued that it could have a fatiguing effect which may lead to accidents.

There are other factors which should be taken into consideration in considering the conspicuity of objects on the road:

- density of traffic. Roads have become so crowded with traffic that the backs of cars provide the foreground more often than not,
- scattering of light by the windscreen, which may be dirty or wet,
- light loss by absorption by a tinted windscreen,
- lighting of the surrounds, which may enhance visual conditions,
- optical guidance provided by direct light from the luminaires. This gives the driver forewarning of the run of the road far ahead, and may be particularly useful in fog,

- age of the road user. Vision becomes less effective with age. For instance, at 60 years of age three times the luminance is needed for the same retinal illuminance as is needed at 20 years of age.[8] In addition, there is more scattering of the light in the older eye, which increases disability glare, and reaction times are slower. There are sufficient data available to allow for these effects, but if we designed for the oldest person likely to be using the road the increase in lighting levels would have to be manifold.

9.4 Some road lighting terminology

Before discussing the silhouette principle, we need to become familiar with some road lighting terms relating to the layout of the luminaires and the light distribution from them.

Figure 9.1 shows the commonly used layouts or arrangements for the luminaires. Which

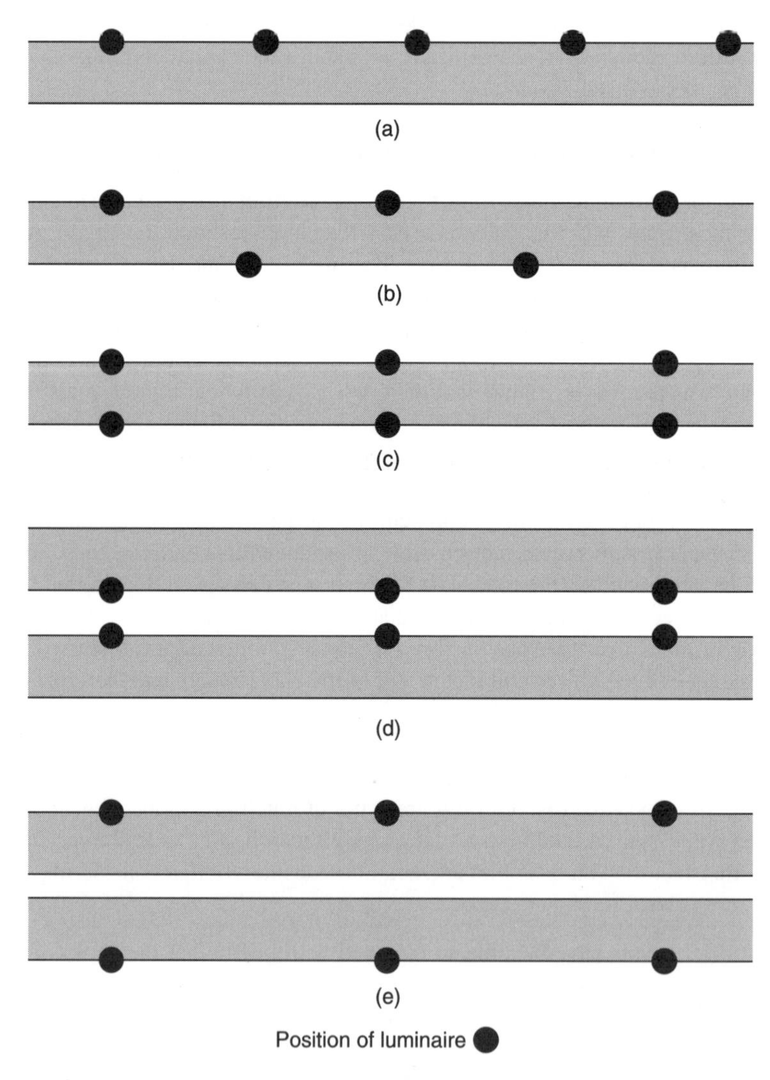

Fig. 9.1 Commonly used arrangements in road lighting. (a) Single side arrangement; (b) staggered arrangement; (c) opposite arrangement; (d) twin central arrangement; (e) opposite arrangement on dual carriageway

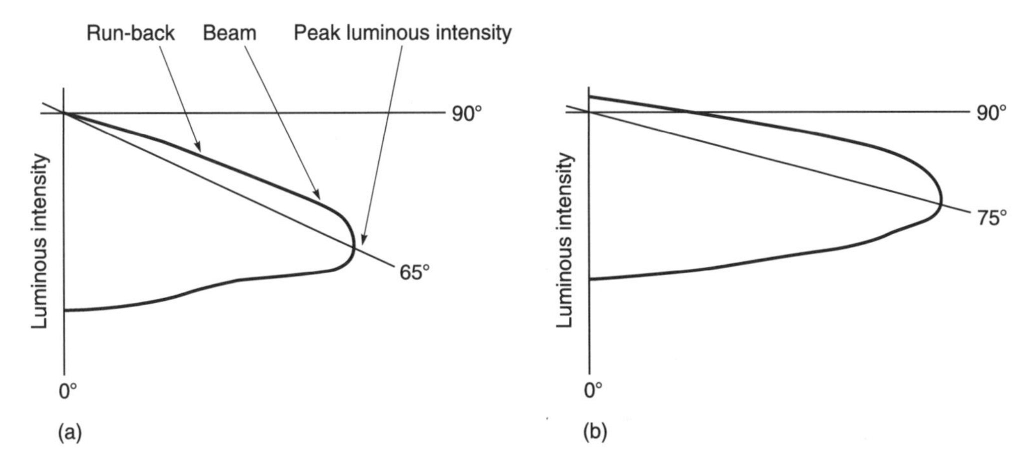

Fig. 9.2 Vertical polar curves through the peak of cut-off and semi-cut-off light distributions. (a) Polar curve of cut-off light distribution; (b) polar curve of semi-cut-off light distribution

arrangement is selected will depend on the light distribution of the luminaires, the width of the carriageway, accessibility for maintenance and economic factors. For instance, as regards factors that are not directly related to the lighting, the single side arrangement is popular outside the UK because of the cheaper cable costs compared with the other arrangements. In addition, maintenance procedures can be carried out more easily. Although the twin central arrangement gives the better lighting effect on dual carriageways, an opposite arrangement may be chosen because of easier accessibility for maintenance, although offset against this is the fact that two cabling runs are needed, on either side of the road.

There are several terms used to describe the general shape of the light distribution from a luminaire. The highest luminous intensity is referred to as the peak luminous intensity or, more simply, as the peak (Figure 9.2).

The part of the light distribution surrounding the peak is known as the beam, and the part of the distribution from the beam to the horizontal is the run-back. Figure 9.2 illustrates the two most common light distributions; cut-off and semi-cut-off. The curves are in the vertical plane through the peak. The cut-off distribution has a peak at 65° or lower and a fast run-back, with a low or zero luminous intensity at 90°. The semi-cut-off distribution has a peak at 75° or lower with a slower run-back and a greater luminous intensity at 90° than the cut-off distribution. A third term, non-cut-off, is sometimes used to describe a distribution where there is no or little pronounced peak and there is little diminution of the luminous intensity in the run-back.

In azimuth, the peak is usually designed to fall between 0° and 15° of azimuth towards the centre of the road. This angle is referred to as the *toe-in* (Figure 9.3). As indicated in this diagram, the light distribution is symmetrical, at least nominally, about the transverse vertical plane ($C = 90°$ and 270°) (see Section 2.2, page 34) through the photometric centre of the luminaire. For situations where the road is wet for a high proportion of the time, it is useful to have a toe-in as great as 25°, but for most general purposes it does not exceed 10°.

Although many of these terms lack precision, and are used in somewhat different ways in different countries, they are useful for descriptive purposes.

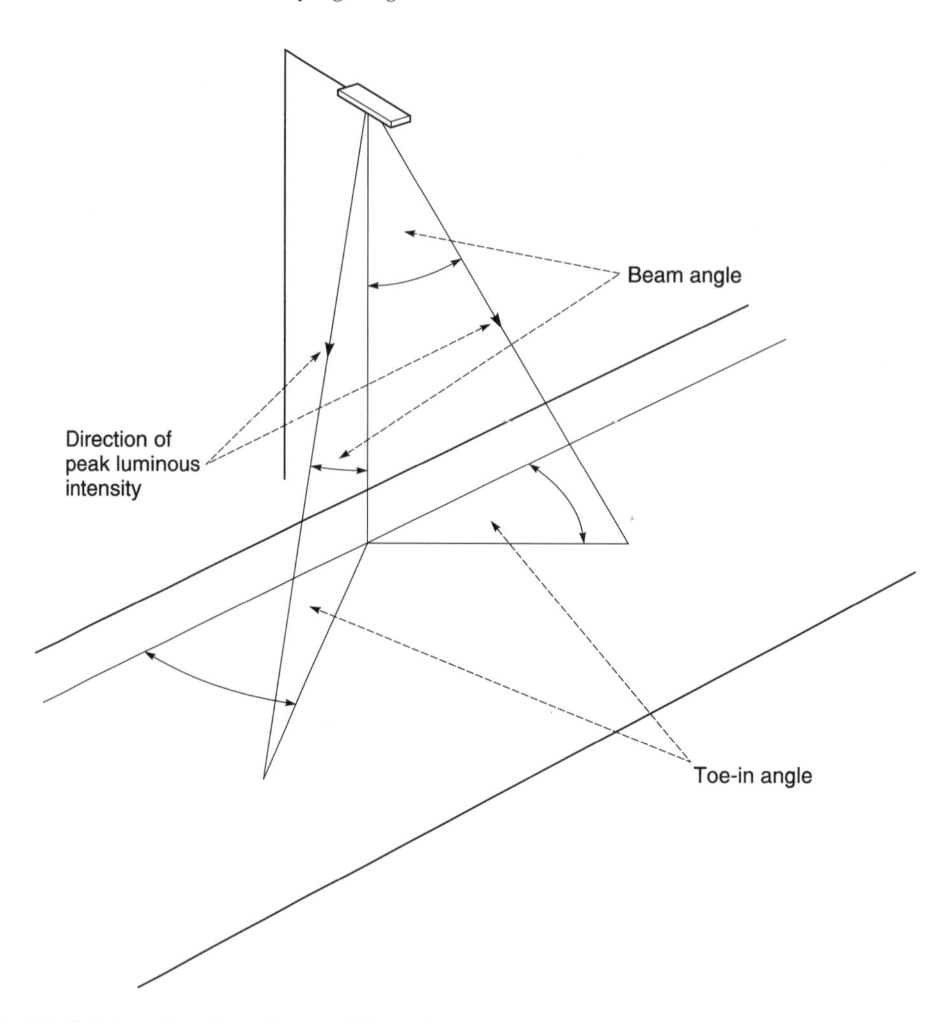

Fig. 9.3 Relation of toe-in to the run of the road

9.5 Lighting the road surface

Each luminaire in the installation produces a bright patch of light on the road and we need to consider the shape and extent of this patch to the approaching driver, and how a number of patches can be fitted together to cover the whole road surface. In practice, this process is carried out by calculation but the following qualitative description gives an understanding of the underlying principles.

As is shown in Figure 9.4(a), schematically, the individual patch of light is T-shaped. The head of the T is produced by mainly diffuse reflection (although the road surface is far from being a uniform diffuser) and the tail of the T is produced by mainly specular or, more precisely, preferential reflection. The maximum extent, away from the observer, of the T is just beyond the transverse line at the nadir of the luminaire. This occurs partly because of the reflection properties of the road surface and partly because of the foreshortening of the head of the T due to the perspective view of the road, which is taken from an eye height of only 1.5 m. The tail of the T is on a line in the vertical plane through the luminaire and the eye of the observer. Its length will

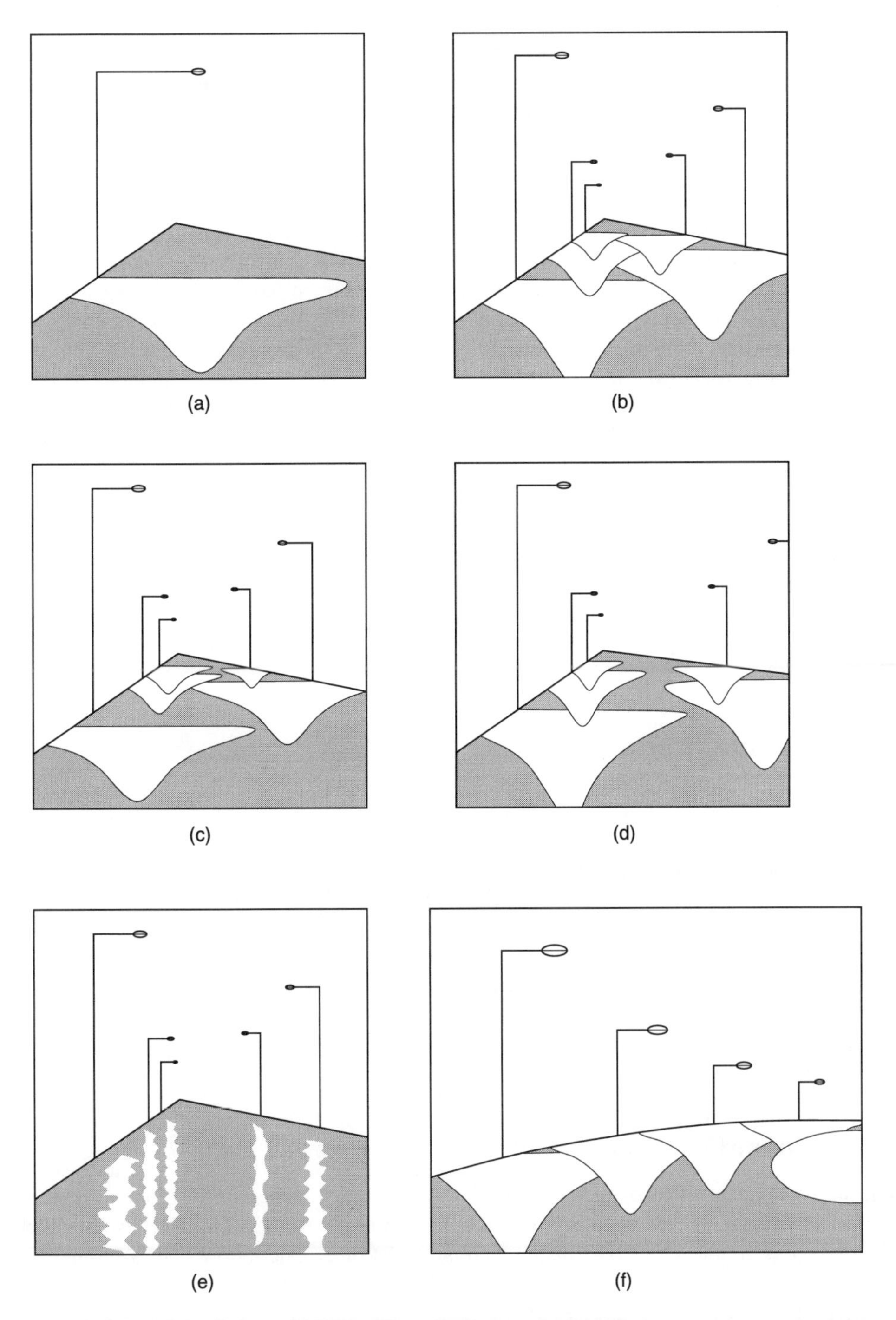

Fig. 9.4 Light patterns produced on the road by different light distributions, road surfaces, and arrangements

depend on the specularity of the road surface and the beam angle of the luminaire. The more specular the road surface the longer the tail, providing the beam angle of the luminaire is high enough to take advantage of the specularity. Anti-skid road surfaces tend to be rough and hence shorten the tail. In this case there is no benefit in having a high beam angle, particularly as it can cause glare. A specular road surface tends to narrow the head of the T. The extreme case occurs in practice when the road is wet. In this condition the T degenerates to a streak of light not much wider than the luminaire. This will extend nearly to the observer.

The remaining diagrams in Figure 9.4 show how the individual patches of light from the luminaires in an installation combine to light the road surface. Figure 9.4(b) shows how the patches in a staggered arrangements are fitted together to cover the road surface. When the lumin-aires have a low beam angle or the surface is very rough, Figure 9.4(c), the tails to the T-shaped patches are shortened. The resultant poor coverage has to be remedied by closing up the spacing between the luminaires, or selecting a luminaire with a more suitable light distribution. In a wider road, Figure 9.4(d), the centre of the road tends to lack coverage, which can be remedied by using an opposite arrangement. When the road is wet, Figure 9.4(e), the T-shaped patches degenerate into streaks, as already stated. As already stated, this can be remedied to a limited extent by having a large toe-in, exceeding 25°. On a bend, Figure 9.4(f), the T-shaped patches are combined by locating the luminaires on the outside of the bend, although it may be necessary to supplement these by luminaires on the inside of the bend on wide roads.

It should be restated that Figure 9.4 is schematic and, in practice, the T-shaped patch consists of graduated luminances, and the parts of the road outside the indicated patches do have some luminance, albeit low. It does not show the effect of increasing the mounting height. This tends to increase the size of the patch but a point is reached where the luminance will be too low for the patch to be recognized as such.

9.6 Quality criteria

Quality criteria, sometimes called the light technical parameters, govern the level and uniformity of road luminance, control of glare, and light falling on the surrounds to the road. They are:

L_{av}, the average road luminance[a];
U_o, the overall uniformity, defined as the minimum to the average luminance;
U_L, the longitudinal uniformity, defined as the minimum luminance to the maximum luminance along a specified line parallel to the direction of the road;
TI, the threshold increment, which is a measure of the disability glare;
G, the glare control mark, which is a measure of discomfort glare;
SR, the surround ratio, which is a measure of the amount of light falling on the surrounds as a proportion of that falling on the road.

9.7 Conventions for installation geometry

CIE and national standardizing bodies have drawn up conventions for positioning the observer, the calculation points, and the position of the luminaires. The national bodies have mostly based their recommendations on CIE 30.2 but there are departures. The following is a summary of the recommendations.

[a] For many years $\bar{L}$ was used as the symbol for average road luminance, but now L_{av} is favoured, perhaps because it is consistent with E_{av}, the symbol for average illuminance.

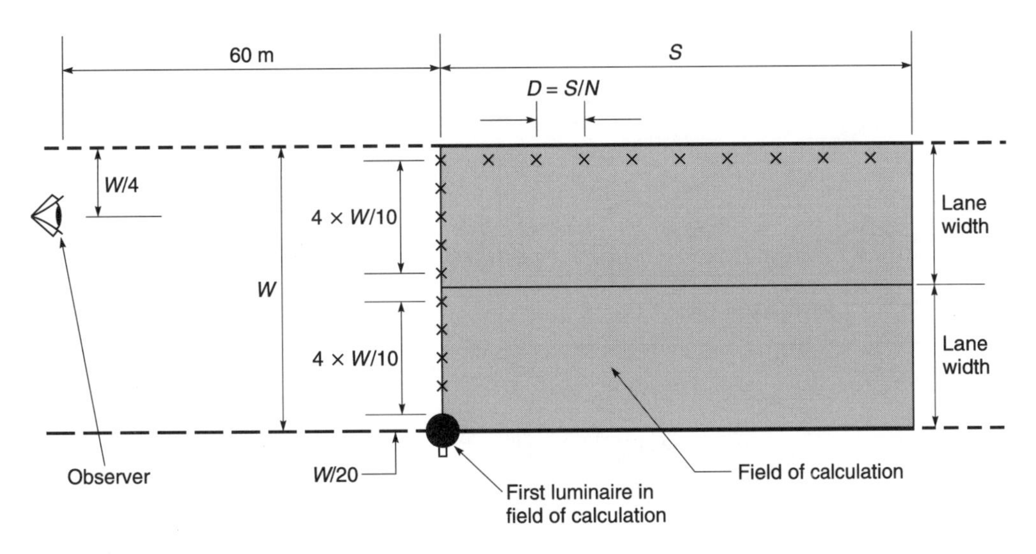

Fig. 9.5 Positions of calculation points for two-lane road

Field of calculation: the field of calculation is the area on the road where points are chosen for the calculation of luminance. It should commence at a luminaire position 60 m ahead of the observer, as shown in Figure 9.5, and it should stretch no further than 180 m from the observer, as this is the limit of applicability of the *r*-table. Within this the field of calculation should cover a section between two luminaires in the same row. Transversely it should be bounded by the edges of the carriageway
Eye height of observer: 1.5 m.
Transverse position of observer: one quarter of the carriageway width across the road for the calculation of L_{av}, U_o, and *TI*. For U_L the observer is on the centre line of each lane. For staggered installations the observer is on the opposite side of the road from the first luminaire.
Number of calculation points longitudinally, *N*: *N* takes on the following values (Figure 9.5):

for $S \leq 50$ m, *N* is 10;
for $S > 50$ m, *N* is the smallest integer giving $D \leq 5$ m, where *D* is the distance between points.

Number of points transversely: CIE recommends five points in each lane but allows three points if the uniformity is good. The number of points must be an odd number as the middle row of points is used to calculate longitudinal uniformity. The half spacing at the edge of the lane should be noted.

UK practice is to use ten points for all roads except motorways, where five points are used in each lane.

In some countries the first row of transverse calculation points is located at a half spacing beyond the first luminaire so that all calculation points can be regarded as being in the centre of rectangles in the calculation field; the usual practice for calculating illuminance both for indoor and outdoor practice.

9.8 Calculation of road surface luminance

Figure 9.6 shows the geometry for the calculation of the luminance at the point *P*. The observer's eye, at *O*, is taken to be 1.5 m above the road surface, by convention. Reflection data

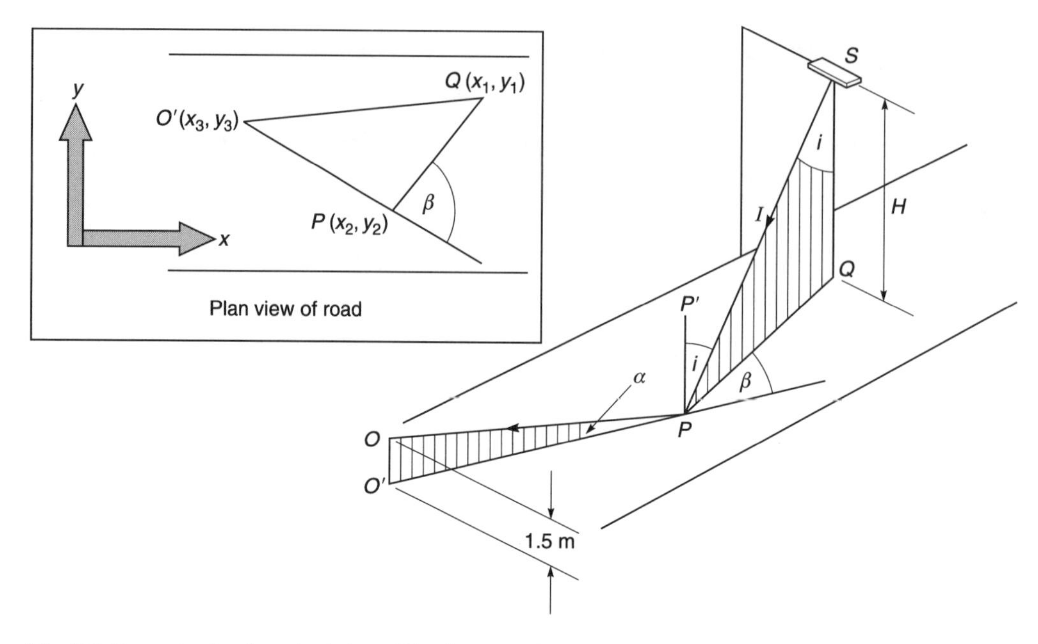

Fig. 9.6 Geometry for the calculation of luminance at *P*

are available in the form of *r*-tables, described in detail later. These are applicable to a viewing angle, α, lying between 0.5° and 1.5°. For an observer height of 1.5 m this means that the area of the road on which calculations can be performed lies between 57 m (= 1.5/tan 1.5°) and 172 m (= 1.5/tan 0.5°) ahead of the observer. It should be noted as a matter of interest that this area occupies only a small part of the visual field (see Figure 9.18, page 386). The luminaire is at *S* and the light ray *SP* emerges from the luminaire at an angle *i* to the downward vertical, which is also equal to the angle of incidence at *P*. At the luminaire, *i* is equal to the angle γ in the (C, γ) system of coordinates when the luminaire is mounted at the attitude at which it was photometrically tested; that is, no rotations have been applied. Correcting for these will be discussed later.

To calculate the luminance of *P* on the road surface (Figure 9.6) as seen by the observer at *O*, the following equation is used:

$$
\begin{aligned}
L &= qE \\
&= \frac{qI \cos^2 i}{H^2} \\
&= (q \cos^3 i) \frac{I}{H^2} \\
&= r \frac{I}{H^2}
\end{aligned}
\tag{9.1}
$$

where

L is the luminance at the point P in candelas per square metre;
q is the luminance coefficient, defined as the ratio of the luminance of an element of surface to the illuminance on it for given angles of viewing and angles of incident light. Methods of measuring it are given in Chapter 15, section 15.10, page 499;
E is the illuminance;
H is the mounting height in metres;
i is the angle of the emergent ray to the downward vertical from the luminaire, which is equal to γ when the luminaire is mounted at the attitude at which it was tested photometrically;
r is the reduced luminance coefficient.

In Equation (9.1), r replaces $q \cos^3 i$ to save having to calculate $\cos^3 i$ each time the equation is used. To enable the correct reduced luminance coefficient to be selected r-tables have been measured for a great variety of road surfaces. In the next section we will examine r-tables in detail before returning to the evaluation of Equation (9.1).

9.8.1 r-TABLES AND THE CLASSIFICATION OF ROAD SURFACES

Table 9.1 shows the format of a typical r-table. The angles β and i are as marked in Figure 9.6. In most published r-tables the left-hand column is labelled γ but, as has already been mentioned, i does not equal γ if the luminaire is rotated in the installation, so i is used in this book to avoid confusion. For convenience, r is multiplied by 10^4 to eliminate decimal points from the figures in the table. The reader will notice that all the values of r in the row for $\tan \gamma = 0$ are the same. This would be expected as the sample is being lit from directly overhead.

Figure 9.7(a) shows the extent of the r-table mapped out on a horizontal surface. The dimensions are in terms of the mounting height H. Figure 9.7(b) shows how this may be used to determine which luminaires should be considered when the luminance at P is being determined. This figure is purely a graphical demonstration for didactic purposes and generally the process is carried out by calculation.

For ease of reference, for calculation purposes, and for classification, it is useful to define four parameters:

- $r(\tan \beta, i)$, — the luminance coefficient at $\tan \beta$ and i,
- average luminance coefficient, — q_o, $Q0$, or Q_o, which we will use in this book,
- specular factor 1 — $S1$, used for classifying surfaces according to their specularity,
- specular factor 2 — $S2$, originally devised for classifying surfaces according to their specularity, but now mainly used simply for reference.

Q_o is found from the equation:

$$Q_o = \frac{1}{\Omega_o} \int_{\Omega_o} q \, d\Omega$$

where Ω_o is the solid angle subtended by the area of integration at the measurement point. This is a rectangle $4H$ towards the observer, $12H$ away from the observer and $3H$ on either side.

In practice, this equation has to be evaluated by numerical integration. This is carried out by multiplying each value in the r-table by the corresponding weighting factor given in Table 9.2, and by $\tan \gamma$. The results are summed and divided by 9.936×10^7.[9,10] Q_o is used as a scaling factor that can be applied to values in an r-table to allow for a change or difference in the

Table 9.1 r-table for $C2$ road surface (values are $r \times 10^4$), $Q_o = 0.07$

tan i	β (deg)																			
	0	2	5	10	15	20	25	30	35	40	45	60	75	90	105	120	135	150	165	180
0.00	329	329	329	329	329	329	329	329	329	329	329	329	329	329	329	329	329	329	329	329
0.25	362	358	371	364	371	369	362	357	351	349	348	340	328	312	299	294	298	288	292	281
0.50	379	368	375	373	367	359	350	340	328	317	306	280	266	249	237	237	231	231	227	235
0.75	380	375	378	365	351	334	315	295	275	256	239	218	198	178	175	176	176	169	175	176
1.00	372	375	372	354	315	277	243	221	205	192	181	152	134	130	125	124	125	129	128	128
1.25	375	373	352	318	265	221	189	166	150	136	125	107	91	93	91	91	88	94	97	97
1.5	354	352	336	271	213	170	140	121	109	97	87	76	67	65	66	66	67	68	71	71
1.75	333	327	302	222	166	129	104	90	75	68	63	53	51	49	49	47	52	51	53	54
2.00	318	310	266	180	121	90	75	62	54	50	48	40	40	38	38	38	41	41	43	45
2.50	268	262	205	119	72	50	41	36	33	29	26	25	23	24	25	24	26	27	29	28
3.00	227	217	147	74	42	29	25	23	21	19	18	16	16	17	18	17	19	21	21	23
3.50	194	168	106	47	30	22	17	14	13	12	12	11	10	11	12	13	15	14	15	14
4.00	168	136	76	34	19	14	13	11	10	10	10	8	8	9	10	9	11	12	11	13
4.50	141	111	54	21	14	11	9	8	8	8	8	7	7	8	8	8	8	10	10	11
5.00	126	90	43	17	10	8	8	7	6	6	7	6	7	6	6	7	8	8	8	9
5.50	107	79	32	12	8	7	7	7	6	5										
6.00	94	65	26	10	7	6	6	6	5											
6.50	86	56	21	8	7	6	5	5												
7.00	78	50	17	7	5	5	5	5												
7.50	70	41	14	7	4	3	4													
8.00	63	37	11	5	4	4	4													
8.50	60	37	10	5	4	4	4													
9.00	56	32	9	5	4	3														
9.50	53	28	9	4	4	4														
10.00	52	27	7	5	4	3														
10.50	45	23	7	4	3	3														
11.00	43	22	7	3	3	3														
11.50	44	22	7	3	3															
12.00	42	20	7	4	3															

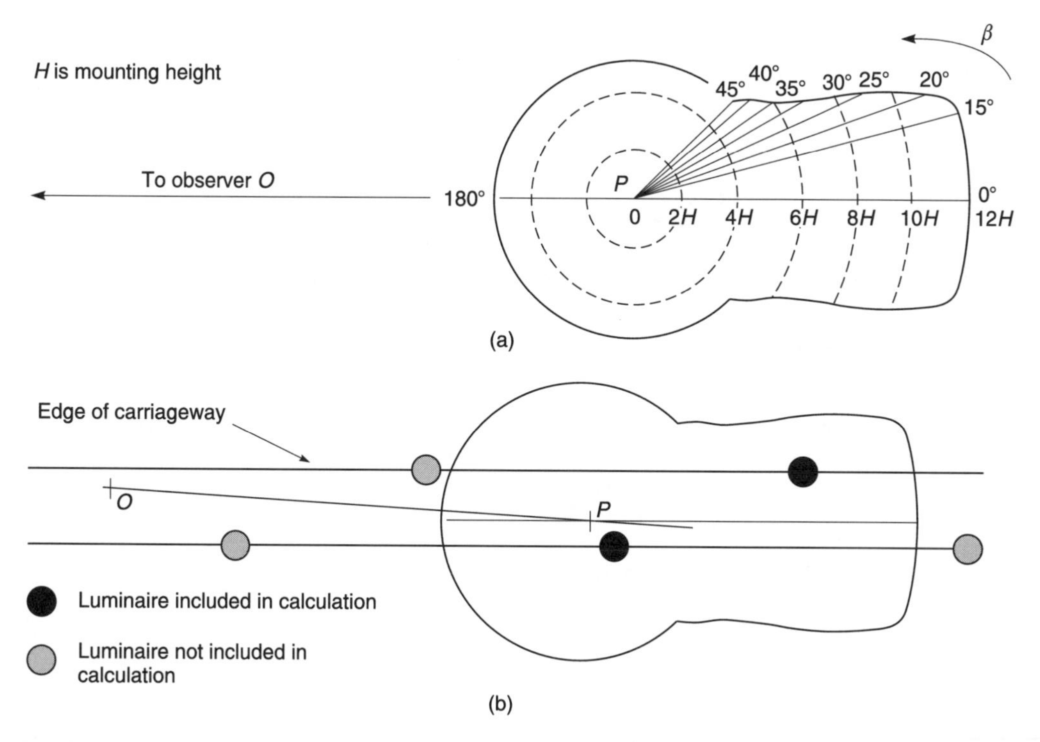

Fig. 9.7 Extent of r-table mapped on the road

overall reflectance of a surface when the specularity as characterized by $S1$ and $S2$, defined below, shows little change.

$$S1 = \frac{r(0, 2)}{r(0, 0)}$$

$$S2 = \frac{Q_o}{r(0, 0)}$$

Worked example 1 Find $S1$ and $S2$ for the road surface given in Table 9.1.

Solution From Table 9.1, $r(0, 0) = 329$, $r(0, 2) = 318$, and $Q_o = 0.07$. Hence:

$$S1 = \frac{318 \times 10^{-4}}{329 \times 10^{-4}}$$

$$= 0.97$$

$$S2 = \frac{0.08}{329 \times 10^{-4}}$$

$$= 2.43$$

Table 9.2 Weighting factors used in the calculation of Q_o

tan *i*	β (deg)																			
	0	2	5	10	15	20	25	30	35	40	45	60	75	90	105	120	135	150	165	180
0.00	8	8	32	22	40	20	40	20	40	25	45	120	60	120	60	120	60	120	60	60
0.25	32	32	128	88	160	80	160	80	160	100	180	480	240	480	240	480	240	480	240	240
0.50	16	16	64	44	80	40	80	40	80	50	90	240	120	240	120	240	120	240	120	120
0.75	32	32	128	88	160	80	160	80	160	100	180	480	240	480	240	480	240	480	240	240
1.00	16	16	64	44	80	40	80	40	80	50	90	240	120	240	120	240	120	240	120	120
1.25	32	32	128	88	160	80	160	80	160	100	180	480	240	480	240	480	240	480	240	240
1.50	16	16	64	44	80	40	80	40	80	50	90	240	120	240	120	240	120	240	120	120
1.75	32	32	128	88	160	80	160	80	160	100	180	480	240	480	240	480	240	480	240	240
2.00	24	24	96	66	120	60	120	60	120	75	135	360	180	360	180	360	180	360	180	180
2.50	64	64	256	176	320	160	320	160	320	200	360	960	480	960	480	960	480	960	480	480
3.00	32	32	128	88	160	80	160	80	160	100	120	510	222	240	180	480	240	480	240	240
3.50	64	64	256	176	320	160	320	160	320	200	120	270	33	0	–30	690	480	960	480	480
4.00	32	32	128	88	160	80	160	80	160	100	60	0	0	0	–75	–30	255	510	222	120
4.50	64	64	256	176	320	160	320	160	320	155	75	0	0	0	0	0	135	372	33	0
5.00	32	32	128	88	160	80	160	105	125	5	–15	0	0	0	0	0	0	33	0	0
5.50	64	64	256	176	320	160	320	170	90	0										
6.00	32	32	128	88	160	80	160	40	0											
6.50	64	64	256	176	320	160	275	35												
7.00	32	32	128	88	160	105	80	–25												
7.50	64	64	256	176	320	170	90													
8.00	32	32	128	88	160	85	45													
8.50	64	64	256	176	320	80	0													
9.00	32	32	128	88	160	40														
9.50	64	64	256	176	320	80														
10.00	32	32	128	88	115	–5														
10.50	64	64	256	176	230	–10														
11.00	32	32	128	113	80	–25														
11.50	64	64	256	186	90															
12.00	16	16	64	69	45															

Worked example 2 By using Table 9.1 it was found that the luminance at a point in an installation is 1.43 cd/m^2. What is the luminance at an equivalent point in an installation with the same geometry having a road surface with $Q_o = 0.05$, and $r(0, 0) = 225$, $r(0, 2) = 227$.

Solution For the unknown road surface:

$$S1 = \frac{r(0, 2)}{r(0, 0)} = \frac{227 \times 10^{-4}}{225 \times 10^{-4}}$$

$$= 1.01$$

$$S2 = \frac{Q_o}{r(0, 0)} = \frac{0.05}{225 \times 10^{-4}}$$

$$= 2.22$$

These values are similar to those obtained in the previous example for the *C*2 surface, and it may be assumed that the surfaces have similar light distribution properties. Q_o may then be used as a scaling factor:

$$\text{luminance at point} = 1.43 \times \frac{0.05}{0.07}$$

$$= 1.02 \text{ cd/m}^2$$

*S*1 is used for the classification of road surfaces (Table 9.3). For dry surfaces three systems are in use, the *N*, *R* and *C*. The *N*-system is generally used for the lighter surfaces, which are commonly used in the Scandinavian countries, where additives are incorporated into the road surfaces to increase the reflectances. The *R*-system is used in other countries but has to some extent been superseded by the *C*-system devised by Burghout,[11] who showed that there was little increase in inaccuracy by having two classes in this system as opposed to four in the other systems. *C*2 is the class of the representative UK road surface recommended BS 5489.[12]

Table 9.3 *R*, *N* and *C* classification of dry road surfaces

System	Class	*S*1 limits	Q_o of standard table
R	*R*1	$S1<0.42$	0.10
	*R*2	$0.42 \leq S1 < 0.85$	0.07
	*R*3	$0.85 \leq S1 < 0.35$	0.07
	*R*4	$0.35 \leq S1$	0.08
N	*N*1	$S1<0.28$	0.10
	*N*2	$0.28 \leq S1 < 0.60$	0.07
	*N*3	$0.60 \leq S1 < 1.30$	0.07
	*N*4	$1.30 \leq S1$	0.08
C	*C*1	$S1 \leq 0.40$	0.10
	*C*2	$S1>0.40$	0.07

For wet road surfaces, a modified parameter, $S1'$, is used in place of $S1$ for classification purposes:[13]

$$S1' = 0.147 \times \text{antilog}\left[\frac{\log\left\{\dfrac{S1\text{-wet}}{0.147}\right\}}{1 - \left\{\dfrac{Q_o\text{-wet}}{0.687}\right\}}\right]$$

where $S1$-wet and Q_o-wet signify that $S1$ and Q_o are measured when the road surface is wet. The reader will note that if $S1$-wet is less than unity the expression in the square brackets becomes negative. $S1$ cannot then be evaluated. CIE 47[14] suggests that, in these cases, $S1$-wet is used in place of $S1'$.

The classes for wet road surfaces are shown in Table 9.4.

Worked example 3 An r-table for a road surface is measured in the wet condition. It is found that $S1$-wet is 6.50 and Q_o-wet is 0.155. Find the W class of the road surface and the scaling factor to enable the standard r-table to be used for calculating luminance in the wet.

Solution

$$S1' = 0.147 \times \text{antilog}\left[\frac{\log\left\{\dfrac{S1\text{-wet}}{0.147}\right\}}{1 - \left\{\dfrac{Q0\text{-wet}}{0.687}\right\}}\right]$$

$$= 0.147 \times \text{antilog}\left[\frac{\log\left\{\dfrac{6.50}{0.147}\right\}}{1 - \left\{\dfrac{0.155}{0.687}\right\}}\right]$$

$$= 19.60$$

The road surface is in class $W2$ for which Q_o-wet is 0.150, from Table 9.4. The scale factor is, therefore, 0.150/0.155 or 0.97.

Table 9.4 *W* classes for wet roads

Class	Limits	Q_o-wet for standard r-table	Q_o-dry
*W*1	$S1<9.6$	0.114	0.088
*W*2	$9.6 \le S1<26.5$	0.150	0.091
*W*3	$26.5 \le S1<73$	0.196	0.097
*W*4	$73 \le S1<200$	0.247	0.104

9.8.2 EVALUATION OF i AND β

To be able to select the appropriate value of r from the r-table and apply Equation (9.1), we have to evaluate the angles i and β. From Figure 9.6 we see:

$$i = \tan^{-1}\frac{PQ}{QS}$$

$$= \tan^{-1}\frac{h}{\sqrt{(x_1 - x_2)^2 + (y_1 - y_2)^2}}$$

$$\beta = 180° - \cos^{-1}\frac{O'P^2 + PQ^2 - O'Q^2}{2 \times O'P \times PQ}$$

$$= 180° - \cos^{-1}\frac{(x_3 - x_2)^2 + (y_3 - y_2)^2 + (x_2 - x_1)^2 + (y_2 - y_1)^2 - (x_1 - x_3)^2 - (y_1 - y_3)^2}{2 \times \sqrt{(x_3 - x_2)^2 + (y_3 - y_2)^2} \times \sqrt{(x_2 - x_1)^2 + (y2 - y_1)^2}}$$

since,

$$O'P^2 = (x_3 - x_2)^2 + (y_3 - y_2)^2$$
$$PQ^2 = (x_2 - x_1)^2 + (y_2 - y_1)^2$$
$$O'Q^2 = (x_1 - x_3)^2 + (y_1 - y_3)^2$$

Worked example 4 In a road lighting installation, the (x, y) coordinates of the observer, luminaire and point of interest are respectively (2, 0), (2, 160), and (8, 149.6077), in metres. The luminaire mounting height is 12 m, and the luminous intensity directed towards the point of interest is 200 cd per klm. If the road surface classification is C, Q_o is 0.07, and the luminous flux of the lamp is 15.5 klm is 0.06, what is the luminance of the point of interest in the direction of the observer, eye height 1.5 m?

Solution Referring to Figure 9.6, page 354, we find:

$$O'P^2 = (x_3 - x_2)^2 + (y_3 - y_2)^2$$
$$= (0 - 149.6077)^2 + (2 - 8)^2 = 22418.4639$$
$$PQ^2 = (x_2 - x_1)^2 + (y_2 - y_1)^2$$
$$= (149.6077 - 160)^2 + (8 - 2)^2 = 144.0000$$
$$O'Q^2 = (x_1 - x_3)^2 + (y_1 - y_3)^2$$
$$= (160 - 0)^2 + (2 - 2)^2 = 25604$$

To confirm that the r-table can be used we have to check that the angle α lies between 0.5° and 1.5°.

$$\tan\alpha = \frac{O'O}{O'P}$$

$$= \frac{1.5}{\sqrt{22418.4}}$$

$$= 0.010$$

so,

$$\alpha = 0.57°$$

As α lies within the specified limits we can proceed with the calculation.

$$\tan i = \frac{H}{\sqrt{(x_1 - x_2)^2 + (y_1 - y_2)^2}}$$

$$= \frac{12}{\sqrt{(160 - 149.6077)^2 + (2 - 8)^2}}$$

$$= 1$$

and

$$\beta = 180° - \cos^{-1} \frac{O'P^2 + PQ^2 - O'Q^2}{2 \times O'P \times PQ}$$

$$= 180° - \cos^{-1} \frac{25600 + 144 - 22418.4404}{2 \times 160 \times 12}$$

$$= 150°$$

r for (150, 1) is 129×10^{-4} for Q_o equal to 0.07. Hence, for Q_o equal to 0.06, r is

$$\frac{0.06}{0.07} \times 129 \times 10^{-4} = 110.6 \times 10^{-4}$$

We can now evaluate the luminance L at P from Equation (9.1).

$$L = r \frac{I}{H^2}$$

$$= 110.6 \times 10^{-4} \times \left(\frac{600 \times 15.5}{144} \right)$$

$$= 0.71 \text{ cd/m}^2$$

This result has been obtained for a single luminaire only. In an installation this procedure has to be repeated for every contributing luminaire (see Figure 9.7, page 357). In addition, the maintenance factor, as will be explained in Section 9.14, page 374, has also to be taken into account. The example has been devised so that it has not been necessary to interpolate when entering the r-table, which, in practice it is necessary to do in nearly every case. Interpolation in the I-table has been described in Chapter 2, but it is now necessary to describe interpolation in the r-table.

9.8.3 INTERPOLATION IN THE r-TABLE

In Chapter 2 we discussed interpolation in the I-table using Lagrange's formula. We use similar procedures in the r-table. Since there are rapid changes of value with tan i and β, particularly for

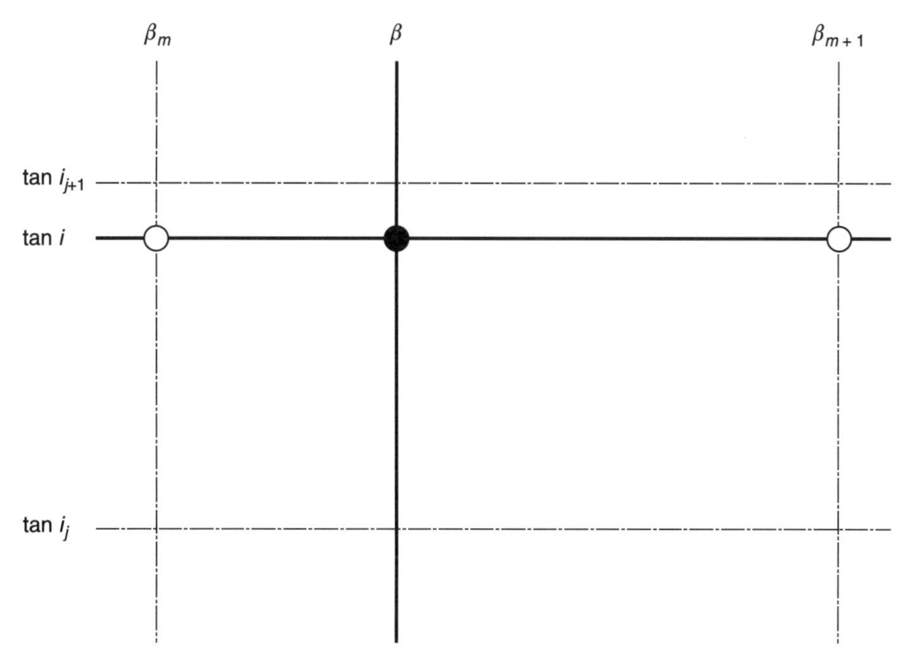

Fig. 9.8 Values required for linear interpolation in the *r*-table

wet surfaces, it is necessary to use quadratic interpolation. At the edges of the table, however, linear interpolation is used as, in certain instances, quadratic interpolation can give rise to negative values. It should be noted that $\beta = 0$ is not at the edge of the table, but in the middle, as the table can be regarded as being symmetrical about $\beta = 0$. We will first describe linear interpolation.

For linear interpolation, the Lagrange formula, put at degree 1, becomes:

$$y = \frac{x - x_2}{x_1 - x_2} y_1 + \frac{x - x_1}{x_2 - x_1} y_2$$

Figure 9.8 indicates the values needed for linear interpolation, which can be applied to either β or tan i first. When it is first applied to β, β is substituted for x in the above equation:

$$x = \beta$$
$$x_1 = \beta_m$$
$$x_2 = \beta_{m+1}$$

From this substitution two constants (K_1 and K_2) can be defined:

$$K_1 = \frac{\beta - \beta_{m+1}}{\beta_m - \beta_{m+1}}$$

$$K_2 = \frac{\beta - \beta_m}{\beta_{m+1} - \beta_m}$$

Since $K_1 + K_2 = 1$ the notation can be simplified by putting $K = K_1$ so $K_2 = 1 - K$.

Substitution of these constants and appropriate values of I for y in Lagrange's equation gives, after simplification:

$$I(\beta, \tan i_j) = I(\beta_m, \tan i_j) + K \times [I(\beta_{m+1}, \tan i_m) - I(\beta_m, \tan i_j)]$$

Similarly,

$$I(\beta, \tan i_{j+1}) = I(\beta_m, \tan i_{j+1}) + K \times [I(\beta_{m+1}, \tan i_{j+1}) - I(\beta_m, \tan i_{j+1})]$$

For interpolation for tan i a similar procedure produces:

$$I(\beta, \tan i) = I(\beta, \tan i_j) + k \times [I(\beta, \tan i_{j+1}) - I(\beta, \tan i_j)]$$

where

$$k = \frac{\tan i - \tan i_j}{\tan i_{j+1} - \tan i_j}$$

In these equations interpolation has been first carried out for β, and then for tan i. If desired, this procedure can be reversed (that is, the interpolation can be first carried out for tan i, followed by β) and the same result obtained.

Figure 9.9 shows the values needed for quadratic interpolation. The suffixes, j and m, refer to the rows and columns of the r-table.

Three values of β and tan i are required for insertion in the interpolation equations. Two values are selected on either side of the value to be interpolated and the third value is selected according to the rules given for interpolation in the I-table.

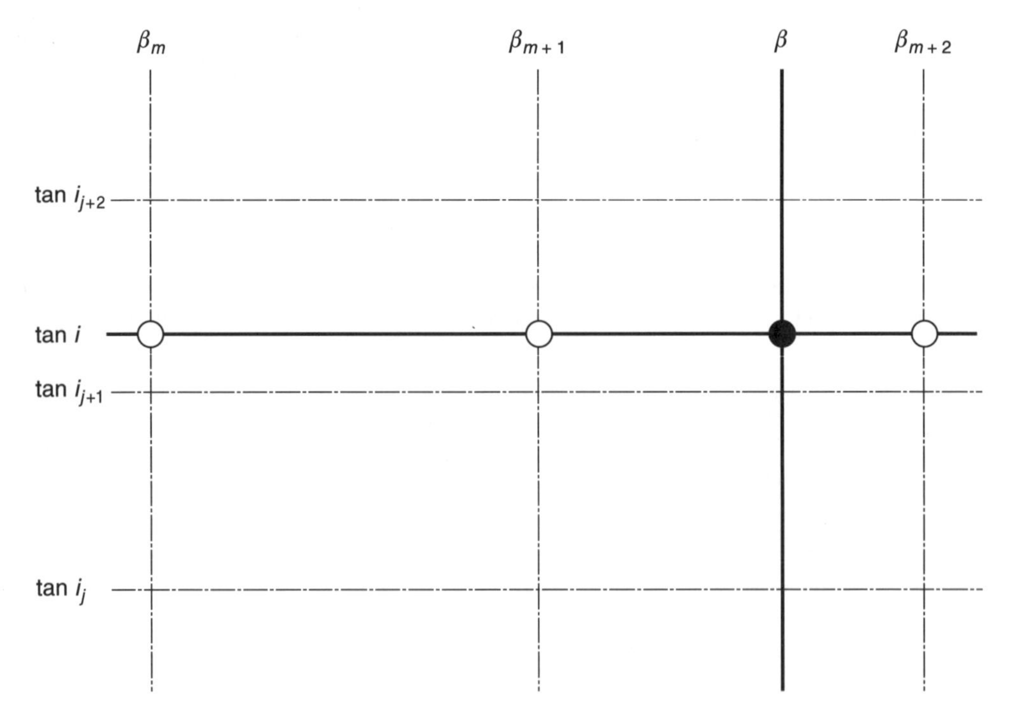

Fig. 9.9 Parameters required for quadratic interpolation in the r-table

This interpolation can be applied to either β or tan i. When it is first applied to β, this parameter is substituted for x in the Lagrange equation:

$$x = \beta$$
$$x_1 = \beta_m$$
$$x_2 = \beta_{m+1}$$
$$x_3 = \beta_{m+2}$$

where β is the angle at which I is to be found by interpolation;

m, $m+1$, $m+2$ are integers indicating the number of the columns in the r-table;

β_m, β_{m+1} and β_{m+2} are values of C for the corresponding column numbers. These are chosen such that $\beta_{m+1}<\beta<\beta_{m+2}$ unless m is zero in which case $\beta_1<\beta<\beta_2$.

From this substitution three constants can be defined:

$$K_1 = \frac{(\beta - \beta_{m+1})(\beta - \beta_{m+2})}{(\beta_m - \beta_{m+1})(\beta_m - \beta_{m+2})}$$

$$K_2 = \frac{(\beta - \beta_m)(\beta - \beta_{m+2})}{(\beta_{m+1} - \beta_m)(\beta_{m+1} - \beta_{m+2})} \tag{9.2}$$

$$K_3 = \frac{(\beta - \beta_m)(\beta - \beta_{m+1})}{(\beta_{m+2} - \beta_m)(\beta_{m+2} - \beta_{m+1})}$$

The reader will notice that $K_1 + K_2 + K_3 = 1$. These constants can be used to set up three more equations:

$$\begin{aligned}
r(\beta, \tan i_j) &= K_1 r(\beta_m, \tan i_j) + K_2 r(\beta_{m+1}, \tan i_j) + K_3 r(\beta_{m+2}, \tan i_j) \\
r(\beta, \tan i_{j+1}) &= K_1 r(\beta_m, \tan i_{j+1}) + K_2 r(\beta_{m+1}, \tan i_{j+1}) + K_3 r(\beta_{m+2}, \tan i_{j+1}) \\
r(\beta, \tan i_{j+2}) &= K_1 r(\beta_m, \tan i_{j+2}) + K_2 r(\beta_{m+1}, \tan i_{j+2}) + K_3 r(\beta_{m+2}, \tan i_{j+2})
\end{aligned} \tag{9.3}$$

Now,

$$k_1 = \frac{(\tan i - \tan i_{j+1})(\tan i - \tan i_{j+2})}{(\tan i_j - \tan i_{j+1})(\tan i_j - \tan i_{j+2})}$$

$$k_2 = \frac{(\tan i - \tan i_j)(\tan i - \tan i_{j+2})}{(\tan i_{j+1} - \tan i_j)(\tan i_{j+1} - \tan i_{j+2})} \tag{9.4}$$

$$k_3 = \frac{(\tan i - \tan i_j)(\tan i - \tan i_{j+1})}{(\tan i_{j+2} - \tan i_j)(\tan i_{j+2} - \tan i_{j+1})}$$

Once again the reader will notice that $k_1 + k_2 + k_3 = 1$. The required value of r is then given by:

$$r(\beta, \tan i) = k_1 r(\beta, \tan i_j) + k_2 r(\beta, \tan i_{j+1}) + k_3 r(\beta, \tan i_{j+2}) \tag{9.5}$$

Worked example 5 Find r for a C2 road surface with a Q_o of 0.06 when β is 17° and tan i is 1.15.

Solution An extract of the relevant part of Table 9.1, page 356, is:

tan i	β (deg) 10	15	20	25	30
0.75	365	351	334	315	295
1.00	354	315	277	243	221
1.25	318	265	221	189	166
1.5	271	213	170	140	121

As the required values do not lie on the borderline of the table, quadratic interpolation is used. The third value of β is 25° and the third value of tan i is 1.5, as these are the higher values. So,

$$\beta = 17$$
$$\beta_m = 15$$
$$\beta_{m+1} = 20$$
$$\beta_{m+2} = 25$$

Substitution of these values in Equations (9.2) gives:

$$K_1 = 0.48$$
$$K_2 = 0.64$$
$$K_3 = -0.12$$

These are substituted in Equations (9.4) to give:

$$k_1 = 0.280000$$
$$k_2 = 0.839999$$
$$k_3 = -0.12$$

Substitution in Equations (9.3) gives:

$$r(\beta, \tan i_j) = 299.32 \times 10^{-4}$$
$$r(\beta, \tan i_{j+1}) = 295.72 \times 10^{-4}$$
$$r(\beta, \tan i_{j+2}) = 246 \times 10^{-4}$$

Substitution in Equation (9.5) gives:

$$r(17, 1.15) = 302.6944 \times 10^{-4} \text{ for a } Q_o \text{ of } 0.07$$

and for a Q_o of 0.06:

$$r(17, 1.15) = 302.6944 \times \frac{0.06}{0.07}$$

$$= 259.5$$

9.9 Calculation of threshold increment

In Section 9.6, page 352, it was mentioned that threshold increment is used as a measure of disability glare. It is now necessary to explain this concept.

Disability glare is caused by light being scattered in the eye. This produces a veil of light over the image of the scene being viewed, which reduces contrasts. Mathematically, this can be shown as follows from the definition of contrast:

$$C_I = \left| \frac{L_o - L_b}{L_b} \right|$$

where

C_I is the initial contrast, without veiling luminance,
L_o is the luminance of the object,
L_b is the luminance of the background.

The equation for C is stated in its usually quoted form; that is, with modulus signs which imply that whether L_o is larger or smaller than L_b is immaterial. However, as Adrian[15] discusses, there is evidence that for the same value of C a target in negative contrast $(L_o - L_b) < 0$ is better seen than one in positive contrast $(L_o - L_b) > 0$.

The equation can then be stated as

$$C_I = \frac{L_o - L_b}{L_b}$$

If there is veiling luminance, L_v, present this will be added on to both L_o and L_b, so the contrast, C_v, will now be:

$$C_v = \frac{L_o - L_b}{L_b + L_v}$$

It is obvious that $|C_v|$ is less than $|C_r|$. The threshold increment is the percentage by which the luminance of the object has to be increased to achieve a contrast equal to that obtained without any veiling luminance.

The veiling luminance can be evaluated from the Stiles–Holladay formula:[16, 17]

$$L_v = K \frac{E_{vert}}{\theta^2}$$

where

E_{vert} is the total illuminance (in lux per 1000 *initial* lamp lumens) produced by new luminaires on a plane at the observer's eye, normal to the line of sight. The observer's eye is taken by convention to be at a height of 1.5 m above road level. In UK practice, it is positioned transversely one quarter the carriageway width from the carriageway edge and longitudinally at a distance in metres of $2.75(H - 1.5)$, where H is the mounting height (in metres), in front of the field of calculation. The line of sight is 1° below the horizontal and in a vertical plane in the longitudinal direction passing through the observer's eye;
θ is the angle in degrees between the line of sight and the centre of each luminaire, where the line of sight is taken to be 1° below the horizontal and directly ahead of the observer. The exponent 2 of θ is only valid for angles from 1.5° to 60° above the horizontal;
K (degrees^2 per steradian) is a constant depending on the age of the observer. For road lighting calculations its value is taken as 10, which is applicable to observers between the ages of 20 and 30 years.

Crawford showed that L_v can be summed for all the contributing sources.[18]

It now remains to relate veiling luminance to threshold increment. This can be done by means of a standard curve of human contrast sensitivity published by the CIE,[19] which enables a relationship between veiling luminance, adaptation luminance L_{av}, and threshold increment to be developed:[20]

$$TI = 65 \frac{L_v \times MF^{0.8}}{L_{av}^{0.8}}$$

$$= 650 \frac{E_{vert} \times MF^{0.8}}{L_{av}^{0.8} \times \theta^2}$$

The adaptation luminance is taken to be equal to the road luminance. During the aging of an installation, the threshold increment is at its greatest when the lamps are giving their greatest output; that is, when the installation is new. It is for this reason that the *maintenance factor* raised to the power of 0.8 is introduced into the formula. In effect it brings L_{av} to the initial condition.

The set-up for selecting the first luminaire is illustrated in Figure 9.10. The eye height is taken to be 1.5 m and the first luminaire is situated in the screening plane provided by the car roof. This is taken to have an elevation of 20°. The observer is situated one quarter of the road width from the nearside edge of the carriageway. This results in the observer being positioned $(\cot^{-1} 20)(H - 1.5)$ or $2.75(H - 1.5)$ from the first luminaire. The veiling luminance is summed for the first luminaire and the luminaires beyond, up to a distance of 500 m. The summation is stopped when a luminaire gives a contribution that is less than 2% of the total veiling luminance

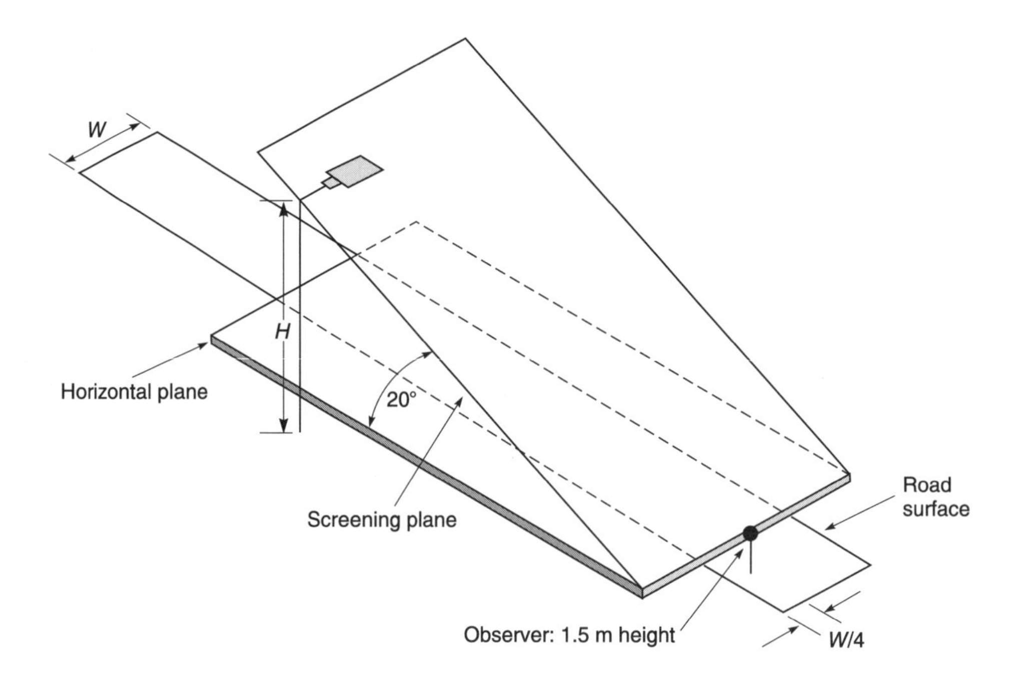

Fig. 9.10 Relationship of the first luminaire included in the calculation of threshold increment to the observer

of the preceding luminaires in the row. Whilst, in present UK practice, the observer is kept stationary, in the practice adopted in many other countries the observer is moved along a longitudinal line and the highest value of threshold increment chosen.

9.10 Glare control mark

Values of the glare control mark, G, range from 1 to 9. 1 implies that the glare is 'unbearable', 3 is 'disturbing', 7 is 'satisfactory', and 9 is 'unnoticeable'. The formula for evaluating glare control mark was developed by de Boer:

$$G = 13.84 - 3.31 \log I_{80} + 1.3 (\log I_{80}/I_{88}) - 0.08 \log I_{80}/I_{88} + 1.29 \log F + 0.97L_{av} + 4.41H' + -1.46 \log p$$

where

G is Glare Control Mark;
I_{80} is the absolute *initial* intensity (cd) at an angle of 80° to the downward vertical, in a vertical plane parallel to the road axis;
I_{80}/I_{88} is ratio of the luminous intensities at 80° and 88° to the downward vertical, in the vertical plane parallel to the road axis;
F is the orthogonally projected flashed area of the luminaire in a direction 76° to the downward vertical, in a plane parallel to the road; $\bar{L}$ is the average road surface luminance (cd/m^2);
L_{av} is the *initial* average road surface luminance (cd/m^2);
H' is the vertical distance (m) between eye level and the luminaire;
p is the number of luminaires per kilometre.

The formula for G is valid for the following ranges of values:

$50 < I_{80} < 7000$ (cd)
$1 < I_{80}/I_{88} < 50$
$0.007 < F < 0.4$ (m^2)
$0.3 < L_{av} < 7$ (cd/m^2)
$5 < H' < 20$ (m)
$20 < p < 100$
number of luminaire rows = 1 or 2

The glare control mark has tended to be little used in recent times. This is because there has been a feeling among users that it does not rate installations in the correct order of discomfort glare. The original work to derive the formula was carried out in the laboratory and this was subsequently validated by field appraisals. However, the fact that the design of luminaires has changed since the work was originally done and there is now more use of high pressure discharge sources rather than low pressure sodium lamps may account for the dissatisfaction with the glare control mark as a glare rating system. In addition, there is a greatly increased use of cut-off luminaires, which provide very good control of both discomfort and disability glare. Field evidence suggests that installations designed according to the threshold increment requirements in Table 9.6, page 372, will have satisfactory control of discomfort glare.

9.11 Surround ratio

The function of the surround ratio (abbreviation SR) is to ensure that sufficient light falls on the surrounds to provide a bright background for objects towards the edge of the carriageway to be revealed. It also helps the driver to anticipate the movements of pedestrians about to cross the road. It is important in curved roads, where the surrounds form the greater part of the background against which objects are seen.

The surround ratio is formally defined as: the ratio of the average illuminance on strips, 5 m wide, or less if space does not permit, which are adjacent to the edges of both sides of the carriageway to the average illuminance on the adjacent strips, 5 m wide or half the width of the carriageway, whichever is the smaller, in the carriageway. For dual carriageways, both carriageways together are treated as a single carriageway unless they are separated by less than 10 m.

The surround ratio is usually determined by calculating the illuminance on a regular array of points on the strips and finding the average.

In situations where the surround ratio is applicable, a value of 0.5 is recommended (see Table 9.6, page 372).

9.12 Lighting classification of roads, and associated quality criteria

To be able to assign quality criteria to roads it is useful to have a system of classifying roads for lighting purposes. This is done in CIE 115, *Recommendations for the Lighting of Roads for Motor and Pedestrian Traffic*. A simplified version of this is shown in Table 9.5.

Some of the terms used in the table need explanation. *At grade* means at the same level as the road, in other words, traffic crossing the road under consideration must be carried by a bridge or

Table 9.5 CIE classification of motor traffic routes for lighting purposes

Description of road	Lighting class
High speed roads with separate carriageways, free of crossings at grade and with complete access control; motorways, express roads.	
Traffic density and complexity of road layout:	
High	M1
Medium	M2
Low	M3
High speed roads, dual carriageway roads.	
Traffic control, such as the presence of signals, and separation of different types of road user into lanes:	
Poor	M1
Good	M2
Important urban traffic routes, radial roads, district distributor roads.	
Traffic control and separation of different types of road user:	
Poor	M2
Good	M3
Connecting less important roads, local distributor roads, residential major access roads. Roads which provide direct access to property and lead to connecting roads.	
Traffic control and separation of different types of road user:	
Poor	M4
Good	M5

in a tunnel. *Complexity of layout* refers to the number of roads entering and leaving the road under consideration.

CIE 115 has not been in existence long enough to judge whether this classification will be adopted by national and international organizations responsible for drafting recommendations. The draft CEN recommendations are loosely based on it but have been adapted to take account of the existing practice in participating European countries.

Table 9.6, page 372, compares CIE and CEN requirements for the various classes. Whilst there are similarities it is evident that CEN has split many of the CIE classes into subclasses. This was done to take account of existing practice in European countries.

CIE 115 has adopted a class of road, or more strictly, a class of situation, from North American practice. This is the conflict area, which is an area where there is an increased potential for collision between road users. Such an area might occur where there is a reduced number of lanes, where the road width is reduced, where the road runs into areas frequented by pedestrians, and so on. A roundabout would be regarded as a conflict area.

For these areas CIE 115 recommends that the lighting standard of the conflict area should be one class higher than that of the roads running into the area (for instance, M3 instead of M4). If the roads running into the conflict area are lit to the highest standard, M1, then the conflict area should also be lit to M1.

In some conflict areas, luminance is not an appropriate criterion since this is calculated for a viewing distance of over 80 m and the operative viewing distance is shorter than this. In these areas illuminance should be used in place of luminance and CIE 115 makes recommendations as to which levels should be used. Since road luminance is not calculated, and in any case the changing viewpoint of the driver makes the adaptation luminance uncertain, it is not possible to calculate *TI*. Instead, disability glare is controlled by limiting the intensity in the run-back zone to 30 cd/klm at 80° and 10 cd/klm at 90° at the azimuth of the directions at which the luminaires are likely to be viewed by the motorist.

It will be noted that the recommended luminances are maintained values, which means that they must not be allowed to drop below these values for the life of the installation. When the installation is new, the values will be greater. As the installation ages the light output of the lamps decreases and dirt accumulates on the luminaires so decreasing the light output. When cleaning the luminaires no longer raises the measured values of luminance sufficiently, it will be necessary to change the lamps. The threshold increment will be at its highest value when the installation is new and it is at this stage that the recommended values must not be exceeded.

9.13 Measures of visibility

As has been stated, most design systems for main road lighting are based on road luminance. However, road luminance is only one factor in determining the visibility of an object on the road. Clearly, it is possible to calculate the luminance of an object on the road surface and to calculate the immediate background luminance, and thereby determine the contrast between the object and the background. From this it is possible to determine whether the object will be seen.

Waldram was the first worker to do this.[21] In his concept of revealing power, he supposed that a square object of 450 mm side was positioned vertically on a transverse line across the road. From a survey, data were available of the frequency distribution of reflectances of pedestrian clothing in the late 1930s, at the time the work was done.[22] At any given point on the road surface this enabled the frequency distribution of luminance of objects to be determined, by multiplying the vertical illuminance at the point by the frequency distribution of reflectances. From this, and a knowledge of the luminance of the road visible immediately above the object, the frequency

Table 9.6 Comparison of requirements for CEN and CIE road lighting standards for motor traffic

CIE Requirements (Maintained values)						CEN Requirements (Maintained values)					
CIE Class	Luminance attributes			*TI*(%) max	*SR*	CEN Class	Luminance attributes			*TI*(%) max	*SR*
	L_{av} (cd/m^2) min	U_0 min	U_L min				L_{av} (cd/m^2) min	U_0 min	U_L min		
M1	2.0	0.4	0.7	10	0.5	ME1	2.0	0.4	0.7	10	0.5
M2	1.5	0.4	0.7	10	0.5	ME2	1.5	0.4	0.7	10	0.5
						ME3a			0.7		
M3	1.0	0.4	0.5	10	0.5	ME3b	1.0	0.4	0.6	15	0.5
						ME3c			0.5		
M4	0.75	0.4	NR	15	NR	ME4a	0.75	0.4	0.6	15	0.5
						ME4b			0.5		
M5	0.5	0.4	0.4	15	NR	ME5	0.5	0.35	0.4	15	0.5
–	–	–	–	–	–	ME6	0.3	0.35	0.4	15	–

NR: No requirement

distribution of contrasts was determined. Work by Dunbar[23] was used to find whether the contrast was sufficient for an object to be visible. At any given point on the road, this enabled the percentage of objects that are visible to be determined. Waldram[24] called this the *Revealing Power*.

Following this procedure meant that the revealing power of a grid of points covering the road surface could be determined. To characterize the performance of a road lighting installation it is very useful to have a single figure of merit. This can be used to compare installations. It was, therefore, very tempting to average the revealing powers found on a stretch of road to produce a single figure. Waldram warned against this. His main argument was that an area of low revealing power could not be compensated for by an area of high revealing power. The area of low revealing power would always present a potential danger for the driver. As can be imagined, the calculation of revealing power is very labour intensive and little work was done on this and kindred concepts until the 1960s, when computers were becoming more generally available.

van Bommel[25] in 1979 developed the concept of *Revealing Power* one stage further by introducing the effect of threshold increment into the calculation. He used this to investigate the trade-off between overall uniformity, the threshold increment, and average luminance of the road surface. For the purpose of this calculation he took the *Revealing Power* at the darkest part of the road as the operative value for an installation.

A related concept, that of *Small Target Visibility* or *STV*, has been developed and investigated in the US in recent years.[26] Straightaway, it has to be pointed out that it differs from *Revealing Power* in that the reflectance of the target is fixed, at 20%. Moreover, it is the weighted average of the *Visibility Level, VL,* of the target over a specified area of the road surface. The Visibility Level is calculated from the formula:

$$VL = \frac{\Delta L_{\text{actual}}}{\Delta L_{\text{threshold}}}$$

where

VL is visibility level;
ΔL_{actual} is the luminance difference between the target and the background;
$\Delta L_{\text{threshold}}$ is the minimum luminance difference needed between a target and its background to make the target visible.

The basis for evaluating *STV* is a method due to Adrian. This allowed the following variables to be taken into account:

- the age of the observer, for the calculation of *STV* this is taken to be 23 years;
- reaction time of the motorist, taken to be 0.2 s;
- size of target, taken to be a square of side 0.18 m;
- luminance of target, evaluated by multiplying the reflectance of the target by its illuminance. The target is assumed to be a uniform diffuser of luminance 0.2 cd/m^2, 83 m ahead of the motorist, vertical, and with its normal parallel to the road axis.

The effect of disability glare is not included, although the Adrian model permits this. A *VL* of 7.5 is recommended for M1 roads grading down to 5.0 for M5 roads.

The visibility concepts discussed indicate that if the overall uniformity of luminance of the road surface is too great, the probability of detecting an object can be decreased. This can be readily understood by considering the extreme case, when the road luminance is perfectly uniform. Here, there is a likelihood of some objects having the same or nearly the same luminance as that of the road, and therefore being invisible. Hence, it appears that a lack of uniformity may be beneficial in making objects on the road conspicuous. Certainly, this has been demonstrated in trial installations. But this is only true for flat targets, which are of necessity of uniform luminance. Lecocq[27] has demonstrated by calculation and in trial installations that *solid* objects will not become invisible, because of the variation of luminance over their surface.

STV is still being developed. There are many aspects of its determination which need to be reconsidered. As mentioned above, a flat target does not seem to relate well to live situations; a solid target is better in this respect. This suggests that it would seem to be more realistic to take into account a range of reflectances of objects as is done in *Revealing Power*. In the calculation of contrast in *STV*, the mean luminance of the parts of the road seen by the observer as being just above and below the target are taken as the background luminance. Work by Ménard and Cariou[28] suggests that it may be better to consider the luminance on either side of the target. Alternatively, consideration should be given to the dynamic situation, where the background to the target is changing as the driver drives along the road. There are other aspects of determination of *STV* open to question. These are the low age of the observer, size of target, and no account being taken of light being reflected from the road surface onto the target.

The future for *STV* and related concepts is uncertain. Work in the US in which an attempt was made to correlate accident rate with *STV* in 50 installations did not produce a positive result.[29] However, road design based on visibility takes into account more variables than road luminance, and may provide the future basis for road lighting design. This would be particularly so if it results in both more effective lighting and a more efficient use of energy.

9.14 Maintenance factors

Maintenance factors are used to allow for the depreciation in the light output of the luminaire and lamp. In general road lighting design practice, no allowance is made for the change in reflection properties of the road surface with time. In the British Standard on road lighting, there are a number of definitions used in connection with maintenance.

Initial luminous flux of lamp: The luminous flux of a lamp in lumens after an initial aging period under stated running conditions. (The lamp is run for the aging period to allow its light output to stabilize, as many discharge lamp types are unstable when freshly off the production line.)

Lamp flux maintenance factor: The proportion of the initial luminous flux of a lamp that is produced after a set time. The data for lamps should be available from lamp manufacturers. Abbreviation: *LLMF* (Lamp lumen maintenance factor).

Luminaire maintenance factor: The light output ratio of a luminaire after a set time in service divided by its initial light output ratio. (Reduction in the light output ratio of a luminaire can be caused by a variety of factors. Perhaps the main one is the deposition of dirt on the light controlling surfaces, but the surfaces themselves may degrade and this may be permanent, or they may move out of alignment owing to continual vibration.) Abbreviation: *LMF*.

Maintenance factor: The product of the lamp flux maintenance factor and the luminaire maintenance factor. The abbreviation used for this is *MF*. Hence $MF = LLMF \times LMF$.

Maintained average light level: The average light level below which the performance of an installation is not allowed to fall. (By average in this definition is meant spatial average and not time average. The implication of the definition is that as soon as, or ideally before, the maintained level is reached, maintenance operations are carried out on the installation.)

Table 9.7 Luminaire maintenance factors

Cleaning interval (months)	Ingress protection number of lamp housing					
	IP5 minimum			IP6 minimum		
	Pollution category					
	High	Medium	Low	High	Medium	Low
18	0.87	0.88	0.91	0.90	0.91	0.92
24	0.84	0.86	0.90	0.88	0.89	0.91
36	0.76	0.82	0.88	0.83	0.87	0.90

Table 9.7 is an abbreviated form of the maintenance factor table from BS 5489. The ingress number refers to the efficacy of the sealing of the lamp compartment of the luminaire. For IP5 the ingress of dust is not totally prevented, but it does not enter in sufficient quantity to interfere with the satisfactory operation of the luminaire. For IP6 there must be no ingress of dust. Full details of IP ratings are given in BS 5490: 1977.

The definitions of the pollution categories are:

- High pollution occurs in the centre of large urban areas and in heavy industrial areas.
- Medium pollution occurs in semi-urban, residential, and light industrial areas.
- Low pollution occurs in rural areas.

In the US, a different system is used. Graphs (Figure 9.11) are given for what is termed the *luminaire dirt depreciation factor* plotted against exposure time in years for different pollution

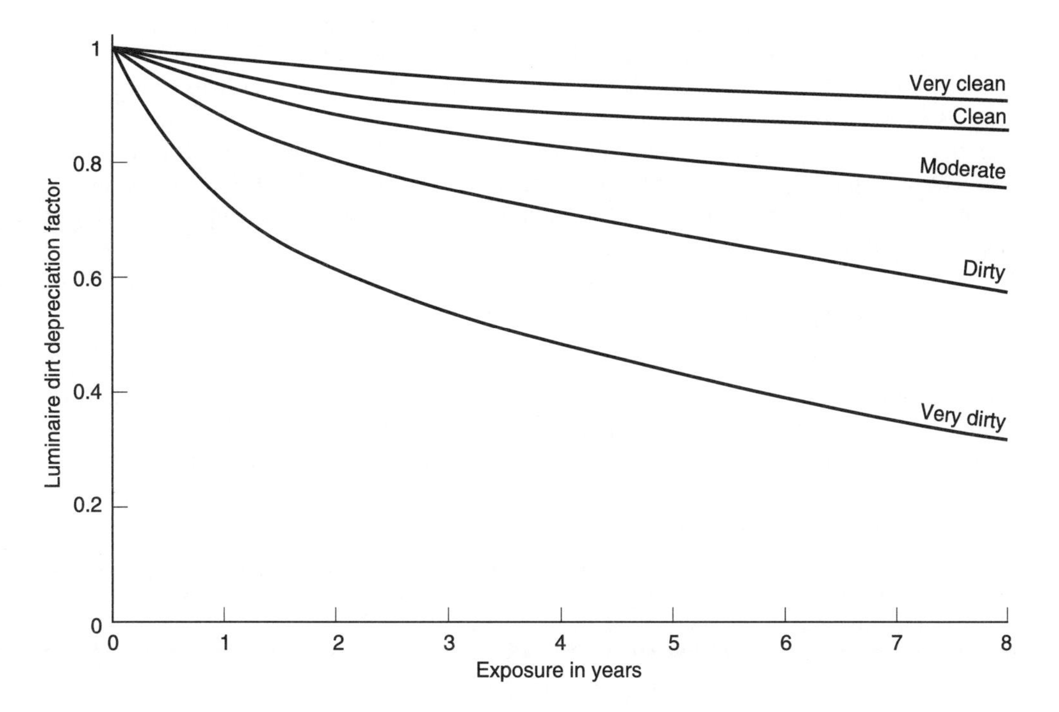

Fig. 9.11 Maintenance factor curves in use in the USA

Table 9.8 Comparison of IESNA and BS maintenance factors for a cleaning interval of three years

IESNA maintenance factor		BS maintenance factor		
	LMF	Pollution category	*LMF* for IP5	*LMF* for IP6
Very clean	0.96	Low	0.88	0.90
Clean	0.91			
Moderate	0.87	Medium	0.82	0.87
Dirty	0.77	High	0.76	0.83
Very dirty	0.56			

categories. Readers should note that the term 'depreciation' in this context has the same meaning as 'maintenance'. No account is taken in these curves of the efficacy of the sealing of the lamp compartment, so it is interesting to compare the results given by the two systems. This is done in Table 9.8 for a three-year maintenance period.

This table would seem to indicate that clean, moderate, and dirty in the IESNA curves correspond roughly to low, medium, and high in the BS table for a three-year maintenance period.

9.15 Tabular and graphical methods of calculation

The adoption of the luminance concept for the regular design of road lighting installations has been made possible by the easy access for designers to powerful computers and easy-to-use programs. However, in the 1960s when the possibility of using luminance design on a day-to-day basis was mooted computers were not generally available and certainly personal computers were a thing of the future. For this reason, tabular and graphical methods were devised to enable luminance and threshold increment calculations to be carried out without the benefit of a computer (although these aids themselves had to be produced by computer). This contrasts with the present situation when computers are universally available, together with very good programs, which has rendered these methods largely obsolete. However, we will make a short review of these methods as they form useful background knowledge for the practising engineer and are referred to in various documents. In addition, they are sometimes useful for assessing the relative performance of a number of luminaires.

9.15.1 TABULAR METHODS

In essence, these methods present the results of calculating the lighting parameters for many commonly used lighting situations. In using them the designer will have to interpolate between results if the exact geometry of the situation of interest does not appear in the table.

In the *Design Table* method of BS 5489: Part 2, the results of calculations on over 400 installations are given. In the standard presentation the road surface used is C2 (Table 9.1, page 356) and a table has to be produced for each mounting height. Table 9.9 shows an extract from such a table.

In the complete table there would be data for opposite, single sided, and twin central geometries, as well as the staggered geometry shown. In addition, data would be shown for wider spacings (S), and more effective widths (W_E).

Effective width W_E is the transverse distance between luminaires, and is sometimes abbreviated as *EW*. For each W_E a value for the spacing index (S_I or *SI*) is quoted. This enables the

Table 9.9 Extract from a road lighting luminaire design table (by courtesy of Thorn Lighting Limited)

Description: Luminaire with aluminium reflector and flat glass
Lamp: 150 W SON-T
Luminaire maintenance category: IP6
Mounting height: 10 m
Design attitude of luminaire: Spigot elevated 0°

S (m)	W_E 8 m, S_I 3.15, S_R 0.63			W_E 9 m, S_I 3.02, S_R 0.61			W_E 10 m, S_I 2.88, S_R 0.59			W_e 11 m, S_I 2.75, S_R 0.59		
	U_0	U_L	V_F	U_0	U_L	V_F	U_0	U_L	V_F	U_0	U_L	V_F
20	0.64	0.76	9	0.62	0.74	9	0.60	0.76	8	0.59	0.76	7
22	0.64	0.73	9	0.61	0.72	8	0.59	0.72	7	0.58	0.72	6
24	0.63	0.69	9	0.60	0.66	8	0.58	0.66	7	0.56	0.67	6
26	0.61	0.64	9	0.57	0.62	8	0.54	0.62	7	0.53	0.62	6
28	0.59	0.63	8	0.55	0.60	8	0.52	0.60	7	0.48	0.60	6
30	0.57	0.63	8	0.53	0.60	7	0.49	0.60	6	0.45	0.58	6
32	0.55	0.62	8	0.51	0.58	7	0.46	0.56	6	0.41	0.53	6
34	0.54	0.60	8	0.48	0.56	7	0.43	0.54	6	0.39	0.52	5
36	0.53	0.58	8	0.47	0.54	7	0.42	0.51	6	0.39	0.49	5

spacing S between luminaires to be found for a required luminance. The equation that enables this to be done is:

$$S = \frac{S_I \times \Phi \times MF}{L_{av}} \tag{9.6}$$

Under S_I in each column is the surround ratio S_R (or SR), which was explained in Section 9.11, page 369. For each spacing, S, U_0 and U_L are quoted together with the *veil factor* V_F (or VF), which enables TI to be calculated from the formula:

$$TI = \frac{V_F \times \Phi}{10\,(L_{av}/MF)^{0.8}} \tag{9.7}$$

This formula calculates the threshold increment for the initial condition of the installation, at which time the threshold increment has its maximum value. This explains why L_{av} is divided by MF.

Worked example 6 What spacing is required to light a road to M2 requirements (Table 9.6, page 372) with a staggered arrangement and a mounting height of 10 m? The design attitude is 0°, the road surface is *C*2, and the lamp to be used is 150 W SON-T, with an initial luminous flux of 17 klm, and a maintenance factor 0.92 for a projected lamp life of 12 000 h (equivalent to three years' burning). The luminaire is IP6, and the pollution category is medium. The effective width W_E is 9m. The cleaning interval is 36 months.

Solution The solution can be broken down into several steps.

(a) Check that SR is greater than 0.5. As it is 0.61, it meets this requirement.
(b) Calculate MF. The luminaire MF for an IP6 luminaire, in a medium pollution environment, with a cleaning interval of 36 months, is 0.87. The MF lamp is 0.92, obtained from the manufacturer. Hence the overall MF is 0.87 × 0.92 or 0.80.
(c) Calculate the spacing to meet the L_{av} requirement of 1.5 cd/m^2. Using Equation (9.6), we obtain:

$$S = \frac{S_I \times \Phi \times MF}{L_{av}}$$

$$= \frac{3.02 \times 17 \times 0.80}{1.5}$$

$$= 27 \text{ m}$$

(d) Check $U_o \geq 0.4$ and $U_l \geq 0.5$. They meet these requirements easily. If they had not it would have been necessary to use a closer spacing.
(e) Check the TI is less than 15%. For this we use Equation (9.7):

$$TI = \frac{V_F \times \Phi}{10(L_{av}/MF)^{0.8}}$$

$$= \frac{8 \times 17}{10(1.5/0.80)^{0.8}}$$

$$= 8.2\%$$

This is well within the upper limit of 15%.

(f) All the lighting requirements have been met and hence a spacing of 27 m is satisfactory.

A variety of what are termed *performance sheets* have been devised by various manufacturers and users working in Europe. These performance sheets differ from the *Design Tables* already described in that the values of the lighting parameters are stated for a variety of situations, ostensibly without the need for any calculation. In practice, some calculation will be needed to take account of the maintenance factor. Table 9.10 is an example of a performance sheet for a luminaire. The complete set of performance sheets for a luminaire would contain data for a variety of effective road widths and arrangements. The reader will notice that values of G are given,

Table 9.10 Example performance sheet
Description: Luminaire with aluminium reflector and flat glass
Lamp: 150 W SON-T   Lamp flux: 17 klm
Luminaire maintenance category: IP6
Design attitude of luminaire: Spigot elevated 5°

			Staggered arrangement											
			Mounting height 10 m						Mounting height 12 m					
			Spacing S (m)						Spacing S (m)					
W_E (m)	Lighting Parameter	Road surface	28	32	36	40	44	48	28	32	36	40	44	48
	L_{av} (cd m^{-2})	R1	1.99	1.74	1.55	1.39	1.26	1.16	1.75	1.54	1.37	1.23	1.12	1.02
		R2	1.47	1.29	1.14	1.02	0.93	0.85	1.31	1.15	1.02	0.92	0.84	0.77
		R3	1.41	1.23	1.10	0.98	0.90	0.82	1.27	1.11	0.99	0.89	0.81	0.74
		R4	1.53	1.34	1.19	1.07	0.97	0.89	1.38	1.21	1.07	0.96	0.88	0.80
	U_0	R1	0.69	0.64	0.59	0.55	0.51	0.46	0.76	0.76	0.72	0.65	0.61	0.56
		R2	0.64	0.58	0.54	0.51	0.48	0.44	0.74	0.71	0.67	0.61	0.58	0.53
		R3	0.62	0.54	0.50	0.48	0.44	0.39	0.71	0.66	0.61	0.55	0.52	0.50
		R4	0.58	0.50	0.47	0.45	0.43	0.38	0.64	0.60	0.56	0.52	0.50	0.48
	U_l	R1	0.74	0.62	0.52	0.44	0.38	0.31	0.83	0.80	0.70	0.56	0.52	0.46
		R2	0.62	0.58	0.53	0.43	0.37	0.31	0.73	0.70	0.66	0.58	0.54	0.53
8		R3	0.65	0.55	0.46	0.42	0.36	0.30	0.74	0.70	0.62	0.57	0.50	0.44
		R4	0.70	0.57	0.46	0.44	0.38	0.38	0.78	0.75	0.62	0.57	0.52	0.51
	G	R1	5.2	5.2	5.3	5.3	5.3	5.3	5.5	5.5	5.6	5.6	5.6	5.6
		R2	5.1	5.1	5.1	5.1	5.2	5.2	5.4	5.4	5.4	5.5	5.5	5.5
		R3	5.1	5.1	5.1	5.1	5.2	5.2	5.4	5.4	5.4	5.4	5.5	5.5
		R4	5.1	5.1	5.2	5.2	5.2	5.2	5.4	5.4	5.4	5.5	5.5	5.5
	TI (%)	R1	3.6	5.1	5.2	5.7	5.8	6.3	3.1	3.3	4.0	4.3	4.4	4.9
		R2	4.6	6.5	6.7	7.3	7.4	8.0	4.0	4.3	5.0	5.4	5.5	6.2
		R3	4.8	6.7	6.9	7.5	7.7	8.3	4.1	4.4	5.2	5.4	5.7	6.4
		R4	4.5	6.3	6.5	7.1	7.2	7.8	3.8	4.1	4.8	5.2	5.3	6.0

which are absent from the *Design Table*, as this parameter is not required for the British Standard. On the other hand, values of *Surround Ratio* are not given as this parameter is not used, or rarely used, on the continent of Europe.

9.15.2 GRAPHICAL METHODS

There are several graphical methods available for calculation of road luminance. Although these are seldom used, no account of road lighting calculations would be complete without a description of them as they are often found in the literature.

Luminance yield curves are used for determining average luminance. Luminance yield factor, η_L, is defined by:

$$\eta_L = \frac{L_{av} \times S \times W}{\Phi \times Q_o}$$

where

L_{av} is the average road luminance,
S is the spacing between luminaires in a row,
W is the road width,
Φ is the lamp flux in lumens,
Q_o is the luminance coefficient.

The curves are obtained by calculating the average luminance for a single luminaire for various road widths. Figure 9.12 shows this function plotted for an example luminaire. The reader will note that there are three curves, which correspond to different positions of the observer. In addition, to comply with the definition of luminance yield factor, the average luminance coefficient is unity. To use the luminance yield curves to find the spacing required to achieve a given luminance, S has to be made the subject of the above equation and maintenance factor MF has to be introduced, giving:

$$S = \frac{\eta_L \times Q_o \times \Phi \times MF}{L_{av} \times W} \qquad (9.8)$$

Worked example 7 A road 12 m wide is to be lighted with the example road lighting luminaire for which the luminance yield curves in Figure 9.12 have been prepared. A staggered arrangement is to be used with a mounting height of 12 m and an overhang of 1 m. The observer is positioned 2.5 m from the kerb. The lamp is a 150 W SON-T with an initial output of 15 000 lm. If the maintenance factor for the lamp and the luminaire is 0.75, Q_o is 0.75, and the average luminance is to be 1.0 cd/m^2, what spacing must be used?

Solution For the kerb side $B/H = 2.5/12 = 0.21$
For the road side $B/H = 7.5/12 = 0.63$

Observer position relative to rows of luminaires:

Left-hand row of luminaires: $1.5/12 = 0.13$ to road side.
Right-hand row of luminaires: $6.5/12 = 0.54$ to road side.

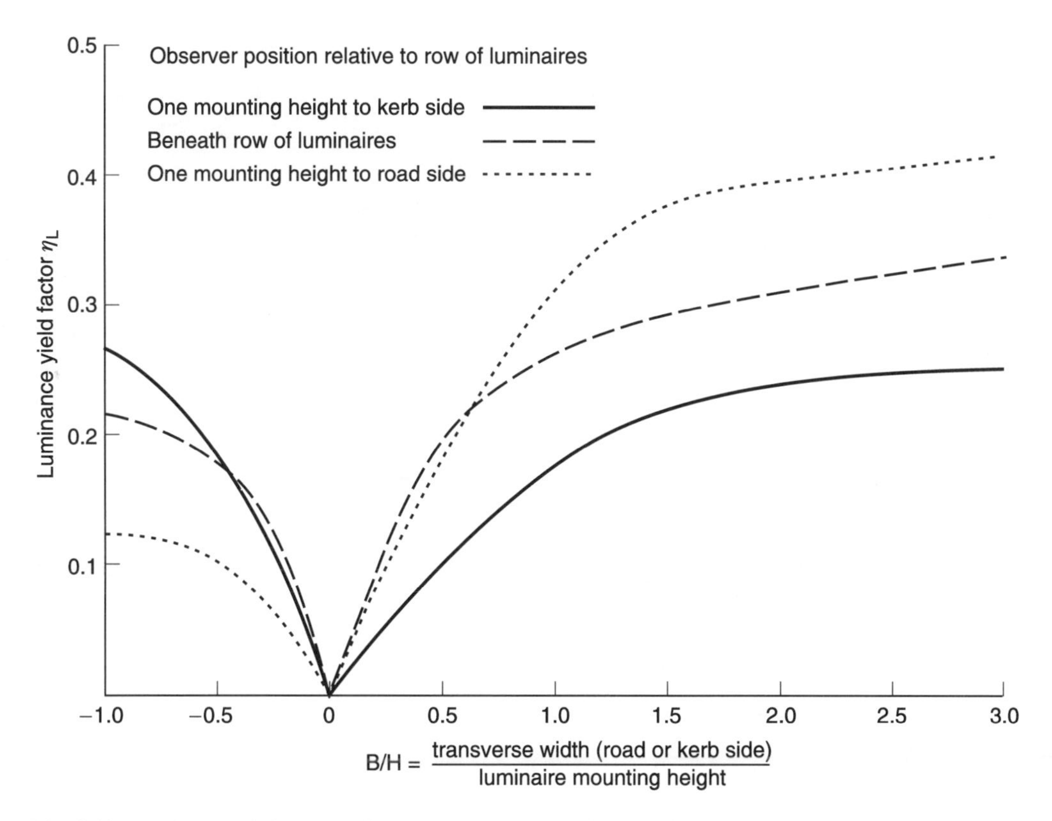

Fig. 9.12 Luminance yield curves for an example road lighing luminaire. Angle of tilt: 0°. Road surface *R*2 with $Q_o = 0$

The luminance yield factors can now be read from the figure. These are:

Kerb side, left-hand row of luminaires: 0.09
Kerb side, right-hand row of luminaires: 0.10
Road side, left-hand row of luminaires: 0.22
Road side, right-hand row of luminaires: 0.23

This gives a total luminance yield factor of 0.64, which can now be substituted in Equation (9.8):

$$S = \frac{\eta_L \times Q_o \times \Phi \times MF}{L_{av} \times W}$$

$$= \frac{0.64 \times 0.075 \times 15\,000 \times 0.75}{1.0 \times 12}$$

$$= 45 \text{ m}$$

For parts of the curves which are steep, a small change in *B*/*H* makes a large difference in the luminance yield factor, so when these parts are used, which is generally the case for most installations, it is difficult to obtain an accurate answer. It should also be noticed that *S* is the spacing

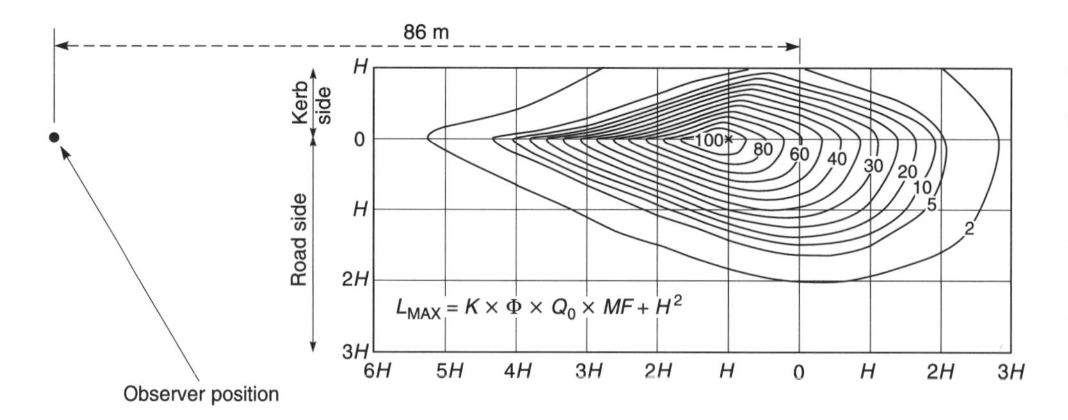

Fig. 9.13 Insoluminance diagram for *Newlight* luminaire for C2 road surface with $Q_0 = 1$

between luminaires in a row, which is different to normal UK practice, where the longitudinal spacing is between luminaires.

9.15.3 ISOLUMINANCE DIAGRAMS

These can be used to find the luminance at a point so that overall and longitudinal uniformity can be determined.

Figure 9.13 shows a typical isoluminance diagram. Distances are expressed in terms of the mounting height *H*. The observer is located to the left of the diagram at a distance of 10 *h* from the luminaire, which is at (0, 0). Luminances are expressed as percentages of the maximum luminance, the formula for which is given on the diagram. The position of the maximum luminance is indicated by the cross in the centre of the diagram.

Two details should be noted about the shape of the contours. First, the contours are asymmetrical about the longitudinal $h = 0$ line. This is because the luminaire directs more light towards the road than the kerb side. Second, when the diagram is viewed at a low angle from the left-hand side it has the T-shape indicated in Figure 9.4, page 351.

The isoluminance diagram is prepared from a luminance grid where the observer is placed in line with the row of luminaires. If in the application for which the diagram is used the observer is outside this line, some inaccuracy will result. However, if it is not necessary to rotate the diagram more than about 5° from the axis of the road the inaccuracy should be acceptable. Otherwise it will be necessary to have isoluminance diagrams for different observer positions.

To use the diagram a drawing of the road to be lit should be prepared scaled in terms of *h*. The isoluminance diagram, on tracing paper, is then placed over the road drawing so that the origin (0, 0) coincides with the position of the luminaire and the longitudinal $h = 0$ line is directed towards the observer. Alternatively, and this may be easier, the road plan can be transferred to tracing paper. The value of the contour which passes through the point at which the luminance is required is noted. Interpolation may be necessary. This is repeated for each luminaire and the total found. To find the luminance this is multiplied by $L_{max}/100$ found from the formula given on the diagram.

Worked example 8 Find the contribution to the luminance of the road surface made by a single *Newlight* luminaire located at a mounting height of 10 m, 150 m ahead of the observer, who is

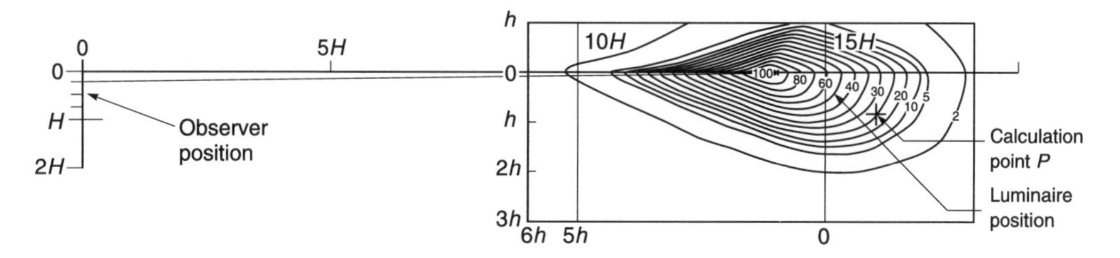

Fig. 9.14 Use of isoluminance diagram for determining luminance at a point

2.5 m from the kerb, on the road side. The point at which the luminance is required is 160 m ahead of the observer and 10 m from the kerb, on the road side. The road surface is *C*2 and Q_0 is 0.07. The lamp flux is 15 000 lumens, and the maintenance factor for the lamp and the luminaire together is 0.75.

Solution The locations of the kerb, observer, luminaire, and calculation point are drawn in plan to the same scale as the isoluminance diagram, as in Figure 9.14. The (0, 0) coordinates of the isoluminance diagram are placed over the position of the luminaire and the diagram rotated so the $h = 0$ longitudinal line passes through the observer. The angle of the rotation is $\tan^{-1}(2.5/150) = 0.95°$, which is small enough for any inaccuracy due to rotation to be ignored. For clarity, some of the contours have been omitted.

Now,

$$L_{\max} = \frac{K \times \Phi \times Q_0 \times MF}{H^2}$$

$$= \frac{0.095 \times 15\,000 \times 0.75 \times 0.75}{100}$$

$$= 8.02 \text{ cd/m}^2$$

The calculation point *P* lies between the 20% and 30% contours at 23%.

So the luminance at *P* is $0.23 \times 8.02 = 1.84$ cd/m^2.

9.15.4 ISOLUMINANCE TEMPLATES

This is the name given to the isoluminance diagrams used in BS 5489: Part 2 for deciding the spacing of luminaires on bends. Their use ensures that the overall uniformity is 0.4. Longitudinal uniformity is not catered for as this parameter has not been defined for curved roads.

The basis of using isoluminance templates relies on the strong correlation between $L_{\text{minimum}}/L_{\text{maximum}}$ and overall uniformity.[30] It was found that when $L_{\text{minimum}}/L_{\text{maximum}}$ is 25% then overall uniformity is 40%. This means that two overlapping isoluminance contours each of 12.5% can be used to find the luminaire spacing to achieve an overall uniformity of 40%.

Figure 9.15 shows such a template. It is recommended that this is drawn to a scale of 1:500 so that it can be used on the maps commonly in use. The reader will notice that there is a circle enclosing the observer position. This circle is placed tangentially to the kerb line so that the observer is in the standard viewing position. Figure 9.16 indicates how the templates are used.

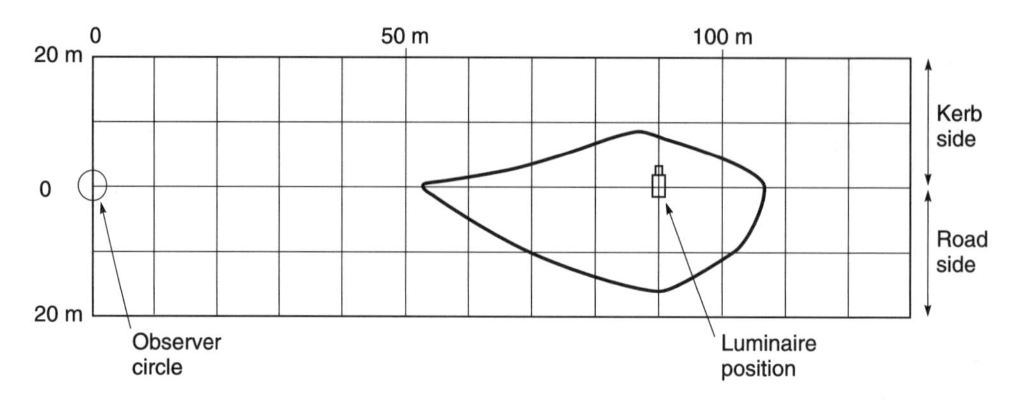

Fig. 9.15 Isoluminance template

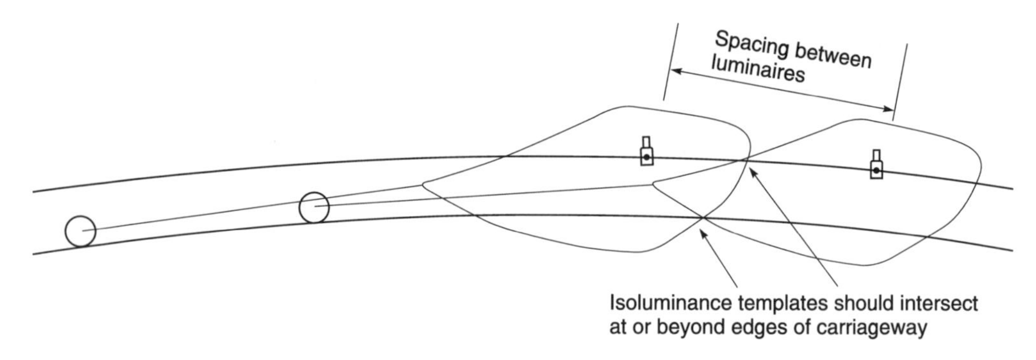

Fig. 9.16 Use of isoluminance templates to find spacing on bends

For a twin central arrangement a second contour is needed (not shown) produced by combining two *I*-tables, one being rotated through 180° of azimuth. Should it be necessary to introduce extra single luminaires into this twin central arrangement, a third contour at 25% of the maximum luminance produced by a single luminaire is required.

Because the isoluminance template is scaled in absolute distances a template is required for each mounting height.

9.16 Perspective view of the road

The lighted view of the road should be represented in perspective view, which is as we see it, rather than in plan, which is the convention for luminance calculations, at least at present. It is not difficult to convert a plan drawing to a perspective one.[31] Figure 9.17 shows how it can be done. *MVU* is the plane of the road and we wish to produce a perspective image of this on the plane *RST*, which is called the picture plane. The road is being viewed from *O*, and all rays of light from the object pass through this point onto the picture plane. The picture plane is vertical, and is perpendicular to *MV*, the longitudinal axis of the road. *f* is the distance of the origin, *O*, from the picture plane, and is equivalent to the focal length in a camera.

From the similar triangles *SOT* and *OVM*,

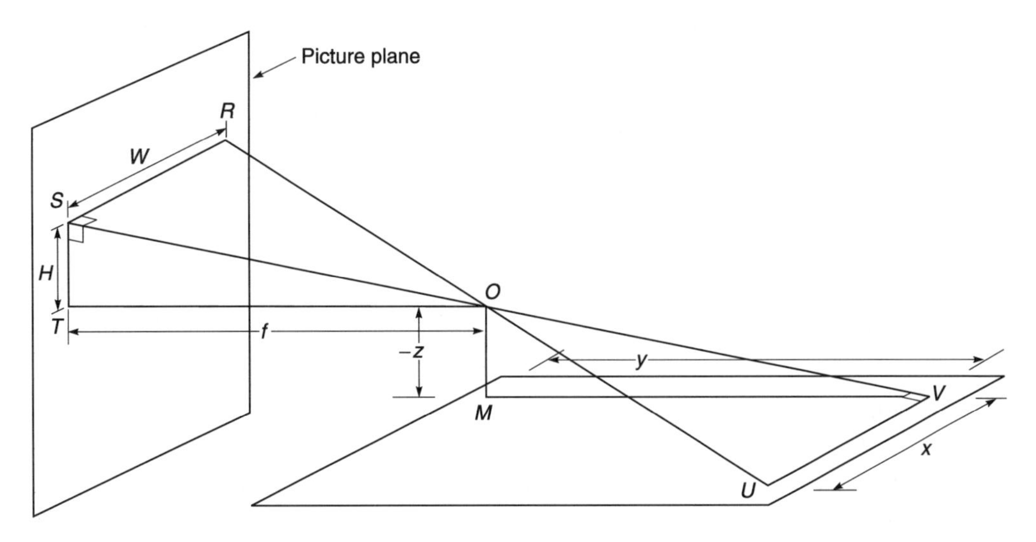

Fig. 9.17 Perspective projection

$$H = \frac{z}{y} f$$

and from the similar triangles *SRO* and *OVU*,

$$W = x\sqrt{\left(\frac{H^2 + f^2}{y^2 + z^2}\right)}$$

From these two equations any point on the road can be mapped onto the picture plane. To simplify drawing it should be noted that straight lines map as straight lines, all longitudinal lines map as lines converging to a point obtained by dropping a perpendicular from *O* onto the picture plane. This is the vanishing point.

Figure 9.18(a) shows a perspective view of the calculation field for a 12 m wide road, and Figure 9.18(b) is an enlarged view of the points in the calculation field. To obtain the correct subjective impression of these views Figure 9.18(a) should be held vertically at 540 mm from the eyes and Figure 9.18(b) should be held vertically at 1860 mm from the eyes. The reader will notice the small proportion of the field of view of the road that the calculation field comprises. In addition, the spacing of the calculation becomes closer as distance increases but is even transversely. There is a case for choosing the position of calculation points so that they are in the centres of equal areas as seen in perspective. This has been tried by some workers and they found that there is very little difference in the values of the luminance measures when this is done, so the present procedure has been retained.

9.17 National variations

There are national variations on nearly every aspect of the design process. We propose here to give a indication of those not already mentioned.

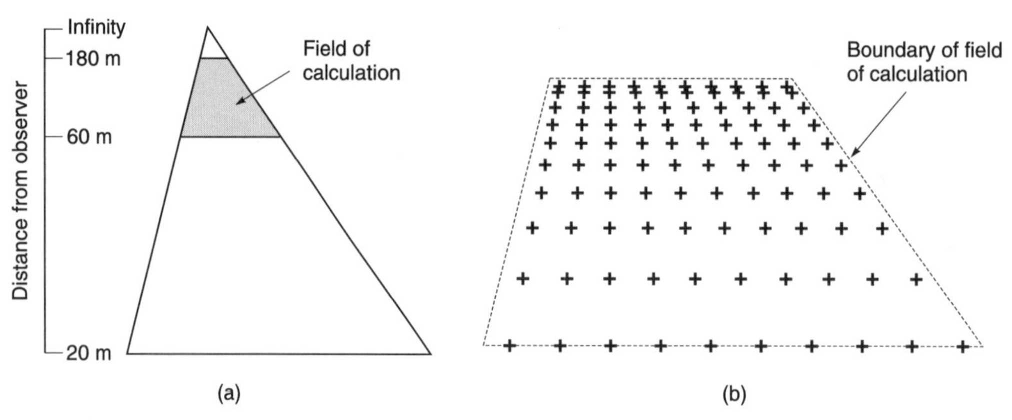

Fig. 9.18 Perspective view of road

In South Africa and Australia, the value of overall uniformity of luminance is 0.33 for all roads, including motorways.[32] Although this seems a small reduction from 0.4 which is used for the great majority of lighting classes in Table 9.6, it usually allows a much greater spacing of luminaires. In countries where there are large distances between towns, this effects considerable economies in running and capital costs, albeit to the detriment of lighting performance.

Reference has been made to the use of a large toe-in to achieve better coverage of the road in wet conditions. This is often used in Nordic countries where the road is wet for a high proportion of the time. An overall uniformity of 0.15 is permitted when the road is wet but when dry the performance of the installation must meet the requirements for the appropriate class of road.

In North American practice, luminance design is the preferred method of design but illuminance criteria can be used if good design judgement is used in their application.[33] For this reason, lighting requirements are given in terms of both illuminance and luminance. However, as the choice of illuminance depends on the reflection properties of the road surface, luminance is indirectly taken into account. The ratio of the veiling luminance to the average luminance is used as a criterion for controlling disability glare in place of threshold increment. The limiting values of this are 0.3 for the more important roads and 0.4 for the other roads. Recommended values of illuminance and uniformity are somewhat lower than those recommended by the CIE.

To help the designer narrow down the selection of luminaires suitable for an application there is a classification system for the light distributions from the luminaires. This uses the spread of light across the road and the throw of light up and down the road, which mainly depend on the angular beam position and the angular width of the beam. The run-back is used to control discomfort and disability glare.

One difference in the calculation of road surface luminance in North American practice to that described here is that the luminance for each longitudinal row of calculation points is calculated for the observer positioned in the line of the row.

9.18 Critique of luminance design

Luminance design for road lighting has been implemented in many countries but there are a number of aspects in which it could be improved.

The perspective area for which the calculation is done is very small and it would be beneficial if the calculation could be extended to parts of the road closer to the driver. This would entail extending the *r*-table so that it embraces a range of viewing angles.

As has already been mentioned, the concept only deals with empty roads, whereas for a significant proportion of the time the view of the road is blocked by traffic.[34] The road surface itself is very often not uniform in its reflection properties. Particularly on light coloured surfaces, a dark coloured longitudinal streak is deposited from exhaust emissions, and the tyre tracks may be of a noticeably different reflectance.

Much of the experimental work that forms the basis of the lighting recommendation was carried out with comparatively young observers. It has already been mentioned that at 60 years of age three times the luminance is needed for same retinal illuminance as is needed at 20 years of age (page 348). Over the same age range threshold increment increases by about 60 per cent, and doubles at 70 years of age.[34] Older drivers are now forming a higher and higher proportion of the driving population.

Road lighting calculations are based on the V_λ photopic spectral response of the eye. This only operates accurately down to a luminance of 2 cd/m^2, below which the mesopic response of the eye operates (see Section 7.1, page 270). This means that the relative efficacy of light sources changes according to the adaptation luminance. For example, whereas the efficacy of metal halide lamps remains fairly stable with decreasing luminance, that of low pressure lamps decreases substantially.[36] At present, no account is taken of this change of efficacy with adaptation luminance.

Modern cars have tinted windscreens, which have a transmission factor of less than 80%. It would be expected that this reduces the effectiveness of the road lighting. In addition, a raster pattern of fine heater wires is incorporated into some windscreens which, it may be conjectured, may produce some scattering of the light and, in effect, increase the threshold increment.

These shortcomings do not mean that designing road lighting by the luminance concept gives bad or misleading results. The system may be regarded as self-regulating since levels and limits are modified from time to time as a result of experience. There is no doubt it is a great improvement on the previous ad hoc practices. In the UK, the full luminance design method was introduced in about 1980 and has had a number of beneficial effects. It has enabled the lighting quantity and quality to be suited to the application in a better way than was previously possible. It has resulted in improvements in the design of luminaires both as regards light distribution and maintenance, thereby effecting economies. An obvious benefit for the driver has been that most present day luminaires have much better glare control, both disability and discomfort, than the luminaires they have replaced. This has been achieved because there has been an incentive for manufacturers to produce luminaires with efficient light distributions and good maintenance characteristics where no incentive existed before.

References

1. Pollard, N. (1994) Sky-glow conscious lighting design. *Lighting Research and Technology*, **26**, 151.
2. CIE 93–1992. *Road lighting as an Accident Countermeasure.*
3. Scott, P. P. (1980) The relationship between road lighting quality and accident frequency. *Laboratory Report LR929* (Crowthorne, Berks: Transport and Road Research Laboratory).
4. Box, P. C. (1973) Freeway accidents and illumination. *Highway Research Record*, **416**, 10–20.
5. Box, P. C. (1979) The relationship between road lighting quality and accident frequency. *Highway Research Board*, USA, **416**, 1–9.
6. Janoff, M. S., Koth, B., McCunney, W., Freedman, M., Duerk, C. and Berkovitz, M. (1977) *Effectiveness of Highway Arterial Lighting*. Report No FHWA-RD—77–37. Washington D.C., Federal Highway Administration, USA.
7. CIBSE (1994) *Code for Interior Lighting.*
8. Weale, R. A. (1968) *From Sight to Light* (Edinburgh and London: Oliver and Boyd) p. 126.
9. CIE 30.2–1982. *Calculation and Measurement of Luminance and Illuminance in Road Lighting.*

10. Sørensen, K. (1974) *Description and Classification of Light Reflection Properties of Road Surfaces*. The Danish Illuminating Engineering laboratory. Report No 10.
11. Burghout, F. (1979) On the relationship between reflection properties, composition and texture of road surfaces. CIE XIX Session Kyoto, pp. 79–65.
12. BS 5489: Part 2 (1992). Road lighting. Part 2. Code of practice for traffic routes.
13. Frederiksen, E. and Gudum, J. (1972) The quality of street lighting under changing weather conditions, *Lighting Research and Technology*, **8**, 90.
14. CIE Publication 47 (1979) *Road Lighting for Wet Conditions*.
15. Adrian, W. (1989) Visibility of targets, model for calculation. *Lighting Research and Technology*, **21**, 4.
16. Stiles, W. S. (1929) *Proc. Roy. Soc.* (London), **104B**, 332.
17. Holladay, L. L. (1927) *J. Opt. Soc. Am.* **14**, 1.
18. Crawford, B. H. (1936) The integration of effects from a number of glare sources. *Proc. Phys. Soc. (London)*, **48**, 35.
19. CIE Technical Report 19 (1972) *A Unified Framework of Methods for Evaluating Visual Performance Aspects of Lighting*.
20. Bell, R. I. and Marsden, A. M. (1975) Disability glare: the relationship between veiling luminance and threshold rise. *Lighting Research and Technology*, **7**(1), 56.
21. Waldram, J. M. (1934) *Ill. Eng. (London)* **27**, 305.
22. Smith, F. C. (1978) Reflection factors and revealing power. *Trans. Illum. Eng. Soc. (London)*, **3**, 196.
23. Dunbar, C. (1938) Values of brightness contrast in artificially lighted streets. *Trans. Illum. Eng. Soc. (London)*, **3**, 187.
24. Waldram, J. M. (1938) The revealing power of street lighting installations. *Trans. Illum. Eng. Soc.* **3**, 173.
25. van Bommel, W. J. M. and de Boer, J. B. (1980) *Road Lighting*. Philips Technical Library – Kluwer Technische Boeken B.V. – Deventer.
26. Lewin, I. (1996) On the road again. *Lighting Design and Application*, May, 66.
27. Lecocq, J. (1994) Visibility and lighting of wet road surfaces. *Lighting Research and Technology*, **26**(2), 75.
28. Ménard, J. and Cariou, J. (1994) Road lighting: Assessment of an installation based on the contrast of a standard target. *Lighting Research and Technology*, **26**(1), 19.
29. (1997) *Safety Benefits of Roadway Lighting Using Small Target Visibility (STV) Design*. Report prepared for Federal Highway Administration, Turner-Fairbank Research Center, Washington, DC.
30. Hargroves, R. A., Simons, R. H. and Simpson, M. D. (1986) Development of a new code of practice for traffic routes. *Lighting Research and Technology*, **18**(4), 143.
31. Waldram, J. M. (1982) A manual of perspective for lighting engineers. *Lighting Research and Technology*, **26**(2), 65.
32. SABS 090: Part 1–1990. Code of Practice for Public lighting Part 1: The lighting of public thoroughfares (South Africa).
33. (1993) *American National Standard for Roadway Lighting*.
34. Narisada, K. (1995) Perception in complex fields. *Lighting Research and Technology*, **27**(3), 123.
35. Adrian, W. and Bhanji, A. (1991) Fundamentals of disability glare. A formula to describe straylight in the eye as a function of the glard angle and age. 1st International Symposium on Glare. pp 185–193. Lighting Research Institute, New York.
36. Hurden, A. (1998) Factors affecting visibility in conditions of low light. *The Lighting Journal* (ILE), **63**(1), 19.

10

Residential Road Lighting

10.1 Introduction

In Chapter 9 on main road lighting, we stated that the lighting is installed chiefly for the benefit of the vehicle driver. In residential road lighting the situation is different; it is installed chiefly for the benefit of the pedestrian, who is mainly interested in seeing objects that are close by, within 5 m, say. These may include the surface of the footway, the surface of the road (in order that it can be crossed safely), and the delineation of the road. In addition, facial recognition of other road users may be important. Speed of reaction is generally not as important as for drivers. It is assumed that motor traffic and pedal cyclists will be using lighting fixed to their vehicles of sufficient quantity and quality to reveal the lie of the road and make their presence known to other road users.

The principle of silhouette vision, as used for main road lighting, which is concerned with seeing objects against a background between some 60 m and 180 m away is, therefore, not applicable to residential road lighting. Adequate lighting of the surfaces of objects close to the observer has to be achieved. The measure mostly used for this is horizontal illuminance at ground level, but in some countries semicylindrical illuminance 1.5 m above ground level , illuminance on a vertical plane 1.8 m above ground level, or hemispherical illuminance at ground level is used as well as, or in place of, horizontal illuminance. The rationale for using these two measures will be discussed in Section 10.3, page 390.

10.2 Lighting and crime

Historically, the lighting of streets and roads was introduced to combat crime.[1] In more recent times the lighting of residential roads has been mainly thought of as an amenity for pedestrians, but since the 1960s there has been a general increase in street crime, which has prompted a number of investigations into crime and lighting. The object of these has been to investigate whether introducing lighting or improving the existing lighting can be shown scientifically to reduce crime. A positive outcome would justify the extra expenditure entailed on equipment, maintenance and power consumption.

The first major investigation was undertaken by Tien *et al.*,[2] in the United States. 103 projects were investigated. The results were disappointing in that the authors found an increase or no change in crime for as many projects as showed a decrease in crime.

Many investigations have been carried out in the UK since the 1980s. These have concentrated more on residents' fear of crime and their perception of the reduction of crime than the actual reduction of crime. There seems to be positive evidence that good road lighting (which we will define later) does reduce fear of crime. This helps to improve morale and civic pride, which leads to the greater use of the area by pedestrians. This in its turn is said to be beneficial because it deters the activities of criminals, who feel that the chance of their being detected is increased.

However, there is evidence from the UK studies that as well as reducing the fear of crime, good lighting does reduce the actual crime rate.[3,4]

It is often argued that reduction of crime in an area merely displaces it to an adjacent area. This was found by Lloyd and Wilson,[5] but they concluded that the combined crime rate for the lit and adjacent areas was reduced. The contrary thesis is being promoted by Painter and Farrington.[6] They suggest that new evidence points to diffusion taking place; that is, the benefits achieved in the relit area diffuse to the adjacent areas.

So far, no studies have been carried out to find the relationship between the quantity and quality of lighting and fear of crime or crime risk.

10.3 Lighting measures

In Section 10.1, horizontal illuminance, semicylindrical illuminance, vertical illuminance, and hemispherical illuminance were enumerated as measures of the quantity of light in residential road lighting. The rationale for the choice of these is not hard to discover. Horizontal illuminance at ground level is an indicator of whether the principal surface with which the pedestrian is concerned is sufficiently lit. However, pedestrians are also concerned with identifying vertical surfaces, and this is presumably why vertical illuminance is used. Semicylindrical illuminance at 1.5 m above ground level would seem to be an appropriate measure of the lighting of these surfaces. 1.5 m was chosen because work by van Bommel and Caminada[7] suggested that recognition of faces is important for pedestrians, both as an amenity and for security. Hemispherical illuminance is mainly used in Denmark, where low mounted lighting is common. This form of lighting gives low horizontal illuminance mid-way between luminaire positions, but nevertheless is judged subjectively to be satisfactory. Hemispherical illuminance gives higher values than horizontal illuminance at these positions. Semicylindrical illuminance does too but was rejected because it gives low or zero values beneath the luminaires.

Investigations by Simons *et al.*[8] into these three measures as indicators of the adequacy of residential lighting showed that horizontal illuminance gives a slightly better correlation with appraisals than the other two measures; that is, for the geometry of installations used in the UK.

10.4 Lighting levels

Table 10.1 summarizes the CIE recommendations.[9] These apply to the whole of the used surface, that is the road and the footways.

The reader will note that the recommended illuminances are maintained values, which means that they must not be allowed to drop below these values for the life of the installation. When the installation is new the values will be greater. As the installation ages the light output of the lamps decreases and dirt accumulates on the luminaires, so decreasing the light output. When cleaning the luminaires no longer raises the measured values of illuminance sufficiently it will be necessary to change the lamps or refurbish the luminaires.

The illuminances relate to situations where the crime risk is negligible. Where the crime risk is high, consideration should be given to using a class one step higher – or where it is severe two steps higher – than would be normally used. Moving a class one step higher in this context means going to a lower class number, for example from P4 to P3. The last three classes in the table should only be used when the crime risk is negligible.

Table 10.1 Summary of CIE recommended illuminances

Lighting class	Description of road or situation	Maintained horizontal illuminance (lx)	
		Average	Minimum
P1	High prestige roads	20	7.5
P2	Heavy night-time use by pedestrians and pedal cyclists	10	3
P3	Moderate night-time use by pedestrians and pedal cyclists	7.5	1.5
P4	Minor night-time use by pedestrians and pedal cyclists	5	1
P5	As P4 but important to preserve village character	3	0.6
P6	As P5 but very minor night-time use	1.5	0.2
P7	Solely visual guidance from the direct light from the luminaires	Not applicable	

Table 10.2 Summary of lighting requirements in BS 5489 Part 3

Category	Description of road or situation	Maintained Horizontal Illuminance (lx)	
		Average	Minimum
3/1	Use by public high, or crime risk high, or traffic usage high	10	5
3/2	Use by public moderate, or crime risk average to low, or traffic usage is of a level equivalent to that of a housing estate access road	6	2.5
3/3	Use by public minor and solely associated with adjacent properties, and crime risk very low, and traffic usage equivalent to that of a residential road	3.5	1

As can be seen from Table 10.2, the UK recommendations in Part 3 of BS 5489 (which were promulgated before the CIE recommendations) correspond very roughly with CIE classes P2, P3, and P4. The degree of crime risk is built into the definition of the class or category. It is worth noting that the prefix 3 used in the designation of the lighting category refers to Part 3 of BS 5489.

As well as giving recommendations in terms of horizontal illuminance, CIE 92, *Guide to the Lighting of Urban Areas*,[10] gives recommendations in terms of semicylindrical illuminance. Generally, for residential areas, the values recommended are roughly equal, numerically, to the horizontal illuminance. The recommendations for semicylindrical illuminance are in terms of the minimum value only and apply in both directions along the run of the road and, as stated previously, 1.5 m above ground level. Table 10.3 is an example of the recommendations for specialized residential areas, which are residential areas with restricted access and may be of high building density.

Because the calculated semicylindrical illuminance under a luminaire is zero (unless its size is taken into account) illuminances within 1 m of the downward vertical through the photometric centre of the luminaire are ignored.

Table 10.4 is condensed from the *IES Lighting Handbook*.[11] It will be seen that the minimum average horizontal illuminances are in the same order of magnitude as those recommended by CIE but the vertical illuminances, which are taken 1.8 m above ground level, are much higher.

Table 10.3 Horizontal and semicylindrical illuminance requirements compared for specialized residential areas (CIE 92)

Situation	Maintained horizontal illuminance		Maintained semicylindrical illuminance
	Average (lx)	Minimum (lx)	Minimum (lx)
High usage	8	4	3
Medium usage	5	2	2
Low usage	3	1	1

Table 10.4 Average maintained illuminances for pedestrian ways (IESNA)

Situation	Minimum average horizontal illuminance (lx)	Average vertical illuminance for pedestrian security (lx)
Roadside sidewalks:		
Commercial	10	22
Intermediate areas	6	11
Residential areas	2	5
Walkways and bikeways distant from roadways	5	5

10.5 Colour of light source

In the UK, the low pressure sodium lamp is commonly used for residential lighting, because of its high efficacy. However, because this source is monochromatic, colours are not rendered. For this reason it is not suitable for those areas where pedestrian activities predominate or where provision for crime risk is a consideration. Instead, in Part 3 of BS 5489,[12] a source emitting substantially 'white' light is recommended. The high pressure sodium lamp would come into this category. CIE 115 makes a similar recommendation.

10.6 Glare

The control of either discomfort or disability glare has not been given the same prominence in residential lighting as it has in main road lighting. This is probably because the visual tasks are not so onerous and, in many cases, direct light from the luminaires is welcomed as providing a cheerful atmosphere and providing optical guidance.

However, there are glare control systems in use that are worthy of description even if their experimental foundations are shaky or non-existent.

In the UK, glare control is achieved by limiting the luminous intensities at 80° and 90° to the downward vertical. The actual restrictions are given in Figure 10.1, page 394. It will be noticed that if the luminaires emit less than 3.5 klm in the lower hemisphere, the luminous intensity limits do not apply. This allows the use of luminaires to provide sparkle.

In the Nordic countries, a formula for Discomfort Glare Rating (*DGR*) is used:

$$DGR = \text{maximum value of } \frac{I}{\sqrt{A}} \quad (\text{cd m}^{-1})$$

where: I is the luminous intensity of the luminaire in candelas; and A is the light emitting part of the luminaire in square metres.

All surfaces are included in the evaluation of A provided that no parts of the light source are visible directly or as broken images. The maximum DGR is found for angles between 80° and 90° to the downward vertical, for a clean luminaire with a lamp emitting its initial luminous flux.

The limiting values are 500 and 1000 for dark and light surrounds respectively. Classes of glare control with higher values of DGR, up to 7000 are proposed by CIE 92.

It is worth recording that the formula $LA^{0.25}$ has been recommended until quite recently by CIE 92. Here, L is the greatest average luminance in the zone between 80° and 90° to the downward vertical from the luminaire. Recommended limiting values vary according to mounting height. One of the problems with the application of this formula is finding the direction of the maximum average luminance.

10.7 Calculation grid

The requirements of Part 3 of BS 5489 apply to the whole of the used surface, which includes the outer edges of the footways as well as the road surface. Points on these outer edges, therefore, have to be included in the search for the minimum illuminance. If the same calculation grid is used for calculating the average and minimum illuminance, points on the edge of the grid have to be given a different weighting from the other points according to the following equation:

$$E_{av} = \frac{\frac{1}{4}E_1 + \frac{1}{2}E_2 + E_3}{1 + \frac{1}{2}C + D}$$

where

- E_1 is the sum of the illuminances at the corner of the grid (lx);
- E_2 is the sum of the illuminances at the boundaries of the grid, excluding the corners (lx);
- E_3 is the sum of the illuminances inside the grid (lx);
- C is the number of points at the boundaries, excluding the corners;
- D is the number of points inside the grid.

The rationale for this equation follows from the fact that if the inside rectangles each have an area of one unit, the corner rectangles will each have an area of one quarter of a unit, and the remaining edge rectangles will each have an area of half a unit.

10.8 Design data

Part 3 of BS 5489 recommends that manufacturers produce a data sheet for the design of installations. This should have the recommended layout (Figure 10.1) to make the comparison of installations easier. Its use for design purposes is tending to be superseded by computer programs, but data sheets are still included in manufacturers' catalogues and are for that reason, described here.

There are three basic elements in the data sheet, which we need to consider in detail. These are the isolux diagram, the utilization factors, and the glare control data.

Luminaire: Resilight

Lamp: 70 W SON-T

Design attitude: Spigot entry elevated 5°

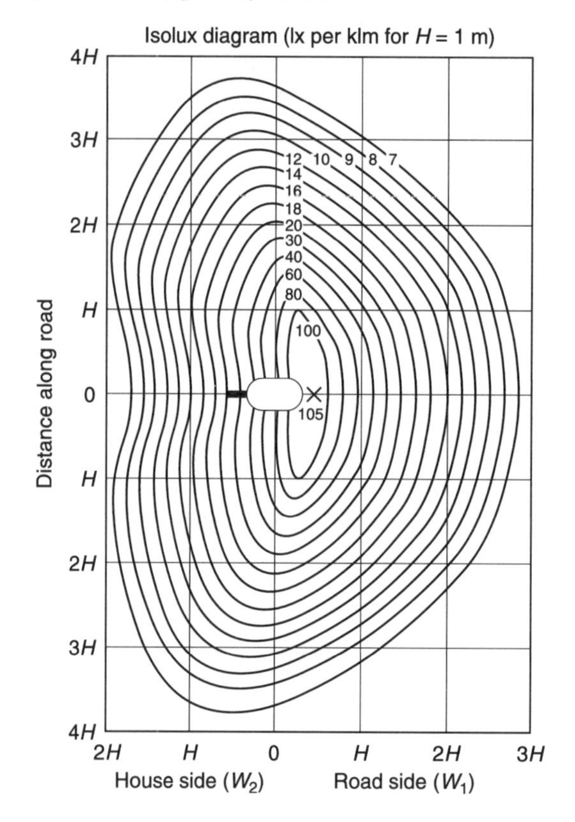

Utilization factors		
W_1/H or W_2/H	Road side U_1	House side U_2
0	0.00	0.00
0.1	0.05	0.04
0.2	0.10	0.09
0.3	0.17	0.11
0.4	0.21	0.14
0.5	0.28	0.16
0.6	0.31	0.17
0.7	0.35	0.19
0.8	0.37	0.21
0.9	0.38	0.22
1.0	0.40	0.23
1.1	0.42	0.25
1.2	0.43	0.26
1.3	0.44	0.27
1.4	0.45	0.28
1.5	0.48	0.29
2.0	0.48	0.32
2.5	0.48	0.32
3.0	0.48	0.32
3.5	0.48	0.32
4.0	0.48	0.32

Glare control data			
Downward light output ratio	**Angle of elevation degrees**	**Maximum intensity**	
		Permitted[1] cd/klm	**Measured** cd/klm
0.85	80	160	145
	90	80	69

[1]No restriction for luminaires emitting less than 3.5 klm in the lower hemisphere

$$\text{Illuminance is given by } \frac{\text{contour valve} \times \text{lamp flux in klm} \times \text{MF}}{(\text{mounting height } H \text{ in m})^2}$$

Fig. 10.1 Design data sheet for residential road luminaire

The isolux diagram. This shows contours joining points of equal illuminance for a horizontal plane under the luminaire (see Section 1.7.5, page 32). Distances are expressed in terms of the mounting height, H. To use it, a plan of the road and the adjacent footpath, if any, drawn to the same scale as the design sheet, is needed. The isolux diagram is placed with its centre (0, 0) over the luminaire from which the illuminance is required. It is then orientated correctly, with the *road side on* the road. The value of the contour passing through the point is then read or estimated if the point lies between two contours. The illuminance can be calculated by using the formula given at the bottom of Figure 10.1. In relation to BS 5489 the isolux diagram is used to find the minimum illuminance to check the design for compliance with the requirements in Table 10.2, page 391. For this purpose the contributions of all the luminaires to the illuminance at a point have to be considered.

The utilization factors (U). These give the fraction of bare lamp luminous flux which reaches infinitely long longitudinal strips. The width of this is expressed in terms of the mounting height. Utilization factors allow the maintained average illuminance over an area to be calculated by use of the following formula:

$$E_{av} = \frac{1000(U_1 + U_2)\Phi MF}{(W_1 + W_2)S}$$

where

- S is the design spacing (m);
- U_1 is the road side utilization factor. This is the fraction of bare lamp flux which reaches the part of the horizontal plane lying between two lines parallel to the road axis, one line being directly under the luminaire, the other being a distance W_1 from it towards the road side;
- U_2 is the house side utilization factor. This is the fraction of bare lamp flux which reaches the part of the horizontal plane lying between two lines parallel to the road axis, one line being directly under the luminaire, the other being a distance W_2 from it towards the house side;
- Φ is the initial luminous flux of the lamp as quoted by the manufacturer (klm);
- MF is the maintenance factor obtained by multiplying the luminaire maintenance factor (LMF) by the lamp lumen maintenance factor ($LLMF$);
- W_1 is the distance between the luminaire and the rear of the far footway (m);
- W_2 is the distance between the luminaire and the rear of the near footway (m).

This equation can be transposed to find the spacing when a specified illuminance is required:

$$S = \frac{1000(U_1 + U_2)\Phi MF}{(W_1 + W_2)E_{av}} \quad (10.1)$$

The glare control data. These allow a check to be made that the light distribution of the luminaire complies with glare control requirements.

The way this data sheet is used is best understood by means of an example.

Worked example A residential road is to be lit to category 3/2 of BS 5489 with the *Resilight* luminaire (data in Figure 10.1). The width of the road surface is 6 m with a 1 m wide footway on either side and the mounting height is 6 m. The environment can be regarded as medium pollution, and the lamp housing of the luminaire has an ingress protection number of IP6. The initial luminous flux of the lamp is 6.3 klm with a lamp lumen maintenance factor ($LLMF$) of

0.92. What spacing is required between luminaires if the arrangement is to be staggered (Section 9.4, page 348).

Solution The solution can be reached in six steps:

Step 1. From Table 10.2, find the relevant requirements. These are an average horizontal illuminance (E_{av}) of 6.0 lx and a minimum (E_{min}) of 2.5 lx.

Step 2. Check that the luminaire meets the glare control requirements. From the data sheet it is obvious that the luminaire meets the requirements. (If there is doubt as to whether a luminaire emits 3.5 klm in the lower hemisphere, this can be found approximately from the utilization factors. The downward light output ratio is approximately equal to the sum of the utilization factors for W_1/H and W_2/H when these are equal to 4, as these nearly represent an infinitely large plane under the luminaire. The sum multiplied by the luminous flux of the lamp is then the flux in the lower hemisphere. However, a caveat should be made. The utilization factors account for the luminous flux falling on an infinitely long plane in the longitudinal direction but only go up to $\tan^{-1} 4$ (or 76°) transversely, so it is possible for some luminous flux not to be included. On the design data sheet shown, the utilization factors appear to converge to limiting values so the assumption made above is probably safe.)

Step 3. Calculate maintenance factor, *MF*. The luminaire *MF* (*LMF*) for an IP6 luminaire, in a medium pollution environment, with a cleaning interval of 18 months is 0.91, from Table 9.7, page 375. The *LMF* is 0.92, obtained from the manufacturer. Hence, the overall *MF* is 0.91 × 0.92 or 0.837.

Step 4. Calculate the spacing which gives E_{av} equal to 6.0 lx. For this we use the utilization factors.

$$\begin{aligned} W_1 &= 6 + 1 \\ &= 7 \\ W_2 &= 1 \end{aligned}$$

so,

$$\begin{aligned} \frac{W_1}{H} &= \frac{7}{6} \\ &= 1.167 \\ \frac{W_2}{H} &= \frac{1}{6} \\ &= 0.167 \end{aligned}$$

By interpolation in the design data sheet (Figure 10.1), we find:

$$\begin{aligned} U_1 &= 0.43 \\ U_2 &= 0.07 \end{aligned}$$

We can now find the spacing to provide an average illuminance of 6 lx:

$$S = \frac{1000(U_1 + U_2)\Phi MF}{(W_1 + W_2)E_{av}}$$

$$= \frac{1000 \times (0.43 + 0.07) \times 6.3 \times 0.837}{(7 + 1) \times 6}$$

$$= 54.9 \text{ m}$$

Step 5. Find the maximum spacing at which the minimum illuminance of 2.5 lx is obtained. For this we use the isolux diagram. Since only two luminaires contribute significantly to a point between luminaires, we need to find the value of the contour that represents half the minimum illuminance, that is 1.25 lx. To do this we rearrange the formula given at the bottom of the design data sheet and substitute in values:

$$\text{contour value} = \frac{\text{illuminance} \times H^2}{\text{lamp flux in klm} \times MF}$$

$$= \frac{1.25 \times 6^2}{6.3 \times 0.837}$$

$$= 8.5 \text{ lx per klm}$$

As shown in Figure 10.2, a plan of the road with the footways is now drawn to the same scale as that of the isolux diagram, which is scaled in terms of H. Two drawings of the interpolated 8.5 lx/klm contour are made on tracing paper on which the position of the luminaire is marked. These are positioned on the plan of the road so that the whole of the road lies within the 8.5 lx/klm contour. When this is done it is found that the spacing is $6.1H$, which, since $H = 6$ m, is equivalent to 36.6 m.

It is possible, although unlikely, that there are points near the intersection of the two contours where the illuminance is less than the required minimum. It is well to check a few points within, say, $0.2H$ of the intersection of the contours to confirm that this does not occur, which is the case in this example. If the minimum illuminance had fallen below the required minimum it would have been necessary to close up the spacing by $0.2H$ and repeat the process.

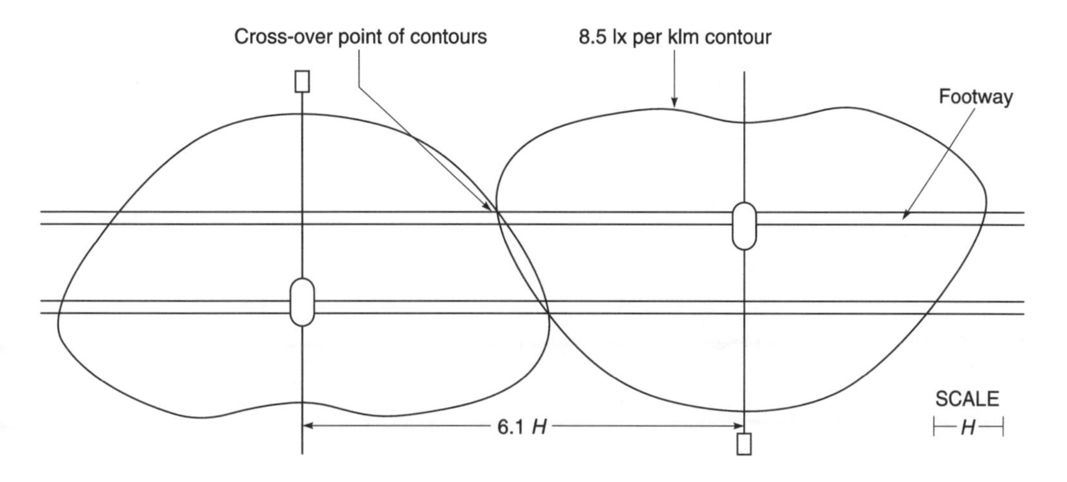

Fig. 10.2 Isolux contours used to find minimum illuminance

Step 6. Determine the final spacing to meet average and minimum illuminance requirements. This is 36 m to meet the requirement for minimum illuminance.

10.9 Derivation of utilization factors

At first sight it might seem possible to calculate the utilization factors by using an illuminance grid. This would, however, have to be infinitely long to collect all the luminous flux for an infinitely long road. To overcome this problem, we consider longitudinal strips provided by the (B, β) system, where the B half-planes intersect the road surface in longitudinal lines, shown in Figure 10.3.

Figure 10.4 shows the angles concerned in more detail. Each longitudinal strip has a width of $0.1H$ and is divided into 2×18 elements, each subtending 5° in the β planes. In the figure these are shown as being rectangular. This is only approximately true: the longitudinal lines are parallel and straight but the transverse lines are parts of parabolas because they are formed by the β angle cones cutting the horizontal plane, which is parallel to their common axis. The fact that the elements are not rectangular does not affect the accuracy of the calculation.

Figure 10.5 shows how the longitudinal strips and elements are portrayed on a web for the (B, β) coordinate system. This is Figure 1.3(c), page 3, viewed along the $(B, \beta) = (90°, 0°)$ axis.

The luminous flux falling on each element is calculated by multiplying the solid angle

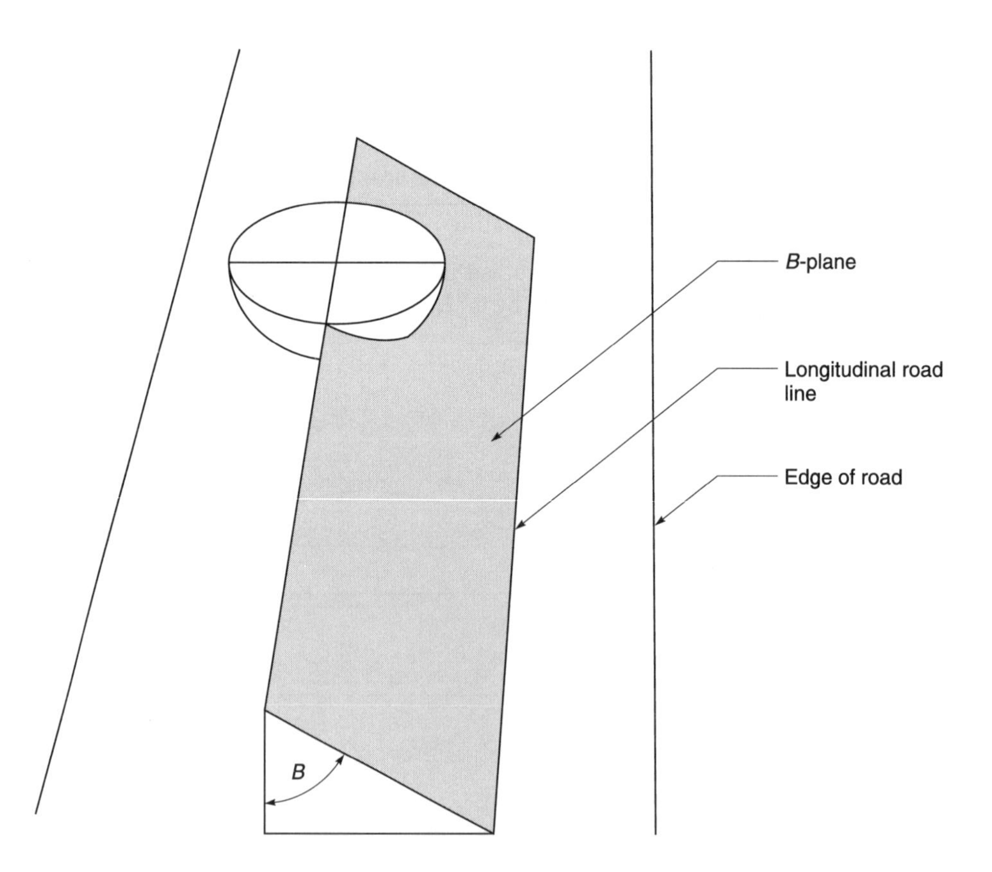

Fig. 10.3 Intersection of B-plane with longitudinal road line

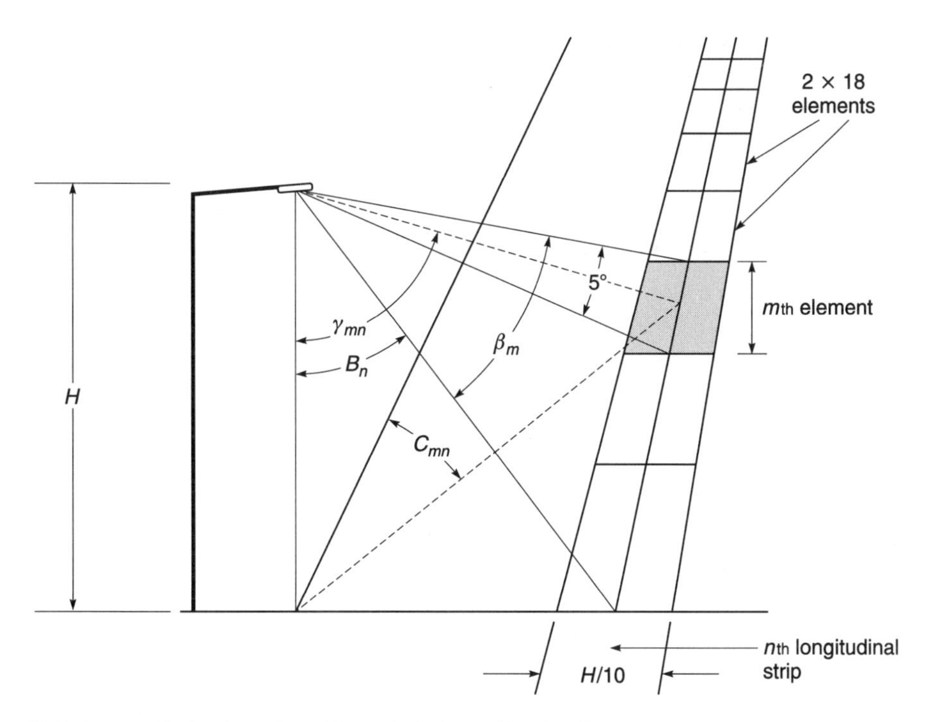

Fig. 10.4 Longitudinal strips of road for calculating utilization factors

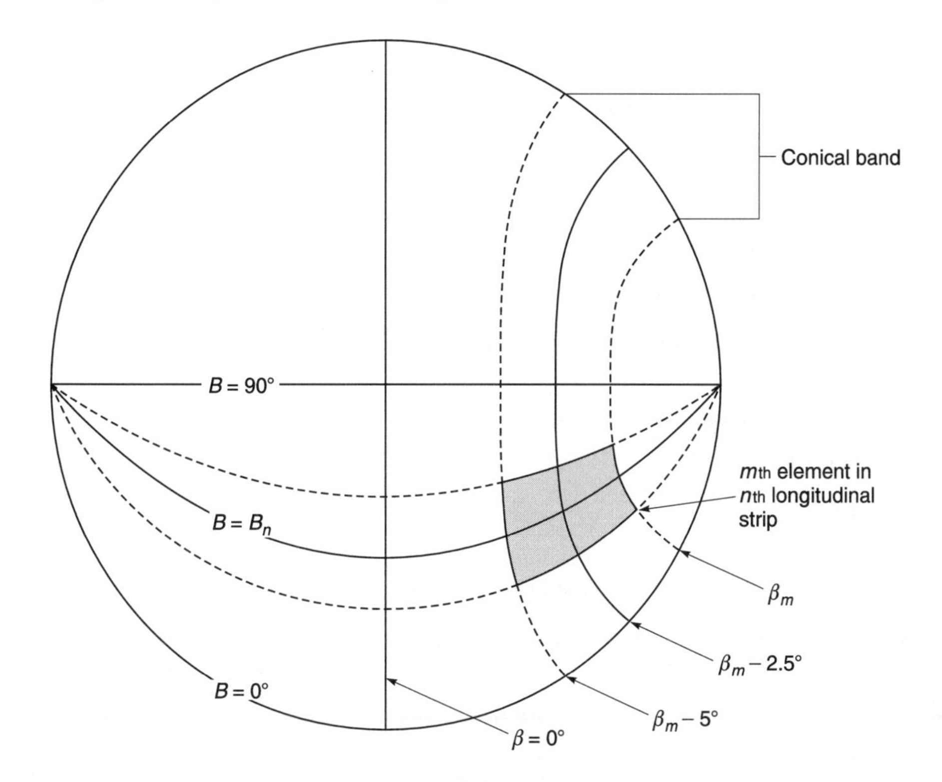

Fig. 10.5 (B, β) coordinate system for the calculation of utilization factors

subtended by each element by the average luminous intensity in the element. This latter is taken as the luminous intensity falling on a point lying on the longitudinal centre-line of the longitudinal strip, and 2.5° from the edge of the strip measured in the β plane, as shown in Figures 10.4 and 10.5. The calculation process can be split down into the following steps.

Step 1. *Solid angle subtended by an element*
For this we can use the formula given in Section 1.7.1, page 25, for the (B, β) system. If ω is the solid angle in steradians between the cones whose half-apex angles are $(\beta_m - 5)$ and (β_m) in degrees of arc, then

$$\begin{aligned}\omega &= 4\pi \sin \tfrac{1}{2}(\beta_2 - \beta_1) \cos \tfrac{1}{2}(\beta_2 + \beta_1) \times \delta B/360 \\ &= 4\pi \sin (2.5) \cos (\beta_m - 2.5) \times \delta B_n/360 \\ &= 2\pi [\sin \beta_m - \sin (\beta_m - 5)] \times \delta B_n/360\end{aligned}$$

where δB_n is the angle in radians subtended by the edges of the nth longitudinal strip. It is calculated from

$$\delta B_n = \tan^{-1} (n/10) - \tan^{-1} [(n - 1)/10]$$

The full equation for the solid angle subtended by an element is then

$$\omega_{mn} = \delta B_n[\sin \beta_m - \sin (\beta_m - 5°)]$$

Step 2. Angle subtended by the centre of an element
The B coordinate of the centre of the nth longitudinal strip is B_n, and the β coordinate of the centre of the mth element is $(5m - 2.5)$, since the width of each element is 5°.

Step 3. Luminous intensity directed to the centre of an element
Since angles in the standard I-table are in (C, γ) coordinates, the coordinates found in Step 2 have to be converted to (C, γ) coordinates by using the formulae given in Section 1.3.3, page 6. These give:

$$C_{mn} = \sin^{-1} \left(\frac{\sin B_n}{\tan \beta_m} \right) \qquad (10.2)$$

$$\gamma_{mn} = \cos^{-1} \{\cos B_n \cos \beta_m\}$$

I_{mn} can now be found from the I-table by using the interpolation procedures given in Section 2.2, page 34.

Step 4. Luminous flux falling on an element
The luminous flux Φ_{mn} falling on the mth element in the nth longitudinal strip is given by

$$\Phi_{mn} = I_{mn}\omega_{mn}$$

Step 5. Luminous flux falling on a longitudinal strip
The luminous flux Φ_n falling on the nth longitudinal strip is given by

$$\Phi_n = 2 \sum_{m=1}^{18} \Phi_{mn}$$

The symbol Σ signifies summation; in this case, over 18 elements. The factor two is introduced here to account for the luminous flux falling down the road as well as up the road.

Step 6. The luminous flux falling on all the longitudinal strips up to the kth
This luminous flux denoted by Ω_k is given by:

$$\Omega_k = \sum_{n=0}^{k} \Phi_n$$

Step 7. Utilization factor
The utilization factor U_k for all the longitudinal strips up to and including the *k*th is then:

$$U_k = \frac{\Omega_k}{\Psi}$$

where Ψ is the bare lamp luminous flux on which the *I*-table is based, usually 1000 lumens.

Problem

1. The *Resilight* luminaire (data in Figure 10.1, page 394) is mounted at 6 m over the kerb of an 8 m wide road with 1 m wide footways. The light output of the lamp is 6.3 klm, *LLMF* is 0.92, and *LMF* is 0.91. What is the maintained illuminance directly under a single luminaire; directly opposite the luminaire, on the edge of the road; and 10 m along the kerb from the vertical line through the luminaire? What spacing is needed between the luminaires in a staggered arrangement for an average maintained illuminance of 6.1 lx to be achieved on the road and footways together?

   Answers: [15.4 lx, 3.1 lx, 1.0 lx, 46 m]

References

1. Hargroves, R. A. (1983) Road lighting. *IEE Proceedings*, **130** Pt A. 420–421.
2. Tien, J. M., O'Donnell, V. F., Barnett, A. and Mirchandani, P. B. (1979) *Phase 1 Report: Street Lighting Projects* (Washington DC: US Government Printing Office).
3. Painter, K. (1989) *Lighting and Crime Prevention: The Edmonton Project* (Middlesex Polytechnic, Centre for Criminology, UK).
4. Painter, K. (1989) *Lighting and Crime Prevention for Community Safety: The Tower Hamlets Study, The First Report* (Middlesex Polytechnic, Centre for Criminology, UK).
5. Lloyd, R. and Wilson, D. (1989) Inner city street lighting and its effects upon crime. *ILE Conference*, Bournemouth.
6. Painter, K. and Farrington, D. P. (in press) Street lighting and crime: Diffusion of benefits in the Stoke-on-Trent project. Chapter to appear in Painter K and Tilley N. *Crime Prevention Studies*, Vol. X (Monsey, NY: Criminal Justice Press).
7. van Bommel, W. J. M. and Caminada, E. (1982) Considerations for the lighting of residential areas for non-motorized traffic. *CIBS Proceedings of National Lighting Conference*. Warwick, UK.
8. Simons, R. H., Hargroves, R. A., Pollard, N. E. and Simpson, M. D. (1987) Lighting criteria for residential roads and areas. *CIE Venice*, Vol. 1, 274.
9. CIE 115–1995. *Recommendations for the Lighting of Roads for Motor and Pedestrian Traffic*.
10. CIE 92 (1992) *Guide to the Lighting of Urban Areas*.
11. (1993) *IES Lighting Handbook*, 8th edition.
12. BS 5489: Part 3 (1992) *Road Lighting: Part 3. Code of Practice for Lighting for Subsidiary Roads and Associated Areas* (amended 1996).

11

Tunnel Lighting

11.1 Introduction

Unlike most other applications, tunnel lighting presents the greatest problems to the designer, not during the hours of night, but during the hours of daylight, and the brighter the daylight is the greater the problem. This is because of what is known as the black hole effect, which exists at the entrance to the tunnel. When the driver approaches the entrance to an unlit tunnel in daylight, he or she sees the opening as a black hole. This is because the eye is adapted to the high luminance of the surrounds of the tunnel entrance (which may include the sky). To make the inside visible the luminance of the first part of the tunnel, known as the threshold zone, has to be a certain fraction of the luminance of the surrounds to the entrance. Based on the work of Schreuder[1] this fraction is given by CIE recommendations the value of 0.1, which means that if the outside luminance is 8000 cd/m^2, as it may well be on a bright day, 800 cd/m^2 will be required in the threshold zone,[a] which is a very high value and costly to achieve. Moreover, when the driver reaches the entrance the eye will take some time to adapt to lower luminances. Hence the high luminances will need to be maintained for some way into the tunnel. The length will depend on the time the eye takes to adapt to low luminances and the speed of the traffic. For this reason, high speed tunnels pose greater lighting problems than low speed tunnels.

At the exit to the tunnel, when the driver emerges into daylight, the problem is ameliorated because the eye adapts much faster from low to high luminances than vice versa.

The outcome of this is that tunnels, particularly those on motorways, may need vast amounts of energy to light them effectively and so the design of the lighting needs careful consideration.

11.2 A diversion: the black hole effect and adaptation level

The black hole effect is important and deserves further explanation.

First, we must introduce the concept of adaptation. It is a fact that when we shine a torch onto a surface at night the surface appears brightly lit (if the torch is working correctly). However, if we do the same thing in sunlight, or for that matter on an overcast day, the extra luminance is not discernible.

To explain this we need to refer once again to the concept of contrast (see Section 9.9, page 366). It was stated that the contrast C of an object of luminance L_o against a uniformly bright

[a] In Japan, a much lower fraction is used; from about 0.02 for 100 km/hr traffic to about 0.008 for 40 km/hr traffic. See Narisada, K. and Yoshikawa, K. (1974) Tunnel entrance lighting – effect of fixation point and other factors on the determination of requirements, *Lighting Research and Technology*, **8**, 9.

background of L_b is given by

$$C = \frac{L_o - L_b}{L_b}$$

We now invoke the Weber–Fechner law, which states that the change in a stimulus that will be just noticeable is a constant ratio of the original stimulus. This 'law' does not hold for the extremes of stimulation and is only approximately true for the stimuli between the extremes, but is, nevertheless, sufficiently accurate for application to tunnel lighting. To apply it to our problem we can regard $L_o - L_b$ as the change in stimulus and L_b as the original stimulus.

For an object to be visible, the contrast C has to be greater than the minimum detectable value or threshold value, which will be constant within the accuracy with which the Weber–Fechner law holds.

In the application of this to tunnel lighting, L_o can be regarded as the luminance of the threshold zone whilst L_b can be regarded as the luminance of the background, also called the adaptation luminance. In addition, we can say that from the equation for contrast,

$$\frac{L_o}{L_b} = C + 1$$

$$= k$$

where k is a new constant.

Hence, for objects to be just visible in the threshold zone they must have a luminance that is higher than a certain proportion k of the adaptation or background luminance. In tunnel lighting, k is chosen so that there is a reasonable certainty of the object being visible.

11.3 Zones of the tunnel

For ease of reference and for design purposes, the tunnel is divided into the four zones illustrated schematically in Figure 11.1. The lighting of these four zones will be discussed separately later in this chapter. In addition to these four internal zones, outside the tunnel there is the access zone, which the driver sees before entering the tunnel.

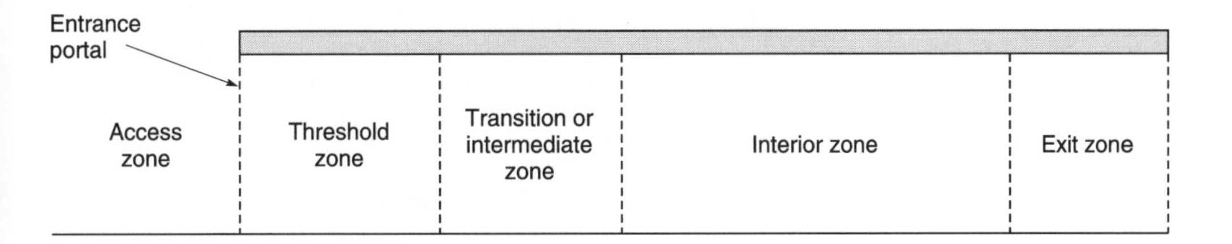

11.4 Types of lighting

Two types of lighting are used – symmetrical and counterbeam (Figure 11.2).

In symmetrical lighting, beams of light are directed with and against the traffic flow, whereas in counterbeam lighting there is only one beam, which is directed against the traffic flow. It follows that symmetrical lighting lights both sides of an object, whereas with counterbeam lighting the side

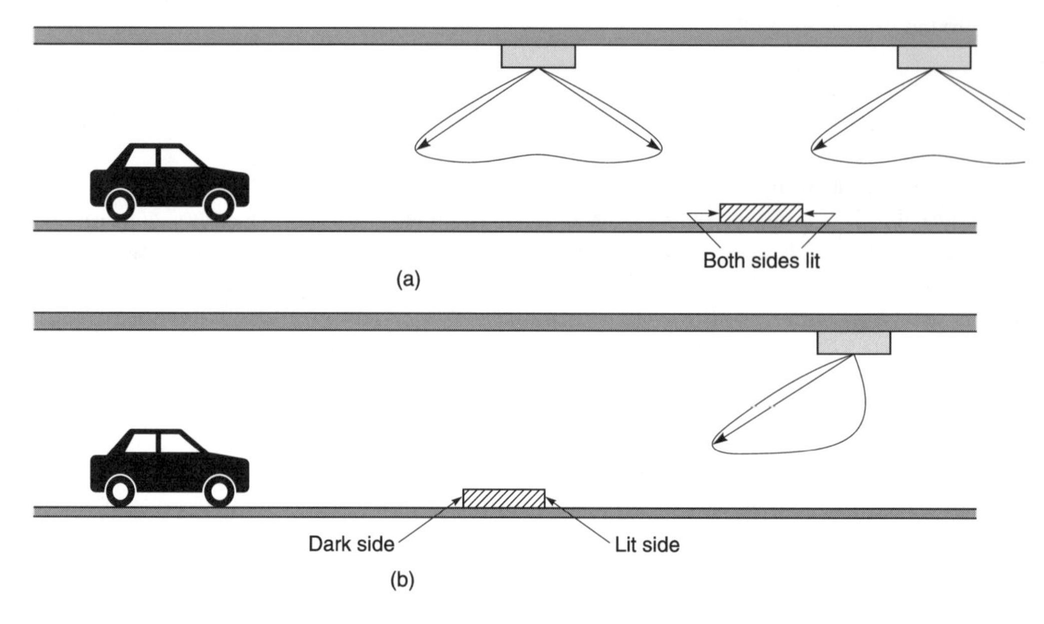

Fig. 11.2 Symmetrical and counterbeam lighting. (a) symmetrical lighting; (b) counterbeam lighting

facing the driver is unlit so the object is seen in silhouette. This increases the contrast between the object and the road, so that less light is needed for an object to be discriminated. However, the objection is often raised that a small vehicle behind a high vehicle may be rendered invisible. Nevertheless, counterbeam lighting, which originated in Switzerland,[2] has been used successfully in a number of countries. The degree to which the light is concentrated in one direction varies between designs.

11.5 Classification of tunnels

Tunnels are classified into two broad groups: long and short. In general, a tunnel is regarded as short if the approaching driver can see through to the end of the tunnel. The lighting requirements for these two types are different.

Long tunnels are further divided into seven classes depending on traffic flow rate, traffic type, optical guidance, and degree of driving comfort. It should be noted that the higher the class number the more onerous are the lighting requirements, which is the reverse of the numbering system adopted for traffic routes (Section 9.12, page 370) and for residential lighting (Section 10.4, page 390).

11.6 Lighting of the entrance to the threshold zone

The starting point for determining the threshold zone luminance is the need to know the adaptation luminance of the driver approaching the tunnel portal. This depends on the luminances in a 20° cone of vision (2 × 10° angle measured from the axis of the cone), the axis of which is aimed at the centre of the entrance portal. The average of these luminances, weighted in a way to be discussed, gives the threshold zone luminance, symbol L_{20}. To determine L_{20} the concept of *stopping distance* (*SD*) has to be introduced. This is the distance required to bring a vehicle to a halt when it is being driven at the design speed for the tunnel. The values are available in national regulations. For instance, the

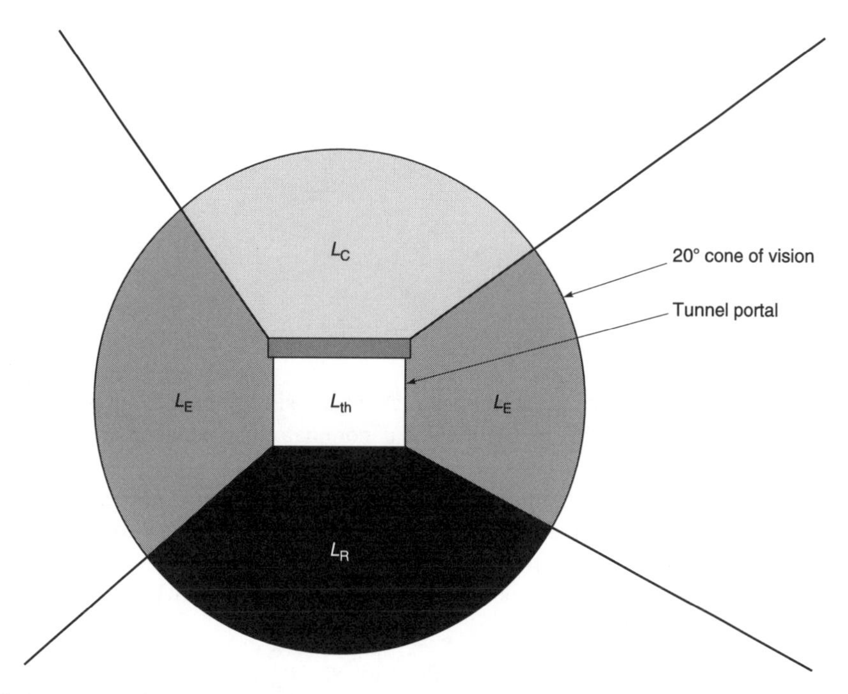

Fig. 11.3 Schematic of the various zones used for determining L_{20}

British Standard BS 5489: Part 7: 1992[3] recommends, for design speeds of 120 km/h and 50 km/h, the most extreme values given, that the *SD* should be 215 m and 50 m.[b] In most circumstances this means that more of the sky will be visible at higher design speeds, which will increase L_{20} so that a higher lighting level will be needed in the threshold zone.

Having determined the stopping distance *SD*, we now need to determine L_{20}. There are three methods available.

(1) This is the most accurate method and assumes that a photograph or drawing is available of the view of the tunnel entrance taken from a distance from the portal that is equal to the stopping distance. The photograph is used to determine the relative projected areas of sky, road, surrounds and entrance portal, as seen in the 20° cone (Figure 11.3). Then

$$L_{20} = \gamma L_C + \rho L_R + \varepsilon L_E + \tau L_{th}$$

where

L_{20} is the access zone luminance (cd/m^2),
L_C is the sky luminance (cd/m^2),
L_R is the road luminance (cd/m^2),
L_E is the surrounding luminance (cd/m^2),
L_{th} is the threshold luminance (cd/m^2),
γ, ρ, ε and τ are the fractions of the circular area being considered occupied by the sky, road, surroundings, and portal respectively.

[b] In fact these figures are for the stopping sight distance (*SSD*), which includes the time for the driver to perceive and to react to an object. They are quoted here because BS 5489: Part 7: 1992 does not give *SD*, as it is based on CIE 61 (1984) and not CIE 88 (1990). *SD* would be shorter than *SSD*.

At this stage L_{th} is not known. Let it be a fraction k of L_{20}, then

$$L_{20} = \gamma L_C + \rho L_R + \varepsilon L_E + \tau[kL_{20}]$$

and by rearrangement of the terms, we obtain

$$L_{20} = \frac{\gamma L_C + \rho L_R + \varepsilon L_E}{1 - k\tau}$$

Also, it will be noted that

$$\gamma + \rho + \varepsilon + \tau = 1$$

so only three of the four fractions need be known. A further simplification can be made since k, in practice, never exceeds 0.1. This means that the denominator can be taken as unity for practical purposes.

Although the above mathematical procedures may seem straightforward there are difficulties in their application. Maximum values of L_C, L_R and L_E are required for insertion into the formula so that the maximum value of L_{20} can be found. These may not occur at the same time of the year. Often the absolute maximum is not used but the maximum that covers 80% of the service time of the tunnel is used instead.

(2) As an alternative to the above, the CEN draft gives a tabular method which, however, is only approximate. The operative value of L_{20} is selected from the table on the basis of the percentage sky visible, whether snow is likely to be present, the stopping distance, and tunnel orientation.

(3) A second approximate method is to compare the access zone geometry with that of a situation for which L_{20} is known. The CEN draft and CIE 88 give a number of sketches that can be used for this purpose.

11.7 Lighting within the threshold zone

As already intimated

$$L_{th} = k \times L_{20}$$

L_{20} has been determined in the previous section so it remains to find k. The value of this constant depends on the lighting class of the tunnel, whether counterbeam or symmetrical lighting is used, and the stopping distance. Table 11.1 shows some typical values for k taking into account these variables.

Table 11.1 *k* values

Lighting Class	Counterbeam lighting		Symmetrical lighting	
	Stopping distance *SD* (m)		Stopping distance *SD* (m)	
	60	160	60	160
7	0.040	0.070	0.050	0.100
5	0.030	0.055	0.035	0.065
3	0.020	0.040	0.025	0.045

Table 11.2 Average road surface luminance for the interior of long tunnels (cd/m^2)

Lighting class	Stopping distance (m)	
	60	160
7	3	10
5	2	6
3	1.5	4

The reader will note the advantage that counterbeam lighting gives. However, which system is more economical has to be assessed on the basis of system efficiency, not solely on the basis of k values, since the counterbeam luminaires may not make as effective use of the available light energy as the symmetrical luminaires.

The threshold zone continues into the tunnel for at least the stopping distance, and for the first half of the zone the threshold L_{th} should be maintained for at least half its length. Thereafter, it is decreased to 0.4 L_{th} in such a way that the luminance of the road does not fall below the value that would be obtained by linear interpolation from L_{th} to 0.4 L_{th} over the second half of the threshold zone. This means that the luminance may be decreased in steps, which would occur in practice since the lighting is provided in stepped packages of luminous flux.

The luminance of the walls of the tunnel should be high enough to reveal objects seen against them. For the most important tunnel lighting classes the luminance of the walls should be similar to that of the road.

11.8 Lighting of the interior zone

The lighting of the transition zone depends on the luminance used in the interior zone, so we have to consider this zone first. Table 11.2 gives an indication of the road luminances used for a variety of lighting classes and stopping distances. The values are minimum maintained.

In UK practice, as stipulated in BS 5489: Part 7, there are only three lighting classes, which are mainly determined by the speed limit. For motorways, where the speed limit is more than 110 km/hr the average maintained road luminance is 10 cd/m^2, for speed limits between 80 and 100 km/hr it is 5 cd/m^2, and for speed limits between 50 to 70 cd/m^2, it is 3 cd/m^2.

11.9 Lighting of the transition zone

We are now in a position to reduce the light level from the end of the threshold zone to the interior zone. This has to be graded so that the adaptation of the eye can keep pace with the reduction of luminance. To enable this to be done CIE 88 and the CEN draft[4] recommend use of the formula:

$$L_{tr} = L_{th}(1.9 + t)^{-1.423}$$

where

L_{tr} is the luminance in the transition zone at a position after the vehicle has travelled for t seconds (cd/m^2);
L_{th} is the luminance at the end of the threshold zone (cd/m^2);
t is in seconds for a vehicle to travel a given distance from the end of the threshold zone.

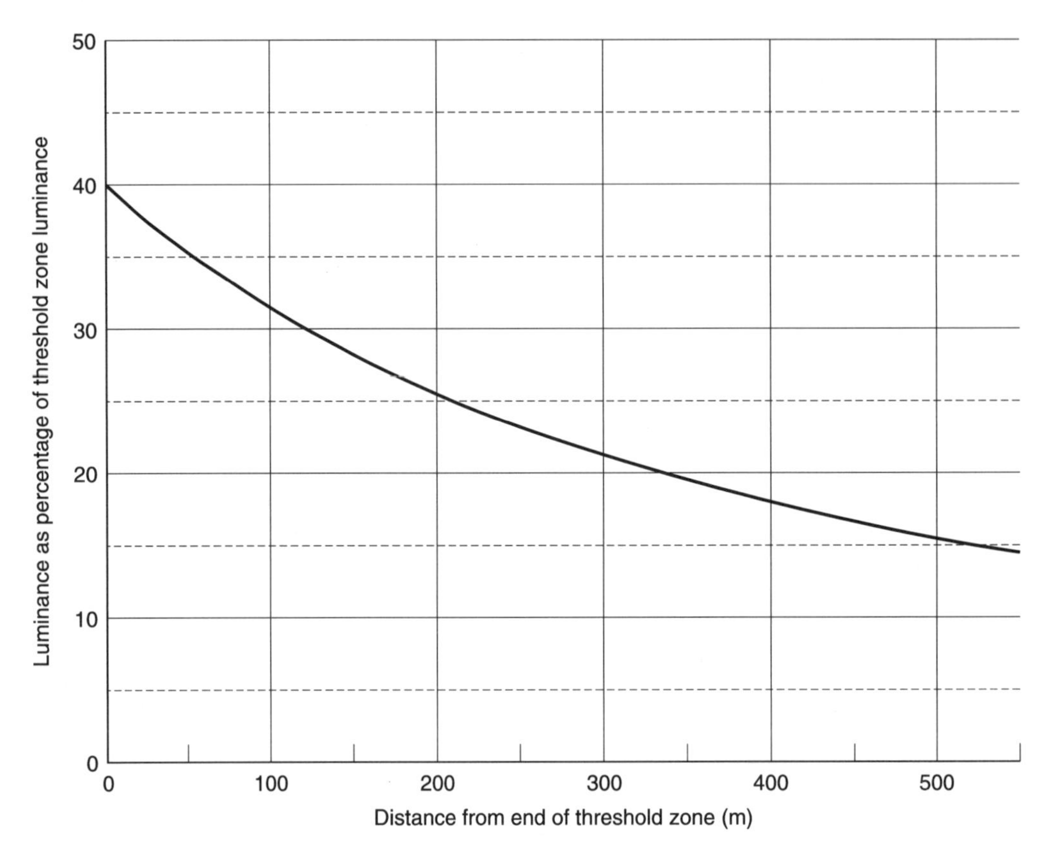

Fig. 11.4 Maximum permissible reduction of road luminance with distance in the transition zone for a vehicle speed of 110 km/hr

This equation is formulated in such a way that when t is zero L_{tr} is 40% of L_{th}, as is required in Section 11.7. Moreover, since t is raised to a negative power, higher speeds require higher values of L_{tr}, as would be expected.

Figure 11.4 is a plot of this equation for a speed of 110 km per hour. In practice it is not possible to grade the lighting to obtain a smooth reduction of the luminance of the road. Instead, the lighting is reduced in steps, but these must be smaller than 3:1.

11.10 Lighting of the exit zone

The adaptation of the eye to increasing luminance takes place so quickly that an increase in luminance is not required for good seeing capability outside the tunnel as the driver approaches the exit in daylight. However, in the highest lighting classes, some increase – maybe as much as five times – in luminance is helpful for revealing small vehicles behind large ones, which may otherwise be rendered inconspicuous because of the glaring effect of the exit in daylight. It also helps with vision in the rear mirror. Another consideration is that if there is the likelihood that the tunnel will be used for traffic travelling in the reverse direction, during periods of maintenance for instance, provision should be made for obtaining the high levels of luminance needed for threshold zone and transition zone lighting.

At night it is recommended that there is a gradual grading off of the lighting if the open road is unlit.

11.11 Other requirements

It is important that the walls are well lighted up to a certain height, 2 m usually being recommended, to help with discrimination of objects in the tunnel. Most recommendations state that the average luminance of the walls should be equal to or greater than the luminance of the road.

The uniformity of luminance of the road is controlled by overall uniformity U_o and longitudinal uniformity U_l as for traffic routes (see Section 9.6, page 352).

Disability glare is controlled in terms of threshold increment *TI* described in Section 9.9, page 366. A variation of the formula quoted in that section is used,

$$TI = 95 \frac{L_v}{L^{0.8}} \quad \text{if} \quad L > 5 \text{ cd/m}^2$$

where L_v is the veiling luminance produced by the initial output of the luminaires visible in a $2 \times 20°$ conical field of view. This is equal to the illuminance produced by these luminaires on an infinitesimal plane, at the driver's eye, which has its normal directed 1° down from the horizontal and parallel to the road axis; and *L* is the luminance of the road for a new installation.

Where discontinuous lines of luminaires are used, unpleasant flicker may result. To avoid this flicker frequencies between 2.5 per second and 15 per second should be avoided except over short lengths.

11.12 Reduction of access zone luminance by screens

If the maximum access zone luminance can be reduced then there can be a reduction in threshold luminance, which can produce a worthwhile economic saving.

Obviously, if the tunnel can be made to run in an east–west direction rather than north–south, the entrance portals will be spared the sun at the brightest part of the day. But this is probably an impracticable suggestion as geographical and other factors dictate the direction in which the tunnel should run.

In some tunnels, daylight screens have been used. These take the form of translucent materials and form a canopy over what would otherwise be the access zone. On the face of it this seems an attractive idea to reduce the lighting needed. However, they are difficult to maintain in terms of their light transmission. They may also drip water onto the road, which may constitute a hazard. When snow falls the transmission of daylight may be reduced, and when it melts it may drip onto the road, and possibly freeze.

Screens, such as louvres, which allow some sunlight to fall on the road are definitely not recommended. This is because they may produce disturbing patterns of light on the road and uncomfortable flicker, as well as having the drawbacks stated for the daylight screens.

Dark walls leading into the access zone and the planting of trees to raise the skyline are helpful.

11.13 Variation of lighting levels with daylight levels

There is a need to vary the lighting levels in the tunnel in accordance with the daylight levels. To do this the average luminance of the access zone is monitored with luminance meters, and variation of the lighting levels in the tunnel is effected either by dimming the lamps, if they are fluorescent, or turning some out. In the latter case, care has to be taken that undue flicker effects do not result.

11.14 Short tunnels

For lighting purposes a tunnel is regarded as short if it is less than 100 m in length and the exit is visible from some distance before the entrance. In these circumstances it may be unnecessary to light the tunnel, providing the daylight penetration is sufficient for the silhouette effect to operate. However, it is possible that the tunnel produces, in effect, a dark frame against which vehicles, cyclists, or pedestrians may be difficult to see. White finishes on the walls are a help in diminishing this effect and some authorities advocate a light-well half-way along the length of the tunnel.

Where lighting is required, the road is lit to a uniform level equal to three times the luminance that would be used in the interior zone of a long tunnel (see Table 11.2, page 407) or to 15 cd/m^2, whichever is the greater.

Bibliography

van Bommel, W. J. M and de Boer, J. B. (1980) *Road lighting.* Deventer, Philips Technical Library.

Schreuder, D. A. (1998) *Road Lighting for Safety* (Thomas Telford Publishing).

References

1. Schreuder, D. A. (1971) Tunnel entrance lighting – a comparison of recommended practice. *Lighting Research and Technology*, **3**, 274.
2. Mäder, F. (1966) Verwendung von besonderen asymmetrischen Leuchten (Schrägstrahlern) zur Beleuchtung von Tunneln. (Application of special asymmetrical luminaires for the lighting of tunnels) *Bull. SEV (Switzerland)*, **60**, 359.
3. BS 5489: Part 7: 1992. *Road lighting. Part 7. Code of practice for the lighting of tunnels and underpasses.*
4. CEN/TC 169 N195 (1997) *Tunnel Lighting* (CEN draft).

12
Floodlighting

12.1 Introduction

By *floodlighting* is meant the lighting of large areas by one or a number of luminaires, the floodlights, which may all have to be individually aimed. The areas so lit may be sports fields, façades of buildings, interiors such as hangers or sports halls, and so on. The luminaires themselves may have beams of very narrow angle, of no more than 5° from the aiming direction, or beams of wide angle, as large as 60°. A number of different calculation methods are employed to meet the illuminance and other requirements. To illustrate these we have chosen two common applications: floodlighting for sports and floodlighting for buildings.

12.2 Floodlighting for sports

The first step in the design of a floodlighting scheme for sports lighting is to decide on the appropriate lighting criteria. Then it is necessary to decide on the general design philosophy for meeting these – height and position of towers or whether a stand roof can be used – choice of light source and luminaire or luminaires, amongst other factors. Conventional photometric data in the form of the *floodlighting diagram* can be of assistance at this stage, especially in making the right choice of luminaire. The next stage is to work out a training scheme for aiming the floodlights. The detailed lighting performance can then be calculated by computer and, if necessary, alterations made to improve the performance. These steps may have to be repeated until a satisfactory result is obtained. Usually, human intervention is needed to judge at the end of every iteration whether the result is an improvement on the last iteration and whether it is acceptable. In practice, the process is not usually as daunting as it might first appear because the designer can draw on previous experience with similar schemes to provide a good starting point. Automatic optimization programs have been written[1] and are in use by some manufacturers.

12.3 Design criteria

The main criteria are horizontal illuminance level and uniformity, vertical illuminance level and uniformity, and glare restriction.

12.3.1 HORIZONTAL ILLUMINANCE

This is generally calculated for ground level. It varies from about 50 lux to 2000 lux depending on the sport and the level of competition; that is, whether it is international, national, regional, local, training, or school sport. Similar criteria are used to decide the illuminance uniformity in terms of minimum to average illuminance, which varies from 0.5 to 0.8. Detailed recommendations are given in CIE[2–6] and CEN[7] publications.

12.3.2 VERTICAL ILLUMINANCE

Where the sport is concerned with vertical surfaces, the illuminance on these surfaces is specified. For instance, for archery, the vertical illuminance specified is 1000 lux for a range of 25 metres and 2000 lux for 50 metres. The recommended uniformity over the target is 0.8.

In addition, vertical illuminance is an important consideration for colour television and filming.[8] Here, levels depend on speed of action and shooting distance – the distance of the camera from the action. For the fastest action and the longest shooting distances, illuminances as high as 1500 lux may be required. Where slow motion recordings are to be made, higher levels than this will be required.

To ensure that the spectators are visible, the vertical illuminance on them should be at least one quarter of the average vertical illuminance on the playing area.

The requirements for uniformity are somewhat complex.

- For vertical planes facing a side-line bordering the main camera area

$$\frac{E_{\mathrm{v\,min}}}{E_{\mathrm{v\,max}}} \geq 0.4$$

  where $E_{\mathrm{v\,min}}$ is the minimum vertical illuminance, and $E_{\mathrm{v\,max}}$ is the maximum vertical illuminance.
- For vertical planes at a calculation grid point (see Section 12.6.2, page 431), these being on the vertical sides of a cube with one side facing a side-line bordering the main camera area (Figure 12.1(a))

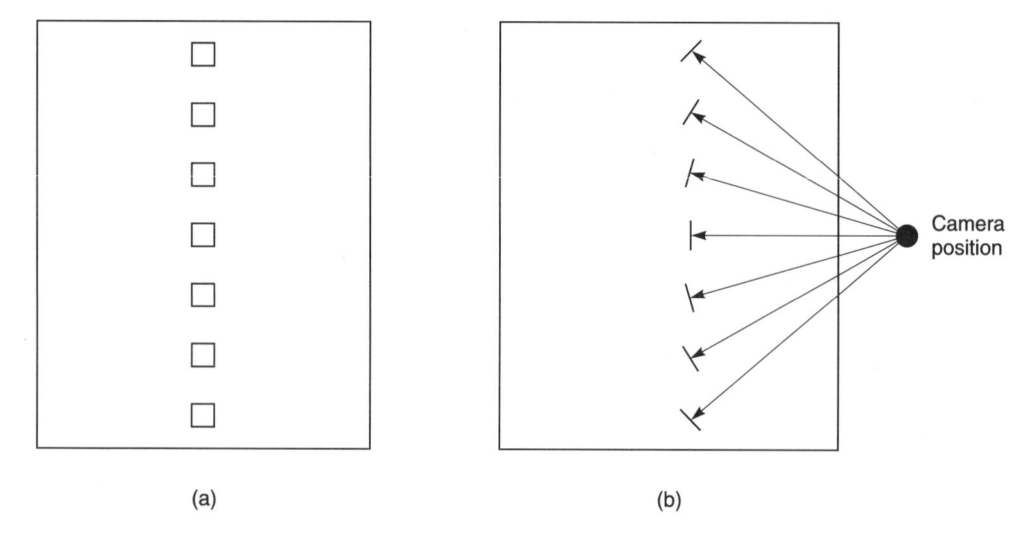

Fig. 12.1 Orientation of vertical planes for calculation and measurement of vertical illuminance

$$\frac{E_{\text{v min}}}{E_{\text{v max}}} \geq 0.3$$

where the symbols have the same meaning as before.

The above requirements apply when the camera may be positioned anywhere outside the border of the playing area. When there is just one camera position, from the centre of one side of the playing area in Figure 12.1(b), only the vertical plane facing the camera need be considered.

- The balance between the average horizontal illuminance $E_{\text{h av}}$ and the average vertical illuminance $E_{\text{v av}}$ on planes facing a side-line bordering the main camera area should be such that

$$0.5 \leq \frac{E_{\text{h min}}}{E_{\text{v av}}} \leq 2$$

- The uniformity of horizontal illuminance should be such that

$$\frac{E_{\text{h min}}}{E_{\text{h max}}} \geq 0.5$$

where $E_{\text{h min}}$ is the minimum horizontal illuminance, and $E_{\text{h max}}$ is the maximum horizontal illuminance.

Whilst we are detailing the requirements for colour television we should mention that the colour temperature of the sources should lie between 4000 K and 6500 K when there is a significant daylight contribution. Otherwise the lower limit can be reduced to 3000 K.

12.3.3 CONTROL OF GLARE

Control of glare in sports lighting has presented a number of problems, not least because of the almost unlimited viewing directions that have to be taken into account, mainly for the players but also for the spectators. Computers have come to the rescue here because they make the calculation of glare ratings in any number of desired directions and at any number of points a practicable possibility. The glare rating method now generally accepted is the CIE *GR* method,[9] which is applicable to floodlighting generally as well as to sports lighting. This is based on the work of van Bommel, Tekelenburg and Fischer,[10] but their formula has been modified so that the higher the rating the more the glare sensation to bring it into harmony with other glare systems. It makes no distinction between discomfort and disability glare.

The CIE formula is

$$GR = 27 + 24 \log \left(\frac{L_{\text{vi}}}{L_{\text{ve}}^{\ 0.9}} \right)$$

where

L_{vi} is the veiling luminance (cd/m^2) produced by the luminaires and is given by

$$L_{\text{vi}} = 10 \sum_{i=1}^{n} \frac{E_{\text{eye } i}}{\theta_i^2}$$

L_{ve} is the veiling luminance (cd/m^2) produced by the environment or background and is given by

$$L_{ve} = \frac{0.035 \times E_{hor\ av} \times \rho}{\pi}$$

or more simply

where

$$L_{ve} = 0.01114 \times E_{hor\ av} \times \rho$$

$E_{eye\ i}$ is the illuminance (lux) in a plane at the observer's eye, normal to the line of sight, given by the *i*th luminaire;
θ_i is the angle between the line of sight and the direction of the *i*th luminaire (degrees);
n is the number of luminaires;
$E_{hor\ av}$ is the average illuminance on the horizontal (lx);
ρ is the reflectance of the horizontal surface, which is regarded as a uniform diffuser

We should pause for a moment and note the family resemblance of this formula to that used for calculation of Threshold Increment in road lighting given in Section 9.9, page 366.

The part used for veiling luminance is the same, based as it is on the Stiles–Holladay formula. In both formulae this is divided by a term for the background luminance, which for road lighting is

$$L_{av}^{0.8}$$

where L_{av} is the luminance of the road surface. For floodlighting it is

$$(0.035 \times L_{av})^{0.9}$$

where, in this formula, L_{av} is the luminance of the horizontal playing surface. The expression for L_{ve} follows from the formula

$$L_{av} = \frac{E_{hor\ av} \times \rho}{\pi}$$

in which it is assumed that the horizontal playing surface is a uniform diffuser of reflectance ρ.

The subjective impression of the glare produced by the installation can be graded on the five-point scale as shown in Table 12.1.

To apply the formula, directions of view have to be decided on as well as observer positions. The application of the formula is restricted to angles of elevation at or below the horizontal, usually at the horizontal because, at this elevation, the maximum *GR* will occur. In azimuth, the

Table 12.1 *GR* assessment scale

Subject Impression of glare	*GR* limits
Unbearable	Greater than 80
Disturbing	Less than 80
Just admissible	Less than 60
Noticeable	Less than 40
Unnoticeable	Less than 20

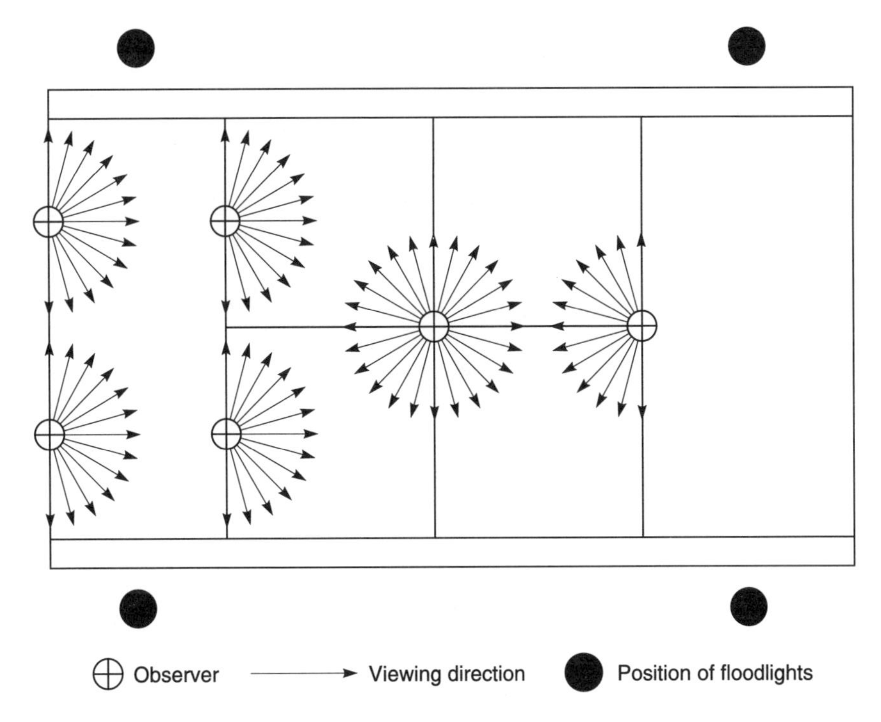

Fig. 12.2 Schematic showing viewing directions for calculating *GR* for a tennis court

GR is evaluated for a range of directions relevant to the lighting application being considered. Figure 12.2 shows a schematic indicating viewing directions for a tennis court. In practice, the observer positions would be taken over a grid covering the tennis court.

In sports lighting, the upper limits for *GR* are 50 for National and International competitions, and 55 for other sports activities, including coaching.

12.4 Training plan

A systematic approach to working out a training plan has been provided by Dovey *et al.*[11] and is illustrated in Figure 12.3. The playing area is lit by four towers *P*, *Q*, *R* and *S* on each of which there are a number of floodlights. Consider the floodlights on tower *S*. Floodlights with narrow beams are directed in an arc furthest from the tower, then come the floodlights with medium spread beams directed further in, with the wide beam flood directed in an arc nearest to the tower.

The angle of aim of the floodlights with narrow beam spread will depend, amongst other things, on the level of glare that can be tolerated by the spectators and also the goal keeper. The spacing of the beams along the arc is determined by superimposing plots of the light distribution of the floodlights, and displacing them sideways until the desired luminous intensity is achieved by trial and error as indicated in Figure 12.4. Care has to be taken that uniformity of illuminance requirements are met. A similar procedure can be adopted for finding the spacing between arcs, but the graphs have to be plotted in terms of illuminance rather than luminous intensity since the aiming angle is being varied.

Allowance has to be made for overlapping of the beams from the four towers. For instance,

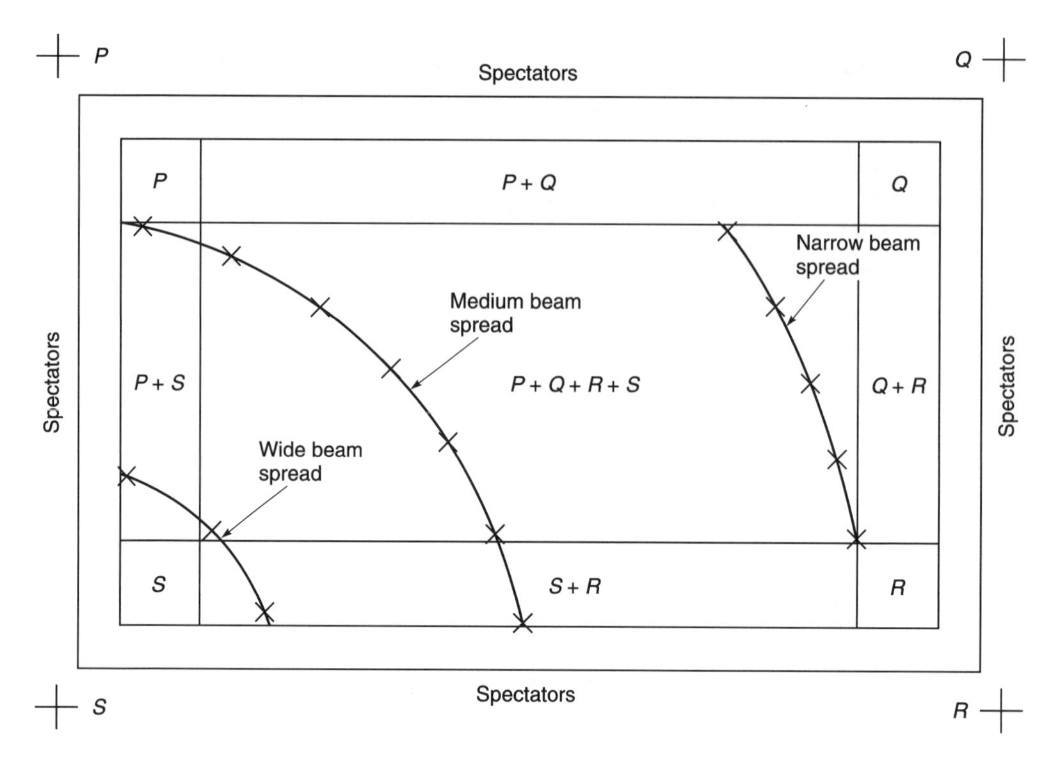

Fig. 12.3 Training plan for tower *S*, and the areas lit from towers *P*, *Q*, *R* and *S*

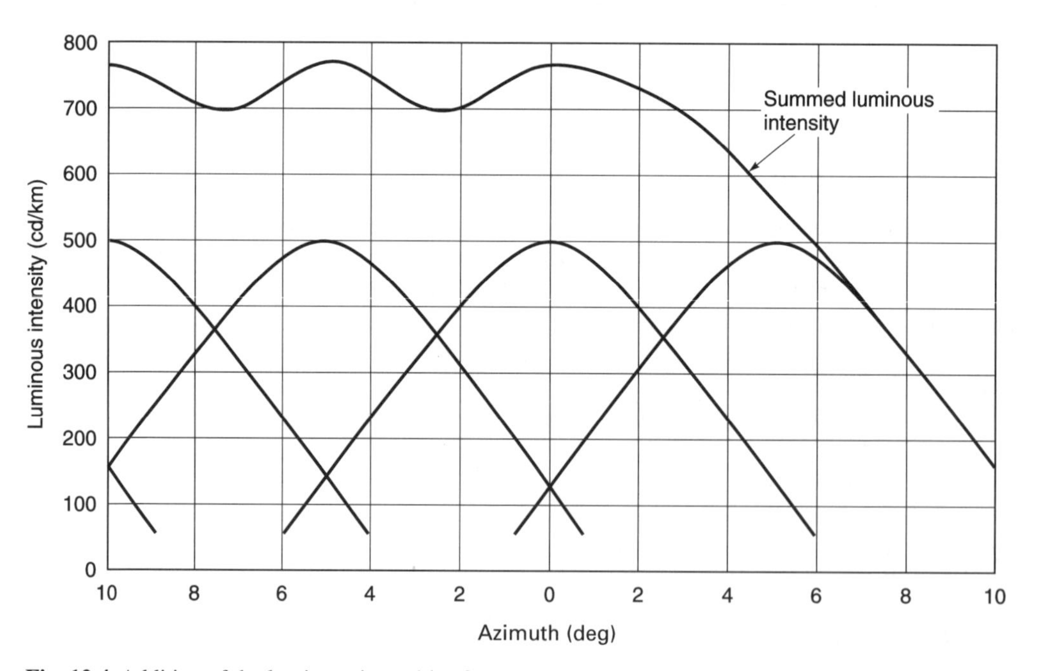

Fig. 12.4 Addition of the luminous intensities from a number of beams

in Figure 12.3 all four towers light the central area (*P*, *Q*, *R* and *S*), whereas only one tower lights each of the corner rectangles (*P*, *Q*, *R* or *S*), and two towers light each of the remaining rectangles (*P* and *Q*, *Q* and *R*, *S* and *R* or *P* and *S*).

A computer method of aiming optimization using a linear programming optimization has been described by Brackett.[12]

12.5 Floodlighting diagram

Figure 12.5 shows the data needed for the manual calculation of schemes. Although a computer program is invariably used for calculating the full photometric data needed for assessing the performance of a scheme, the floodlighting diagram is useful – or even indispensable – for selecting the floodlighting luminaires and determining the training plan.

The data sheet consists of two main elements, the beam data and the floodlighting diagram.

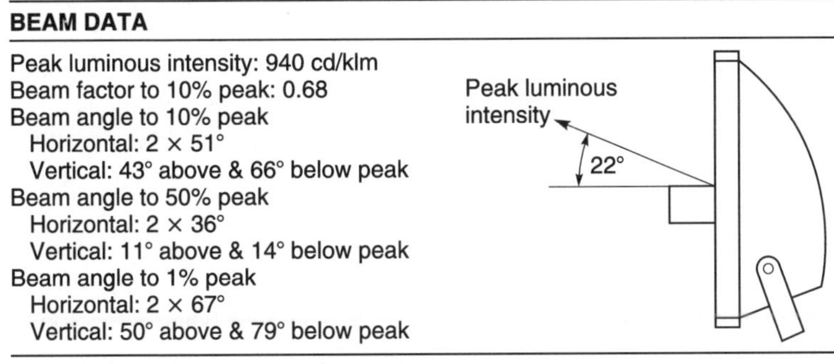

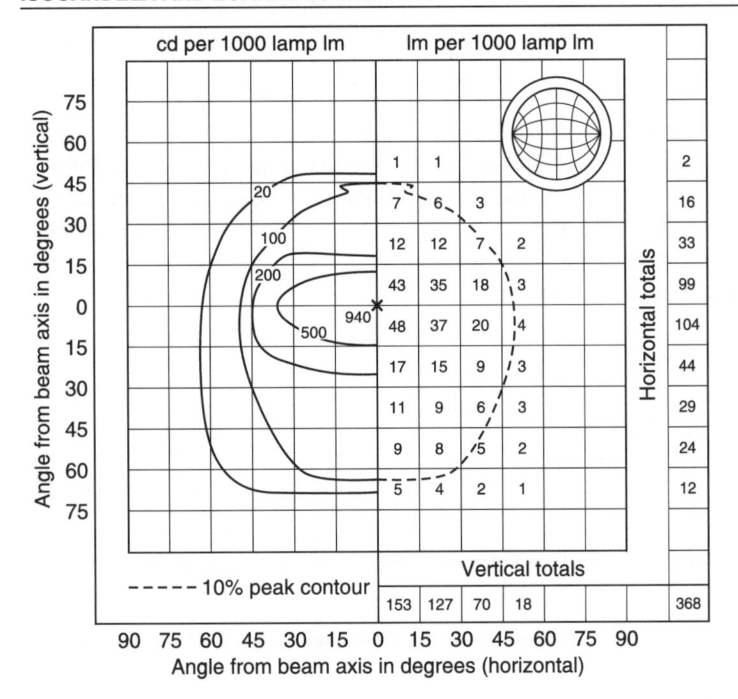

Fig. 12.5 Floodlighting diagram for *SONPAK* floodlight (by courtesy of Thorn Lighting Limited)

The beam data help in the selection of the floodlight. The peak luminous intensity will indicate whether the desired illuminance can be achieved at the salient points. The beam factor enables the luminous flux in the beam to be determined by multiplying by the light output of the bare lamp. The beam widths to various percentages of the peak luminous intensity are also useful indicators of the suitability of the floodlight. In practice, the beam angle to 1% of the peak luminous intensity gives the useful angular extent of the floodlight, as light outside this will not add significantly to the illuminance, but may cause glare or atmospheric pollution.

The coordinate system used for the floodlighting diagram itself is generally the (B, β) and not the (C, γ) used in most other applications (see Section 1.2, page 2). To remind the reader of this, a thumbnail sketch is shown of this coordinate system on the figure. The choice eases the calculations as will be apparent later, but a word of caution is needed. In at least one country, a floodlighting diagram similar to that shown in the figure, but on the (C, γ) system of coordinates, is used. It is conventional to label the B angles as vertical angles (V) and the β angles as horizontal angles (H), which is more descriptive, and the system is sometimes referred to as the (V, H) system, especially in IESNA practice (see Section 2.6.2, page 60). The direction of the peak luminous intensity is taken as the origin (0°, 0°) of the system. The direction of the peak luminous intensity with respect to a mechanical feature of the floodlight, such as its mouth or the front glass, is shown on a diagram to permit the floodlight to be aimed when the scheme is set up in the field.

The left-hand side of the figure is an isocandela plot, which enables the illuminance to be found at any point, although the accuracy with which this can be done depends on the closeness of the contours and how large the diagram is.

The right-hand side of the diagram gives the luminous flux enclosed by B planes and β cones. The reader should notice that the total luminous flux emitted by the floodlight is 2×368 or 736 lumens per 1000 lamp lumens. In practice, as will be apparent when we come to the worked example, the fraction of bare lamp luminous flux that is utilized on the lit area is much less than this figure would seem to indicate. It is usually of the order of 30%.

It is assumed that the light distribution from the floodlight is symmetrical about the $H = 0°$ vertical plane.

Suppose we know the (x, y) coordinates of a point on an area to be floodlit and we want to find this point on the diagram. We need to find the corresponding (V, H) coordinates, for which purpose Figure 12.6 can be used. In this, L is the luminaire mounted at a height of h over F, with its peak luminous intensity directed at P. R is the point whose (V, H) coordinates we wish to find in terms of x, y, h and FP. θ is the angle of incidence and will be used later in the determination of the illuminance.

From the diagram it is apparent that

$$V = \angle FLS - \angle FLP$$

$$= \tan^{-1} \frac{FS}{FL} - \tan^{-1} \frac{FP}{FL}$$

$$= \tan^{-1} \frac{y}{h} - \tan^{-1} \frac{FP}{h}$$

and

$$H = \tan^{-1} \frac{x}{\sqrt{y^2 + h^2}}$$

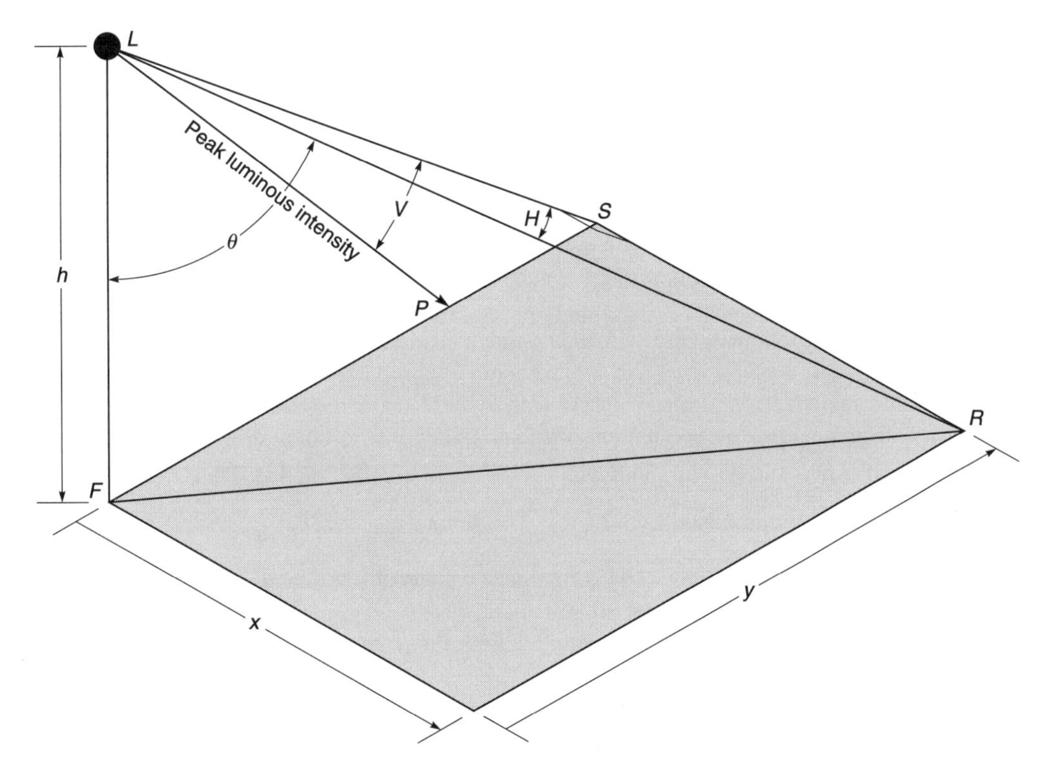

Fig. 12.6 Relationship of angles and distances used in floodlighting

The illuminance at R is given by

$$E = (I \times \cos^3 \theta \times \tau) \times \left[\frac{\Phi \times LLMF \times LMF}{h^2} \right] \quad (12.1)$$

where

I is the luminous intensity (cd/klm);
θ is the angle of incidence given by

$$\cos \theta = \frac{LF}{LR}$$

$$= \frac{L}{\sqrt{(x^2 + y^2 + h^2)}}$$

τ is the attenuation due to atmospheric losses. This can be surprisingly high. It is of course very variable, depending as it does on weather conditions, and the length of the light path.[13] It certainly becomes important when the mounting towers are over 30 m high, when it may be as much as 15% on what might appear to the uninitiated to be a clear night. It can be calculated from Bouguer's law, namely

$$\tau = e^{-\sigma R} \tag{12.2}$$

where

e is the base of natural logarithms, 2.7183 approximately;
σ is the extinction coefficient (m^{-1});
R is the length of the light path (m).

LLMF is the maintenance factor for the lamp, usually obtainable from the manufacturer; *LMF* is the maintenance factor for the luminaire.

In the equation for E, the part in the round brackets varies with the position of the illuminated point, whereas the part in the square brackets remains constant if h does not vary. It may, therefore, be convenient to evaluate these separately when a number of floodlights are used to light a given area.

To find the average illuminance E_{av} on an area, the boundary of the area is marked out on the floodlighting diagram and the luminous flux on the enclosed area totalled. Then

$$\text{luminous flux within boundary} = \Phi_{boundary} \times \Phi \times MF_{lamp} \times MF_{luminaire} \times \tau$$

is determined for each floodlight. The luminous flux for. all the floodlights is summed and divided by the area to give the average illuminance.

We can now consider how the floodlighting diagram may be used in a calculation.

Worked example 1 The area *ABCD* shown in Figure 12.7 is to be lit by four floodlights L_1, L_2, L_3 and L_4 (not shown) positioned over F_1, F_2, F_3 and F_4 at a mounting height of 10 m and aimed with their peak luminous intensity directed towards P_1, P_2, P_3 and P_4 respectively. Their photometric performance is given in Figure 12.5, page 417. Find the average horizontal illuminance and the horizontal illuminance at each corner and in the centre of the area. Take the lamp luminous flux as 28.5 klm, the lamp lumen maintenance factor (*LLMF*) as 0.9, the luminaire maintenance factor (*LMF*) as 0.85, and the atmospheric transmission as 0.95.

Answer As all the floodlights are symmetrically placed with respect to *ABCD* we need only consider one floodlight L_1 initially. The first task is to find the (V, H) coordinates of the points A, S, B, E, C, R, D and P. These can then be marked on Figure 12.8 to find the boundary of the lit area on the isocandela diagram.

V and H are calculated in the spreadsheets in Tables 12.2 and 12.3. The short time spent in preparing spreadsheets as opposed to using a hand calculator is well justified as, inevitably, some re-aiming is necessary and spreadsheets enable this to be done quickly and with less risk of error. At the same time, θ is calculated as it will be needed for the calculation of point illuminances. It should be noted that *AGB* and *DJC* lie on curves but an accuracy commensurate with that to which the diagram can be read is obtained by using straight lines.

Negative values of V arise when V is below the peak luminous intensity and negative values of H arise when x is negative. From Figure 12.8 it is apparent that 292 lm per 1000 lamp lumens falls within the boundary. Hence

$$\begin{aligned}\text{luminous flux within boundary} &= \Phi_{boundary} \times \Phi \times MF_{lamp} \times MF_{luminaire} \times \tau \\ &= 292 \times 28.5 \times 0.90 \times 0.85 \times 0.95 \\ &= 6050 \text{ lm}\end{aligned}$$

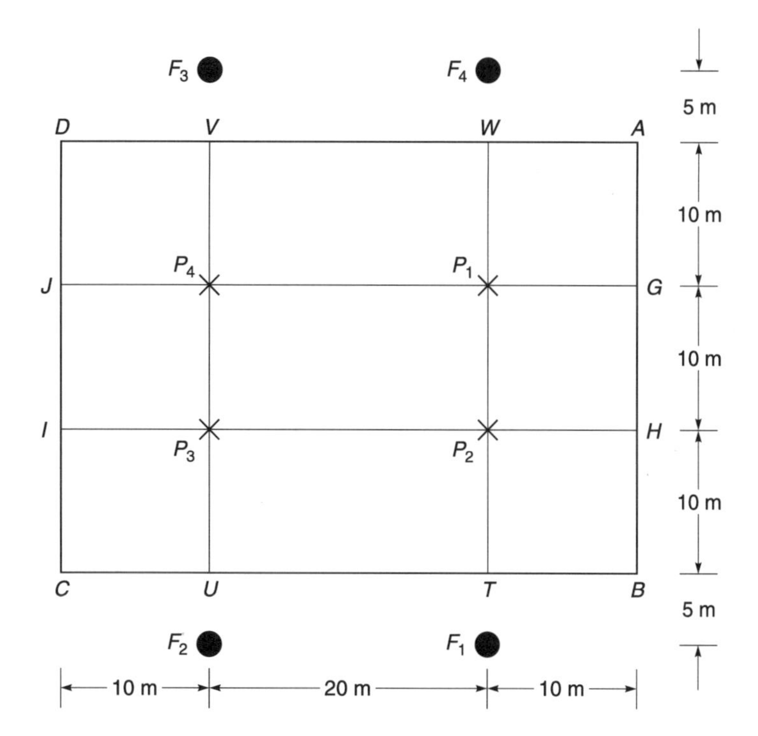

Fig. 12.7 Plan of lit area

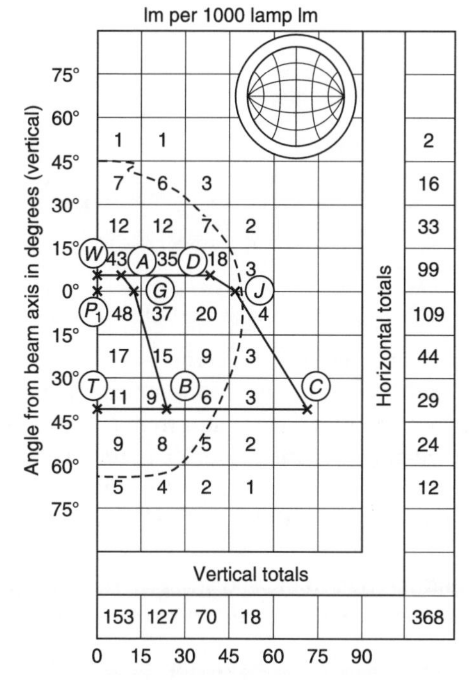

Luminous flux on left side (*lm per 1000 lamp lm*)	Luminous flux on right side (*lm per 1000 lamp lm*)
29	15
107	43
44	20
20	14
0	0
Total 200	Total 92

Grand total: 292 lm per 1000 lamp lm

Fig. 12.8 Lit area plotted onto isocandela diagram

Table 12.2 Calculation of V for luminaire L_1

Line	F_1P_1 (m)	y (m)	h (m)	$\tan^{-1}\frac{F_1P_1}{h}$ (deg)	$\tan^{-1}\frac{Y}{h}$ (deg)	$V = \tan^{-1}\frac{Y}{h} - \tan^{-1}\frac{FP}{h}$ (deg)
CTB	25	5	10	68.2	26.6	– 41.6
JP_1G	25	25	10	68.2	68.2	0.0
DWA	25	35	10	68.2	74.1	5.9

Table 12.3 Calculation of H and θ for luminaire L_1

Point	x (m)	y (m)	h (m)	$H = \tan^{-1}\frac{x}{\sqrt{y^2+h^2}}$ (deg)	$\cos\theta = \frac{h}{\sqrt{x^2+y^2+h^2}}$
B	5	5	10	24.1	0.816
C	–30	5	10	–69.6	0.312
G	5	25	10	10.5	0.365
J	–30	25	10	–48.1	0.248
A	5	35	10	7.8	0.272
D	–30	35	10	–39.5	0.212

Since the floodlights are symmetrically placed with respect to the boundary, this result can be multiplied by four to find the total luminous flux incident within the boundary, and divided by the area to give the average illuminance.

$$\text{average illuminance} = \frac{\Phi_{\text{boundary}} \times \text{number of floodlights}}{\text{area within boundary}}$$

$$= \frac{6050 \times 4}{40 \times 30}$$

$$= 20 \text{ lx}$$

The second part of the question is to find the illuminances at the corners of the area. We can do this by finding the illuminances at A, B, C and D from floodlight L_1 and adding the result. This is justified because of the symmetrical layout; the illuminance at B from L_2 is the same as the illuminance at C from L_1, and so on for the other points. First we need to find the luminous intensity in these directions for which we return to the isocandela part of the floodlighting diagram (Figure 12.9) and plot these points on it making use of the (V, H) angles previously determined.

The values of luminous intensity are read and inserted in the spreadsheet, Table 12.4. Since we are taking the atmospheric attenuation as constant at 0.95, we can modify Equation (12.1), to read

$$E = (I \times \cos^3\theta) \times \left[\frac{\Phi \times MF_{\text{lamp}} \times MF_{\text{luminaire}} \times \tau}{h^2}\right] \tag{12.3}$$

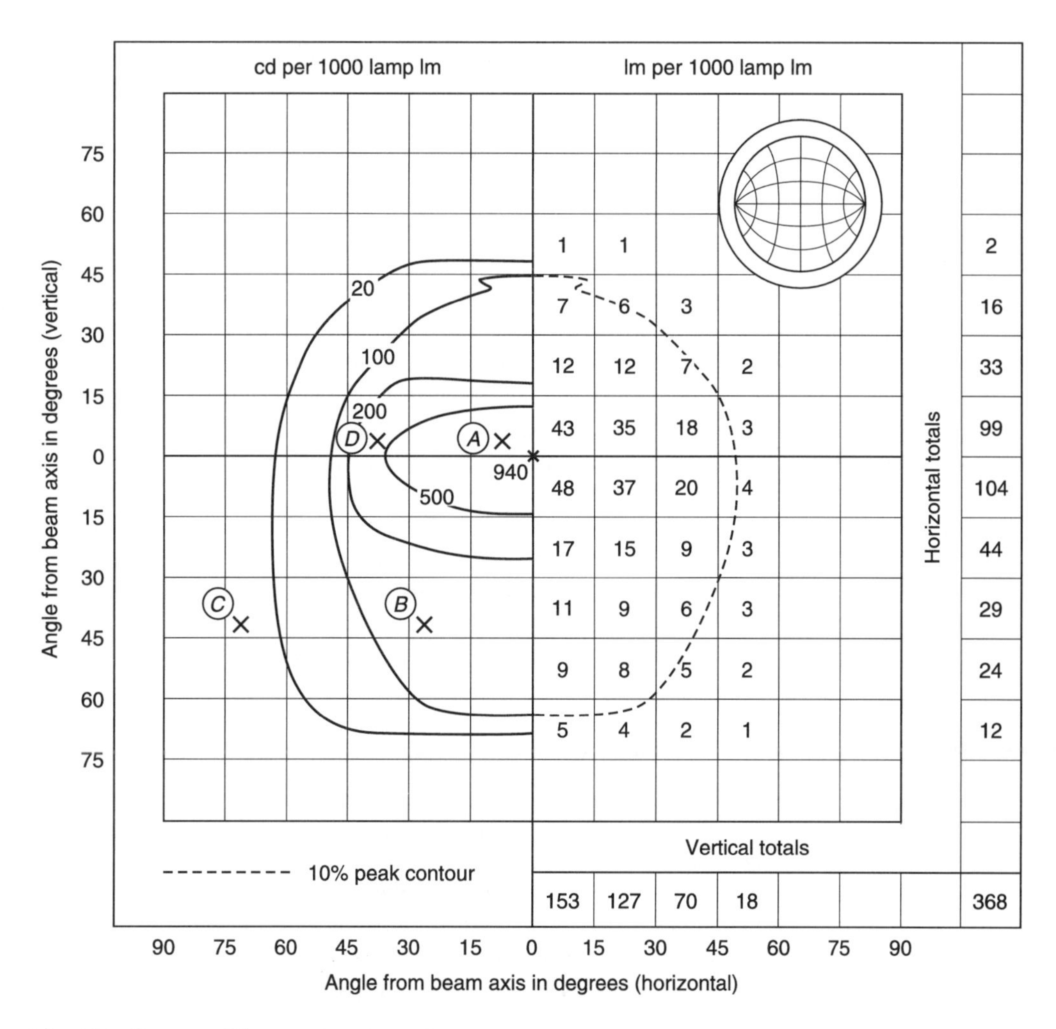

Fig. 12.9 Corners of lit area plotted onto isocandela diagram

In this, for ease of calculation, we replace the part in square brackets by a constant K such that

$$K = \frac{\Phi \times LLMF \times LMF \times \tau}{h^2}$$

$$= \frac{28.5 \times 0.90 \times 0.85 \times 0.95}{10^2}$$

$$= 0.207 \text{ lm/m}^2$$

Usually the minimum illuminance is obtained at or near the corner of the illuminated area so the indications are that excellent uniformity is achieved (minimum illuminance ÷ average illuminance).

Table 12.4 Calculation of the illuminance at the corners of the lit area

Point	I (cd/klm)	$\cos\theta$	K (lm m^{-2})	$E = I \times \cos^3\theta \times K$ (lx)
A	700	0.272	0.207	2.92
B	130	0.816	0.207	14.63
C	0	0.312	0.207	0.00
D	400	0.209	0.207	0.76
			Total	18.30

Comments on Worked example 1

1. To demonstrate the use of the method, the reduction of illuminance due to attenuation was taken as a constant. If the extinction coefficient is known, a more accurate allowance can be made using Equation (12.2), as will be demonstrated in Worked example 2.
2. Interpolation between the isocandela lines tends to be inaccurate especially where there are large steps in luminous intensity between them and they are widely spaced on the diagram. Similarly, estimation of the luminous flux in the angular boxes may also be inaccurate when they are cut by the boundary, as the luminous flux is not spread evenly in each of the boxes.
3. It is interesting to work out the fraction of bare lamp luminous flux utilized.

   Total bare lamp luminous flux: $4 \times 28.5 = 114$ klm.

   Total luminous flux falling on area: $6.048 \times 4 = 24$ klm.

   Utilization: $24/114 = 0.21$ with maintenance factor and attenuation losses, or $0.21/(0.90 \times 0.85 \times 0.95) = 0.29$ when these losses are discounted. This agrees well with the 0.3 figure quoted earlier in this chapter.
4. The floodlights are aimed at two-thirds of the distance along the lit area. Experience shows that this is a good starting point for meeting photometric requirements. In addition, the floodlights are set back 5 m from the lit area. By reducing the horizontal angle H this helps to increase the luminous intensity and therefore the illuminance at the corners.

Worked example 2 Calculate the attenuation of luminous intensity for distances of 50 m and 100 m when the extinction coefficient is 1.1×10^{-3} m^{-1}, which represents the figure for a clear day.

Answer We use Equation (12.2), page 420,

for $R = 50$ m,

$$\begin{aligned}\tau &= e^{-\sigma R}\\ &= 2.7183^{-1.1 \times 10^{-3} \times 50}\\ &= 0.95\end{aligned}$$

and for $R = 100$ m,

$$\begin{aligned}\tau &= e^{-\sigma R}\\ &= 2.7183^{-1.1 \times 10^{-3} \times 100}\\ &= 0.90\end{aligned}$$

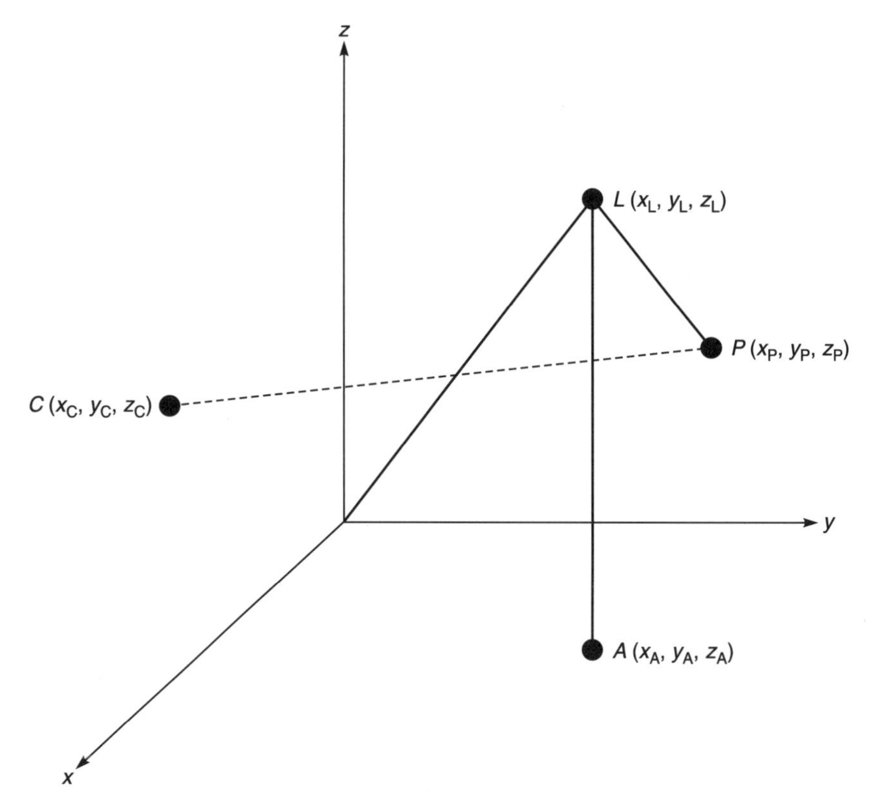

Fig. 12.10 x, y, z coordinate system for floodlighting

12.6 Illuminance in complex situations

Whilst the floodlighting diagram is useful for situations where there are few floodlights and the illuminances at only a few points on a single surface are required, it is not suitable for more detailed calculations. For this, an *I*-table has to be used and a computer program will be required.

To tackle the calculation in a systematic way, a coordinate system in three dimensions (x, y and z) is used.

Figure 12.10 displays the coordinate system. L is the luminaire, A its aiming point, and P the point at which calculations are required. C is the camera position.

Let $\boldsymbol{n}_x$, $\boldsymbol{n}_y$, $\boldsymbol{n}_z$ be the unit vectors parallel to the x, y and z axes (see Section 3.9, page 112). Then

$$\boldsymbol{LA} = (x_A - x_L)\boldsymbol{n}_x + (y_A - y_L)\boldsymbol{n}_y + (z_A - z_L)\boldsymbol{n}_z$$
$$\boldsymbol{LP} = (x_P - x_L)\boldsymbol{n}_x + (y_P - y_L)\boldsymbol{n}_y + (z_P - z_L)\boldsymbol{n}_z \quad (12.4)$$

Squaring and taking the square root of these give

$$LA = \sqrt{(x_A - x_L)^2 + (y_A - y_L)^2 + (z_A - z_L)^2}$$
$$LA = \sqrt{(x_P - x_L)^2 + (y_P - y_L)^2 + (z_P - z_L)^2} \quad (12.5)$$

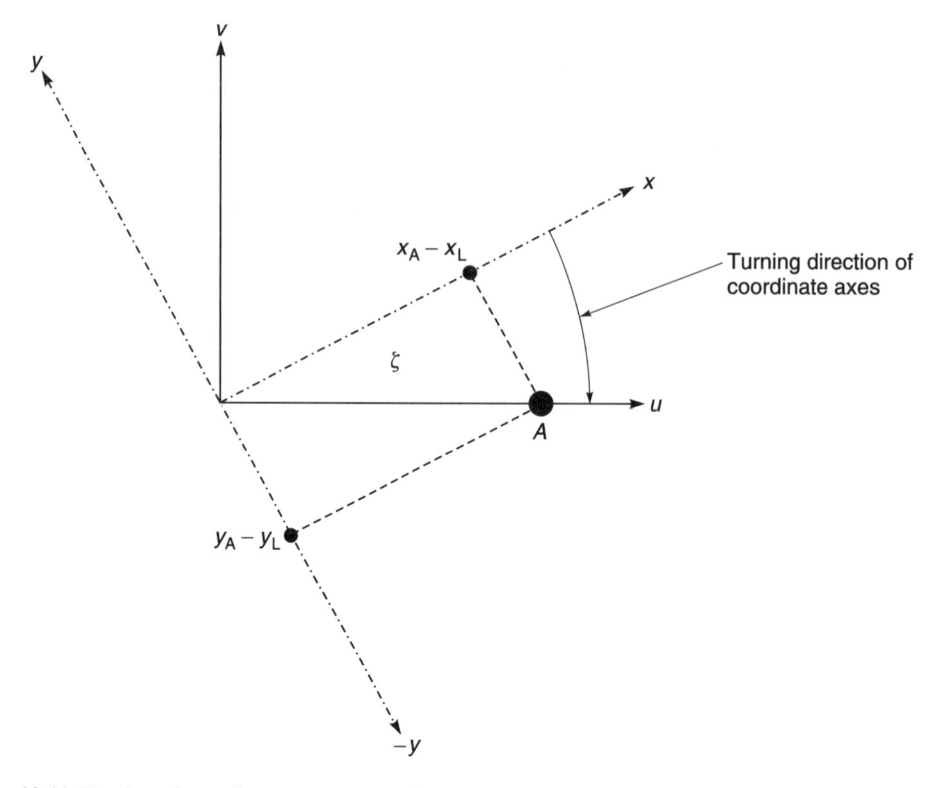

Fig. 12.11 Turning of coordinate system to make the abscissa coincide with *A*

To find *B* and *β* we have to relate the *x*, *y*, *z* coordinates to a new coordinate system in which the *x*, *z* plane passes through *A* and *L*. This is illustrated in Figure 12.12 and is designated the *u*, *v*, *w* coordinate system. To effect the translation we first have to move the *x*, *y*, *z* system linearly so that the *z* axis passes through *L*, and then rotate it so that *A* (the aiming point) lies in the *x*, *z* plane or in the *u*, *w* plane in the new coordinate system.

The linear movement of the coordinate system is effected by subtracting the coordinates for *L* from the coordinates for *A* and *P*. The rotation is effected by using a matrix similar to that in Equation (2.21), page 48, but we first have to find the angle of rotation. Let this be ζ as indicated in Figure 12.11.

From the figure it is apparent that

$$\cos\zeta = \frac{x_A - x_L}{\sqrt{(x_A - x_L)^2 + (y_A - y_L)^2}}$$

$$\sin\zeta = \frac{y_A - y_L}{\sqrt{(x_A - x_L)^2 + (y_A - y_L)^2}} \tag{12.6}$$

Matrix multiplication then gives the transformed coordinates

$$(u, v, w) = (x \cos\zeta + y \sin\zeta,\ x \sin\zeta - y \cos\zeta,\ z) \tag{12.7}$$

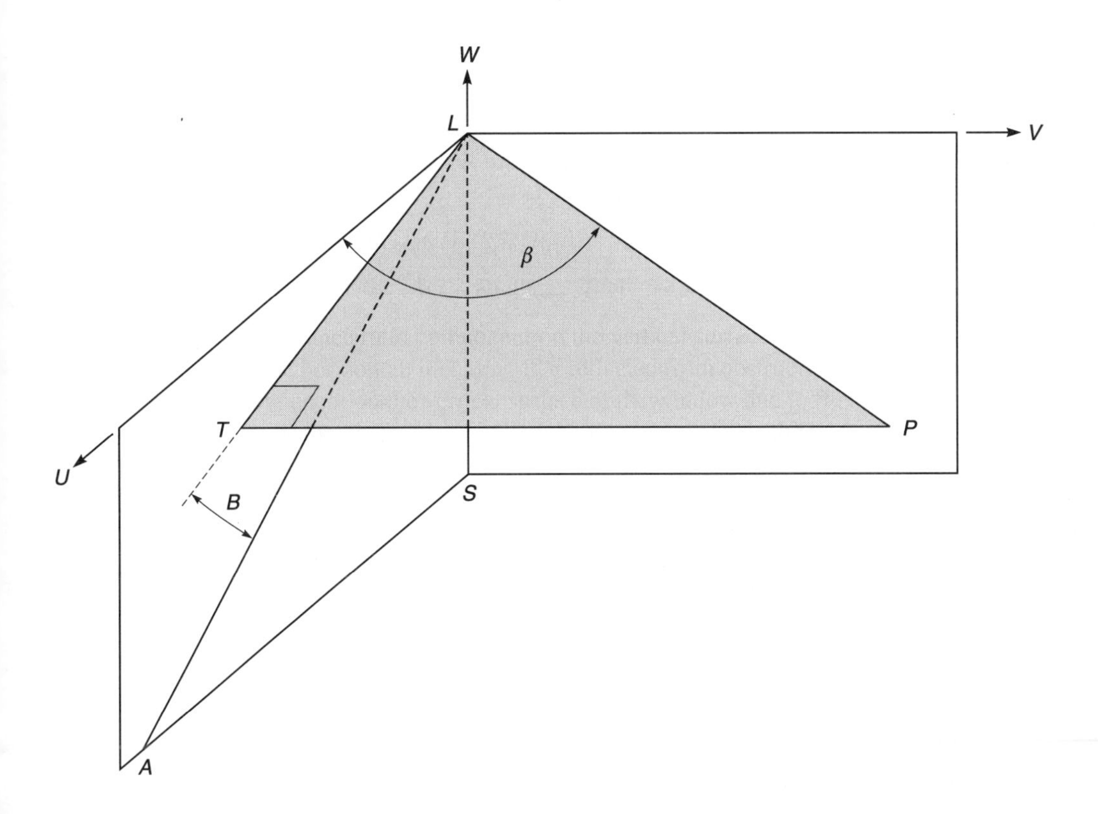

Fig. 12.12 Transformed coordinate system

Figure 12.12 shows the floodlight L on the u, v, w coordinate system with the angles B and β inserted. To find these the coordinates of T have to be determined, where T is the orthogonal projection of P on the u, w plane. They are $(u_P, 0, w_P)$.

Let $\boldsymbol{n}_u$, $\boldsymbol{n}_v$, $\boldsymbol{n}_w$ be the unit vectors parallel to the u, v and w axes. Then

$$\begin{aligned} \boldsymbol{LT}_{uvw} &= u_T\boldsymbol{n}_u + w_T\boldsymbol{n}_w \\ &= u_P\boldsymbol{n}_u + w_P\boldsymbol{n}_w \\ \boldsymbol{LA}_{uvw} &= u_A\boldsymbol{n}_u + w_A\boldsymbol{n}_w \\ \boldsymbol{LP}_{uvw} &= u_P\boldsymbol{n}_u + v_P\boldsymbol{n}_v + w_P\boldsymbol{n}_w \end{aligned} \tag{12.8}$$

where the suffix uvw indicates that the vector relates to the uvw coordinate system and the suffices on u, v and w indicate the point to which the coordinate refers. Squaring the equation for $\boldsymbol{LT}_{uvw}$ and taking the square root gives

$$LT = \sqrt{u_P^2 + w_P^2} \tag{12.9}$$

B can be found from the scalar product of $\boldsymbol{LT}$ and $\boldsymbol{LA}$ but this gives no indication of whether B is positive or negative. For this reason it is better to use the equation

$$\begin{aligned}\cos^{-1} \mathrm{B} &= \cos^{-1}\angle TLS - \cos^{-1}\angle ALS \\ &= \cos^{-1}\left(\frac{w_{\mathrm{T}}}{LT}\right) - \cos^{-1}\left(\frac{w_{\mathrm{A}}}{LA}\right) \\ &= \cos^{-1}\left(\frac{w_{\mathrm{P}}}{LT}\right) - \cos^{-1}\left(\frac{w_{\mathrm{A}}}{LA}\right)\end{aligned} \quad (12.10)$$

since $w_{\mathrm{P}} = w_{\mathrm{T}}$.

Alternatively, an equation based on that for V in Table 12.2 could be used.

In addition

$$\begin{aligned}\sin \beta &= \frac{TP}{LP} \\ &= \frac{v_{\mathrm{P}}}{\sqrt{u_{\mathrm{T}}^2 + v_{\mathrm{T}}^2 + w_{\mathrm{P}}^2}} \\ &= \frac{v_{\mathrm{P}}}{\sqrt{u_{\mathrm{P}}^2 + v_{\mathrm{P}}^2 + w_{\mathrm{P}}^2}}\end{aligned} \quad (12.11)$$

Alternatively we could use

$$\begin{aligned}\cos \beta &= \frac{LT}{LP} \\ &= \frac{\sqrt{u_{\mathrm{P}}^2 + w_{\mathrm{P}}^2}}{\sqrt{u_{\mathrm{P}}^2 + v_{\mathrm{P}}^2 + w_{\mathrm{P}}^2}}\end{aligned} \quad (12.12)$$

the sign of which does not change depending on which side of the u, w plane P lies. Since most I-tables do not distinguish between positive and negative values of β, this may be the more convenient formula.

12.6.1 CALCULATION OF ILLUMINANCE

In the calculation of illuminance three quantities have to be determined, the luminous intensity, the distance between the luminaire and the illuminated point, and the angle of incidence.

1. Luminous intensity

The determination of the B, β coordinates, in Equation (12.10) and (12.11) or (12.12), enables the luminous intensity to be found from the I-table, with interpolation if necessary.

2. Distance between luminaire and calculation point

In Figure 12.12 this is LP, which has already been found in Equation (12.5), page 425.

3. Angle of incidence

This is the angle, θ, between $\mathbf{PL}$ and the normal to an illuminated facet at the calculation point. To use the scalar product to determine θ it is essential we adopt a consistent sign convention for the direction of the two vectors. We will consider the vectors as emanating outwards from the illuminated facet. ***LP*** has already been determined in Equation (12.5), and we require ***PL***, which is the negative value of this.

We have to consider a number of cases for representing the normal to the illuminated facet as a vector.

(a) *Normal to facet aligned with the x axis*
The vector for a facet facing towards the positive direction of the x axis is $\boldsymbol{n}_x$.
The vector for a facet facing towards the negative direction of the x axis is $-\boldsymbol{n}_x$.
(b) *Normal to facet aligned with the y axis*
The vector for a facet facing towards the positive direction of the y-axis is $\boldsymbol{n}_y$.
The vector for a facet facing towards the negative direction of the y-axis is $-\boldsymbol{n}_y$.
(c) *Normal to facet aligned with the z axis*
The vector for a facet facing towards the positive direction of the z-axis is $\boldsymbol{n}_z$.
The vector for a facet facing towards the negative direction of the z-axis is $-\boldsymbol{n}_z$.
(d) *Normal to facet directed towards camera*
Let the coordinates of the camera position C be (x_C, y_C, z_C), then the vector ***PC*** is given by

$$\boldsymbol{PC} = (x_C - x_P)\boldsymbol{n}_x + (y_C - y_P)\boldsymbol{n}_y + (z_C - z_P)\boldsymbol{n}_z \tag{12.13}$$

The angle of incidence is found by scalar multiplication of the normal to the facet with ***PL***. For instance, for the facet normal to the camera,

$$\cos\theta_C = \frac{\boldsymbol{PC}\cdot\boldsymbol{PL}}{PL \times PL}$$

$$= \frac{[(x_C - x_P)\boldsymbol{n}_x + (y_C - y_P)\boldsymbol{n}_y + (z_C - z_P)\boldsymbol{n}_z]\cdot[(x_L - x_P)\boldsymbol{n}_x + (y_L - y_P)\boldsymbol{n}_y + (z_L - z_P)\boldsymbol{n}_z]}{\sqrt{(x_C - x_P)^2 + (y_C - y_P)^2 + (z_C - z_P)^2}\ \sqrt{(x_L - x_P)^2 + (y_L - y_P)^2\ (z_L - z_P)^2}}$$

$$= \frac{(x_C - x_P)\ (x_L - x_P) + (y_C - y_P)\ (y_L - y_P) + (z_c - z_P)\ (z_L - z_P)}{\sqrt{(x_C - x_P)^2 + (y_C - y_P)^2 + (z_C - z_P)^2}\ \sqrt{(x_L - x_P)^2 + (y_L - y_P)^2\ (z_L - z_P)^2}} \tag{12.14}$$

To find the horizontal illuminance on a surface facing upwards we take the normal as $\boldsymbol{n}_z$, then

$$\cos\theta_{H+} = \frac{\boldsymbol{n}_z\cdot\boldsymbol{PL}}{PL}$$

$$= \frac{\boldsymbol{n}_z\cdot[(x_L - x_P)\boldsymbol{n}_x + (y_L - y_P)\boldsymbol{n}_y + (z_L - z_P)\boldsymbol{n}_z]}{PL}$$

$$= \frac{z_L - z_P}{\sqrt{(x_L - x_P)^2 + (y_L - y_P)^2 + (z_L - z_P)^2}} \tag{12.15}$$

where the H+ suffix on θ indicates that the normal to the facet faces upwards.

If cos θ is negative then the facet does not receive light, as explained in more detail in Section 3.9, page 112.

All the elements have now been found that enable the illuminance to be determined at a point.

Worked example 3 A floodlight is positioned at L = (60, 0, 30) and aimed at A = (120, –45, 0). For P at (90, –135, 0) calculate B, β, the angle of incidence θ_C of the light on a plane normal to the camera direction, and the angle of incidence θ_{H+} on the upper surface of the horizontal plane. All dimensions are in metres.

Answer The work is set out in two spreadsheets. These are connected so that the data in one are automatically transferred to the other when needed. The work could be done in one spreadsheet with a considerable gain in speed of calculation but presentation in this book would be difficult. An explanation of the mathematics involved is given after each spreadsheet.

Table 12.5 Spreadsheet for the calculation of ζ, B and β. All linear dimensions in metres

1	2	3	4	5	6	7	8	9	10
X_A	X_P	X_L	$X_A–X_L$	$X_P–X_L$	LA	$\cos \zeta$	U_A	U_P	LT
120.0	90.0	60.0	60.0	30.0	80.78	0.800	75.00	105.00	109.2
Y_A	Y_P	Y_L	$Y_A–Y_L$	$Y_P–Y_L$	LP	$\sin \zeta$	V_A	V_P	B(°)
–45.0	–135.0	0.0	–45.0	–135.0	141.51	–0.600	0.0	90.0	–5.9
Z_A	Z_P	Z_L	$Z_A–Z_L$	$Z_P–Z_L$			W_A	W_P	β(°)
0.0	0.0	30.0	–30.0	–30.0			–30.00	–30.00	39.5

Spreadsheet	**Column number**	**Explanation and reference to equations**
Table 12.5	1–3	The coordinates for A, P and L are entered.
	4 & 5	The coordinates for L are subtracted from those for A and P. In effect this moves the origin to L so that the coordinate system can be rotated about the Z axis.
	6	LA and LP are calculated from Equation (12.5), page 425.
	7	$\cos \zeta$ and $\sin \zeta$ are calculated from Equation (12.6), page 426. When the aiming point A is in the x, z plane the denominator of the expressions for $\cos \zeta$ and $\sin \zeta$ is zero. To make provision for this an IF statement is needed to force the result $\cos \zeta = 1$ and $\sin \zeta = 0$.
	8 & 9	Equation (12.7), page 426 is applied to the coordinates in columns 4 and 5. Note that $y_A = 0.0$, which is the desired result of turning the coordinate system through ζ.
	10	LT is found from Equation (12.9), page 427. B and β are found from Equations (12.10) and (12.12), page 428.

Table 12.6 Spreadsheet for calculating $\cos \phi_C$ and $\cos \phi_{H+}$, All linear dimensions in metres

1	2	3	4	5	6	7	8	9
X_A	X_P	X_L	X_C	$X_A–X_L$	$X_C–X_P$	$X_L–X_P$	$\cos \theta_C$	$\cos \theta_{H+}$
120.0	90.0	60.0	3.0	60.0	–87.0	–30.0	0.0072	0.212
Y_A	Y_P	Y_L	Y_C	$Y_A–Y_L$	$Y_C–Y_P$	$Y_L–Y_P$	θ_C(°)	θ_{H+}(°)
–45.0	–135.0	0.0	6.0	–45.0	141.0	135.0	89.6	77.8
Z_A	Z_P	Z_L	Z_C	$Z_A–Z_L$	$Z_C–Z_P$	$Z_L–Z_P$		
0.0	0.0	30.0	9.0	–30.0	9.0	30.0		

Spreadsheet	Column number	Explanation and reference to equations
Table 12.6	1–3	These values are transferred from the first spreadsheet.
	4	The camera coordinates are entered.
	8	$\cos \theta_C$ is calculated from Equation (12.14), page 429.
	9	$\cos \theta_{H+}$ is calculated from Equation (12.15), page 429.

12.6.2 GRID SIZE

The number of points taken for calculation and measurement has to be tailored to the size of the lit area. CIE and CEN recommend the use of the following formula to determine the grid size

$$p = 0.2 \times 5^{\log_{10} d}$$

where p is the grid size (m), and d is the longer dimension of the area (m).

The number of points along the longer dimension is given by the odd number that is closest to d/p. From this, the distance between grid points along the longer dimension is determined and in the shorter dimension the spacing between grid points should be chosen to give grids that are as square as possible, with the proviso that there should be an odd number of points. Odd numbers of points are recommended to allow values both for calculation and measurement to be taken along the centre-lines of the enclosing rectangles. This means that the spacing from the boundary is half that between points. For measurements, the number of points may be reduced to keep the measurement effort within practicable limits but the points should be evenly distributed throughout the area. The average illuminance is simply obtained by summing the values and dividing by the number of points. Where an area is used for a particular activity a sub-grid may be used.

The formula has been constructed in such a way that a reasonable number of points is taken where large distances are involved, as indicated in Table 12.7.

The draft CEN recommendations for sports lighting include, in tabular form, the number of recommended grid points, based on the above formula. However, the formula is useful where a number of activities take place in the same area and in general floodlighting.

12.7 The floodlighting of buildings

The term 'floodlighting' is less appropriate when it is applied to buildings than it is to sports fields or to car parks; for 'Flooding' a building with light is not the main aim. The main aim of

Table 12.7 Grid spacing versus size of playing area

d (m)	p (m)	Calculated number of points	Number of points adjusted to be odd
1	0.2	5	5
10	1.0	10	11
100	5.0	20	21
1000	25.0	40	41

lighting a building is to display that building after dark. Only attractive or notable buildings usually merit being lit up after dark. There are, of course, other reasons for lighting the exterior of a building after dark, such as for security reasons, and, perhaps most commonly, for publicity – to attract the public to the building for commercial reasons.

It is relatively easy to flood the exterior of a building with light and examples of overlit buildings are common. The main reason for this overlighting is probably the installer's fear that the client might otherwise consider the building underlit and be dissatisfied. This is especially true of buildings lit for commercial reasons.

12.8 Revealing the building after dark

When the purpose of the exterior lighting is to reveal the architectural excellence of the building, then the shadows produced by the lighting are as important as the illumination produced on the building's surfaces.

It is the arrangement of the light sources to produce appropriate light and shade that is the essence of most design. The disposition of the light sources, the intensity distribution of the floodlights, the light output of the lamps and the colour of the light all have an important part to play in a successful design.

The designer's choice of these aspects of the design will depend upon the architectural form of the building, the materials from which the exterior is constructed, in terms of colour and texture, and the brightness of the locality in which the building is situated.

Although calculations have a part to play in the design, in this instance they are not as critical as they would be in the design of the lighting for a sports stadium or a general office interior.

12.9 Lighting levels and design calculations

Many notable buildings, such as ancient churches, are often best floodlit with just a few floodlights aimed in such a way as to allow the variation of the orientation of the surfaces and their varying distances from the floodlights to produce the light and shadow patterns that best reveal the architectural features of the building. Detailed calculations, in such a case, would be difficult. Even with the aid of a computer, the inputting of the structural information would be time-consuming.

Two types of calculation usually provide sufficient information for a competent designer:

(1) an average illuminance calculation over a 'window' outlining the area to be lit at an appropriate distance from the floodlight.
(2) sufficient point by point calculations to ensure that the designer's lighting level and distribution objectives are being met by the design.

It is necessary to decide on the appropriate average lighting levels to be provided before such calculations can be undertaken in a meaningful way.

Since the effect required is aesthetic and not performance-related, the lighting level is not as critical as in other types of design, for example as in an office. It is also not so cost-related, since it is often possible to change the size of a floodlight without greatly increasing the cost of the installation. To these factors must be added the fact that floodlight outputs are the product of the lamp wattage and these have relatively large steps; for example 70 W, 150 W, 250 W and, say, 400 W. However, since the inverse-square law applies to floodlighting

Table 12.8 Approximate average illuminance (lux) for preliminary design

ρ	*L* Rural	*M* Suburban	*H* Town centre
0.8	15	25	40
0.6	20	35	60
0.4	30	50	85
0.3	40	65	110
0.2	60	100	170

calculations, the effects of moving a floodlight a relatively short distance can double or halve illuminances.

Table 12.8 gives a set of illuminances, as a design guide, to enable the approximate number and size of floodlights required to be determined. The values are based upon experience and are related to building reflectance. An estimate of the surface reflectance can be obtained on site by measuring the illuminance on an unobstructed part of the building under overcast daylight and then measuring the reflected illuminance from that surface at a distance close enough for the surface to be considered as an approximation to an infinite area source without the photocell casting a shadow.

Then,

$$\rho \approx \frac{E(\text{reflected})}{E(\text{direct})}$$

Worked example 4 A small church is to be floodlit. The elevation and plan of the church are shown in Figure 12.13(a) and (b). The church lies on open ground in the centre of a village and can be seen from all directions.

Answer The proposal is to emphasize the tower and to light the whole church so that the building preserves its unity. The intention is to light the south and north sides of the aisle–nave areas each with two floodlights and to light the top of the tower with narrow angle floodlights from the ground, and the bottom of the tower and the adjacent part of the aisles with wide-angle low wattage floodlights. The chancel will also be lit with a wide-angle low wattage floodlight. The different distances of the aisle and clerestory areas from the large floodlights will provide different illuminance levels and so avoid a flat appearance.

The design is supported by the following calculations.

The side of the church, excluding the tower, fills a 'window' 22 m × 12 m. The main body of the building is constructed from brick with a 50% reflectance. From Table 12.8 an average illuminance value of 25 lux was selected.

For the initial calculation, a utilization of 0.3 is assumed and in this clean area a maintenance factor of 0.8 is also assumed.

$$E_{av} = \frac{F_L \times UF \times MF}{A}$$

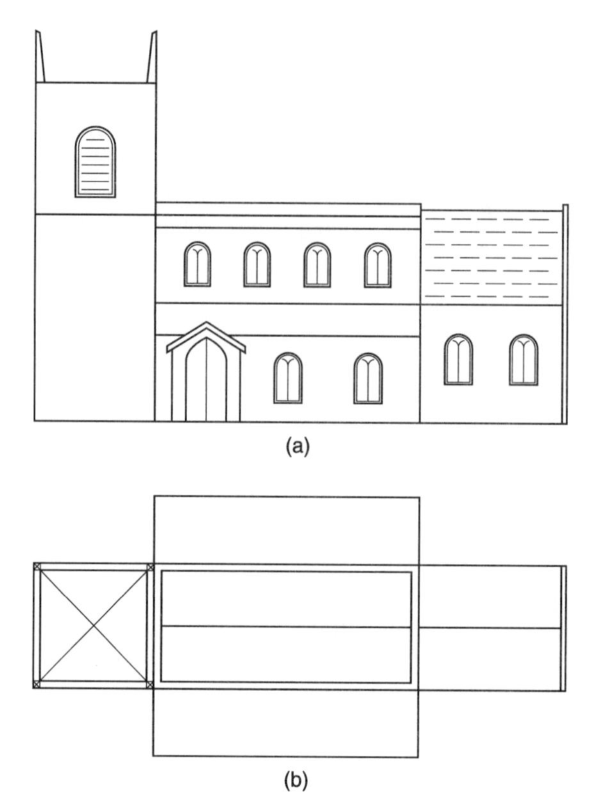

Fig. 12.13 Church to be floodlit

$$F_L = \frac{E_{av} \times A}{UF \times MF}$$

$$= \frac{25 \times 22 \times 12}{0.3 \times 0.8}$$

$$= 27\,500 \text{ lm}$$

Two 150 W high pressure sodium lamps would give 2 × 14000 = 28000 lumens. The floodlights chosen had a total vertical beam angle of 44° (to 50% of maximum luminous intensity) and a total horizontal beam angle of 61°.

For calculation purposes these floodlights were placed symmetrically on a line 11 m from the aisle wall and parallel to it. The main beam of each floodlight was directed at 9 m above the ground level; that is, aimed at the clerestory. The computer program was designed to treat the building as a box and so to deal with the fact that the clerestory window walls, which are above the aisles, lie 4 m further away from the floodlights than the aisle walls, two rectangular plots were employed at different distances from the floodlights.

Two illuminance grids were obtained; one at 9 m, which was the distance of the aisle walls from the floodlights and the other at 13 m, which was the horizontal distance of the plane of the clerestory from the floodlights. These are shown in Figure 12.14(a) and (b).

(a) Aisle

	0.5	1.5	2.5	3.5	4.5	5.5	6.5	7.5	8.5	9.5	10.5	11.5	12.5	13.5	14.5	15.5	16.5	17.5	18.5	19.5	20.5	21.5
11.5	8.1	9.0	9.8	10.4	10.9	11.2	11.6	12.0	12.5	13.1	13.5	13.5	13.1	12.5	12.0	11.6	11.2	10.9	10.4	9.8	9.0	8.1
10.5	9.2	10.4	11.4	12.2	12.8	13.2	13.5	13.8	14.2	14.7	15.2	15.2	14.7	14.2	13.8	13.5	13.2	12.8	12.2	11.4	10.4	9.2
9.5	10.4	11.9	13.1	14.2	14.9	15.3	15.7	15.7	16.0	16.3	16.7	16.7	16.3	16.0	15.7	15.7	15.3	14.9	14.2	13.1	11.9	10.4
8.5	11.5	13.5	15.1	16.4	17.3	17.8	18.1	18.1	17.9	17.9	18.1	18.1	17.9	17.9	18.1	18.1	17.8	17.3	16.4	15.1	13.5	11.5
7.5	12.5	15.1	17.3	18.9	20.0	20.6	20.9	20.6	19.9	19.5	19.3	19.3	19.5	19.9	20.6	20.9	20.6	20.0	18.9	17.3	15.1	12.5
6.5	13.3	16.6	19.3	21.4	22.8	23.5	23.7	23.0	21.9	20.6	19.8	19.8	20.6	21.9	23.0	23.7	23.5	22.8	21.4	19.3	16.6	13.3
5.5	13.9	17.6	20.9	23.5	25.3	26.1	26.2	25.2	23.5	21.3	19.8	19.8	21.3	23.5	25.2	26.2	26.1	25.3	23.5	20.9	17.6	13.9
4.5	14.2	18.4	22.1	25.0	27.0	28.0	28.0	26.8	24.7	22.0	19.6	19.6	22.0	24.7	26.8	28.0	28.0	27.0	25.0	22.1	18.4	14.2
3.5	13.9	18.6	22.6	25.8	28.0	29.1	29.0	27.7	25.4	22.3	19.0	19.0	22.3	25.4	27.7	29.0	29.1	28.0	25.8	22.6	18.6	13.9
2.5	13.1	17.8	22.1	25.5	27.8	28.9	28.8	27.4	24.9	21.6	18.2	18.2	21.6	24.9	27.4	28.8	28.9	27.8	25.5	22.1	17.8	13.1
1.5	12.0	16.3	20.5	24.2	26.6	27.6	27.5	25.9	23.1	20.1	17.1	17.1	20.1	23.1	25.9	27.5	27.6	26.6	24.2	20.5	16.3	12.0
0.5	10.7	14.3	18.1	21.5	23.8	24.8	24.6	23.1	20.6	18.1	15.9	15.9	18.1	20.6	23.1	24.6	24.8	23.8	21.5	18.1	14.3	10.7

(b) Clerestory / Chancel

	0.5	1.5	2.5	3.5	4.5	5.5	6.5	7.5	8.5	9.5	10.5	11.5	12.5	13.5	14.5	15.5	16.5	17.5	18.5	19.5	20.5	21.5
11.5	7.8	8.4	9.0	9.5	10.0	10.5	11.1	11.8	12.5	12.9	13.2	13.2	12.9	12.5	11.8	11.1	10.5	10.0	9.5	9.0	8.4	7.8
10.5	8.6	9.4	10.1	10.6	11.1	11.6	12.3	13.0	13.7	14.2	14.5	14.5	14.2	13.7	13.0	12.3	11.6	11.1	10.6	10.1	9.4	8.6
9.5	9.4	10.3	11.1	11.8	12.3	12.7	13.5	14.1	14.8	15.5	15.8	15.8	15.5	14.8	14.1	13.5	12.7	12.3	11.8	11.1	10.3	9.4
8.5	10.2	11.3	12.2	12.9	13.5	14.0	14.6	15.3	16.0	16.8	17.1	17.1	16.8	16.0	15.3	14.6	14.0	13.5	12.9	12.2	11.3	10.2
7.5	10.9	12.2	13.3	14.1	14.7	15.2	15.7	16.4	17.0	17.8	18.2	18.2	17.8	17.0	16.4	15.7	15.2	14.7	14.1	13.3	12.2	10.9
6.5	11.5	13.0	14.2	15.1	15.8	16.3	16.8	17.3	17.8	18.5	19.0	19.0	18.5	17.8	17.3	16.8	16.3	15.8	15.1	14.2	13.0	11.5
5.5	11.9	13.5	14.9	15.9	16.7	17.2	17.6	18.0	18.3	18.9	19.4	19.4	18.9	18.3	18.0	17.6	17.2	16.7	15.9	14.9	13.5	11.9
4.5	12.0	13.8	15.2	16.4	17.2	17.8	18.1	18.3	18.6	19.0	19.5	19.5	19.0	18.6	18.3	18.1	17.8	17.2	16.4	15.2	13.8	12.0
3.5	12.0	13.7	15.2	16.4	17.3	17.9	18.2	18.3	18.5	18.8	19.4	19.4	18.8	18.5	18.3	18.2	17.9	17.3	16.4	15.2	13.7	12.0
2.5	11.8	13.5	15.0	16.2	17.1	17.7	18.0	18.0	18.1	18.4	18.9	18.9	18.4	18.1	18.0	18.0	17.7	17.1	16.2	15.0	13.5	11.8
1.5	11.2	12.9	14.4	15.6	16.5	17.1	17.3	17.3	17.4	17.5	17.8	17.8	17.5	17.4	17.3	17.3	17.1	16.5	15.6	14.4	12.9	11.2
0.5	10.4	12.0	13.5	14.7	15.6	16.1	16.4	16.4	16.3	16.2	16.4	16.4	16.2	16.3	16.4	16.4	16.1	15.6	14.7	13.5	12.0	10.4

Fig. 12.14 Illuminance grid

17.4	19.6	30.4	43.9	60.8	46.3	30.5	19.0
16.2	23.7	39.3	59.7	68.6	46.0	29.9	18.0
15.0	28.2	50.4	63.1	66.9	43.4	28.2	16.1
13.8	30.8	50.8	61.1	59.0	38.5	24.5	12.8
12.6	27.8	48.9	54.2	50.9	32.0	17.6	9.0
11.4	25.9	43.3	45.8	37.5	20.3	13.1	7.8
10.2	23.1	31.5	28.1	25.6	16.2	9.8	6.9
9.0	17.3	21.9	21.8	15.5	12.4	9.2	6.9
7.8	15.7	17.2	17.2	14.7	11.7	9.4	7.7
6.6	16.1	16.8	16.6	14.8	12.4	11.0	9.5
5.4	19.8	20.2	19.6	18.2	16.3	14.1	11.5
4.2	25.4	26.5	26.1	22.5	18.1	14.3	11.3
3.0	25.2	25.5	23.4	19.8	17.4	14.6	11.9
1.8	27.3	27.3	25.2	22.1	18.7	15.3	12.2
0.6	22.7	23.0	20.7	17.8	15.0	12.2	9.9
	0.4	1.3	2.1	3.0	3.9	4.7	5.6

(c)

Fig. 12.14 *continued*

In Figure 12.14(b) it is the clerestory and the chancel walls and roof that are the relevant areas. On this plot, the aisle area is crossed through. Comparing the clerestory plot with the aisle plot shows the difference in illuminance produced by the set back (that is, the different distances of the upper and lower walls from the floodlight positions) and the consequent emphasis of the building shape.

The top of the tower is lit on three sides by 150 W narrow angle floodlights. These have a horizontal beam angle of 12° (to 50% of maximum intensity) and a vertical beam angle of 10°. The bottom of the tower is lit by a 70 W wide angle floodlight, having a horizontal beam angle of 60° and a vertical beam angle of 40°. Figure 12.14(c) shows an illuminance grid for these two floodlights.

The back of the tower (east side) would be lit from the roof on the nave by a 150 W floodlight similar to those described above, but with a specially specified diffuse finish to the reflector to double the beam angle, since it will be mounted to illuminate the tower more squarely and so a wider spread of light is required to avoid a very high and patchy illuminance.

The calculations have simply ensured that the lighting level will be adequate and that the tower has been emphasized as intended. However, all such installations should be preceded by a trial to determine the best position for each floodlight, since small movements of the floodlights can greatly affect the shadows produced.

The lighting levels vary considerably, but a variation of 20:1 is not uncommon or unacceptable as long as it is not too sudden.

For simplicity, the computer calculations for the nave and chancel areas have been carried out assuming that the floodlights are aimed directly at the building, at right angles to the walls. However, in practice, during a trial it is usually found that moving the floodlights a short distance to one side (in our example, probably towards the chancel end of the church) and then angling the floodlights towards the walls gives a better rendering of the building texture and features because of the shadowing that this produces. Figure 12.15 gives the final position of all the floodlights.

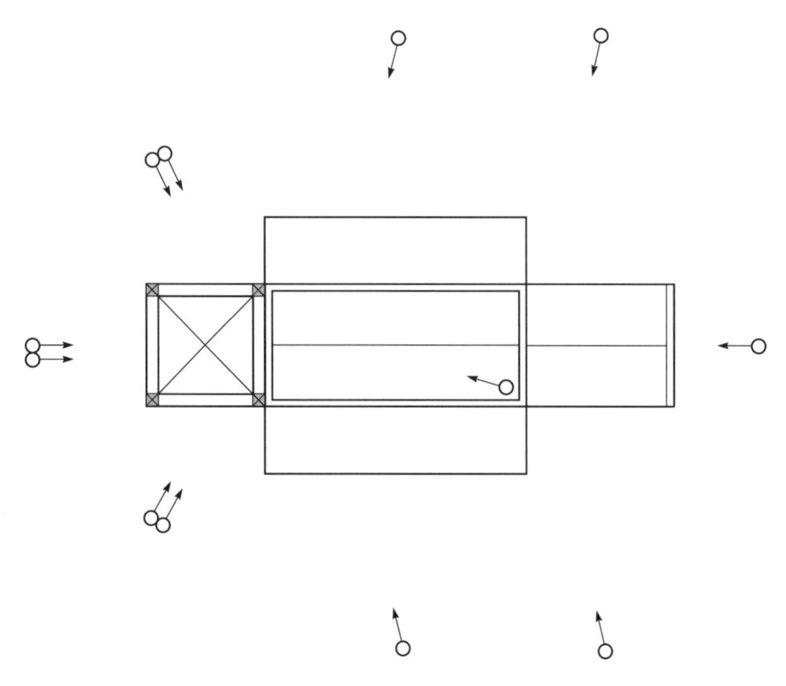

Fig. 12.15 Final position of floodlights

12.10 Public buildings and statues

In Section 12.9 our example featured the floodlighting of a church. There, the different distances of parts of the structure from the floodlights produced the variation in illuminance that preserves an interesting appearance. It is a tribute to the designers of our ancient churches that they look so pleasing when lit from below, when the builders so long ago expected their buildings to be lit only from above.

Other public buildings present different problems. For example, many have pillared porticos. The problem with pillars is that it is almost always best to light them at an angle to the direction of view and never, if possible, *from* the direction of view. Therefore, when a portico has multiple pillars it presents a real problem as to where the floodlights should be sited. This is especially true of, say, a public building in the centre of a major city, where there are severe restrictions on the positions of the floodlights.

One very successful solution is to floodlight the part of the building behind the pillars, so that the pillars themselves are seen in silhouette. The floodlights can then be mounted behind the pillars.

It is often commented that lighting is an art as well as a science, and nowhere is this more true than in the floodlighting of notable buildings, monuments and, particularly, statues.

It is always wise to carry out a trial when designing a floodlighting scheme, but it is essential when lighting a statue or similar monument or sculpture.

The lighting level will depend on the ambient brightness of the surroundings, which may be high or low. In addition, the viewing distance must be taken into account. The further the viewing distance, the higher the illuminance level required. The size of the object or statue being lit must also be taken into account; for the smaller the statue the more brightly it will need to be lit.

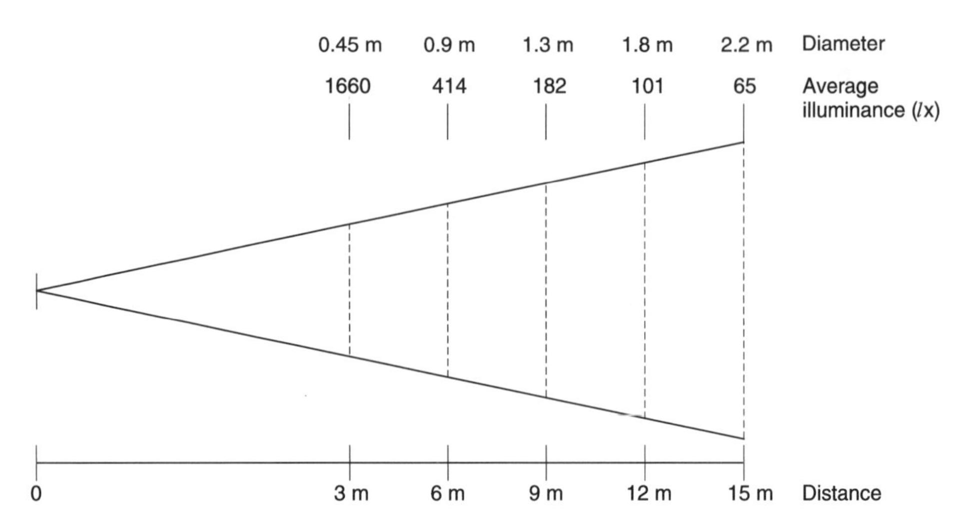

Fig. 12.16 Cone diagram

To this fact must be added the special features of the statue or monument itself; the material from which it is constructed and its surface finish.

The starting point is to assume that whatever the recommended average illuminance for floodlighting a building, the illuminance on the statue will need to be considerably higher than that, if it is to be seen as a special feature in its own right. Therefore, calculations have only a part to play in this type of lighting. One important feature is the location of the floodlight or floodlights; and another is the floodlight distribution, since glare and stray light becomes very important. The calculational device that is often most useful is the 'cone' diagram, in which the average illuminance is specified at different distances from the apex of the cone, where the floodlight is assumed to be located. Such a diagram is shown in Figure 12.16.

These diagrams enable the appropriate floodlight to be chosen to give the required illuminance at the mounting distance of the floodlight from the statue or the mounting distance required using that floodlight to achieve the desired illuminance. An illuminance level of three times, or more, than that for floodlighting a building, in the same area, could easily be required.

In some cases, lighting the statue from two opposing directions and using lamps of a similar type, but different colour temperatures, can give a very pleasing effect on a statue. As would be expected, modelling is a major consideration with statues and, just as with the lighting of pillars, it is important not to light the statue from the main direction of view because this gives a flat appearance.

Problems

1. Recalculate Worked example 2 (page 424) for an extinction coefficient of $1.3 \times 10^{-3}\ \text{m}^{-1}$.
   Answer: [0.94 at 50 m, and 0.88 at 100 m.]
2. In Worked example 1 (page 420) find the illuminance at P_1, P_2, P_3 and P_4 from floodlight F_1 and the average illuminance in the rectangle enclosed by the four points.

   Answer: [10 lx at P_1, 21 lx at P_2, 3 lx at P_3 and 0.8 lx at P_4. Average illuminance in rectangle from the one floodlight is 7.4 lx. Note, because of the uncertainty of interpolation these are approximate answers.]

3. In a floodlighting scheme, the coordinates of a floodlight are (20, 10, 20), and it is aimed at (35, 8, 0). What are the (B, β) coordinates of a point P with coordinates (30, 15, 0)? What is the angle of incidence and the illuminance of the light on a horizontal surface at P if the luminous intensity from the floodlight is 60 000 cd? What is the illuminance at P on a surface normal to the direction of the camera, which is positioned at (32, 17, 3)? All dimensions are in metres.

   Answers: [$(B, \beta) = (12.3°, 15.9°)$, $\theta_{H+} = 29.2°$, $E_{H+} = 99.8$ lx, $\theta_C = 77.5°$, $E_C = 24.8$ lx.]

Bibliography

CIBSE Lighting Guide 4: 1990. Sports

CIBSE Lighting Guide L6: 1992. The Outdoor Environment

Coaton, J. R. and Marsden, A. M. (1999) *Lamps and Lighting* (Arnold) 445–478.

References

1. Brackett, W. E. (1978) Automatic floodlight aiming optimization. *JIES*, 223–233.
2. CIE Publication 42 (1978) *Lighting for Tennis.*
3. CIE Publication 45 (1979) *Lighting for Ice Sports.*
4. CIE Publication 57 (1983) *Lighting for Football.*
5. CIE Publication 58 (1983) *Lighting for Sports Halls.*
6. CIE Publication 62 (1984) *Lighting for Swimming Pools.*
7. CEN draft European Standard (1997) *Lighting Application: Sports Lighting*. CEN/TC 169/WG 4 N 109E.
8. CIE Publication 83 (1989) *Guide for the Lighting of Sports Events for Colour Television and Filming.*
9. CIE Publication 112 (1994) *Glare Evaluation System for Use Within Outdoor Sports and Area Lighting.*
10. van Bommel, W. J. M., Tekelenburg, D. and Fischer, D. (1983) A glare evaluation system for outdoor sports lighting and its consequences for design practice. *Proceedings CIE 20th Session*, Amsterdam, CIE Publication Number 56. Paper D505, pp. 1–4.
11. Dovey, G. M., Peirce, M. W. and Price, W. A. (1964) *Trans IES (London)*, **29**, 29.
12. Brackett, W. E. (1978) Automatic floodlight aiming optimization. *JIES* 223–233.
13. Heard, F. W., Stone, F. H. S and Jewess, B. W. (1976) Effect of atmospheric attenuation on exterior lighting design. *Lighting Research and Technology*, **8**, 151–156.

13

Specific Applications: Airfield Lighting and Emergency Lighting

13.1 Airfield Lighting

13.1.1 INTRODUCTION

Airfield lighting differs from the other lighting systems described in this book in that, typically, the lights are not used to illuminate surfaces but are used as signals; that is, the luminaires are viewed directly. Perhaps the nearest case to this so far described is in road lighting where the road lighting luminaires are used to give visual guidance, but even here their main function is to provide a bright road surface. The main reason for not lighting the surface of the runway is that no part of the luminaires is allowed to protrude more than 150 mm from the runway, so as not to obstruct the aircraft. As a consequence of this low mounting height, the light would be incident at grazing angles. This would tend to exaggerate the appearance of small irregularities in the runway and, in any case, would give a very patchy appearance. Moreover, the lit surface would not have the high luminance needed for the airfield to stand out at a distance from other lit surfaces, such as roads.

The most critical functions of airfield lighting are to enable the pilot to identify the airfield and to provide a lighting pattern to help in landing the aeroplane safely. It should then guide the pilot to the apron for unloading. The lighting also has the converse task of guiding the pilot from the apron to the runway, giving a signal to indicate whether the runway is clear for take-off, and finally providing guidance for the take-off.

Whilst a number of airports have provision for blind landing of aircraft there is still a place for visual landing aids even at these airports; visual monitoring gives the pilot confidence that the landing is proceeding safely.

For the provision of visual aids, airports are classified according to the visibility conditions in which they can be used. These conditions are measured in terms of runway visual range (RVR) and are used to create four categories.

(1) Unrestricted visibility.
(2) Category I: RVR down to 800 m.
(3) Category II: RVR down to 400 m.
(4) Category III: RVR down to 200 m for Category IIIA, RVR down to 50 m for Category IIIB, and zero for Category IIIC.

In the UK, an airport would have to be lit to Category III to enable landings to take place in all the weather conditions that are likely to occur. The situations in other parts of the world might well be different.

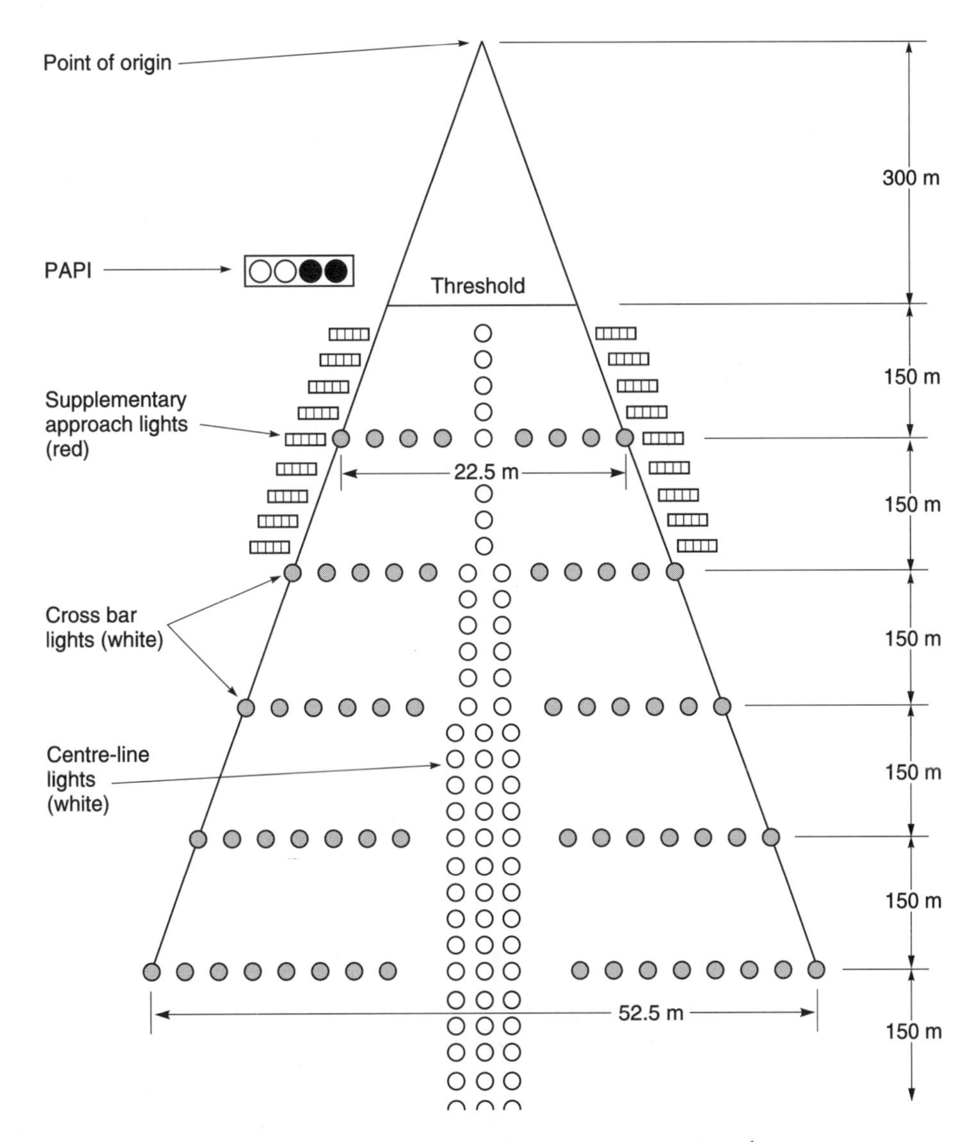

Fig. 13.1 Perspective of approach lighting pattern (after Bridgers and Richards[1])

13.1.2 THE APPROACH LANDING LIGHTS

Figure 13.1 shows a typical configuration of landing lights in the approach zone.[1,2] These are usually 200 W or 300 W tungsten halogen lamps in paraboloidal reflectors. Although ideally mounted at ground level, they are usually mounted on poles to take account of the ground contours. They all give white light signals.

Apart from the angle of approach indicators, which will be discussed separately, there are two essential elements in the pattern. These are the centre-line lights and the cross bar lights. Calvert,[3] of the Royal Aircraft Establishment, produced the theory, the parafoveal streamer theory, on which this pattern is based. This makes use of the fact that if an observer is moving in a straight line towards a point, the point of origin in Figure 13.1, all other points appear to stream from that point even if the origin is not visible to the pilot. The point of origin, the reader should note, does not mark the end of the runway. It is produced by the progressive narrowing of the cross bars as indicated on the figure.

The pilot, then, maintains a symmetrical streamer position. The rate at which the radial streamers in the field of view varies with the height and the rate of descent are two essential items of information. In addition the centre-line lights enable the pilot to line up the aircraft with the runway, and the cross-bar lights enable the pilot to keep the aircraft horizontal from wing tip to wing tip; that is, to keep the roll at zero. To do this the cross-bar lights are kept level with the horizontal edges of the windscreen. They must also be at right angles to the centre-line otherwise the aircraft is displaced to one side of the runway.

The cross-bars give some indication of the angle of descent. They should disappear below the windscreen at a constant rate. If they disappear above the windscreen the rate of descent is too steep. However, as already mentioned, more precise information is given by the angle of approach indicators.

The supplementary approach lights each consist of four lighting units grouped together. They are coloured red to provide contrast with the threshold lights, which are green, and serve as an additional warning to the pilot not to land short.

13.1.3 THE RUNWAY LIGHTING

The runway touchdown starts at the threshold lights. All the luminaires (Figure 13.2) are recessed except for those at the edge, which may protrude. The centre-line lights are colour

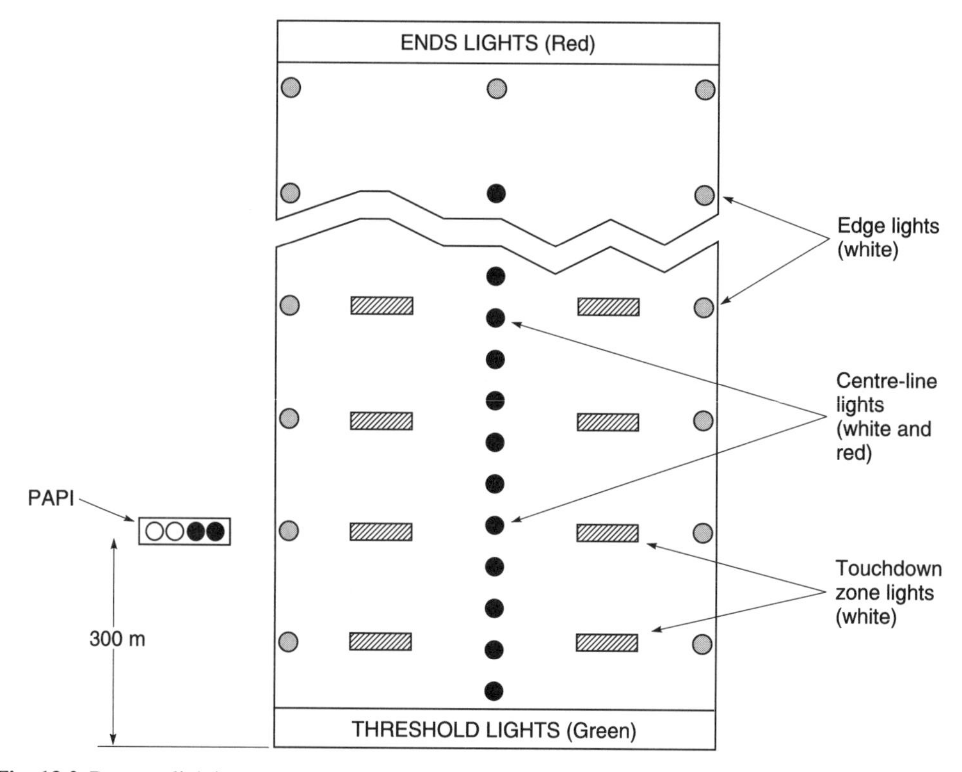

Fig. 13.2 Runway lighting

coded to provide *distance-to-run* information. Nearest the threshold they are white, they then alternate with stretches of red and white lights and end with red lights towards the end of the runway.

The requirements for the runway lights are onerous optically, thermally and mechanically and their design is described in Section 13.1.5.

13.1.4 ANGLE OF APPROACH INDICATORS AND THEIR EVOLUTION

The evolution of the design of angle of approach indicators makes an interesting story because it illustrates how their performance has been improved in the course of time and the method of indication simplified.

The first indicators were, in effect, projectors with three sectors: the top being yellow, the middle green, and the bottom red. An indicator was located to one side of the touchdown point and the pilot had to follow the yellow sector down. There were two main drawbacks. First, as the pilot approached touchdown the yellow sector became too narrow to follow and he or she had to ignore it. Second, it was not sufficiently bright to distinguish it easily from the other airfield lights at night and, in any case, could not be used during daytime.

In about 1957 this was replaced by the VASI (Visual Approach Slope Indicator) developed by Calvert[4] and Sparke.[5] Basically, this provides a parallel corridor down which the pilot flies the plane as illustrated in Figure 13.3.

The corridor is formed by two lines of units, six on either side of the runway as in Figure 13.4.

Each unit or indicator gives a sector of white light and a sector of red light separated by a narrow zone of pink light subtending an angle of only $\frac{1}{2}°$. When the pilot is in the landing corridor he or she sees the red sector from the rear line of units and the white sector from the front line of units. If the pilot is too low, both lines of units appear red, if too high both appear white.

The original design of the units is shown in Figure 13.5. The size of the filter in relation to the size of the slot and their separation would indicate that the pink zone should extend over $\tan^{-1}$ (50/1500) or 2°. However, owing to the white sector being much brighter than the red sector, the pink zone is only visible as such over $\frac{1}{2}°$. Three or four 200 W lamps are used in each indicator.

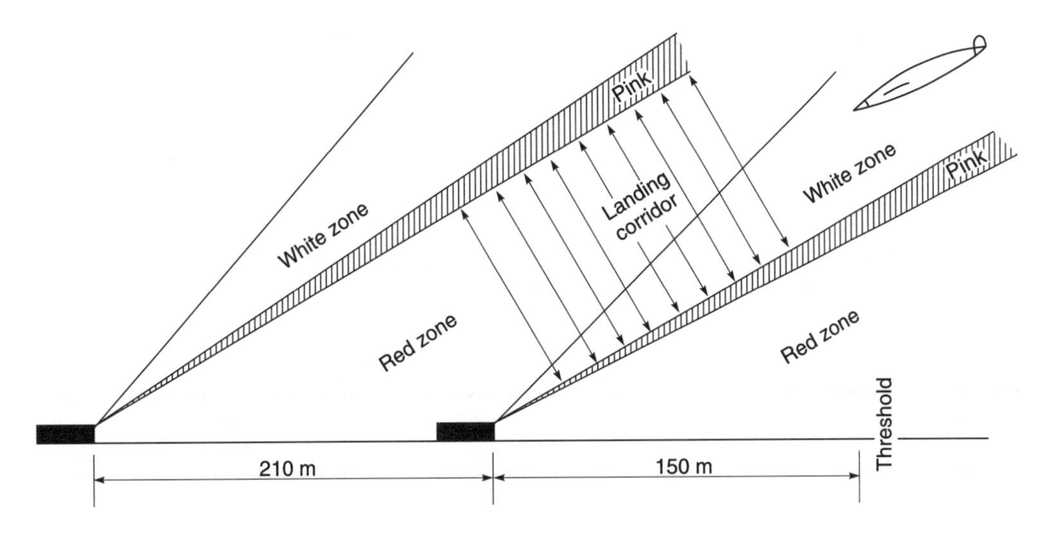

Fig. 13.3 Landing corridor formed by VASI units

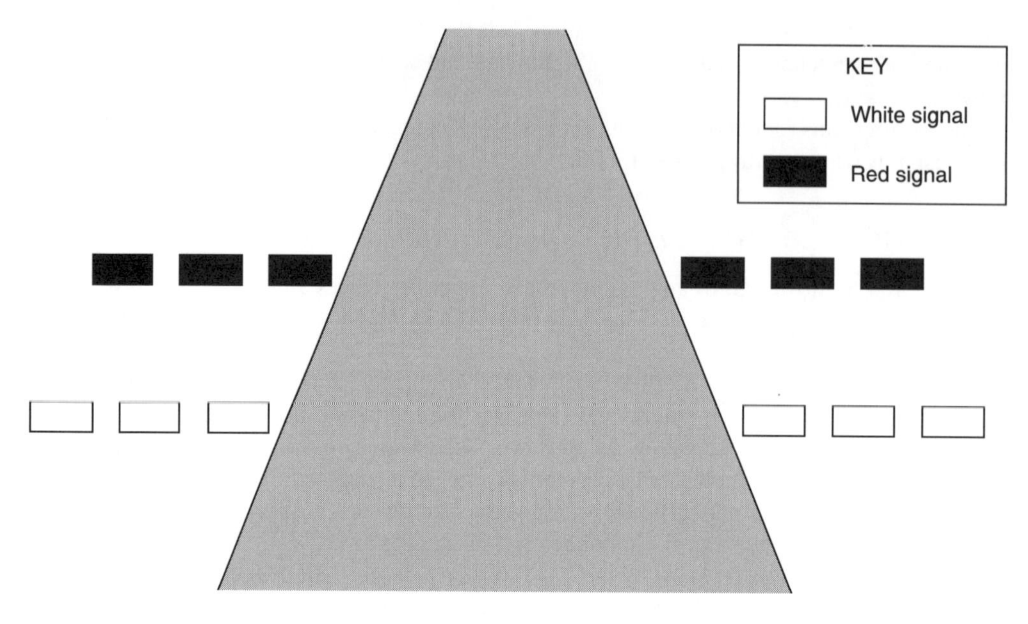

Fig. 13.4 Pilot's view of VASI when at the correct glide slope

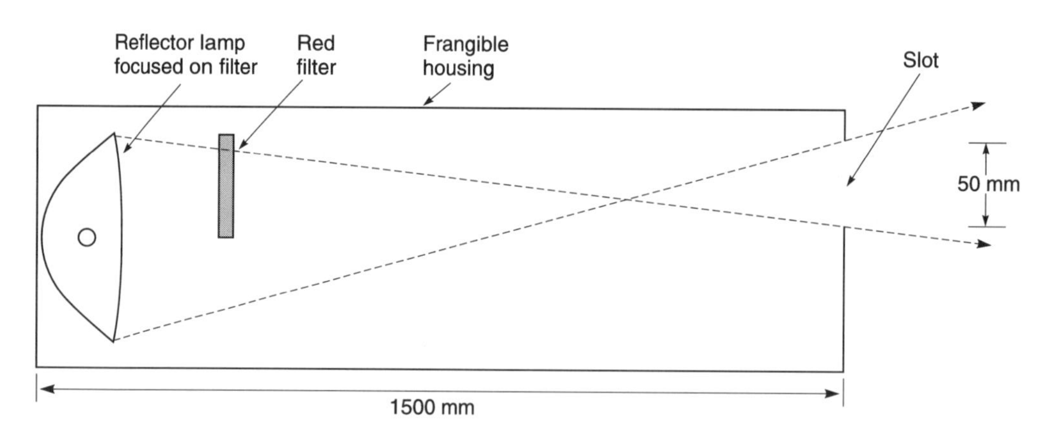

Fig. 13.5 Optical construction of VASI unit

The manufacture of this type of unit is straightforward in that alignment of the principal components, the slot, the filter, and the lamp, can be simply achieved. However, it is large (1500 mm by 1500 mm by 500 mm high) and, as a consequence, is susceptible to movement by wind and jet-blast. Moreover, it is cumbersome to transport. As a result, it was replaced by a projector system.[6] This allows the unit to be much smaller (typically 400 mm by 300 mm by 230 mm high) and to provide a sharper transition from red to white, which is required in the PAPI (Precision Approach Path Indicator) system to be described next.

Figure 13.6 shows the view the pilot has of the PAPI system according to the angle of approach. There are four separate PAPI units on the left of the runway, set at slightly different angles to each other in the way to be described. The angles on the diagrams refer to the angle of the aircraft's approach path in relation to the correct approach path, which is normally at an elevation of 3°. When the aircraft is on the correct approach path two red and two white signals are visible. If the approach path is too high, progressively three and then four white signals come into view. Conversely, if the approach path is too low, progressively three then four red signals come into view. The two main advantages of this system are that it gives the rate information to the pilot, that is it tells him or her by how much the aircraft is straying from the correct approach path, and, because there is only a single line of units, it can be lined up exactly with the ILS (the instrument landing system). In addition, only four units are needed in contrast to the 12 needed for the VASI system.

Figure 13.7 shows a cross-section through the projection system. The 200 W tungsten halogen lamp is housed in an ellipsoidal reflector that focuses light onto the red filter. An image of this is formed at infinity by the front lenses.

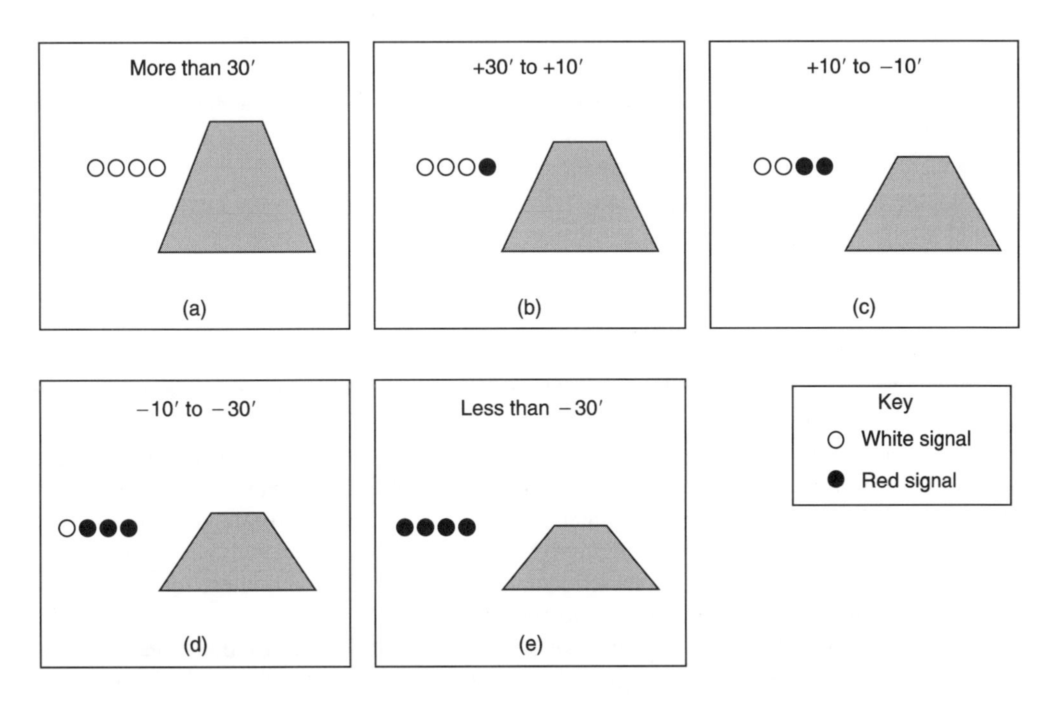

Fig. 13.6 Pilot's view of the PAPI. (a) Much too high; (b) too high; (c) correct glide path; (d) too low; (e) much too low

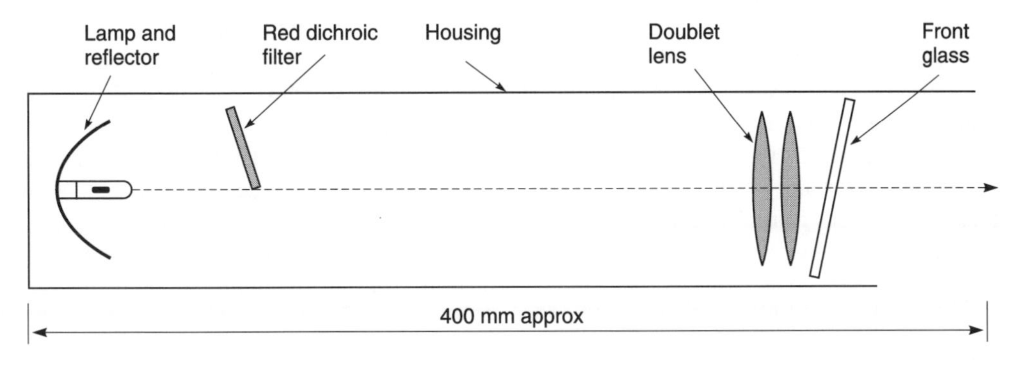

Fig. 13.7 Optical system of a PAPI unit

The angular width of the transition from red to white needs to be as sharp as possible, about three minutes of arc, which can be obtained by a good quality optical system. Two components in particular are crucial; the filter and the front lens.

The filter should be as thin as possible. Normally red filters are made from through-coloured borosilicate glass, the colouring agent being selenium. The glass has to be borosilicate to withstand the heat generated by the radiation focused on it. This type of filter is generally made from a blown cylinder and tends to be slightly curved and somewhat variable in thickness. Moreover, the colouring agent may not be homogeneously dispersed in the glass. It is also worth mentioning that the glass tends to become less red as it heats up, although not to the extent that it no longer complies with the relevant colour signal specification. A dichroic interference filter, which is a thin layer vacuum-deposited on a substrate of only 0.5 mm thick borosilicate glass for this application overcomes these problems, but brings one difficulty in its train.[7] Light that is reflected by the filter is green, and tends to dilute the signal if it is misdirected. This is the reason for the filter being tilted (exaggerated in the figure).

To obtain the required luminous intensity the front lens has to work at an aperture of *f*2,[a] and be fully flashed at this aperture. Because of spherical aberration, a single lens is incapable of giving the desired performance and a doublet lens is needed.

The front glass is tilted down to direct reflected light from the sky and elsewhere onto the ground.

13.1.5 INSET LUMINAIRES

As the name implies, inset luminaires are set into the pavement and are used for both the runways and taxiways. Typically they should be no more than 300 mm in diameter and protrude no more than 17 mm above the pavement. The restriction in diameter is to keep civil engineering costs to a minimum, and the restriction in height is to obviate damage to aircraft tyres, and to allow snow ploughs to run over them (in countries where this is applicable), as well as the use of arrester hooks. For runway inset luminaires they must be able to withstand the impact and vibration caused by aircraft landing on them. Moreover, the heat generated must be limited to obviate damage to aircraft tyres. All these requirements are in addition to those for photometric performance and are indeed onerous, but are possible to meet.

Figure 13.8 shows the outside appearance of a typical unit. This is a bidirectional unit and therefore has two windows facing in opposite directions, although not necessarily diametrically opposite since the beams may need to be displaced in azimuth to one side of the centre-line. Omnidirectional luminaires have a complete glass ring to allow the light to emerge in all azimuths. The body casting is sunk into the pavement and bolted onto a permanent base. To allow for easy servicing, which has to be done quickly so as to keep the runways and taxiways in use, the body casting can easily be released by undoing the bolts to allow the optical unit to be removed and replaced by a serviced unit. The removed unit, which is typically only 100 mm in diameter and 30 mm deep, can then be serviced in a workshop at leisure. Besides allowing fast maintenance, this type of design saves having to transport heavy and bulky parts. As part of the lens lies below the surface of the pavement, drainage has to be provided to prevent the gully in front of the lens from filling with water.

Figure 13.9 shows a section through the optical system for a bi-directional unit. The lamp is a 24 V tungsten halogen with its filament configured to give the maximum luminous intensity towards the lenses. Whilst a discharge lamp would be more efficacious in terms of lumens per

[a] The ratio of the focal length of the lens to its diameter. See page 239.

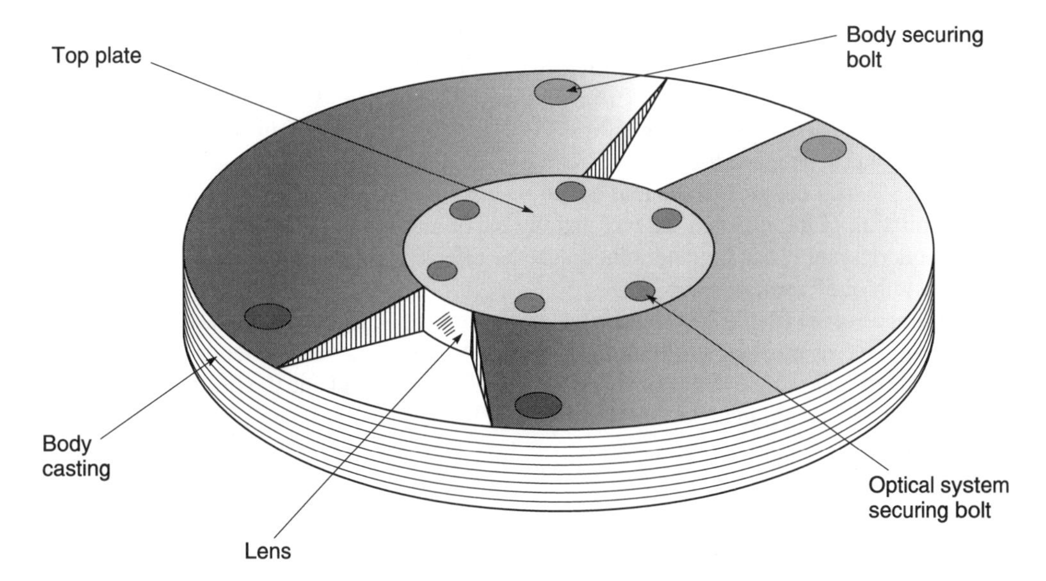

Fig. 13.8 Typical inset unit

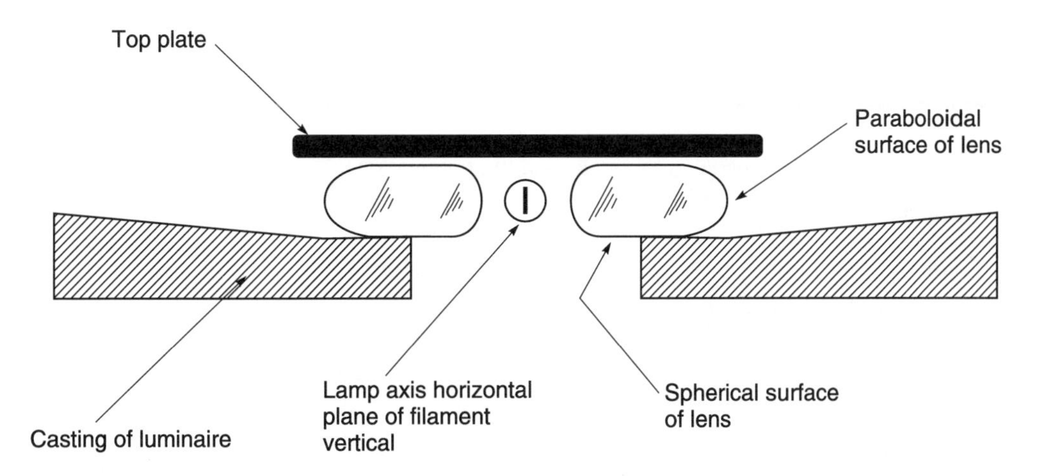

Fig. 13.9 Optical system of a bi-directional runway inset luminaire

watt, the need for instant restriking presents a problem as does dimming. The maximum power consumption by the lamp is limited to 200 W because of restrictions on the temperature of the top plate, as already mentioned.

The outer surface of the lens is paraboloidal to correct for spherical aberration, which would otherwise occur as the lens is required to work at a high *f* number. This allows the maximum possible luminous intensity to be achieved. Where a coloured signal is required, a suitable filter is inserted in the system.

Photometric requirements for these luminaires are laid down by the International Civil Aviation Authority.[8] The bi-directional inset runway luminaire requires two beams at 180° in azimuth, each with a peak luminous intensity of 7500 candelas above the horizontal. From the complete specification for the light distribution it is possible to calculate that 400 lumens are required in each beam. These data allow the size of the window to be estimated.

The forward luminous intensity of the lamp is 420 candelas. Its luminance is 18 cd/mm^2, hence the required area of the lens is 7500 ÷ 18 = 417 mm^2. This is the minimum figure because an allowance has to be made for transmission losses and incomplete flashing of the lens. If these losses are 20% then the area of the lens has to be increased to 500 m^2 approximately. From the height limitation of the unit and the fact that the maximum luminous intensity is required at 3° above the horizontal it can be determined that the effective height of the lens will be 10 mm, giving a width of 50 mm.

The distance of the filament from the back of the lens can be calculated by taking into account the fact that the lens must emit 400 lm in the beam. A loss of 20% in the lens has already been used but an extra allowance has to be made for the light falling outside the beam, probably raising the losses to 30%. From this we find that 400 ÷ (1 – 0.30) = 570 lm will have to be collected.

Now

$$\begin{aligned}\Phi &= I \times \omega \\ &= I \times \frac{A}{d^2}\end{aligned}$$

where

Φ is the luminous flux collected by the lens (lm);
I is the luminous intensity of the lamp directed towards the lens (cd);
A is the area of the lens (mm);
d is the distance of the filament to the back of the lens (mm);
ω is the solid angle subtended by the lens at the filament (sr).

From this, we have

$$\begin{aligned}d &= \sqrt{(I \times A) \div \Phi} \\ &= \sqrt{(420 \times 500) \div 570} \\ &= 19 \text{ mm}\end{aligned}$$

This is very approximate for the reasons we have stated, and because the filament is regarded as a point source of light in order to calculate the solid angle ω and because the luminous intensity is regarded as constant in the direction of the lens. However, these calculations enable the feasibility of achieving the required light distribution to be assessed and give a good starting point for the detailed design work.

For an actual inset luminaire, the achieved utilization of luminous flux in each beam has been measured and found to be of the order of 12%.[9]

13.1.6 OTHER AIRFIELD LUMINAIRES

Lights are provided to light the taxiways between the runway and the apron where the aircraft are loaded and unloaded. These are either green centre-line lights or blue edge luminaires, which may be elevated.

The apron may be equipped with visual aids to enable the pilot to dock his or her plane accurately in position. These consist of luminaires where two light sources have to be lined up with each other.

Amongst the assortment of other lights the wig-wags must be mentioned, if only because of their appealing name. These are used at holding areas leading to and from the runway. They consist of two lights that are alternately switched on and off. They give a signal that cannot be missed because of an apparent three-dimensional effect. There are three colours: red, amber, or clear, which signal respectively: do not pass (on entrance to runway), hold until cleared (on entrance to runway), and exit runway to apron.

13.2 Emergency lighting

13.2.1 THE NEED

The function of emergency lighting is to provide sufficient light for the occupants of a building to escape when the normal lighting fails. The reason for failure may be that the mains supply has been interrupted by an outside agency or, in the worst case, it may be due to fire. The lighting should be designed to prevent 'panic', which is the word most used when a disaster is being investigated.[10]

The emergency lighting is used in conjunction with emergency lighting signs that guide people along the designated escape route and constitute an essential part of the escape system.

In Europe, a CEN Standard has been prepared on this subject.[11]

13.2.2 PROVISION OF EMERGENCY LIGHTING

There are three ways of making provision for emergency lighting.

(1) Stand-by generators. These provide an almost immediate return to power and duration is almost limitless. They may be the only solution when power is needed for running other essential services, as may be the case in a hospital. They need good maintenance and, of course, space.
(2) Central battery systems. These have the same advantages as stand-by generators and, similarly, they need dedicated space, but with special ventilation. They need constant attention to ensure that they are kept in working order.
(3) Self-contained emergency lighting luminaires. These are luminaires with their own battery packs which come into operation when the mains fails, and are therefore independent of a central distribution system. There are different types of such luminaires:

- non-maintained luminaire. This has a lamp that only comes into operation when triggered;
- maintained luminaire. This has a lamp that is in operation when required for normal purposes and when the mains fail;
- sustained luminaire. This contains two or more lamps, at least one of which is used for normal operation and one for emergency conditions.

13.2.3 TYPES OF EMERGENCY LIGHTING AND THEIR REQUIREMENTS

(1) Escape route lighting. As the name implies, this is lighting along the escape route. The average level recommended in the UK is 0.2 lux,[12] based on the work of Simmons.[13] However Boyce's[14] investigations suggest that this should be treated as a minimum and 1 lux allows faster and surer movement. In fact, other European countries feel that 0.2 lux is inadequate and want it raised to 1 lux. The central band of the escape route should be illuminated to at least half this value. 0.2 lux may be retained for escape routes without hazards.

(2) Anti-panic area lighting. This applies in large open areas such as halls and offices without defined escape routes. The lighting should enable occupants to move safely towards the escape routes. The proposed minimum illuminance is 0.5 lux, with a ratio of the maximum to the minimum illuminance not greater than 40:1.
(3) High risk task area lighting. This applies to areas where hazardous activities are taking place and the aim is to allow those people engaged in these to close down the operations safely and then escape. The lighting level proposed is 15 lux or 10% of the normal lighting level, whichever is the greater. As for anti-panic lighting, the ratio of the maximum to the minimum illuminance should not be greater than 40:1.

There are other lighting requirements besides the ones stated above. Disability glare is controlled by limiting the maximum luminous intensity of the luminaire in directions from 60° to 90° to the downward vertical. The limit increases with mounting height. The illuminance must reach 50% of the required illuminance within 0.5 seconds and its full value within 60 seconds for escape route lighting and anti-panic lighting. For high risk task area lighting the required illuminance must be reached within 0.5 seconds. The colour rendering index of the lamps R_a should be at least 40 to allow colours to be identified reasonably well.

The CEN draft Standard states that the lighting should be from above. However, there are low-mounted way-guidance systems, which have a mounting height of no more than about one metre. These can take a number of forms; a track of miniature tungsten filament lamps, a track of electroluminescent lamps, or a strip of photoluminescent material, and may consist of the following:[15]

- escape route marking;
- exit door frame marking;
- exit door signs;
- immediate direction signs;
- low mounted lighting.

Investigation of way-finding systems appears to show that they are at least as effective as the conventional forms of emergency lighting, and may have some advantages in smoke conditions.[16] They are at present used in specialized situations such as aircraft.

References

1. Bridgers, D.J. and Richards, H. (1974) Airfield lighting. *Lighting Research and Technology,* **6**(3), 144.
2. Johnson, B. (1984) *Airport lighting systems. CIBS National Lighting Conference*, pp. 44–52.
3. Calvert, E.S. (1950) *Trans. IES (London)*, **15**, 6.
4. Calvert, E.S. (1957) The theory of visual judgements in motion and its application to the design of landing aids for aircraft. *Trans IES (London)*, **22**(10), 271–297.
5. Sparke, J.W. (1958) *Royal Aircraft Establishment – Technical Report EL 160.*
6. Bridgers, D.J. (1978) *Veni; Vidi; VASI? IES – CIBS.* National Lighting Conference, pp. D1–D6.
7. Halstead, M.B., Jewess, B.W., Simons, R.H. and Wilkinson, B.S. (1984) *Colour problems in airfield landing slope indicators. CIBS National Lighting Conference*, pp. 53–68.
8. (1971) *International Standards for Aerodromes.* Annex 14, Convention on International Civil Aviation, 6th Edition.

9. Richards, M.J. (1970) The development of a taxiway light. *Lighting Research and Technology,* **2**(3), 186.
10. McKenna, B. and Morin, T. (1997) *Emergency lighting.* SANCI-CIE International Conference – Lighting in Developing Countries, pp. 106–108.
11. CEN Standard. EN 1838 (1999) *Lighting Applications – Emergency Lighting.*
12. (1986) Code of Practice for the emergency lighting of premises. *BS 5266 Part 1.*
13. Simmons, R.C. (1975) Illuminance, diversity and disability glare in emergency lighting. *Lighting Research and Technology*, **7**(4), 125.
14. Boyce, P.R. (1985) Movement under emergency lighting: the effect of illuminance. *Lighting Research and Technology*, **17**(2), 51.
15. (1993) Draft CEN Standard CEN TC 169 WG3B N7. *Emergency Lighting*: *Low Mounted Way Guidance Systems – Electrically Powered.*
16. Aizlewood, C.E. and Webber, G.M.B. (1995) Escape route lighting: comparison of human performance with traditional lighting and wayfinding systems. *Lighting Research and Technology*, **27**(3), 133.

14
Daylight Calculations

14.1 Introduction

'Daylighting' presents what might seem at first glance to be a set of impossible problems; one of which is that daylight is continuously variable and another is that, when the sky is clear, the position of the sun is the major determining factor, which itself changes both throughout the day and the year. When the sky is overcast it has a variable luminance, both in distribution and magnitude.

Much study and thought has been given to these factors and appropriate assumptions and approximations have been adopted, so that daylighting of buildings can be quantified.

In countries such as the UK, where completely overcast skies are a common feature, the criteria are based upon the relative illuminance inside a room to that under the unobstructed overcast sky at the same moment in time. The ratio of these illuminances is called the daylight factor (*df*) and is expressed as a percentage. The daylight factor includes interreflected light and can be for a point, a surface or the average for the room.

In this chapter, we will consider first the completely overcast sky, since this is the basis of most UK daylight design.

14.2 The overcast sky

The CIE have published an agreed empirical formula, which is in general use for an overcast sky. This is:

$$L_\theta = \frac{1}{3} L_z(1 + 2 \sin \theta)$$

where L_θ is the luminance of the sky at an angle θ to the horizontal and where L_z is the luminance at the zenith, in cd/m^2.

Figure 14.1(a) shows a representation of a sky dome of radius r. The illuminance at P due to an element of the sky is given by:

$$\delta E = \frac{L_{\theta\alpha} r\delta\theta r\delta\alpha \cos \theta \sin \theta}{r^2}$$

$$= L_{\theta\alpha} \sin \theta \cos \theta\, \delta\theta\delta\alpha$$

where $L_{\theta\alpha}$ is the luminance at an angle θ to the horizontal and at the azimuth angle α, measured from a specified direction.

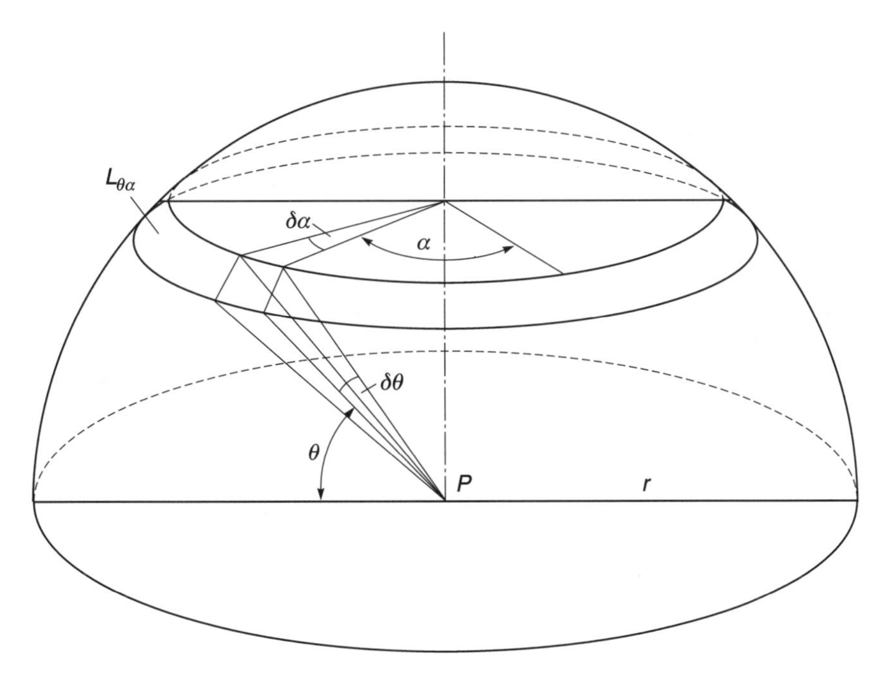

Fig. 14.1(a) Calculation of illuminance from an overcast sky

If the luminance of the overcast sky is assumed to be given by the CIE formula, then

$$E = \int_{\alpha_1}^{\alpha_2} \int_{\theta_1}^{\theta_2} \frac{1}{3} L_z (1 + 2\sin\theta) \sin\theta \cos\theta \, d\theta \, d\alpha$$

$$= \int_{\alpha_1}^{\alpha_2} \int_{\theta_1}^{\theta_2} \frac{1}{3} L_z (\sin\theta\cos\theta + 2\sin^2\theta\cos\theta) \, d\theta \, d\alpha$$

$$= \frac{1}{3} L_z \left[\frac{1}{2}\sin^2\theta + \frac{2}{3}\sin^3\theta \right]_{\theta_1}^{\theta_2} \times \int_{\alpha_1}^{\alpha_2} d\alpha$$

$$= \frac{1}{3} L_z \left(\frac{\sin^2\theta_2 - \sin^2\theta_1}{2} + \frac{2\sin^3\theta_2 - 2\sin^3\theta_1}{3} \right) (\alpha_2 - \alpha_1)$$

for the whole sky $\theta_2 = \pi/2$, $\theta_1 = 0$ and $\alpha_2 = 2\pi$, $\alpha_1 = 0$

giving,

$$E_{sky} = \frac{7\pi}{9} L_z \text{ lux}$$

and the sky component of the daylight factor due to any element at $\delta\theta$, $\delta\alpha$ at θ, α

$$= \frac{1}{3} L_z(1 + 2 \sin \theta) \sin \theta \cos \theta \, \delta\theta\delta\alpha \div \frac{7\pi}{9} L_z$$

$$= \frac{3}{7\pi} (1 + 2 \sin \theta) \sin \theta \cos \theta \, \delta\theta\delta\alpha$$

Note It is sometimes considered difficult to conceive of the sky as having a radius (r in the earlier equation) and so it is necessary to translate the problem into one where the distance in question is simple to determine.

If we consider how the sky luminance is measured we have, in the simple case, an aperture through which the sky is illuminating a photocell at a known distance, see Figure 14.1(b). The aperture and the photocell are separated by a distance d such that the inverse square law can be applied, and are opposite and parallel.

Then

$$E = \frac{I}{d^2}$$

where I is the intensity at the aperture produced by the patch of sky seen through the aperture and a is the area of the aperture.

Then

$$E = \frac{L \times a}{d^2}$$

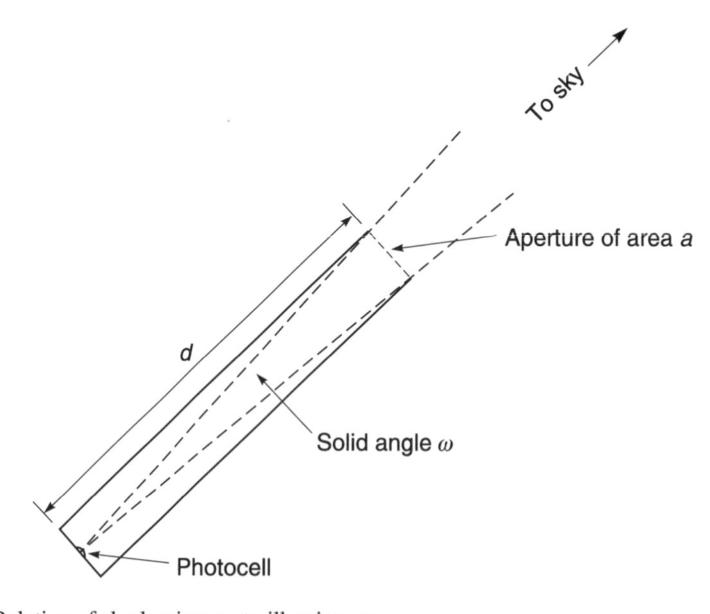

Fig. 14.1(b) Relation of sky luminance to illuminance

If the aperture is small enough a/d^2 may be taken as equal to ω, the solid angle subtended at the photocell by the element of sky considered.

So, $E = L\omega$ or $L = E/\omega$.

If we consider the hemisphere not as a sky dome but as a means of calculating ω and make $r = 1.0$ we arrive at $\omega = a/r^2 = a$ or $\omega = \cos\theta\ \delta\theta\delta\alpha$ and continue as before.

14.3 Window area

At an early stage in the design, it is very useful to have an estimate of the window area that will be required for a given daylight factor. This estimate can be based on a simple calculation of the average daylight factor for all the room surfaces, using a modification of the formula for an integrating sphere (see Section 5.2, Chapter 5).

From Chapter 5,

$$E_{av} = \frac{F_L}{A(1-\rho)} \text{ lux}$$

where F_L in the input luminous flux (lumens), A is the internal surface area (m^2) and ρ the reflectance of the inside of the sphere, and E_{av} the average illuminance on the inside of the sphere. In the case of a room lit by a window, F_L would be the flux that enters via the window.

In a room where the room surfaces had different reflectances, it would be necessary to use the average reflectance of all room surfaces in place of ρ. Let this average reflectance be R then the formula becomes

$$E_{av} = \frac{F_L}{A(1-R)}$$

where

$$R = \frac{\rho_C A_C + \rho_F A_F + \rho_W A_W}{A_C + A_F + A_W}$$

and where ρ_C, ρ_F, ρ_W are the ceiling, floor, wall reflectances, etc., A_C, A_F, A_W are the ceiling, floor, wall areas, etc.

The term F_L now requires consideration.

If F_L is replaced by $TW(\theta/2)$ where T is the transmittance of the glazing, W is the area of the glazing and θ is the vertical angle in degrees of unobstructed sky measured from the middle of the window (see Figure 14.2) the formula becomes

$$df = \frac{TW\theta}{2A(1-R)} \% \qquad (14.1)$$

where df is the average daylight factor for the room.

This expression depends upon the fact that the value $\theta/2$ has been found to be a reasonable approximation to the ratio of the illuminance on the vertical face of the window to that on the horizontal from an unobstructed overcast sky, as a percentage.

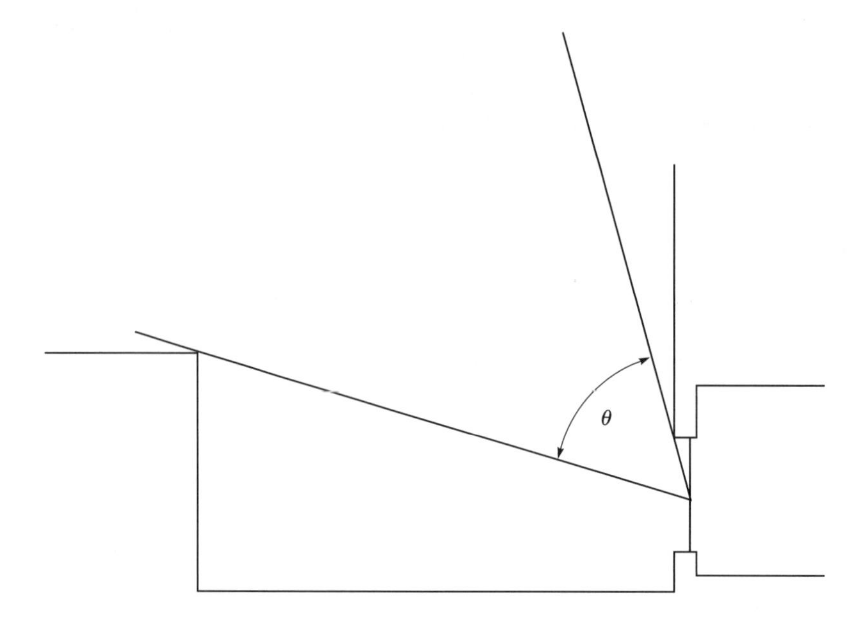

Fig. 14.2 The vertical angle θ of unobstructed sky measured from the middle of the window

Thus,

$$TW\frac{\theta}{2} \times \frac{E_{sky}}{100}$$

is equal to the luminous flux passing through the window into the room.

This gives

$$\frac{E_{av}}{E_{sky}} \times 100\% = \frac{TW\dfrac{\theta}{2}}{A(1-R)} = \text{average daylight factor within the room}$$

The relationship between the illuminance on the vertical window surface and the angle of unobstructed sky was pointed out by Lynes.[1]

The illuminance ratios (C) used in the Lynes comparison were those obtained by measurements from Waldram Diagrams for glazed windows (see Section 14.5.3), assuming that the obstructions were horizontal bands of luminance one-tenth of the mean sky illuminance.[2] These ratios do not include light incident from below the horizontal. If the average luminance of the ground is assumed to be one-tenth of the mean sky luminance then, also assuming the ground to form a half-infinite plane, the increase in this vertical illuminance will be 5%. The total illuminance ratio on the vertical plane is then $C + 5$.

Table 14.1 shows the good agreement between this value and $\theta/2$ in degrees, apart from $\theta = 10°$.

The simple expression in Equation (14.1) gives the average daylight factor for all the room surfaces. The formula could be transposed to give a value for W, the window area, needed to

Table 14.1 Comparison between the $C + 5\%$ and the $\theta/2$ values

θ	Angle of obstruction	C (%)	$C + 5$ (%)	$\theta/2$
90°	0°	39	44	45
80°	10°	35	40	40
70°	20°	31	36	35
60°	30°	25	30	30
50°	40°	20	25	25
40°	50°	14	19	20
30°	60°	10	15	15
20°	70°	7	12	10
10°	80°	5	10	5

achieve a specified value of daylight factor. However, recommendations for daylight factors are usually given in terms of the average daylight factor required on the working plane; that is, at desk or work bench height above the floor. Crisp and Littlefair have produced a modified version of the Lynes formula to give a better estimate of the daylight factor for the working plane.[3] This was done by considering the relationship between the average illuminance for all the room surfaces and that for the working plane. Selecting a square room, half as high as it is wide and considering a range of reflectance values, it indicates that an approximate relationship is given by

$$E_{av} = E_{WP}\left(\frac{1 + R}{2}\right)$$

where E_{WP} is the average working plane illuminance. If this relationship is accepted then we may write

$$\frac{TW\theta}{2A(1 - R)} = df_{WP}\left(\frac{1 + R}{2}\right)$$

where df_{WP} is the average daylight factor on the working plane giving

$$df_{WP} = \frac{TW\theta}{A(1 - R^2)}\ \%$$

Transposing this formula to give the required window area,

$$W = \frac{df_{WP} \times A(1 - R^2)}{T\theta}\ \text{m}^2$$

As a simple example, let us estimate the window area required to give a 2% daylight factor at the working plane. Let us assume that the vertical angle of unobstructed sky is 45°, T is 0.8, R is 0.5 and the room area 64 m^2.

$$W = \frac{df_{WP} \times A(1 - R^2)}{T\theta}\ \text{m}^2$$

$$= \frac{2 \times 64(1 - 0.25)}{0.8 \times 45}$$

$$= 2.7 \text{ m}^2$$

(Note: in this example no allowance has been made for dirt on the windows.)

From the above, it can be seen how easy this formula is to use as a first step in window design and it has found much favour. Figure 14.3 shows the results obtained when this formula is used to estimate the average working plane daylight factor for a model room.

Also shown in Figure 14.3 are the results obtained when using the Longmore development from the BRS split-flux formula for the working plane daylight factor.[4] The method is termed 'split-flux' beause it is based on the assumption that the light arriving at the window from above the horizontal is assumed to be incident upon the lower part of the room surfaces and flux arriving from below the horizontal on the upper surfaces of the room.

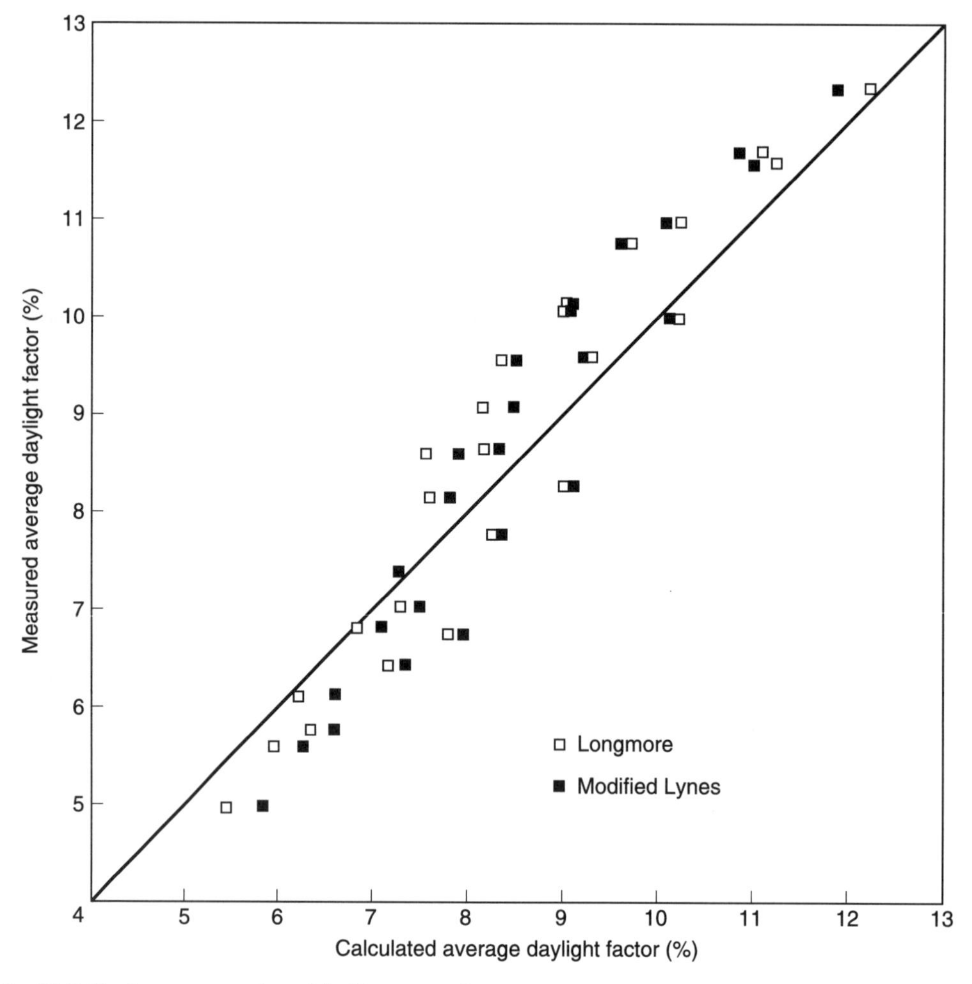

Fig. 14.3 The Longmore and modified Lynes results

The Longmore formula is

$$df_{WP} = 0.85W\left(\frac{C}{A_{FW}} + \frac{CR_{FW} + 5R_{CW}}{A(1-R)}\right)\%$$

If the assumptions regarding the glass transmittance and the ground luminance are removed, this expression may be written:

$$df_{WP} = TW\left(\frac{C}{A_{FW}} + \frac{CR_{FW} + D_G R_{CW} R_G}{A(1-R)}\right)\%$$

Here, C is the illuminance ratio component on the vertical surface of the window due to flux incident from above the horizontal, including that reflected from obstructions and $D_G \times R_G$ is the illuminance ratio component on the vertical surface of the window due to flux reflected from the ground and the parts of the obstructions below the horizontal. D_G is the ratio of half the ground illuminance to the sky illuminance.

A_{FW} is the area of the floor and lower parts of the walls below the mid-height of the window (not including the window wall).

R_{FW} is the average reflectance of A_{FW}.

R_{CW} is the average reflectance of the ceiling and upper walls above the mid-height of the window (not including the window wall).

A is the total area of all the interior surfaces.

R, as before, is the average reflectance of all the interior surfaces.

R_G is the average reflectance of the ground.

The Longmore formula used the values for C obtained by measurement from a Waldram diagram, as mentioned earlier. It is of interest now to seek to develop an analytical expression for the terms C and D_G in the expanded Longmore-type formula given above. The approach used is that of Tregenza.[5]

14.4 Development of the coefficients *C* and D_G

The coefficient C has two components:

(1) that due to the direct illuminance from the unobstructed part of the sky, and
(2) the light reflected onto the window from the obstructing building.

(1) *Direct illuminance from the sky*

Consider Figure 14.4(a), a vertical window placed at P will be illuminated by half the sky dome, unless it is obstructed. Two types of obstruction are common.

First, there is obstruction by buildings in front of the window plus obstruction from any canopy over the window. This type of horizontal obstruction can be denoted by two angles: θ_L the angle from the horizontal at the mid-height of the window subtended by obstructing buildings, and θ_H the angle also measured from the horizontal denoting the upper extent of the unobstructed sky.

The second type of obstruction is due to tall vertical obstructions, or deep window reveals.

The extent of the obstructions in the azimuth plane are denoted by α_L and α_R. Then

$$E_W = \int_{\theta_L}^{\theta_H}\int_{\alpha_L}^{\alpha_R} L_{\theta\alpha} \cos^2\theta \cos\alpha \, d\alpha \, d\theta$$

where E_W is the window illuminance and, as before,

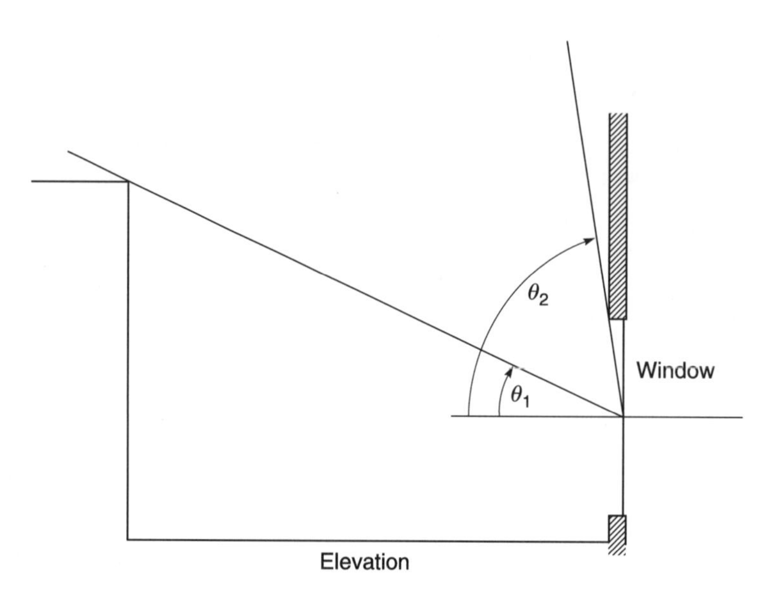

Fig. 14.4(a) The angles relating to horizontal obstructions

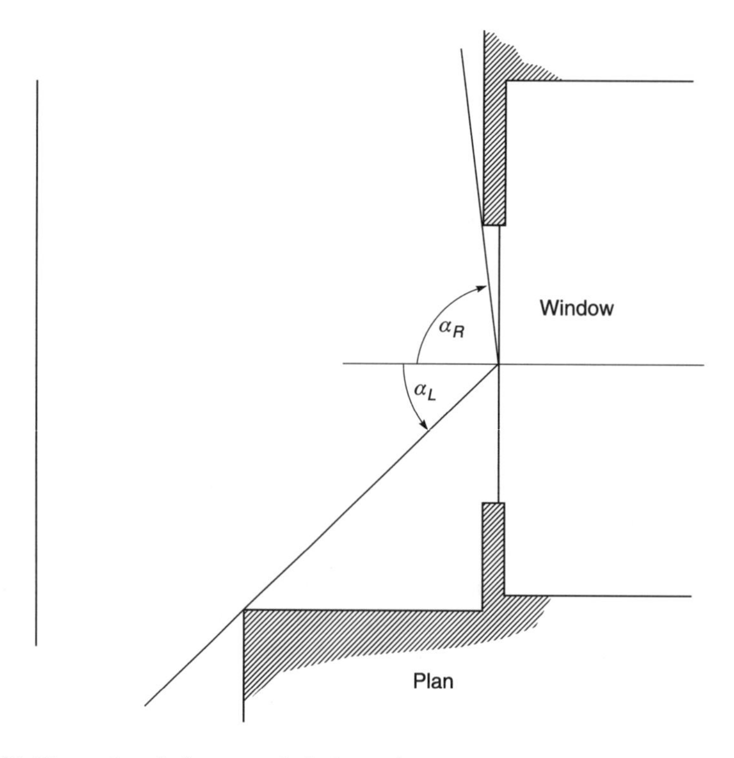

Fig. 14.4(b) The angles relating to vertical obstructions

$$L_{\theta\alpha} = \frac{1}{3} L_z(1 + 2 \sin \theta)$$

So,

$$E_W = \int_{\theta_L}^{\theta_H} \int_{\alpha_L}^{\alpha_R} \frac{1}{3} L_z(\cos^2 \theta + 2 \sin \theta \cos^2 \theta) \cos \alpha \, d\theta \, d\alpha$$

Now,

$$\cos^2 \theta = \frac{1}{2} (\cos 2\theta + 1)$$

giving

$$\int \cos^2 \theta \, d\theta = \frac{1}{2} \int \cos 2\theta \, d\theta + \frac{1}{2} \int d\theta$$

$$= \frac{\sin 2\theta}{4} + \frac{\theta}{2}$$

also,

$$\int 2 \sin \theta \cos^2 \theta \, d\theta = \frac{-2 \cos^3 \theta}{3}$$

and

$$\int \cos \alpha \, d\alpha = \sin \alpha$$

Thus,

$$E_W = \frac{1}{3} L_z(\sin \alpha_L + \sin \alpha_R) \times \left(\frac{\theta_H - \theta_L}{2} + \frac{\sin 2\theta_H - \sin 2\theta_L}{4} - \frac{2 \cos^3 \theta_H - 2 \cos^3 \theta_L}{3} \right) \tag{14.2}$$

Note that the two terms in the first bracket are both positive, since the angles are measured in opposite senses (Figure 14.4(b)).

(2) *Reflected illuminance*
It is now necessary to allow for the light reflected onto the windows from the obstructing buildings.

If we assume that the illuminance on the obstructing building has the same value as the illuminance on the window itself we should not overestimate the reflected contribution to the final window illuminance.

In addition if the building luminance is $E_W R_B/\pi$ (for uniform diffusion), then we can calculate the contribution to the window illuminance of reflection from the building. This is done, in the first instance, by subtracting the illuminance that would be produced on the obstructed window by a sky of uniform luminance from the illuminance that would be produced on an unobstructed window by a sky of uniform luminance. In this calculation, we use the luminance of the obstructing building

$$L_B = \frac{E_W R_B}{\pi} = \frac{L_Z f R_B}{\pi}$$

The illuminance on the window from an obstructed sky of luminance L_B is given by

$$\int_{\theta_L}^{\theta_H} \int_{\alpha_L}^{\alpha_R} L_B \cos^2\theta \cos\alpha \, d\alpha \, d\theta$$

$$= L_B(\sin\alpha_L + \sin\alpha_R)\left(\frac{\theta_H - \theta_L}{2} + \frac{\sin 2\theta_H - \sin 2\theta_L}{4}\right)$$

So, the illuminance on the window by reflection from an obstructing building of luminance L_B is

$$E_{WB} = L_B\left[\frac{\pi}{2} - (\sin\alpha_L + \sin\alpha_R)\left(\frac{\theta_H - \theta_L}{L} + \frac{\sin 2\theta_H - \sin 2\theta_L}{4}\right)\right] \tag{14.3}$$

To conclude the calculation of coefficient C, it is necessary to consider interreflection in the cavity between the building and the ground. If we treat it as a simple cavity we may write

Total incident flux on window = direct flux + reflected flux within the cavity

Assuming the flux to be distributed uniformly we may divide by the area of the window and the result may be expressed as

$$E_{WT} = E_{WB} + E_{WT} R_O$$

So,

$$E_{WT} = \frac{E_{WB}}{1 - R_O} \tag{14.4}$$

where E_{WT} is the total component of reflected illuminance on the window, E_{WB} is the component reflected from the obstructing building and R_O is the effective reflectance of the cavity.

R_O is given by

$$R_O = m\frac{R_B + R_G}{2}$$

where m is the fraction of reflected flux not lost out of the cavity; for example, to the sky.

Since we are dealing with low reflectances and a second-order effect, a simple mean of R_B and R_G has been taken, assuming equal areas of building façade and the ground. If we continue this simplifying assumption, we arrive at $m = 1 - 0.5$ as the fraction of flux that remains after each reflection (see Section 4.9.1).

If it is assumed that half the reflected flux remains within the cavity then

$$m = 0.5 \text{ and } R_O = \frac{R_B + R_G}{4}$$

In Section 14.2 the value of illuminance at a point on the horizontal plane from the whole sky was calculated as

$$E_{sky} = \frac{7\pi}{9} L_z \text{ lux}$$

So,

$$C = \frac{E_W + E_{WT}}{E_{sky}} \times 100\%$$

$$= \frac{9}{7\pi} f \left(1 + \frac{R_B g}{\pi(1 - R_O)}\right) \times 100\% \text{ (see Equations (14.2), (14.3) and (14.4))}$$

where

$$f = \frac{1}{3}(\sin\alpha_L + \sin\alpha_R)\left(\frac{\theta_H - \theta_L}{2} + \frac{\sin 2\theta_H - \sin 2\theta_L}{4} - \frac{2\cos^3\theta_H - 2\cos^3\theta_L}{3}\right)$$

and

$$g = \left[\frac{\pi}{2} - (\sin\alpha_L + \sin\alpha_R)\left(\frac{\theta_H - \theta_L}{2} + \frac{\sin 2\theta_H - \sin 2\theta_L}{4}\right)\right]$$

This completes the calculation of the coefficient C.

Although the ground reflectance was included in the above calculation, the fact that the flux that was reflected had first struck the buildings above the horizontal, before being reflected, places it correctly in the evaluation of coefficient C.

The coefficient D_G can be evaluated as follows.

Assume that the ground outside the building containing the window has a luminance given by $R_G E_G/\pi$ and that this may be considered to be a semi-infinite plane.

Then

$$E_{WG} = \frac{R_G E_G}{2\pi} \times \pi$$

where E_G is the ground illuminance and E_{WG} is the illuminance on the window due to light reflected from the ground. Within a built-up area, the ratio of E_G/E_{sky} can be assumed to be of the order of 0.5 giving

$$\frac{E_{WG}}{E_{sky}} = \frac{R_G E_G}{2E_{sky}}$$

or

$$\frac{E_{WG}}{E_{sky}} \% = R_G D_G \%$$

where

$$D_G = \frac{E_G}{2E_{sky}} \times 100\% \qquad (14.5)$$

$$= \frac{0.5}{2} \times 100\%$$

$$= 25$$

If the ground reflectance is assumed to be 0.2 then this gives rise to the term $5R_{CW}$ in the Longmore formula.

Table 14.2 gives values of $C/2$ for $\theta_H = 90°$ taking the reflectance of the ground and the obstructing buildings as 0.2.

Table 14.2 Calculated values of $C/2$ for $\theta_H = 90°$ taking the reflectance of the ground and obstructing buildings as 0.2

θ_L	ϕ_L or ϕ_R, half-sky angle in azimuth								
	10°	20°	30°	40°	50°	60°	70°	80°	90°
0°	3.8	7.3	10.5	13.2	15.6	17.4	18.7	19.5	19.8
10°	3.2	6.3	9.1	11.6	13.6	15.3	16.5	17.2	17.5
20°	2.6	5.1	7.4	9.4	11.1	12.5	13.5	14.1	14.3
30°	1.9	3.8	5.5	7.0	8.3	9.4	10.2	10.6	10.8
40°	1.3	2.5	3.7	4.7	5.6	6.3	6.8	7.1	7.2
50°	0.7	1.4	2.1	2.7	3.2	3.6	3.9	4.1	4.2
60°	0.3	0.7	1.0	1.2	1.5	1.7	1.8	1.9	1.9
70°	0.1	0.2	0.3	0.4	0.5	0.5	0.6	0.6	0.6
80°	0	0	0	0.1	0.1	0.1	0.1	0.1	0.1

Example Consider Figure 14.5. This shows a window 5 m wide by 2.5 m high. The window is set in the side wall of a 6 m wide square room, 3 m high. The window is obstructed by a continuous row of buildings to 20° above the horizon. The room reflectances are $\rho_C = 0.7$, $\rho_F = 0.2$ and $\rho_W = 0.5$. Assume the obstructing buildings and the ground have a reflectance of 0.2. Calculate the average daylight factor.

Fig. 14.5 The obstructed window related to the example in Section 14.4

The average reflectance of the walls below the centre height of the window (assuming it terminates at the ceiling height and excluding the window wall) and the floor is given by

$$R_{FW} = \frac{\rho_F A_F + \rho_W A_{LW}}{A_F + A_{LW}} = \frac{0.2 \times 6 \times 6 + 0.5 \times 1.75 \times 6 \times 3}{6 \times 6 + 1.75 \times 6 \times 3}$$

$$= 0.34$$

where A_{LW} is the area of the lower walls.

The average reflectances of the walls (excluding the window wall) and the ceiling above the centre height of the window is given by

$$R_{CW} = \frac{\rho_C A_C + \rho_W A_{UW}}{A_C + A_{UW}} = \frac{0.7 \times 6 \times 6 + 0.5 \times 1.25 \times 6 \times 3}{6 \times 6 + 1.25 \times 6 \times 3}$$

$$= 0.62$$

where A_{UW} is the area of the upper walls

$$R_O = \frac{R_B + R_G}{4} = \frac{0.2 + 0.2}{4} = 0.1$$

The average reflectance of all surfaces

$$R = \frac{\rho_C A_C + \rho_F A_F + \rho_W A_W}{A_C + A_F + A_W} = \frac{0.7 \times 6 \times 6 + 0.2 \times 6 \times 6 + 0.5 \times 3 \times 6 \times 4}{6 \times 6 + 6 \times 6 + 3 \times 6 \times 4}$$

$$= 0.475$$

$$C = \frac{9}{7\pi} f \left(\frac{1 + R_B g}{\pi(1 - R_O)} \right) \times 100\%$$

$$f = \frac{1}{3} (\sin \alpha_L + \sin \alpha_R) \times \left(\frac{\theta_H - \theta_L}{2} + \frac{\sin 2\theta_H - \sin 2\theta_L}{4} - \frac{2 \cos^3 \theta_H - 2 \cos^3 \theta_L}{3} \right)$$

In this case $\alpha_L = \alpha_R$, $\theta_H = 90°$, $\theta_L = 20°$

$$f = \frac{1}{3} (2) \left[\frac{1}{2} \left(\frac{\pi}{2} - \frac{20}{180} \times \pi \right) + \frac{0 - 0.643}{4} - \frac{0 - 1.659}{3} \right]$$

$$= 0.669$$

$$g = \left[\frac{\pi}{2} - (\sin \alpha_L + \sin \alpha_R) \left(\frac{\theta_H - \theta_L}{2} + \frac{\sin 2\theta_H - \sin 2\theta_L}{4} \right) \right]$$

$$= \left[\frac{\pi}{2} - (2)(0.785 - 0.174 - 0.161) \right]$$

$$= 0.671$$

If we take $D_G = 25$ as assumed previously (Equation 14.5).
Total area of all room surfaces

$$A = 2 \times 6 \times 6 + 3 \times 6 \times 4 = 144 \text{ m}^2$$

Let us assume the glass has a transmittance of 0.85 and an allowance for dirt is made at 0.8 then the effective value of $T = 0.85 \times 0.8 = 0.68$.

Let us also assume that the window frame reduces the net area of glazing by 20%, giving $W = 5 \times 2.5 \times 0.8 = 10 \text{ m}^2$.

Then

$$df_{WP} = TW\left(\frac{C}{A_{FW}} + \frac{CR_{FW} + D_G R_{CW} R_G}{A(1-R)}\right)\%$$

$$= 0.68 \times 10\left(\frac{28.7}{76.5} + \frac{28.7 \times 0.34 + 25 \times 0.62 \times 0.2}{144(1-0.475)}\right)\%$$

$$= 4.0\%$$

In the above example, α_L and α_R were each taken as 90°. If the window had deep reveals or if there were vertical obstructions to one or both sides, these angles would be less and the term $(\sin \alpha_L + \sin \alpha_R)$ would be reduced in value, which would reduce the value of C.

Let us calculate the value for the daylight factor that would have been obtained using the modified Lynes formula.

$$df_{WP} = \frac{TW\theta}{A(1-R^2)}\%$$

$$= \frac{0.68 \times 10 \times 70}{144(1-0.475^2)}\%$$

$$= \frac{476}{144 \times 0.774}\%$$

$$= 4.3\%$$

14.5 Daylight factor at a point

The previous section dealt with the calculation of the average daylight factor for the working plane or area below the working plane and this is of considerable value in the early stages of the daylight design.

Once the project has reached a more detailed stage, more detailed information about the distribution of the daylight within the space is often required. This information can be obtained from point by point daylight factor calculations and could be displayed in terms of daylight factor contours.

There are manual and graphical methods that can be used to calculate the daylight factor at a point, but such detailed calculations are most appropriately carried out by computer.

The most popular method of calculating arrays of daylight factors for the working plane is to use daylight coefficients.[6] The concept is based on dividing the sky up into a large number of small elements and calculating the effect of each element separately, hence the appropriateness of using a computer.

In Section 14.2, the following expression relating to Figure 14.1 was introduced:

$$\delta E = \frac{L_{\theta\alpha} r\delta\theta \; r\delta\alpha \cos\theta \sin\theta}{r^2}$$

where δE is the illuminance at P due to an element of sky. In this equation $(r\delta\theta r\delta\alpha \cos\theta)/r^2$ represents the solid angle subtended by the element of the sky at point P. This simplifies to

$$\delta\Omega = \delta\theta\delta\alpha \cos\theta$$

So that $\delta E = L_{\theta\alpha}\delta\Omega \sin\theta$.

If the transmittance of the glazing is $T_{\theta\alpha}$, then

$$\begin{aligned}\delta E &= L_{\theta\alpha}\delta\Omega \sin\theta \, T_{\theta\alpha}\\ &= L_{\theta\alpha}\delta\Omega D_{\theta\alpha}\end{aligned}$$

where $D_{\theta\alpha} = T_{\theta\alpha} \sin\theta$, which is the daylight coefficient for direct light at point P for this element of sky.

Since $D_{\theta\alpha}$ is independent of $L_{\theta\alpha}$ the daylight coefficients do not need to be re-evaluated if the luminance distribution is changed.

The daylight coefficient can be developed to include an interreflection calculation for each element and so a general definition of the daylight coefficient is

$$D_{\theta\alpha} = \frac{\delta E_{\theta\alpha}}{L_{\theta\alpha}\delta\Omega_{\theta\alpha}}$$

From the above,

$$E = \int_0^{2\pi}\int_0^{\pi/2} D_{\theta\alpha} L_{\theta\alpha} \cos\theta \; \mathrm{d}\theta \; \mathrm{d}\alpha$$

It is sometimes convenient to split this equation into two sections by separating the direct and reflected components of $D_{\theta\alpha}$:[6]

$$D_{\theta\alpha} = D_{\mathrm{d}\theta\alpha} + D_{\mathrm{i}\theta\alpha}$$

where d notes the direct daylight coefficient and i the interreflected daylight coefficient

$$E = E_{\mathrm{d}} + E_{\mathrm{i}}$$

where E_{d} is the direct illuminance at the point and E_{i} is the interreflected component.

So,

$$E = \int\int_{\text{window area}} D_{\mathrm{d}\theta\alpha} L_{\theta\alpha} \cos\theta \; \mathrm{d}\theta \; \mathrm{d}\alpha + \int_0^{\pi}\int_0^{\pi/2} D_{\mathrm{i}\theta\alpha} L_{\theta\alpha} \cos\theta \; \mathrm{d}\theta \; \mathrm{d}\alpha$$

The first part of the equation deals with the direct light at the point which must be received from the direction of the window. The second part of the equation deals with the interreflected light and this can enter the room from an entire half of the sky and be interreflected.

The double integrals generated by this type of daylighting calculation can usually only be evaluated by numerical integration.

There are a number of ways to calculate the detailed interreflected components, such as Ray Tracing,[7] the Monte Carlo Method[8] or the radiosity method used in this book (Chapter 5).

14.5.1 TRANSMITTANCE $T_{\theta\alpha}$

The transmittance of clear glass can be taken as 0.8 to 0.85 for diffuse transmittance in, for example, the formulae for average daylight factor (double glazing would be 0.70 to 0.75).

For daylight coefficient calculations, the value of transmittance varies with the angle of incidence and values of $T_{\theta\alpha}$ are required.

Littlefair has developed the following equation for this purpose[9]

$$\begin{aligned} T_{\theta\alpha} = {} & 0.623 + 0.3 \cos \alpha - 0.137 \cos^2 \alpha + 0.51 \cos \theta \\ & - 0.66 \cos \theta \cos \alpha + 0.346 \cos \theta \cos^2 \alpha \\ & - 0.285 \cos^2 \theta + 0.427 \cos^2 \theta \cos \alpha - 0.246 \cos^2 \theta \cos^2 \alpha \end{aligned}$$

A useful document relating to Daylight Algorithms is published by the DTI-sponsored Energy Technology Support Unit (ETSU).[10]

14.5.2 DIRECT SUNLIGHT

The daylight coefficient method introduced in Section 14.5 can be easily adapted to the calculation of the direct sunlight illuminance at a point within a room.[6] Since, in this method, the sky is treated as an array of point sources, the daylight coefficient formula for a single point source may be modified as follows:

$$\begin{aligned} E_S &= D_{\theta\alpha} L_{\theta\alpha} \cos \theta \, d\theta \, d\alpha \\ &= D_{\theta\alpha} \frac{E_{NS}}{\omega} \cos \theta \, d\theta \, d\alpha \\ &= D_{\theta\alpha} E_{NS} \end{aligned}$$

where E_{NS} is the direct normal solar illuminance outdoors and $D_{\theta\alpha}$ is the daylight coefficient for the element at the solar attitude and azimuth. E_{NS} may be calculated from the formula given below.[11]

$$E_{NS} = 128 e^{-0.5 \operatorname{cosec} \gamma} \text{ klx}$$

where

$$\gamma = \sin^{-1}(\sin \phi \sin \delta - \cos \phi \cos \delta \cos 15t)$$

and

ϕ = geographical latitude (positive in the northern hemisphere);
δ = solar declination (see Table 14.3 – north is positive, south negative);
t = hours since midnight (the term $15t$ is in degrees).

14.5.3 THE WALDRAM DIAGRAM

The Waldram Diagram was mentioned in Section 14.3 and this is a graphical method for solving the double integrals encountered in evaluating daylight factors.

Table 14.3 Solar declination δ at different times in the year

Date	δ
June 22	23° 27'N
May 21/Jul 24	20° N
Apr 16/Aug 28	10° N
March 21/Sept 23	0°
Feb 23/Oct 20	10° S
Jan 21/Nov 22	20° S
Dec 22	23° 27'S

In its simplest form, using the results of Section 14.2, it consists of a network with the abscissae proportional to α and the ordinates proportional to

$$\left(\frac{1}{2}\sin^2\theta + \frac{2}{3}\sin^3\theta\right)$$

This gives δx proportional to $\delta\alpha$ and δy proportional to $(\sin\theta\cos\theta + 2\sin^2\theta\cos\theta)\delta\theta$. The whole diagram represents half the sky. Areas on this diagram are proportional to the sky component of the daylight factor (see Figure 14.6). If the angles from the point, for which the daylight factor is required, to the window boundaries are calculated, the outline of the window can be plotted on the diagram as shown. The area of this plot, divided by twice the total area of the diagram, gives the value of the sky component of the daylight factor when expressed as a percentage. The diagram shown in the figure is for an unglazed window.

Notice how, on this diagram, horizontal lines on the window become curved because, as the

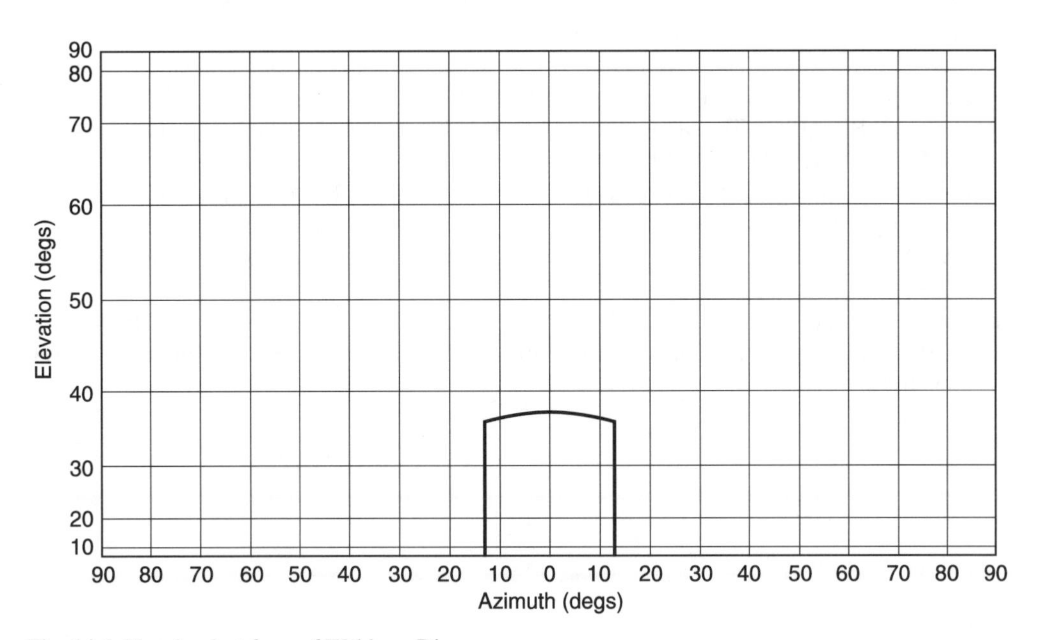

Fig. 14.6 The simplest form of Waldram Diagram

point on the horizontal line moves away from the centre of the diagram, it subtends a smaller and smaller angle at the illuminated point and therefore moves down the diagram. In addition, a glazed window will have a transmittance that falls with the angle of incidence and Waldram Diagrams are sometimes constructed to take both of these effects into account. The diagrams consist of 'droop' lines superimposed on the basic diagram so that horizontal obstructions can be plotted easily and window transmission taken into account.[12]

Obviously, another form of the diagram could be produced from the equations developed in Section 14.4 and related to the illuminance on vertical surfaces.

An alternative method is to use the BRS Sky Component Tables published by BRE.[12] These tables give values of the sky component in terms of window size and position relative to the illuminated point on the working plane.

A worked example using these tables is given in the 1994 CIBSE Code (see also *Window Design Applications Manual*, AMI (London) CIBSE (1987)).

References

1. Lynes, J.A. (1979) A sequence for daylighting design. *Lighting Research and Technology*, **11**(2), 106.
2. Hopkinson, R.G., Longmore, J. and Petherbridge, P. (1954) An empirical formula for computation of indirect component of daylight factor. *Trans. Illum. Eng. Soc. (London)* **19**(7), 201–219.
3. Crisp, V.H.C. and Littlefair, P.J. (1984) Average daylight factor prediction. *CIBSE National Lighting Conference.*
4. Longmore, J. (1975) Daylighting; a current view. *Tt. Ltg.*, **68**, 113.
5. Tregenza, P.R. (1989) Modification of split-flux formulae for mean daylight factor and internal reflected component with large external obstructions. *Lighting Research and Technology*, **1**(3), 125–128.
6. Littlefair, P.J. (1992) Daylight coefficients for practical computation of internal illuminances. *Lighting Research and Technology*, **24**(3), 127–135.
7. Ward, G. and Rubinstein, F.M. (1988) A new technique for computer simulation of illuminated spaces. *J. Illum. Eng. Soc.*, **17**(1), 80–91.
8. Tregenza, P.R. (1983) The Monte Carlo Method in Lighting Calculations. *Lighting Research and Technology*, **15**(4), 163–170.
9. Littlefair, P.J. (1982) Effective glass transmittance under a CIE sky. *Lighting Research and Technology*, **14**(4), 232–235.
10. (1993) *Daylighting Algorithms, ETSU S 1350*, Energy Technology Support Unit (ETSU), DTI.
11. Coaton, J.R. and Marsden, AM. (eds) (1997) *Lamps and Lighting*, Chapter 19 (Arnold) 367.
12. (1986) Estimating daylight in buildings: Part I, *BRE Digest 309* (Garston: Building Research Establishment).

15

Measurements

15.1 General

Basically two types of measurement are required, laboratory and field. Generally, laboratory measurements are required to provide data for the prediction of the performance of a material in a luminaire, or a luminaire in an installation. In the field, measurements are required to establish whether the required or predicted performance has been achieved. The object of this chapter is to describe and discuss the various methods of carrying out these lighting measurements.

15.2 Photoelectric cells

The photoelectric cell is used in all modern photometric apparatus. This produces an electric current that varies according to the illuminance falling on it. Before the advent of the photocell, measurements had to be made by visual comparison, which was not only time consuming but produced results that depended on the skill of the observer. The photocells used in most apparatus are silicon diode cells. These have tended to displace the selenium photocell because of their better performance as regards aging, stability, fatigue and linearity. For some applications, such as luminance meters and the collimating photometer, photomultipliers are used. These have greater sensitivity than the silicon diode but need a high voltage for their operation. The characteristics that are important in the selection of photocells are as follows.

- Spectral response to radiation. The spectral sensitivity of the photocell should closely match that of the eye. As shown in Figure 15.1, the selenium photocell has an advantage here in that its spectral sensitivity curve peaks at approximately the same wavelength as that of the eye, so that correction can be obtained by a commercially available filter. The spectral sensitivity of the silicon diode, on the other hand, is very different from that of the eye and needs a specially computed filter. Two types of filter are available. One is made from a mosaic of different colour filters fitted together side by side. The other is a matrix consisting of different colour filters of the same size as the photocell and used in parallel. By measuring the response of the individual photocell before correction, a very close approximation to the sensitivity of the eye can be obtained by varying the sizes of the components in the case of the mosaic filter and the thickness of the components in the case of the matrix filter. As can be imagined, these are expensive processes. Correction for the various mesopic sensitivity functions can also be obtained. Care has to be taken that the response of the photocell to infra-red and ultra-violet radiation is minimal. The matrix filter has the advantage over the mosaic filter that the whole of the cell surface is uniformly corrected, which for some applications is essential.

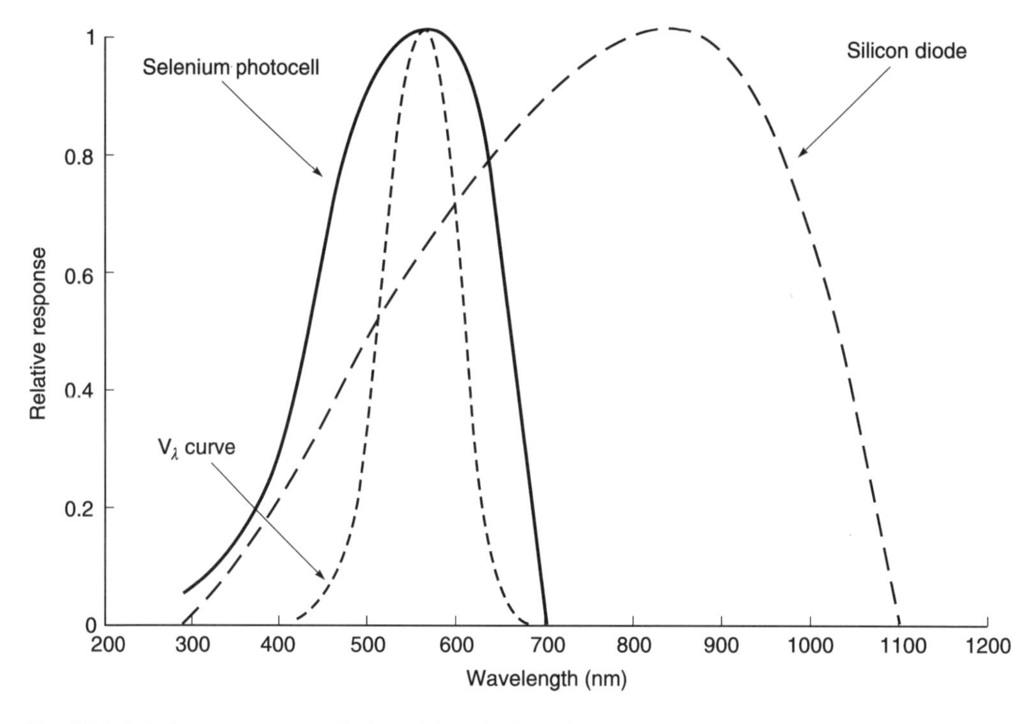

Fig. 15.1 Relative response to radiation of the selenium photocell and the silicon diode

In the relative method of photometry, where the light distribution of a luminaire is compared with that of the bare lamp, spectral sensitivity correction need not be as stringent as with other forms of photometry, provided the luminaire does not alter the spectral light distribution of the lamp or lamps significantly. When photometry has to be performed on lights where the radiation is concentrated over a relatively narrow band, such as signal lights, it is prudent to obtain a correction factor for that particular source.

- Cosine response. Illuminance varies as the cosine of the angle of incidence of the light. However, the illuminance measured with a bare photocell falls away faster than this because its surface reflects an increasingly higher proportion of the incident light as the angle of incidence is increased (see Section 6.8.12, page 265). As explained in Section 15.11.1, page 501, correction generally needs to be made for this, although in distribution photometers, where the angle of incidence is not greater than 20°, correction is not necessary.
- Linearity of response. The output from the photocell and its associated circuit must be proportional to the illuminance on the photocell. This can be checked on a photometric bench but it is more convenient to use a specially constructed linearity tester (see page 490).
- Fatigue. By this is meant the reduction of output of the photocell when it is exposed to light for a fairly long period, 30 minutes say. It may only be manifest when the photocell is exposed for a period, covered and re-exposed. In addition it may be more pronounced at higher illuminances. During the dark periods it should be checked that the reading returns to, and remains at, zero.
- Response time. The photocell and its associated circuit will take a finite time to produce the final reading. If the photocell or light source is moved while readings are being taken, as is often the case in goniophotometry (to be described in the next section), a trial run should be

made to check that substantially the same readings are obtained for a moving photocell or light source as are obtained when both the photocell and light source are stationary at each measurement point.

- Temperature dependence. The output of the photocell and its associated circuit will be affected by the ambient temperature. In relative photometry, this should not affect the final result provided that the temperatures at the photometer head, when measurements are taken on the bare lamp, and the luminaire agree within close limits. How close the limits need to be can be determined by varying the temperature when a steady illuminance is falling on the photocell and recording the variation in the signal. In absolute photometry, keeping the temperature constant is more important. Photocells with thermostatic control are commercially available.
- Non-uniform sensitivity over surface of a photocell. Usually there is some variation of sensitivity over the surface of a photocell. This is because current generated near the centre of the photocell has further to travel to the collection ring at the periphery of the photocell than current generated elsewhere. For small photocells the effect is slight, but for large photocells it may be significant. Colour correction by means of a mosaic filter will produce non-uniform sensitivity. In certain applications, such as goniophotometry, non-uniform sensitivity may not be important but, in other applications, such as luminance measurement, it may well be of paramount importance.
- Sensitivity to polarized light. At angles near Brewster's angle (see Section 6.8, page 237) light polarized in certain planes will be reflected from the surface of the photocell.
- Evaluation of modulated light. All light sources that run on alternating current produce a modulated waveform. This is more pronounced with light sources where the emission is directly from an arc tube than when the emission is from the intermediary of a phosphor (fluorescent lamp) or from a filament (tungsten lamp). Photocells obey Talbot's law so that the current generated is equal to the average illuminance, provided that the characteristics of the circuit are such that the indication is steady. There is a departure from this at high values of circuit resistance and also at a low frequency of interruption.

The allowable variation of these characteristics is given in CIE 121.[1] In addition it is wise to check that the sensitivity of the photocell and its associated circuit in terms of the electrical signal generated is sufficient for the application for which it will be used.

15.3 Light distribution photometry

A more exact, if more ponderous, term for this is luminous intensity distribution photometry. It describes how the luminous intensity of a luminaire is distributed in space according to one of the coordinate systems described in Chapter 1. The apparatus is often dignified by the name goniophotometer, but more usually referred to as a polar curve apparatus, polar curve photometer, or simply photometer. In its simplest form, the goniophotometer consists of a photocell that can be moved relative to the luminaire so that the luminous intensity can be measured in any direction in space. There is a great variety of configurations for the apparatus, as will be seen. The main considerations governing these are as follows.

- Length of the optical path in relation to the space available. For the inverse square law to apply to an accuracy acceptable in practice, it is usually reckoned that the length of the optical path should be at least five times that of the largest dimension of the largest luminaire to be measured. This is the ratio recommended by the CIE 121, with two provisos. First, the light distribution should be approximately cosine in planes passing through the lamp axis. Second, the ratio of the path length to the largest dimension of the luminaire normal to the lamp axis should be at least 15. For certain types of luminaire, such as floodlights, where the light is focused into a beam, this distance may have to be much longer. This will be discussed in Section 15.6, page 487. However, for general purpose luminaires, the five times rule can be held to apply. This means that to measure a luminaire of length 2400 mm (the longest luminaire generally encountered) an optical path of 12 m is required. Accommodating this distance obviously has problems, particularly in a building that is not purpose built. However, as will be seen, there are ways of reducing or eliminating these problems.

These stipulations as regards test distance apply when the requirement is to measure luminous intensity. To measure the total light output ratio, much shorter distances can be used. This can best be understood by regarding the readings as illuminance measurements taken over the surface of a sphere. If these illuminance values are multiplied by their respective areas the total flux will be found, since illuminance times area equals luminous flux.

The same is true of the luminous flux above and below the horizontal, provided that the depth of the luminaire is not great compared with the test distance. The fact that the light output ratio can be measured at a short distance can be used to advantage where a family of luminaires has to be measured and the different members of the family differ only in length. This means that accurate luminous intensities can be measured on one of the luminaires of relatively short length. The light output ratios of the other members of the family can then be measured and used as scaling factors to be applied to the luminous intensity data measured on the short luminaire.

However, it should be noted that even though the total, up, and down LOR may be measured accurately at short distances, the luminous flux in zones is not. In Figure 15.2 it is evident that the light will fall onto different zones as the sphere size is increased.

Dunlop and Finch[2] carried out experimental work in which they analysed photometric data

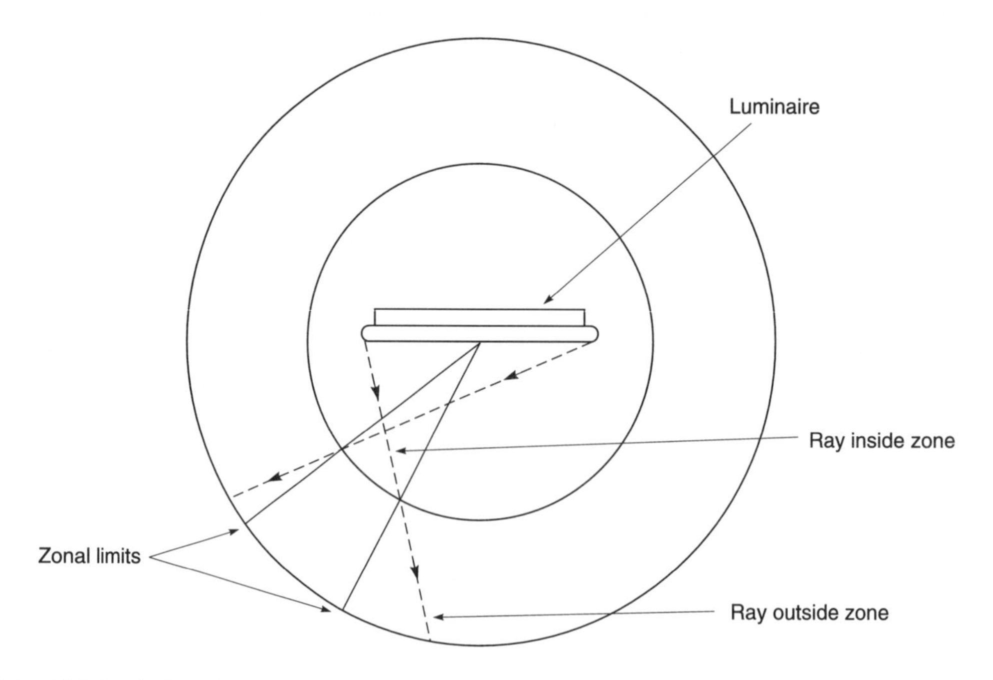

Fig. 15.2 Vertical section through concentric spheres enclosing luminaire

derived from measurements taken at 3000 mm and 12 000 mm on 2400 mm and 1200 mm luminaires. In the worst case, the 2400 mm luminaire tested at 3000 mm, the utilization factors did not differ by more than one unit in the second figure from those derived from the 12 000 mm data. However, there were quite considerable differences in zonal luminous flux, as would be expected. In an extension of this work, Riggs and Lampert,[3] who also compared measurements taken at 3000 mm and 12 000 mm on 2400 mm luminaires, showed that the accuracy of the transverse polar curve was good in most cases. Exceptions occurred at angles above 75°. The axial polar curve was invariably markedly inaccurate, sometimes by as much as 30%. Their conclusion was that 3000 mm photometry gives light output ratios and utilization factors that are accurate enough for practical purposes. More recently these findings have been confirmed by Frost.[4]

An argument sometimes put forward as a justification of short distance photometry is that since we are usually concerned with short lengths in actual installations, short distance photometry is the more meaningful.[5] This is erroneous, for all calculations depend on the use of true luminous intensity. Thus, data derived by short distance photometry cannot be used to calculate accurate values of illuminance.

- Movement of the luminaire. In some luminaires, parts, such as louvres, are held in place by gravity. Obviously if these luminaires have to be rotated about their own axis during a test then these parts will have to be secured, which can be time consuming. A more serious problem to overcome is that the light output of some lamps changes with the temperature of the air surrounding them. This means that if an open luminaire, for instance, is turned upside down, the lamp may well operate at a lower temperate and increase its light output. Fluorescent lamps behave in this way, so that it is desirable that the luminaire be operated in its working attitude or position. It may still be necessary to rotate the luminaire about its vertical (first) axis. If this is done too rapidly it is possible that the layers of warm air surrounding the lamp are disturbed, thereby changing the light output. Rotation of certain high pressure discharge lamps may alter their light output. There may not be a marked change if the rotation is about the axis of the arc tube but otherwise the change may be very marked.
- Reduction of stray light. By stray light is meant light that may reach the photocell other than directly from the source under test. Ideally, stray light should be entirely eliminated but this is impossible in practice. There are many places from which the stray light may come. Inevitably, the photocell sees the background to the luminaire (Figure 15.3). This background will change if the angle of view of the photocell changes. It is, therefore, necessary that all these surfaces, and they may be extensive in area, are painted matt black. Matt black paints vary in their reflectivity and one should be chosen that has a low reflectance. Matt black carpet can be a good material for absorbing light as the light is interreflected and absorbed in the tufts. It is essential that these surfaces are kept free of dust, which has a surprisingly high reflectance. Other surfaces that may reflect light are the bevelled edges of mirrors, and supporting structures of the photometer. These latter may be troublesome if the photocell 'views' them at glancing angles of reflection and the light is incident at the same angles.

15.4 Basic components for a light distribution photometer

(1) Light measuring head. This takes the form of a photocell or photomultiplier in a housing. The housing narrows the field of view of the photocell to reduce the amount of stray light arriving at the photocell as already noted; however, the photocell can still see the background to the luminaire (Figure 15.3). Figure 15.25, page 506, shows how the baffles may be located. It is essential that the photocell, together with its associated circuit, has linearity of response

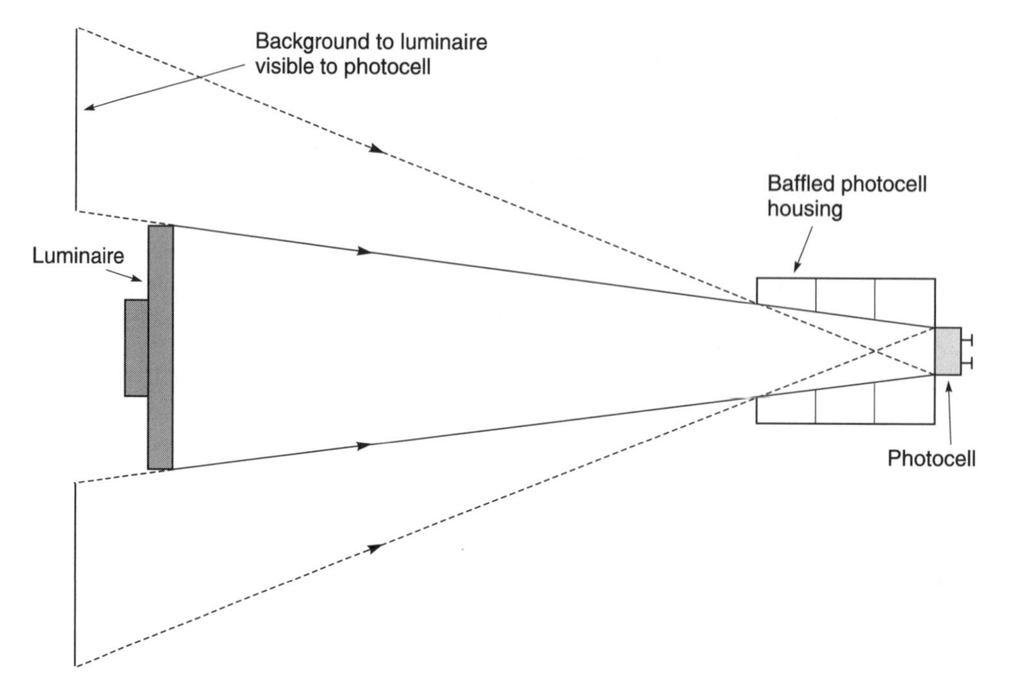

Fig. 15.3 Baffled photocell housing to reduce the amount of stray light reaching the photocell

(Section 15.2, page 472). If the photometer is calibrated each time a test is performed, in terms of the bare lamp (that is, relative photometry is used), then it need not be calibrated in lux, but for absolute photometry it must be calibrated in lux. Correction to the V_λ photopic spectral response of the eye, or any other appropriate response (scotopic or mesopic) is essential. Correction for oblique light incidence, the cosine correction (page 472) is usually not important as the maximum angle of the light incidence is small. In fact, cosine correction may be undesirable as it may decrease the effective sensitivity of the photocell.

(2) Mount for luminaire. This usually takes the form of a bar from which the luminaire can be suspended or a spigot for mounting road lighting luminaires. Provision should be made for mounting the luminaire securely so there is no untoward movement during a test. For road lighting luminaires, it is helpful if the angle of the mounting spigot can be adjusted easily. A gantry may be necessary where the luminaire has to be lifted to a height that cannot be reached easily. If one is used it should be checked that the luminaire maintains its test attitude when in its test position. Attention to ease of mounting is repaid in time saved.

(3) Means of rotation and angular indication. Sturdy and vibration-free means of rotation is required. Indication is usually remote by electronic means, typically shaft encoders, especially if any degree of automation is required.

(4) Temperature control. The light output of some lamps, notably fluorescent ones, as has already been mentioned, is sensitive to temperature. When luminaires using these lamps are being measured, it is essential that the temperature is kept constant. CIE 121 and various national standards recommend an ambient temperature of 25 ± 1°C. This is difficult to achieve in the large space that a goniophotometer may occupy. Air conditioning has to be used with caution because it can cause draughts, which will waft away the warm air surrounding the luminaire or bare lamp. The room housing the goniophotometer should preferably be in the bowels of the building.

15.5 Light distribution goniophotometers for the (C, γ) coordinate system

The (C, γ) is the coordinate system that is generally used for interior luminaires and there is pressure by some lighting organizations, national and international, to have it adopted as the universal system applicable to all luminaire types. The designs of goniophotometer using the (C, γ) coordinate system are best considered in order of complexity. A great variety of designs is possible and, whilst the following descriptions are not exhaustive, they should provide sufficient background for solving particular problems.

15.5.1 SINGLE MOVING PHOTOCELL, NO MIRRORS

Figure 15.4 shows a simple photometer with a moving photocell on the end of a rotating arm. For convenience of mounting the luminaire, a gantry is used. If the arm is mounted on the gantry, only half the head room is needed. Luminous intensity measurements below the horizontal can be taken with the gantry in its upper position, and measurements above the horizontal can be taken with the gantry in its lower position. This saves head room but is inconvenient, and there may be a temperature difference at the two heights.

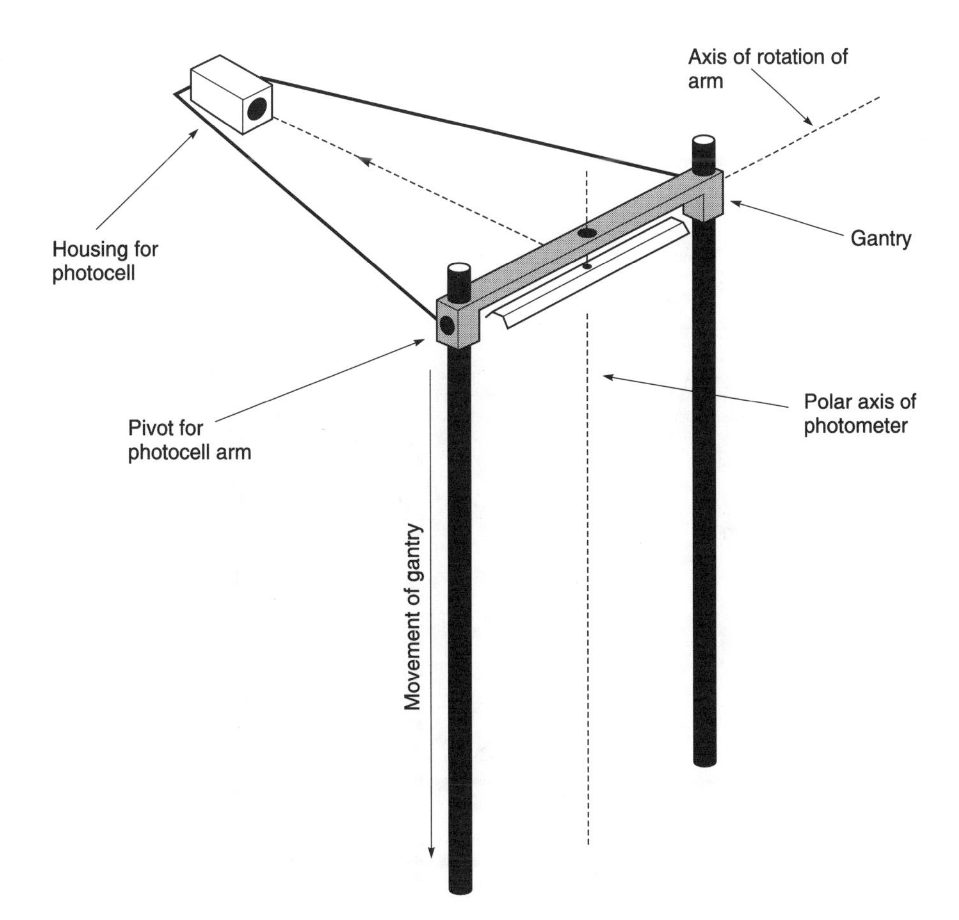

Fig. 15.4 Revolving arm photometer

The photometer illustrated in Figure 15.5 employs a track along which the photocell can be moved by a cable. Whilst this overcomes the problem of sagging and torsion in the arm, a large diameter track is difficult to construct and advantage cannot be taken of being able to test the luminaire at two different heights.

Figure 15.6 shows a revolving beam photometer. The photocell housing and luminaire are mounted at either end of a revolving beam. By means of the pulley arrangement shown, the luminaire is kept in its correct working position as the beam rotates; the central pulleys are kept stationary, but the end pulleys are free to rotate about their own axes. Sometimes gear wheels are used instead of pulleys, but over the long distances used in luminaire photometry (as opposed to lamp photometry) many or large wheels would be needed to span the distance. This apparatus has the advantage that it is possible to have an optical path nearly equal to the height of the laboratory and to make an uninterrupted sweep around the luminaire. The luminaire can be mounted on the apparatus when the beam is at its nadir of rotation.

There are two problems that have to be overcome with this apparatus. First, the top of the laboratory is usually warmer than the bottom so care has to be taken that the difference is not so

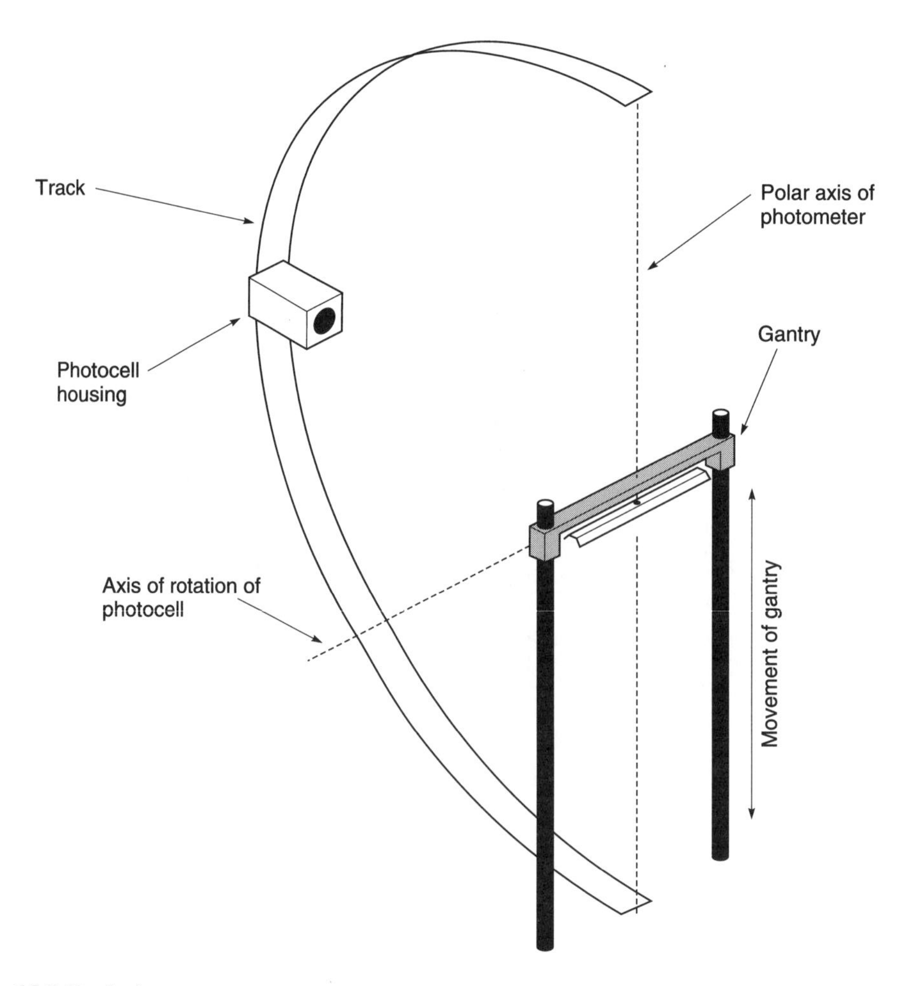

Fig. 15.5 Track photometer

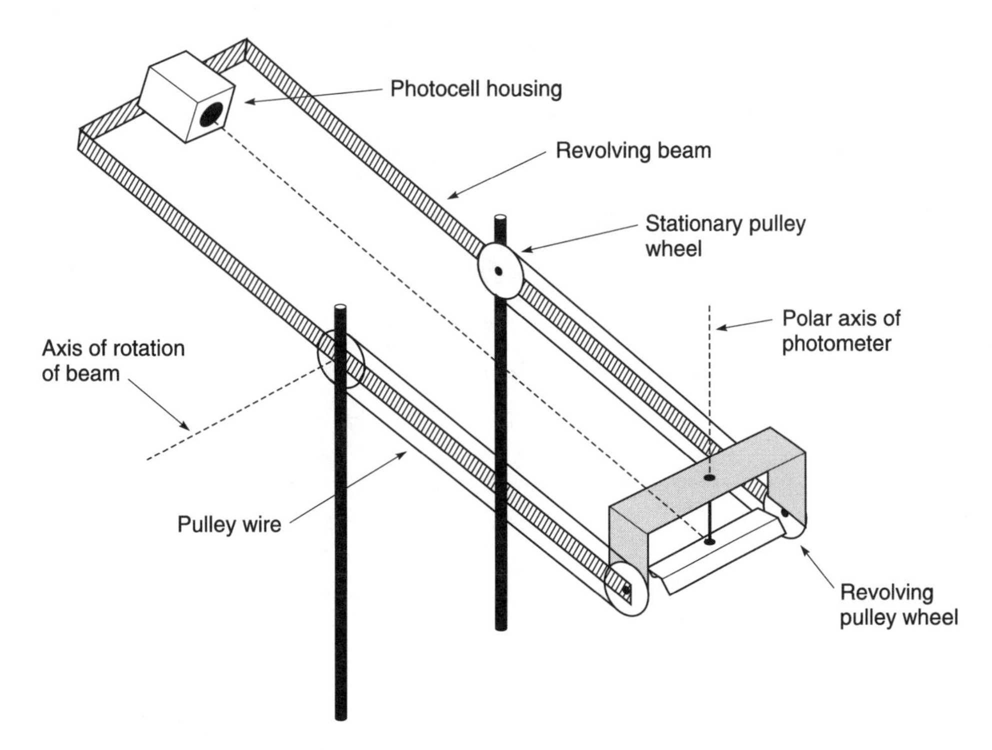

Fig. 15.6 Revolving beam photometer

great as to affect the measurements significantly. Second, air is wafted through the luminaire as the beam is rotated, so its speed of rotation has to be restricted to minimize any cooling effect.

Figure 15.7 shows a photometer suitable for use when headroom is very limited. Its use involves turning the luminaire out of its normal working position by means of the elbow joint shown in the figure. The apparatus is calibrated with the bare lamp in its normal position, a photocell at the nadir being used. The luminaire, also in its normal position, is then mounted on the apparatus and another reading taken at the nadir. The ratio of these two readings provides a correction factor for when the luminaire is turned so that its axis is in the vertical plane. At any given azimuth the measurement at the nadir of the luminaire provides a monitoring reading for correction of all the readings at that azimuth. When the azimuth is changed, an adequate time should be given for the luminaire to stabilize.

Very often with the designs so far described, the optical path cannot be made long enough for measurements on the full range of luminaires. Mirrors can be used to overcome this problem by folding the optical path onto itself and so make the apparatus more compact. Photometers employing these will be described next followed by photometers employing other ingenious solutions.

15.5.2 MIRROR PHOTOMETERS

When mirrors are rotated, flexing takes place so that the surface may become cylindrical. This will alter the area of the image of the luminaire, and, therefore, the measured luminous intensity. The rule used for astronomical telescopes to reduce flexing is that the thickness of the mirror should be one sixth of its diameter. Following this rule would be very expensive, moreover, the

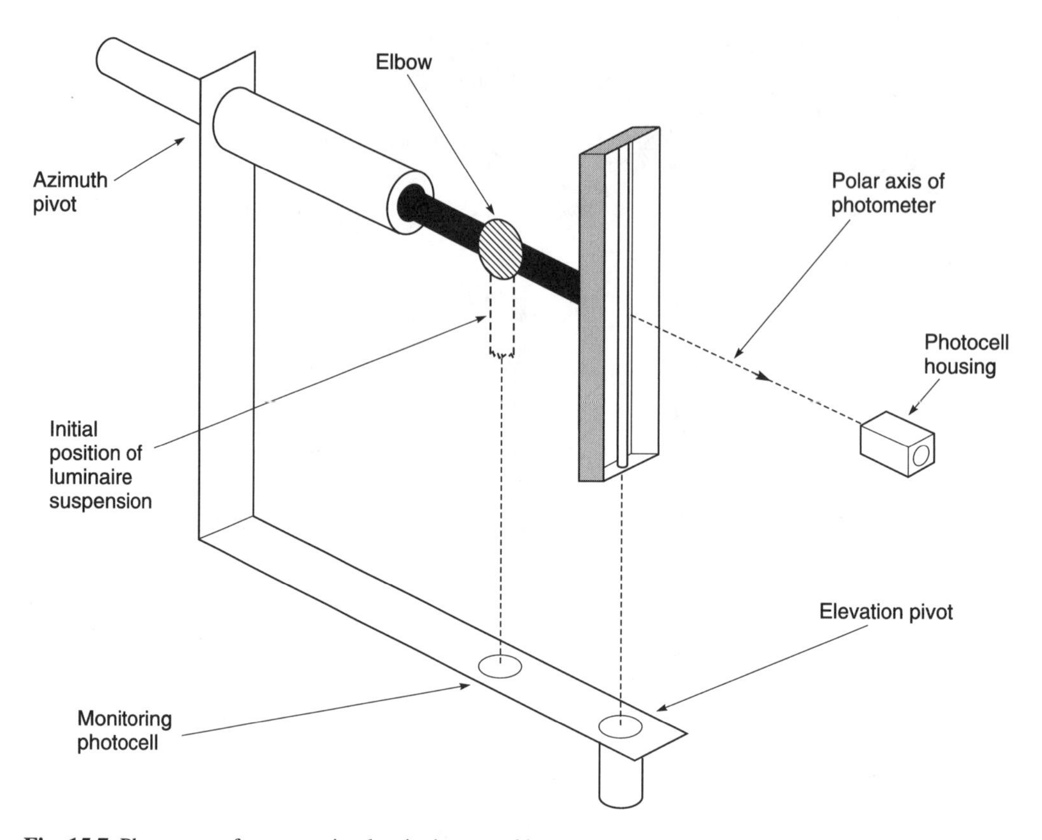

Fig. 15.7 Photometer for measuring luminaire out of its normal working attitude

supporting structure would have to be correspondingly increased in strength. In practice, it has been found that mirrors with a thickness of only 6 mm for 700 mm diameter are satisfactory, although small errors, albeit of no practical consequence, are incurred. Flexing of the mirror can be overcome by cementing it to a lightweight supporting structure. A good quality coating is required, which must be evenly reflective over the surface of the mirror. If the mirror is coated on the back surface, the use of white glass, and not tinted glass, tinted perhaps because of the presence of iron or other salts, is to be preferred.

Figure 15.8 shows a single mirror distribution photometer. In this, both the mirror and the photocell housing are fixed to the arm and rotate round the luminaire. As with the apparatus shown in Figure 15.4, the movement of the gantry can be used to halve the headroom needed. If this facility is not required, then in the interests of mechanical stability it is probably better to have the pivot of the arm held to a structure that does not move with the gantry.

To save headroom, the mirror can be brought nearer to the luminaire and the optical path brought out sideways, as in Figure 15.9, but the mirror must not reflect light onto the luminaire.

Figure 15.10 shows various other configurations for mirrors. If, as in Figure 15.10(a), the photocell is placed in the axis of rotation and at right angles to this axis, it need not move with the arm since the angle of incidence stays constant. However, the light is incident from different directions as the arm rotates, and care has to be taken that the photocell is equally sensitive to light incident in those different directions. Alternatively, two (Figure 15.10(b)) or three (Figure 15.10(c)) mirrors can be used to direct into the axis of rotation of the arm, so that the light is

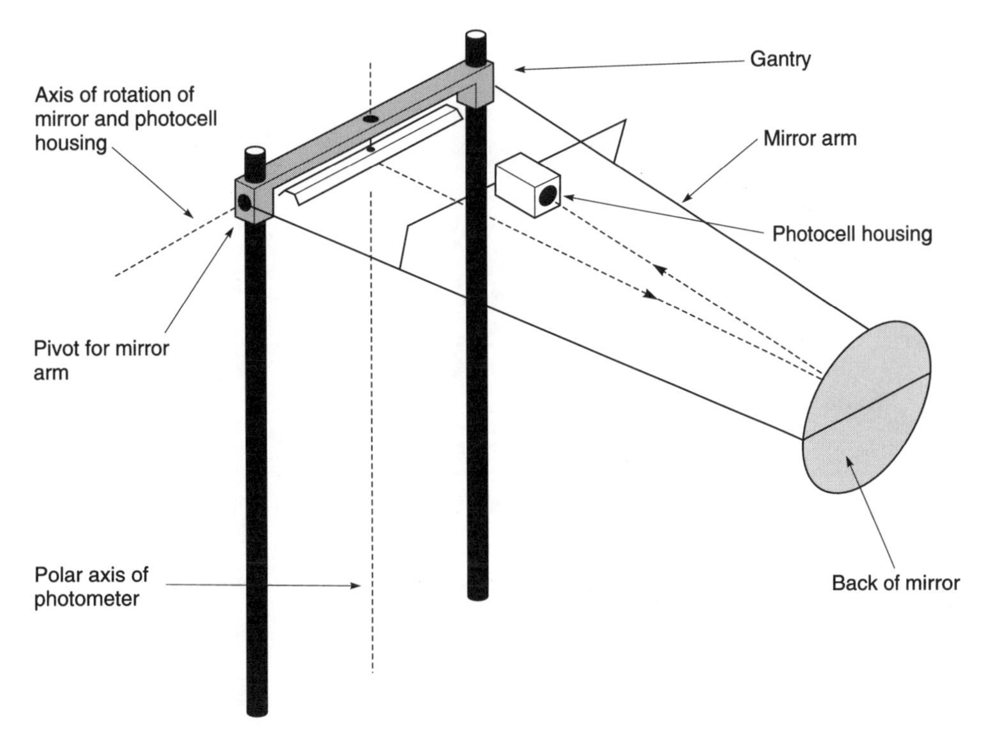

Fig. 15.8 Single mirror photometer

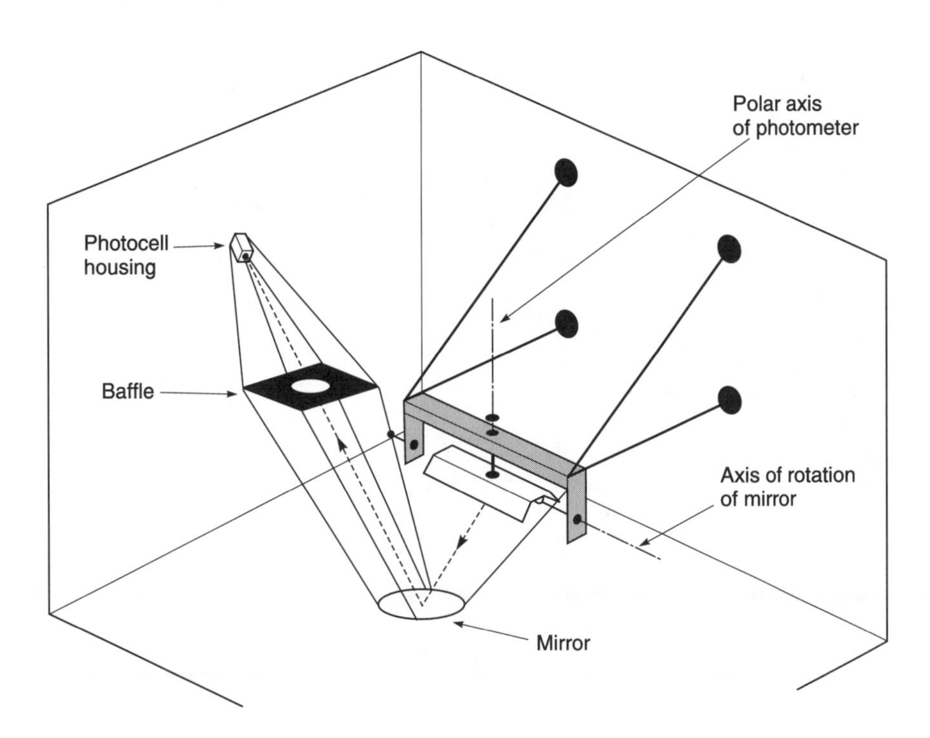

Fig. 15.9 Single mirror photometer – light path brought out sideways

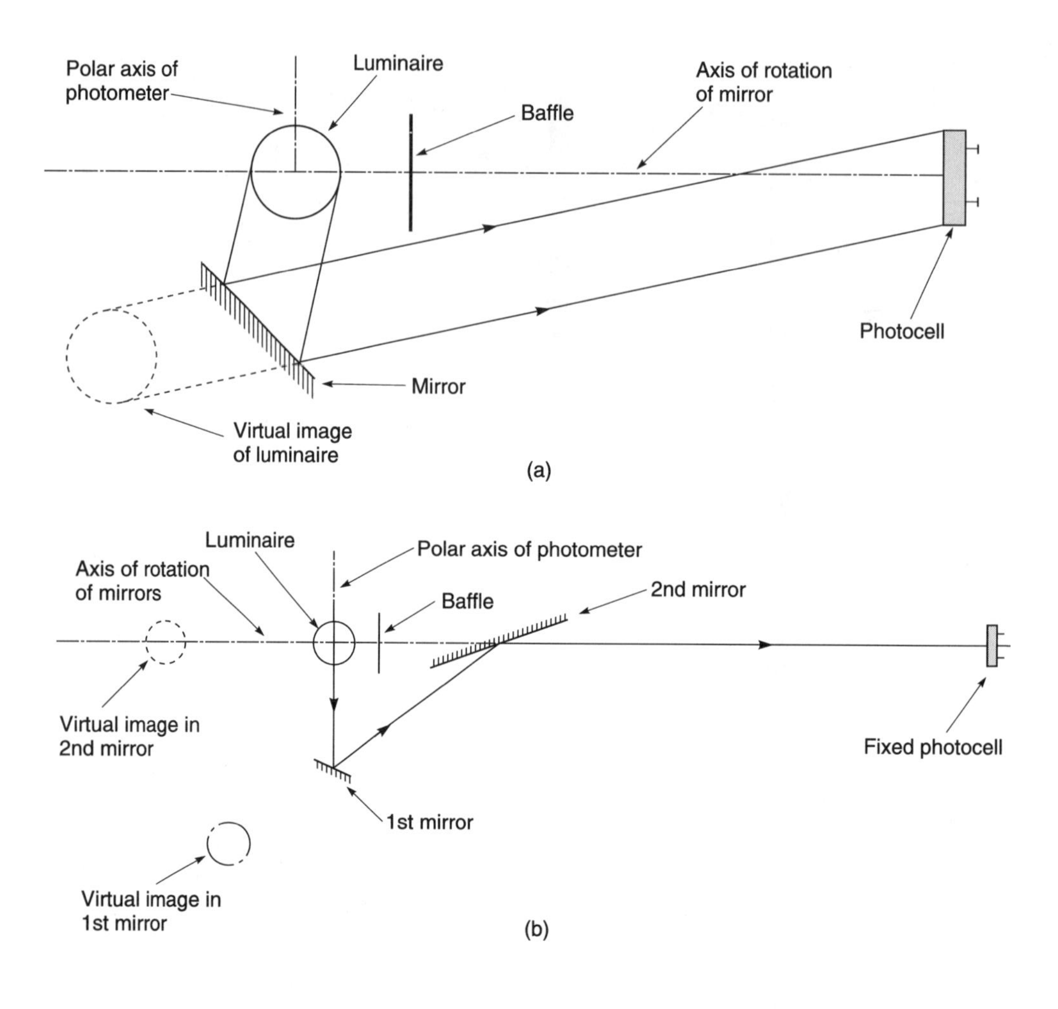

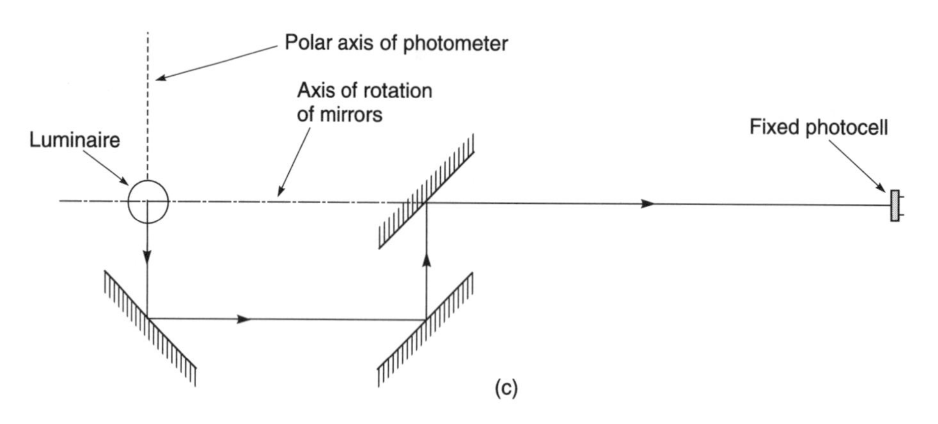

Fig. 15.10 Various configurations for mirror photometers

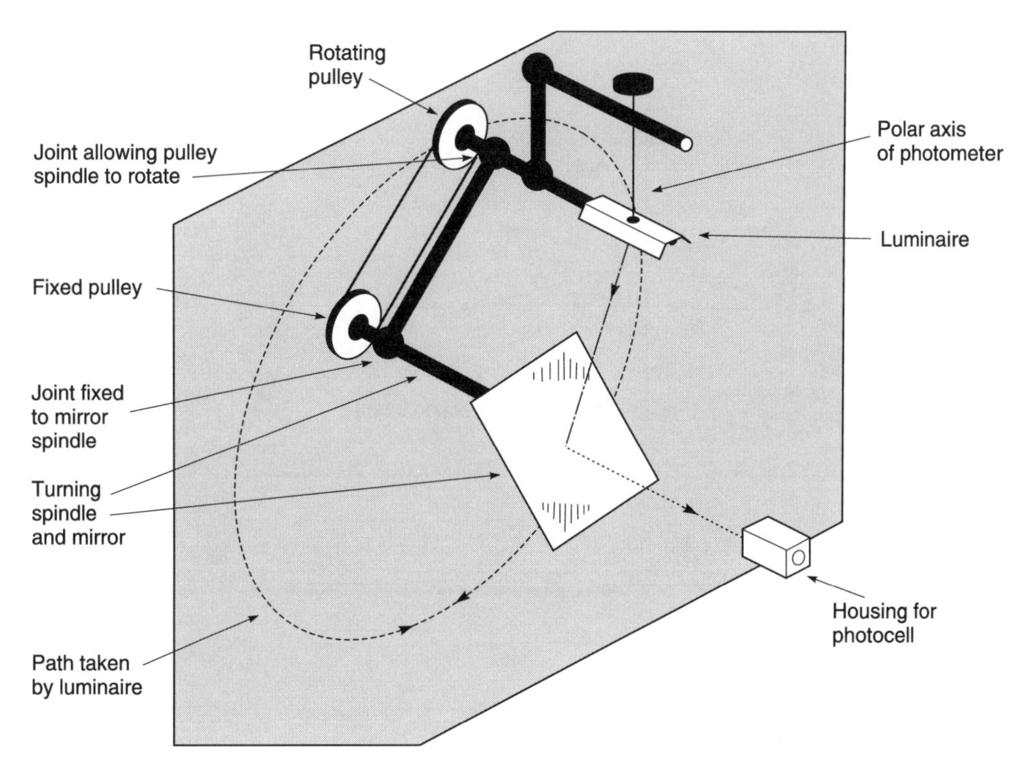

Fig. 15.11 Single mirror arrangement with fixed photocell; rotating light source

always normal to the surface of the photocell. These arrangements also have the advantage that the photocell can be taken as far back from the goniophotometer as the available space will allow, provided the mirrors are large enough to allow a complete image of the luminaire as seen from the photocell.

In an adaptation of the beam goniophotometer, the luminaire rotates around a centrally placed rotating mirror, which reflects the light into the axis of rotation. This is shown in Figure 15.11. It effects an economy in the number of mirrors required, but suffers from the same drawbacks as the beam goniophotometer.

15.5.3 COLLIMATING PHOTOMETERS

In this method, a lens, a parabolic mirror, or a mechanical means is used to select those rays of light that can be considered as contributing to the luminous intensity in a given direction. The advantage of this method is that the length of the optical path is considerably shortened. Some authors refer to this method as 'inverse collimation' but since the elements used are referred to as 'collimators' the term collimation has been adopted here. The method may be most readily understood by reference to some examples.

In Figure 15.12(a) light from the source is focused at the aperture in the screen by the collimating lens, so that only a narrow beam of light is accepted. The aperture, which is set at the focus of the collimating lens, is imaged by the second lens onto the surface of the photocell. The collection angle of the system is determined by the diameter of the aperture in relation to the focal length of the collimating lens; it is made narrower by decreasing the aperture and increasing the

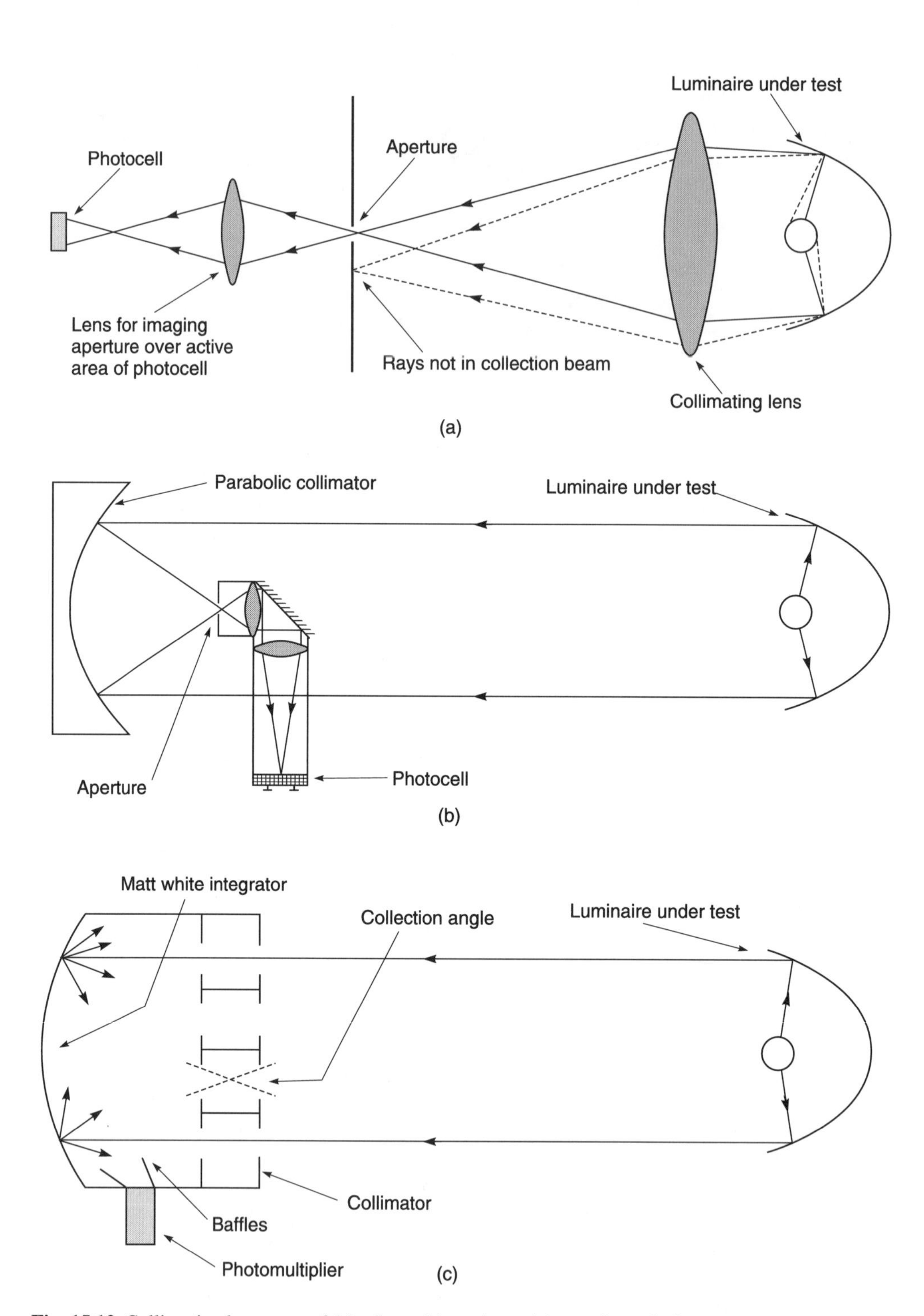

Fig. 15.12 Collimation by means of (a) a lens; (b) a mirror; (c) a perforated plate

focal length. Errors due to aberrations in the collimating lens can be reduced by making the focal length of the lens much greater (at least six times) than its diameter. This also reduces the error due to the light from the lens being incident obliquely on the photocell. It is essential that the collimating lens has a diameter somewhat greater than that of the source so that all the light in the direction of measurement is collected. To reduce interreflections the spacing between the lens and the source should be as great as possible. A system using a parabolic mirror, instead of a lens, can also be used (Figure 15.12(b)). This allows a greater source diameter to be measured (because it is easier to make a large diameter mirror than a lens) but it should be noted that an error is introduced by the obstruction of the pick-up elements that direct light on to the photocell.

Even a parabolic mirror would be too expensive for measurements on a 1500 mm or 2400 mm luminaire. To overcome this, an ingenious system devised by Frederiksen[6] uses perforated plates placed one behind the other to restrict the collection angle to 4° (Figure 15.12(c)). Behind these a matt white surface integrates the light, the illuminance from which is measured by a photomultiplier. The sensitivity of a photomultiplier is required because the collimator allows very little light to pass. It has to be baffled in such a way that it produces a reasonably equal response to a given quantity light no matter where this enters the collimator. Complete equality of response, or nearly complete equality, is then obtained by blocking some of the holes.

The collimator itself is made of injection-moulded plastic plates 1000 mm by 1000 mm by 10 mm thick. Six of these plates are stacked behind each other to form an element of the collimator (only two plates are shown in Figure 15.12(c)). Since the collimator is about 1800 mm in diameter, support may be necessary to prevent it from sagging.

As for the other collimating photometers, the diameter of the collimator, d, has to be somewhat greater than the largest linear dimension, D, of the luminaire to be tested (Figure 15.13):

$$d = D + 2r \tan \varepsilon$$

where r is the distance between the luminaire and the surface of the collimator, and ε is the collection angle.

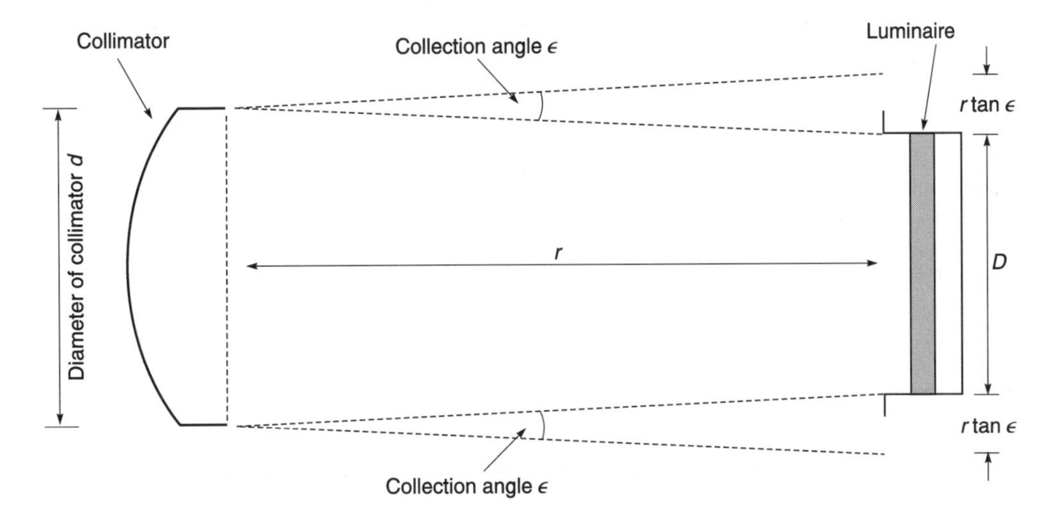

Fig. 15.13 Size of collimator in relation to luminaire

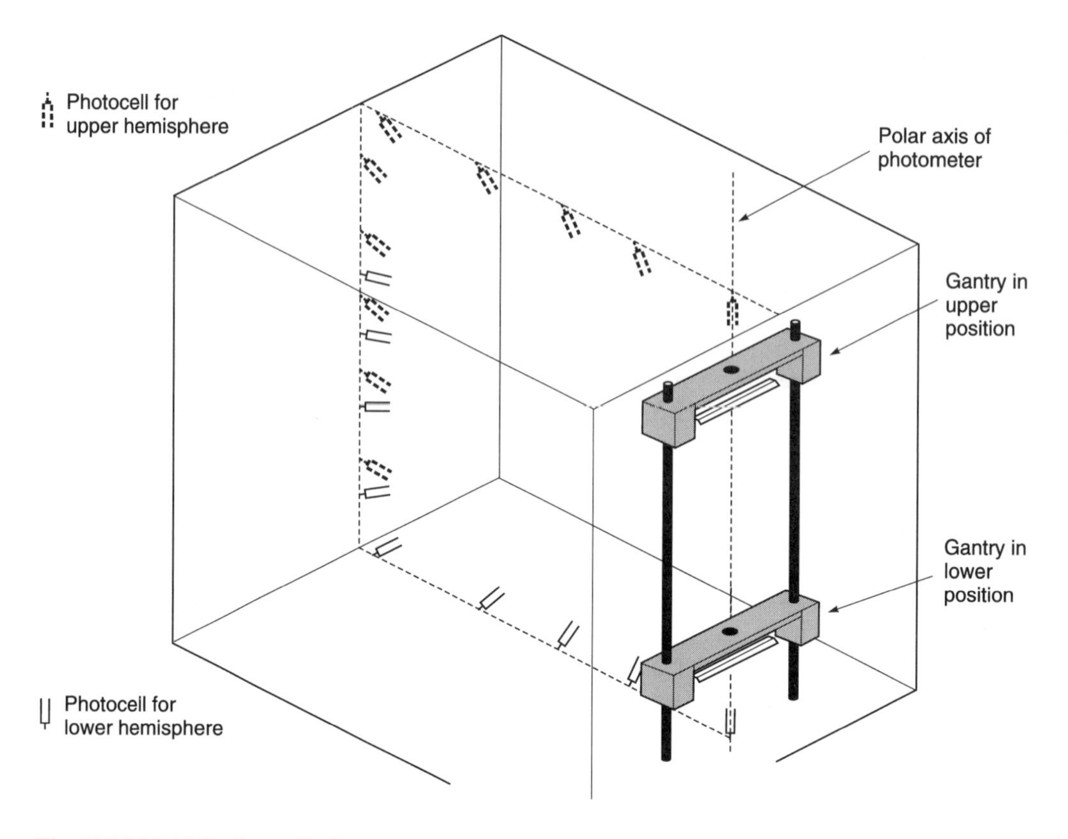

Fig. 15.14 Multiple photocell photometer

15.5.4 MULTIPLE PHOTOCELL PHOTOMETER

In this method, a separate photocell is used to receive the light at each of the required angles. The advantages are that the measurements can be taken very rapidly, the number of moving parts is minimized, and the photocell at the nadir can be used as a monitoring photocell. The disadvantage is the great number of photocells needed, which are costly and need to be kept in calibration. For measurements at 2.5° intervals in the vertical plane at least 73 photocells are needed.

Such a photometer has been described by Pritchard and Simons.[7] The layout of the photocells is shown in Figure 15.14. There are two sets of photocells: one for the measurements in the lower hemisphere, used when the luminaire is in its upper position; and another for measurements in the upper hemisphere, used when the luminaire is in its lower position. To ensure that the scaling of the luminous intensities is the same in both hemispheres, a, extra photocell is positioned to take readings at 75° when the luminaire is in its lower position on the gantry. The photocells at 90° for both the hemispheres could be used but very often the luminous intensity at this angle is nearly zero, or is zero.

The photocells are calibrated relative to one another by means of a fluorescent tube, which is rotated about its own axis so that each photocell 'sees' the same part of the tube. During a test, all luminous intensity measurements are related to the measurement at the nadir. This monitoring eliminates fluctuations due to voltage and temperature variations. In addition, where only the shape of the light distribution is required, as in some development work, the warming up time for the lamp can be eliminated.

For reliability all switching is by electronic means and not by mechanical means, which can be unreliable.

15.6 Goniophotometers for floodlights and projectors

Floodlights and projectors have to be tested at much greater distances than luminaires for indoors and road lighting. This is because the receptor should be beyond the cross-over point (see Figure 6.13, page 215) so that it 'sees' the whole of the projector flashed. The test distance may be miles in length for very parallel beams, as from searchlights, but with floodlights used for sports stadia, 30 m is generally sufficient.

A goniophotometer is required for aiming the luminaire in the desired direction. The goniophotometer in Figure 15.15(a) uses the (B, β) coordinate system, whereas the goniophotometer in Figure 15.15(b) uses the (C, γ) coordinate system. If desired, both coordinate systems can be combined in the same apparatus. The apparatus needs to be robust as floodlights can be very heavy. Moreover, a worm drive should be provided so that degrees of arc can be divided into parts, say tenths, accurately.

Both these photometers suffer from the drawback that the luminaire is moved out of its working position as the test proceeds, which may affect the light output from the lamp. It may be possible to overcome this problem by means of a monitoring photocell which has to be arranged so that it always 'looks' at the luminaire in the same direction. This photocell needs to 'look' at the whole luminaire, and not just part of the arc tube of the lamp, as the distribution of luminance of the arc tube may vary as it is rotated. This may rule out the use of fibre optics which, otherwise, would provide a very convenient solution to the problem as they cause little obstruction to

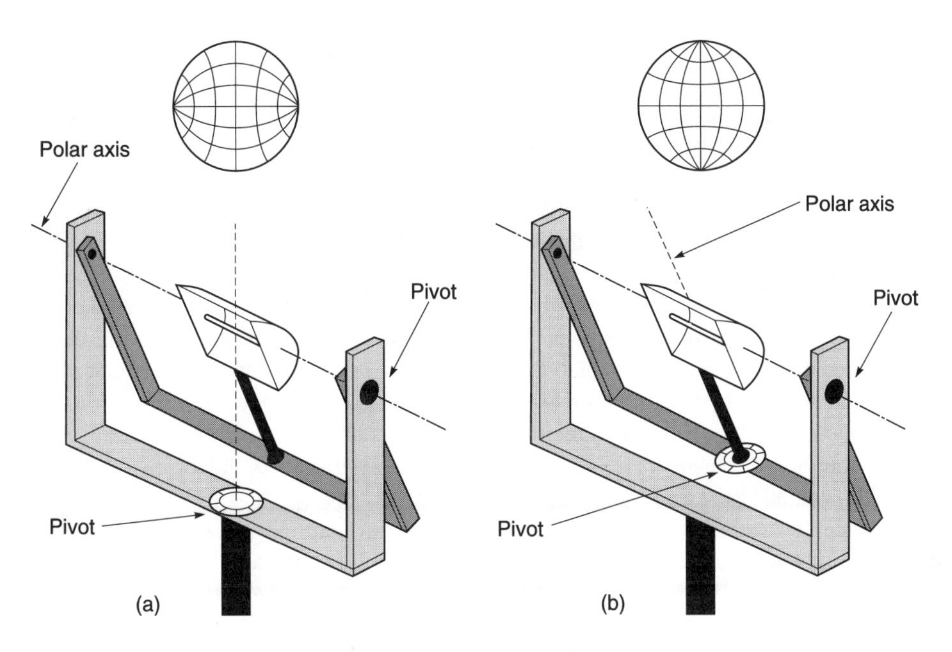

Fig. 15.15 Two forms of goniophotometer for testing floodlights

the light and enable the monitoring photocell to be housed away from the heat of the lamp. Another solution is to keep the luminaire at its correct aiming angle during the test and to rotate mirrors about the luminaire, as in Figure 15.10, page 482. Strictly, a different set of data is required for each aiming angle.

Because, as already stated, the test may need to be conducted with the photocell at a large distance from the luminaire, readings taken on the bare lamp for calibration purposes may be too small to achieve good accuracy. Provided the photocell is sensitive enough, it may be possible to overcome this problem by the use of different ranges on the photocell current measuring or indicating device. Alternatively, provision can be made for bringing the photocell closer to the goniophotometer. If a large receptor is used, a check should be made that its angle of subtense at the luminaire is not too great.

An alternative approach is to use an accurate illuminance meter. The luminous flux of the test lamp should be measured so that all readings can be multiplied by a factor to correct for any deviation from the nominal light output of the lamp.

15.7 Checking the alignment of a goniophotometer

The procedures described here apply to a goniophotometer using the (C, γ) system of coordinates, but they can be adapted for goniophotometers using the other coordinate systems. They can also be used in setting up a new photometer. A laser, which throws a bright spot of light at a distance, eases the task considerably.

1. The first check should be for the verticality of the polar axis. If the goniophotometer uses a gantry the test should be done with the gantry at the height at which the luminaire is tested. If it is used at more than one height then the test has to be repeated at each height.

   A plumb bob is dropped from the centre of the spindle used for turning the luminaire through C angles and its position marked on the floor (Figure 15.16(a)). If the gantry is used at more than one height, all the nadirs should be coincident. Moreover, when the gantry is loaded with the heaviest luminaire to be tested, there should be no significant deviation from the position marked on the floor when the gantry is unloaded.

   A laser is then fixed to the spindle and directed to the nadir. The laser does not have to be in line with the spindle for this test, it can be to one side. When the spindle is rotated, the spot should remain stationary. If it does not it will describe an ellipse, which will be very nearly a circle. The centre of this should be marked and the laser directed towards it (Figure 15.16(b)). The spot should remain stationary. The distance between the nadir found with the plumb bob and this spot indicates the amount that the spindle is out of vertical. An adjustment should be made accordingly.
2. The next step is to ensure that the vertical plane in which the measurements are taken is truly vertical and passes through the polar axis. For goniophotometers with rotating arms, this means checking that the axis of rotation is horizontal and passes through the polar axis.

   This can be done by attaching the laser to the arm used for rotation and using it to find the extended axis of rotation on a suitable surface, in a similar way to that which was used for the vertical spindle. A string is then tightly stretched between this point and the centre of the spindle used for rotating the arm. This should be horizontal – one way to check this is with a water level (basically water in a transparent flexible tube). In addition it should pass through the polar axis of the goniophotometer. If it does not then the necessary adjustments have to be made.

   The point where the two axes cross is the centre of the goniophotometer and some means should be provided for marking this position. Since this is in mid-air it is convenient to use

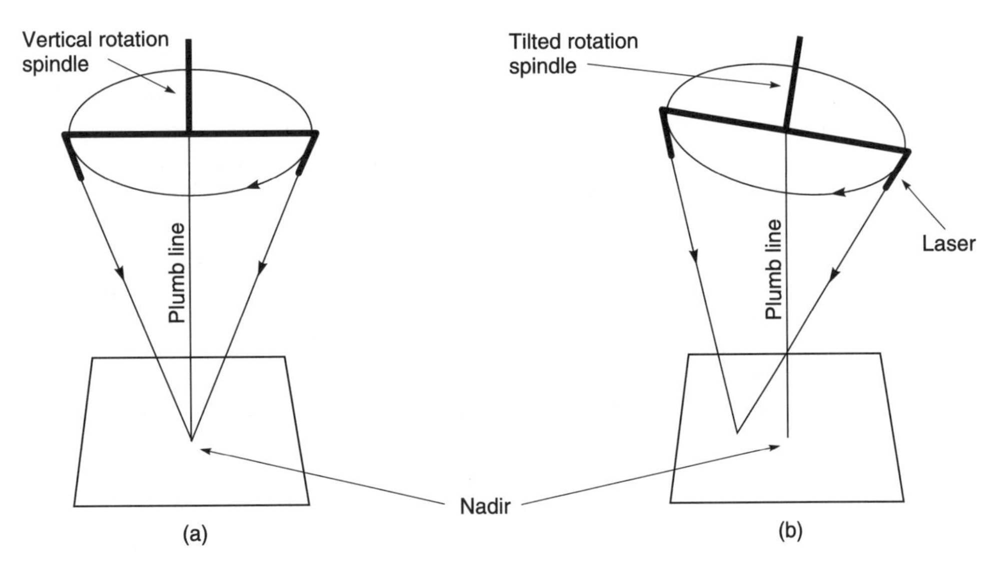

Fig. 15.16 Testing verticality of polar axis

an optical device such as a laser fixed to a wall, at the same height as the centre. Then, when it is switched on, its spot of light can be seen shining on the luminaire. It is also a good idea to mark permanently on a wall or other convenient surface the nadir and the point indicating the extended horizontal axis. These can then be used for future reference.

3. It is now necessary to find the position of the $C = 0°$ vertical plane. This can be done most conveniently by using a device that allows the laser to be mounted coaxially on the vertical goniophotometer spindle and permits movement of the laser through measured angles in the vertical plane. This device is turned so the laser beam is coaxial with the horizontal axis of the goniophotometer. The laser is then turned through a right-angle in the horizontal plane. Its beam will now lie at $C = 0°$, $\gamma = 90°$, allowing the angle indicators on the goniophotometer to be set. If the device for holding the laser is suitably calibrated, it permits the accuracy of γ indication to be checked at a number of angles.
4. The last step is to set the photocell box so that the laser beam is directed to the centre of it. If there are mirrors, these will have to be tilted to achieve this. A check should be made that when the laser beam is directed downwards and the goniophotometer arm is moved into the nadir position, the laser spot still lies in the centre of the photocell. Any straying indicates that there is bending in the arm or mirror. CIE 121 and most national standards allow an error of 0.5° in measurement.

15.7.1 OTHER CHECKS ON THE GONIOPHOTOMETER

CIE 121 lists a number of performance checks which should be carried out.

1. Stray light. An observer should view the luminaire mounted on the goniophotometer from the position of the photocell. A careful search for light reflected from the supporting structure and the bevelled edges of mirrors should be made. In addition, the background to the luminaire should be examined. Ideally this should be done from a number of angular positions; that is, at different values of γ.

2. Reflectance of mirrors. For this test a lamp is selected that exhibits little change of luminous intensity over a reasonably large angular zone of ±10°, say. An opal tungsten lamp is suitable. The goniophotometer arm is positioned at the nadir and the light source is positioned on a horizontal bar attached to the azimuth spindle. When the bar is rotated, the luminous intensity should stay constant for any position of the lamp on the bar. In addition, when the light source is moved along the bar, the reading should remain constant. Care should be taken that the voltage is kept constant as the output of tungsten lamps varies as the power 3.3 of the voltage. This means that a 1% variation in voltage produces a 3.3% variation in light output. If this source of error is made insignificant, variations in measured luminous intensity are most likely due to variations in the reflectance of the mirror or mirrors. Other possibilities that should be eliminated are stray light, the photocell not being at the nadir, and flexing in the luminaire mounting during its rotation.
3. Linearity of response of photocells. This can be checked on a photometric bench but it is more convenient to use a specially constructed linearity testing box. As shown in Figure 15.17 this consists of a box housing a number of light sources, usually GLS lamps run at a slightly lower voltage than the nominal voltage to increase life. These should be of different outputs, roughly proportional to the numbers marked on the diagram, so that a range from 1 to 20 in nearly equal steps of illuminance can be obtained by exposing different combinations of lamps. The range can be extended by adding lamps of higher power or bringing the box nearer to the photocell. The different combinations of lamps for exposing to the photocell are obtained either by operating shutters or by switching. Shutters are probably preferable as repeated switching of the lamps might affect their output, and the switches themselves might introduce inaccuracies due to resistance at the contacts. It is used by exposing a number of lamps to the photocell at once, and then checking that the reading so obtained is equal to the sum of the readings when the lamps are exposed separately. The discrepancy is a measure of lack of linearity plus any experimental error due to voltage fluctuations.

   An advantage of this method over using a photometric bench is that photocells can be tested *in situ*, which is important when they are incorporated into the apparatus.
4. Constancy of nadir reading. During a test when measurements are taken in a number of *C* half-planes, the nadir reading should stay constant. If there is variation, it can be due to the causes mentioned in point 2. This is a check that can be carried out every time a full test is done. The print-out of the *I*-table can, with advantage, be programmed to warn of any undue variation.

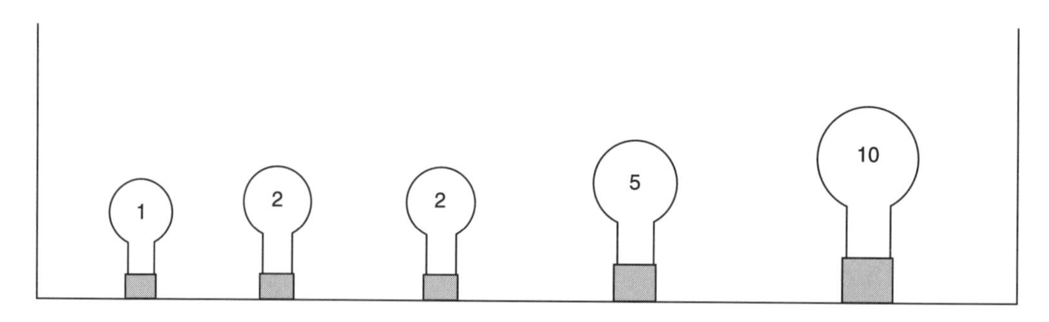

Fig. 15.17 Linearity of response testing box

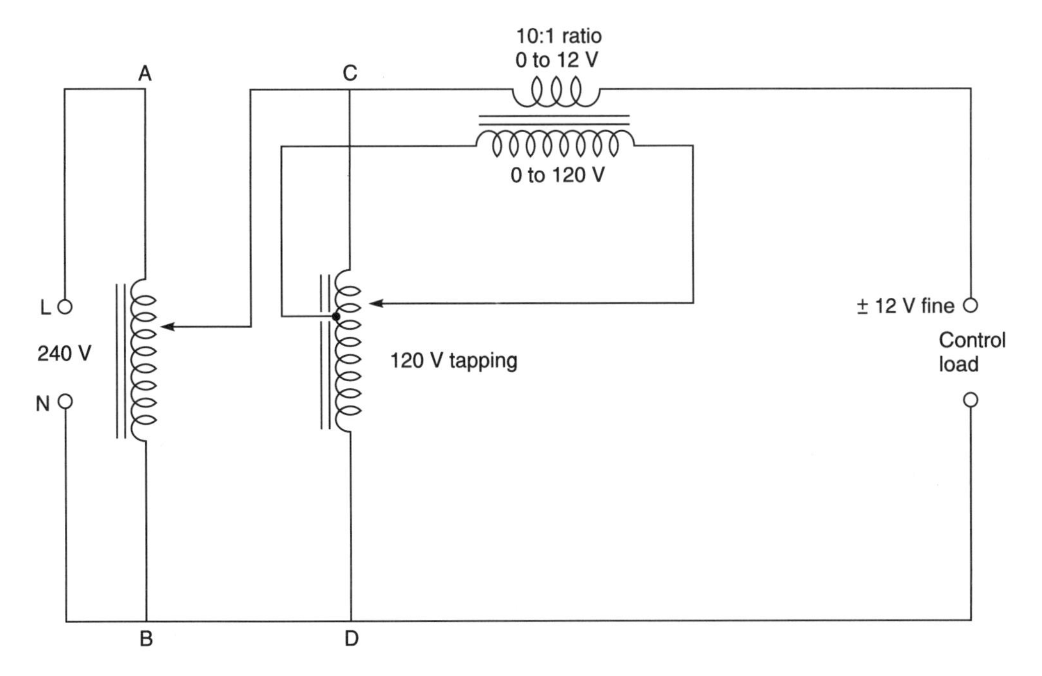

Fig. 15.18 Buck-boost fine voltage adjustment

15.7.2 ANCILLARY EQUIPMENT

Many hundreds of readings are needed to produce an *I*-table and some degree of automation is required to save time and overcome the tedium of doing the work manually. For this, a remote indication of angle is required, which is generally achieved by means of shaft encoders. These are capable of measuring to the nearest 0.1° of arc without having to resort to gearing, which is necessary with magslips. Also, they can be used to feed the signal to a computer.

A good voltage stabilizer is required, preferably one that does not give generous harmonics. It should be remembered, as already mentioned, that the output of filament lamps varies as the power 3.3 of the voltage. Hence, it is imperative that the voltage is monitored during a test, unless the light output itself is monitored. The task of doing this is eased if the measurements can be completed quickly.

A method of fine adjustment of the voltage will be required. A variable auto-transformer used by itself will not permit fine enough control. The buck-boost circuit shown in Figure 15.18 can be used for this purpose.

AB is a variable auto-transformer capable of taking the full load current. CD provides the fine adjustment through the fixed transformer. It need only be capable of taking one tenth of the full load current.

Where low voltages are required, a step down transformer fed from a variable auto-transformer can be used as an alternative to the buck-boost circuit.

15.8 Determination of light output ratios by integrators

An integrator is a device such as a sphere with a white internal finish that collects all the luminous flux from a light source placed in it. Provided that certain conditions, to be described, are fulfilled, the illuminance on the integrator wall is proportional to the total luminous flux emitted. The devices are used in some laboratories for two reasons. They provide an easy method of measuring light output ratios because there are fewer readings needed than with distribution photometry, as few as four for a single lamp luminaire. This cuts down on the work required, but distribution photometry may still needed. Another advantage is that an integrator needs less room (or volume) than a light distribution photometer so it is easier to control the temperature, which is an important consideration with temperature sensitive lamps. There is no doubt that they are useful for development work where maximizing the light output ratio is an important consideration. However, it has to be said that, because of automation, distribution photometry is becoming so fast and reliable that the case for using integrators is less strong than it used to be.

15.8.1 THE INTEGRATING SPHERE

As will be explained, various shapes of integrator are used, but only the sphere has the fundamentally correct shape. The validity of its use depends on the principle that the illuminance received on one area of a sphere from another part is independent of the relative positions of the two parts. This fact has already been proved in Section 3.6.2, page 76, in connection with determining the total luminous flux emitted by a uniform diffuser. Making use of it, we can relate the illuminance received by a window to the total luminous flux emitted by a source, provided that the window is shaded from direct light by a baffle. A typical arrangement is shown in Figure 15.19. The luminaire or lamp to be tested is suspended at or near the centre of the sphere, and measurements of light output are made by the photocell, which is baffled so that no light from the source can reach it directly. The auxiliary lamp with its baffle is used to compensate for absorption of interreflected light by the method explained later. The inside of the sphere is coated

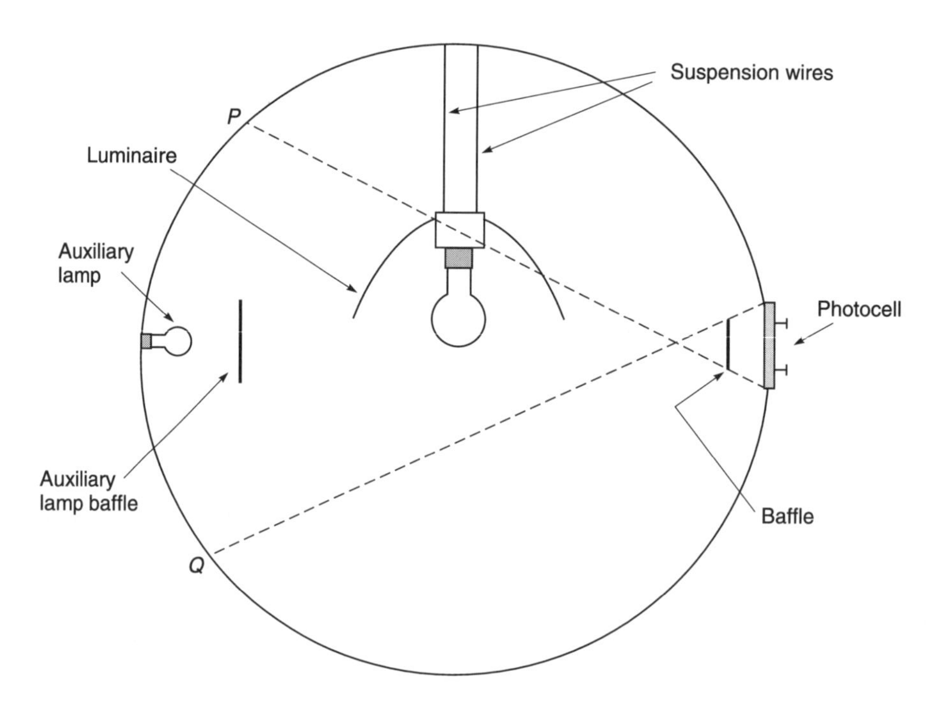

Fig. 15.19 Integrating sphere

with paint of reflectance ρ, and is assumed to behave as a uniform diffuser. In general, the direct illuminance will vary over the surface of the sphere.

If A is the area of the sphere, the total luminous flux F from the source is given by:

$$F = \frac{\pi}{\rho} \int L \, dA \tag{15.1}$$

where L is the initial luminance of an element dA of the sphere wall.

The light reflected after the first reflection will illuminate all parts of the sphere, including the photocell, evenly.

Hence, the illuminance due to the reflected light from dA received by all parts of the sphere, including the photocell, will be $(\pi/A)L \, dA$, and the illuminance due to light reflected from all parts of the sphere will be $(\pi/A) \int L \, dA$.

From (15.1) this equals $(\rho/A)F$.

Similarly, the illuminance due to the second reflection is $(\rho^2/A)F$. This process can be carried out indefinitely so that the final illuminance E is given by:

$$F = \frac{\rho}{A} F + \frac{\rho^2}{A} F + \frac{\rho^3}{A} F + \ldots$$

$$= \frac{\rho F}{A} \left(\frac{1}{1 - \rho} \right)$$

Hence, E is independent of the distribution of light from the source or luminaire.

Theoretically, if the reflectance of the sphere surface is known, it is possible to calculate F by measuring E. In practice, however, the assumptions made in the theory cannot be satisfied and so the sphere is only used for comparing sources; that is, for substitution photometry. The bare lamp is placed in the integrator and the reading R_B obtained, this is replaced by the luminaire and the reading R_L obtained. The light output ratio is then R_L/R_B. This has to be multiplied by a correction factor to allow for obstruction, as given.

There are three main departures from the ideal sphere which must be considered.

Effect of baffle

Luminous flux falling on the baffle will produce less illuminance on the photocell than if it fell on part of the sphere wall 'visible' by the photocell. In addition, if direct light falls on the part of the sphere wall shaded by the baffle (PQ) it will also be less effective in producing illuminance on the photocell. Hence, the source should be positioned, if possible, in such a way that the minimum luminous flux falls on the baffle or shaded part of the sphere wall. In addition, the baffle should be made as small as possible, consistent with all light emitting parts of the bare lamp and luminaire being shaded from view.

To reduce the error due to shading by the baffle, some authorities advocate the use of one that is translucent, the degree of translucency being determined by experiment. Other authorities argue that such a correction will only apply to one particular position of the source and screen, and if these are materially altered, a greater error than before may result. Another suggestion for reducing the error is to paint the baffle with paint of the highest reflectance available, so reducing the amount of luminous flux absorbed.

It can be shown by experiment that the position of the screen is not critical. Obviously it should not be too close to the photocell, since in this position it would reduce the view of the photocell. On the other hand, the closer it is placed to the source the more luminous flux it will absorb. For a small source, Walsh[8] recommends placing it at a third of the sphere radius from the source. Since a range of sizes of luminaire may have to be catered for, a good compromise is to place it half-way between the end of the luminaire and the photocell.

Diffusion properties of paint

Errors will arise because the paint does not behave as a uniform diffuser, but these have not been investigated, and all that can be done is to ensure that the finish is as matt as possible. BS 354[9] makes recommendations for paint formulation, as does CIE 84.[10]

Obstruction

The luminaire and baffle interfere with the interreflection of the light in the sphere. Large luminaires may absorb a considerable amount of luminous flux, so introducing a large error. This may be minimized by making the sphere as large as possible, by not using paint of the highest reflectance (less than 80%), and by using the auxiliary lamp method.

In the latter, an incandescent lamp is fixed on part of the sphere wall that is not visible from the photocell position, that is *PQ* in Figure 15.19. Considerations of symmetry suggest the best position for this is on the line that passes through the photocell and the centre of the sphere. As shown, the auxiliary lamp is screened to prevent direct light from it reaching the source or luminaire, since its function is to correct for absorption of the interreflected light. To correct for absorption by the lamp to be used in the luminaire, the following three readings are taken with the auxiliary lamp on:

(1) R_1 with the sphere empty,
(2) R_2 with the unlit lamp in the sphere.
(3) R_3 with the unlit luminaire in the sphere.

The measured output of the lamp should, therefore, be increased by R_1/R_2. Similarly, the output of the luminaire should be increased by R_1/R_3. The overall correction factor to be applied to the light output ratio is then equal to:

$$\frac{R_1}{R_3} \times \frac{R_2}{R_1} = \frac{R_2}{R_3}$$

So only the two readings R_2 and R_3 are required.

Other shapes of integrator

Owing to the expense of constructing a sphere, the difficulty of access inside it, and perhaps lack of headroom, other shapes of integrator are often used. If the integrator is used for comparing sources with a similar light distribution, then probably the shape is not critical. In luminaire photometry the light distribution from the lamp and the luminaire are, with very few exceptions, quite different. Keitz[11] investigated various types of integrator by finding how the photocell reading changed when a spotlight was directed onto various parts of the integrator wall. Even with the sphere, the reading fell by as much as 15% when the spotlight was directed onto the baffle or the part of the wall masked by the baffle. As would be expected, the greatest readings were registered when the spotlight was directed onto the parts of the wall the photocell could 'see' without obstruction. With two polyhedra, one 14 sided the other 19 sided, the maximum error increased to 20%, the error occurring over a greater proportion of the wall area than in the sphere.

These figures confirm one practical detail we have mentioned already; that is, the luminaire or lamp should be positioned so that as little of the luminous flux falls on the baffle and masked wall area as possible. They also indicate it is important that before integrator measurements of light output ratio are accepted as reliable, they should be compared with those obtained by using light distribution photometry, which, as stated previously, is a more fundamental method. It has been found in practice that it is possible to obtain better agreement between the two methods than the figures quoted above might indicate. Possibly this is due to the errors obtained with the bare lamp and luminaire tending to cancel each other out, particularly in the case of linear fluorescent luminaires, where the shape of the axial light distribution is usually little modified, except where louvres are used.

Construction of the integrator

The materials that have most often been used in the construction of the integrator are wood, aluminium, mild steel, and fibre-glass.[12]

Access to integrators for painting and suspending the luminaire presents a problem for integrators with a height greater than about 2 m. A common solution is to use a travelling platform which goes through a trap-door half-way up the integrator.

To provide platforms for the task of repainting the integrator when it is higher than about 3.5 m scaffolding may need to be erected. Provision, such as sockets in the wall of the integrator, should be made for this. An alternative approach (Figure 15.20) is to have the integrator in two halves that can be drawn apart horizontally. Part of the top of the integrator can be left

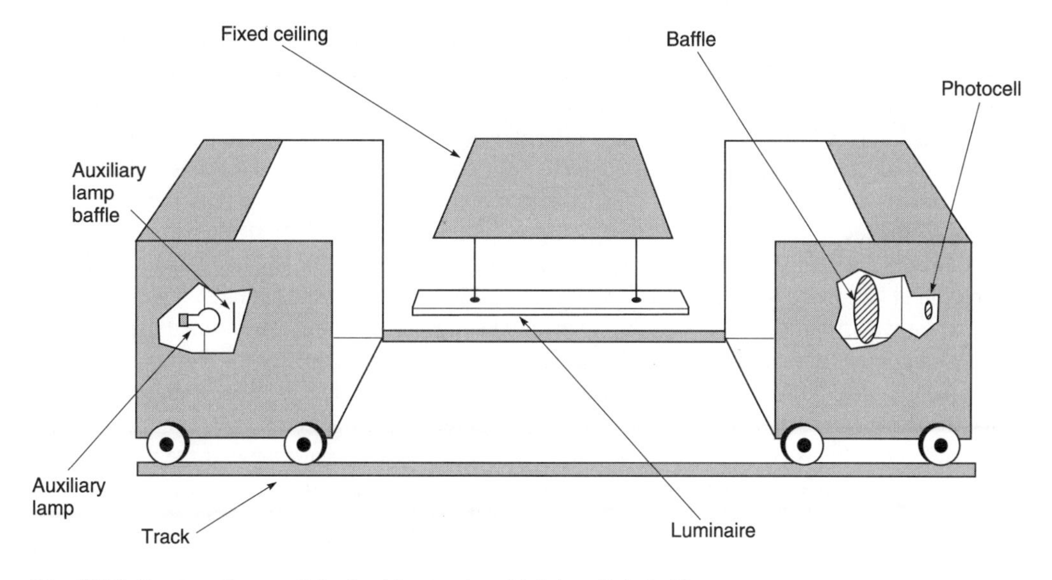

Fig. 15.20 Rectangular parallelepiped integrator which is split in half

stationary for suspension of the luminaire or lamp. This has a number of advantages. Besides making access to the integrator relatively easy for repainting, it allows the two halves to be moved well clear of operations when a luminaire is being suspended. This not only makes the process easier but reduces the risk of soiling the bottom. As most luminaires direct the greater proportion of their light downwards, even a small amount of soiling can affect the measurement of light output ratio appreciably. In addition, the two halves arrangement allows the luminaire to stabilize thermally in a less confined space, which is an important consideration with lamps having a luminous output sensitive to temperature.

15.9 Practical procedures for testing luminaires

The practical procedures for testing luminaires are described in detail in CIE 121 and various national documents such as BS 5225[13] and the US *Lighting Handbook*,[14] so there would be no point in our repeating these. Instead we will discuss some of the basic problems involved.

15.9.1 CALIBRATION OF LUMINOUS INTENSITY DATA

Luminous intensity data may be calibrated in one of two ways, by the absolute method or the relative method.

In the absolute method, the light distribution photometer is calibrated by means of a standard lamp of known luminous intensity in a known direction, or by means of an accurately calibrated illuminance meter and application of the inverse square law. After this is done, the luminaire to be tested is mounted on the photometer and readings converted into candelas. This method is sometimes used in the testing of floodlights, but the data so produced are of limited use for design data. This is because the designer requires data based on the nominal or notional output of the lamp used in the luminaire and not the output of the particular lamp used for the test. In other words, the designer requires data that can be scaled to be representative of the actual light output of the lamp to be used in the lighting installation.

The relative method of photometry enables this to be done. The usual base adopted for the light output of the lamp is 1000 lumens. To apply the method, the photometrist has first of all to choose lamps that have the correct nominal dimensions and electrical characteristics. These are then aged for the periods recommended in the various standards. The next step is to determine on the photometer the ratio of the luminous intensity in a convenient direction to the luminous flux of the lamp (cd/lm). Since this is a ratio, its value depends only on the shape of the light distribution of the lamp and not on its absolute light output in lumens. This means that absolute measurements are only taken once, when the nominal output of the lamp type is determined, usually by the lamp manufacturer.

As this ratio depends only on the shape of the light distribution from the lamp, tungsten lamps can be run at a reduced voltage to prolong their life. Fluorescent lamps maintain a very constant shape of light distribution from one lamp to the next and Baumgartner[15] determined that the ratio of the luminous flux output to the luminous intensity at right angles to the axis is 9.25. Theoretically, if the tube were a uniform diffuser the figure would be π^2. However, tempting as it may be to save time by using a readily available figure, better practice is to determine the ratio for each lamp. This is because the ratio varies with the length of the tube and the thickness of the fluorescent powder coating, which may not be constant circumferentially.

To illustrate the relative method, consider the following example. We photometer the lamp and determine that it gives 10 candelas per 1000 lumens in a particular direction. So with the photometer head in this direction, we set the luminous intensity recording instrument to 10 candelas. We then put this same lamp in the test luminaire, and the recording instrument will be indicating candelas per 1000 lumens.

If an integrator is used to determine the light output ratio (*LOR*), the procedure is slightly different. The luminous flux of the luminaire is determined in arbitrary units (*A*) by using one of the methods in Section 1.7, then;

$$\text{Scale factor} = \frac{A}{LOR \times 1000 \times \text{number of lamps}}$$

This is applied to the luminous intensity values as a multiplier so that they represent candelas per 1000 lumens.

15.9.2 PHOTOMETRIC CENTRE OF LUMINAIRE

The photometric centre of the luminaire is the point at which the inverse square law applies most closely. The position of this point may change according to direction. This may be understood by considering a box-shaped diffuser. In the direction of the nadir, the inverse square law would apply most closely from the centre of the bottom panel, whereas normal to one of the side panels, it would apply from the centre of that side panel. Hence, we have to make a compromise that gives the least overall error. It would be impracticable to determine this for each luminaire; instead certain conventions are adopted in CIE 121 and national standards. Some of these are indicated in Figure 15.21.

15.9.3 TEMPERATURE CONTROL

CIE 121 states that the ambient temperature should be 25±1°C during a test, which is a very close tolerance for the large volume that a distribution photometer occupies. The lamps that are most sensitive to temperature change are fluorescent, and when these – or luminaires incorporating them – are being tested it is important to keep the temperature within the specified limits. Moveover, as has already been mentioned, there should be no draughts as these will alter the temperature of the air surrounding the lamp or luminaire. Most other lamps are relatively insensitive to temperature fluctuations. However, it is possible that temperature fluctuations will affect the output from the photocell or photocells, or their associated amplifiers, unless these have provision for thermostatic control (see Section 15.2, page 471).

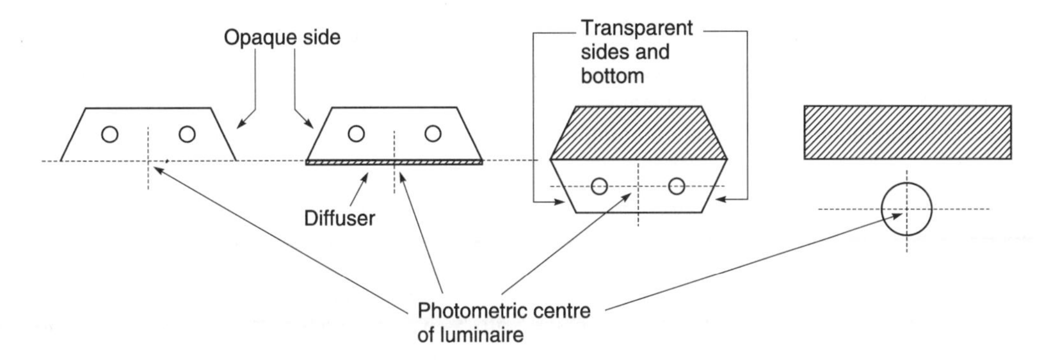

Fig. 15.21 Examples of photometric centres of luminaires

If a luminaire is to be used at a temperature significantly different from 25°C, then its output should be measured at this temperature. Alternatively, temperature measurements that will enable a correction to be made must be taken in the luminaire. The bare lamps should still be measured at the same temperature (usually 25°C) as that at which the nominal light output is measured.

15.9.4 ELECTRICAL MEASUREMENTS

The lamp, in the case of filament lamps, or the control gear, in the case of discharge lamps, should be run at rated voltage. To do this accurately, leads – which are separate from the power supply leads – should be taken back to the voltmeter. This eliminates the error arising from the volt drop in the power leads and is particularly important with distribution photometers where the leads may be of considerable length.

15.9.5 MULTI-LAMP LUMINAIRES

Multi-lamp luminaires use more than one lamp of the same type. This is distinct from blended lamp luminaires (Section 15.9.7), which use different lamp types in the same luminaire.

The lamps for multi-lamp luminaires should be photometrically matched to within 5% when used on the same ballast. This can conveniently be done in an integrator. Ballasts should also be matched to within 5% of the corresponding reference ballast.

15.9.6 ANGULAR SPACING OF LUMINOUS INTENSITY MEASUREMENTS

The angular spacing of the luminous intensity readings has to be such that interpolation procedures can be carried out to give the accuracy required for the practical application of the data.

For general purpose luminaires, measurements are generally taken at 5° in the vertical (*C*) planes and at 45° or 30° in azimuth. However, as automation of photometry is now commonly used, measurements closer than these, particularly in azimuth, may be taken to provide greater accuracy in calculation of illuminance and glare.

For road lighting luminaires, the spacing of the measurements is graded in CIE 30.2 according to the likely rate of change of luminous intensity with angle and the importance of the luminous intensity in affecting the road surface luminance. For instance, the angular steps are closer together in the direction of the main light beam than in regions under the luminaire. However, new standards being prepared recommend even angular spacing, certainly in the lower hemisphere.

In conclusion, it can be said that with the advent of automation a high proportion of the time spent in testing a luminaire is in preparation and warming up. It is well, therefore, to make the angular spacings as close as is practicable so that full data are on hand for any demanding calculation that may arise in the future.

15.9.7 BLENDED LAMP LUMINAIRES

In these luminaires, two different types of lamp are used. For instance, in some industrial floodlights for high-bay lighting, a high intensity discharge lamp and a tungsten lamp are used in the same luminaire. If we want to test the luminaire with both lamps on at the same time we would have to make sure that the outputs of the lamps are in proportion to their nominal outputs, which may be difficult to achieve. However, this problem can be overcome by testing the luminaire with only one lamp alight at a time, but with the other lamp in position. In the relative method of photometry this yields two *I*-tables, each based on 1000 lumens.

Let Φ_A and Φ_B be the nominal luminous flux outputs, in kilolumens, of the lamps *A* and *B*. In a given direction, let the corresponding luminous intensity measurements be I_A and I_B, per kilolumen. Then the luminous intensity in the chosen direction is $I_A\Phi_A + I_B\Phi_B$.

When an integrator is used, the light output ratios LOR_A and LOR_B, determined separately for the two lamps *A* and *B*, are combined in the following formula to give the *LOR* for the combination:

$$LOR = \frac{(LOR_A \times \Phi_A) + (\mathrm{LOR}_B \times \Phi_B)}{\Phi_A + \Phi_B}$$

This method is valid only if the heat generated by one lamp does not affect the light output from the other lamp.

15.10 Measurement of *r*-tables

The reflection properties of road surfaces need to be measured to enable the road surface luminance to be calculated for design purposes, as described in Section 9.8, page 353. Figure 15.22 shows the general layout of the apparatus used for these measurements. There are three main elements:

(1) a turntable on which the sample is mounted and can be rotated about a vertical axis,
(2) luminance meter, which views the sample along the reference axis (usually road axis) of the sample, at an angle (90 – *α*)° to the axis of rotation of the sample,

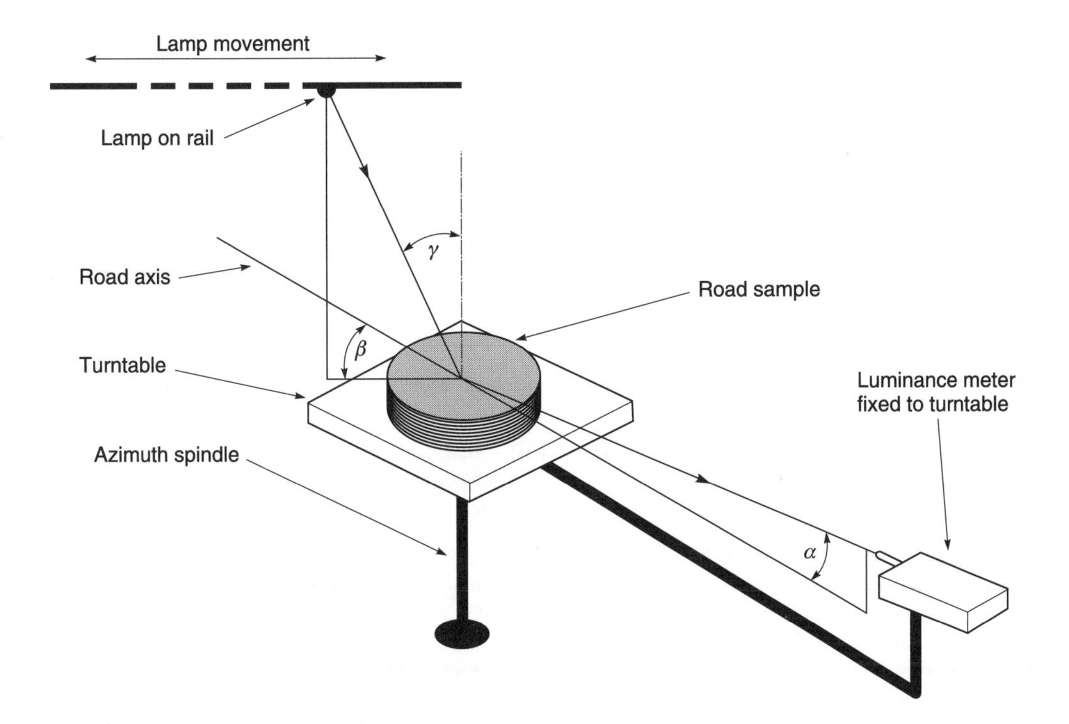

Fig. 15.22 Apparatus for measuring an *r*-table

(3) a light source, which illuminates the sample and can be moved to vary the angle γ, the angle of incidence.

The turntable should be sufficiently sturdy to turn the sample without vibration or deflection. Adjustments should be provided to enable the top surface of the sample to be brought to the horizontal. Since the surface is often rough, achieving this is problematical. CIE 30-2[16] recommends placing a 5 mm to 10 mm foam pad on the sample and surmounting this with a flat mirror, which can be checked for horizontality with a suitable optical device.

The luminance meter is rigidly fixed to the turntable so that it rotates with the sample and always views it in the reference axis of the road. The luminance meter should have a field of view of five minutes of arc in the vertical plane and ten minutes in the horizontal plane. The distance between the luminance meter and the sample should be sufficiently great for the whole of the lens to be above the level of the plane of measurement of the sample.

The angle γ can be varied either by moving the light source along a horizontal rail, as illustrated, or at the end of an arm. The former method is preferred in CIE 30-2 because, it is argued, this is more in accord with what happens in practice – the angle subtended by the source at a point on the road sample diminishes the further the source is from the sample. In addition, it is very difficult to achieve the required angular accuracy with an arm, especially for parts of the r-table (Section 9.8.1, page 355) where values of tan γ are high. For instance, r values are required for tan γ equal to 11.5 and 12, which correspond to angles of 85.03° and 85.23°. However, the lamp rail requires a long laboratory – more than 12 times the height of the rail above the sample – which may not be available.

The sample is usually cut with a diameter of about 200 mm and a thickness of 100 mm. Since many road surfaces may flex and flow with the passage of time, the samples should be mounted on an inflexible base. The road axis should be marked on the sample, although it is generally assumed that road surface samples are isotropic. The reflection properties of a road surface may change with time, even in the laboratory. It is, therefore, important to take measurements as soon as possible after cutting the sample.

In some instances, r-tables for wet surfaces are required, especially in the Nordic countries, where the road surface may be wet for a significant proportion of the time.[17] For the purpose of obtaining the standard wet condition, the sample is sprayed under the defined conditions in CIE 30-2, and measurements are taken 30 minutes after turning off the spray. Distilled water should be used otherwise there may be a deposit of salts left on the surface of the sample.

15.11 Illuminance measurements

Basically, an illuminance meter consists of a photometric head connected to a meter for indicating the current generated. The photometric head incorporates a photocell together with a means of correcting its spectral sensitivity to that required, and a means of correcting the directional response.

For convenience, it is better to have the photometric head on a lead connected to the meter than have the head connected directly to the meter. The remote indication allows the operator to take readings without obstructing the incident light. Generally, the instrument has a number of scales that enable a wide range of illuminances to be measured accurately. Provision for connection to a recorder is helpful where many readings have to be taken. A tripod may be necessary to achieve the correct height, although its use may be rendered impossible by the presence of furniture and other impedimenta. Gimbals for correcting the levelling of the photometric head may be helpful in dealing with uneven surfaces such as may occur in sports stadia.

Photometric heads for measuring different types of illuminance are now described.

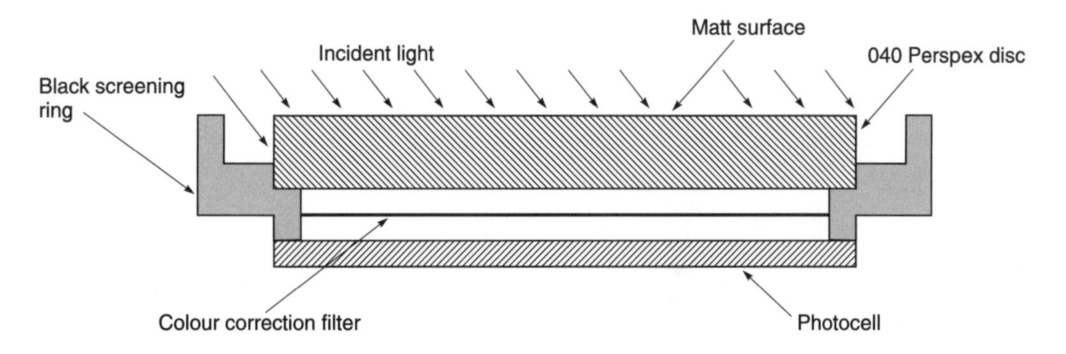

Fig. 15.23 Correction device for oblique light incidence

15.11.1 PLANAR ILLUMINANCE

Planar illuminance is illuminance on a plane that is coincident with the surface of the photocell. The surface of the photocell, whether selenium or silicon diode, tends to be glossy. The result is that a high proportion of the light impinging at angles of incidence of over about 60° tends to be reflected and, consequently, does not reach the sensitive surface of the photocell. At 60° to the normal, the error is about 10%, which increases to 40% at 85°. For a hemispherical sky the error is from 8% to 10% for a cell calibrated with normally incident light.[16] A matt finish provides only partial correction.

Figure 15.23 shows an accurate method of correction.[18] The sides of the matt finish opal disc overcompensate for the light incident at high angles, and this is corrected by the screening ring. The top surface of this must be level with the top surface of the opal disc so that light incident at 90° does not reach the opal disc. It is imperative that the top of the disc is kept clean.

When only one light source is being measured, as is usually the case in a laboratory, the effect of the reflection loss can be eliminated by tilting the photocell so that the light is incident normally to its surface. The readings are then multiplied by the cosine of the angle of tilt. However, this method is only valid if the source can be regarded as a point.

15.11.2 OTHER MEASURES OF ILLUMINANCE

Figure 15.24(a) shows the head for measuring cylindrical illuminance. Basically, it consists of a matt opal cylinder placed over a horizontal photocell. Owing to the preferential reflection of nearly vertical light rays, this head may be subject to error. An elaborate design to achieve greater accuracy has been developed by Gooding *et al.*[19] in which light travelling in a nearly vertical direction is directed onto the photocells by reflectors. The design for measuring semicylindrical illuminance (Figure 15.24(b)) is similar but half the cylinder is masked. The hemispherical photometer head (Figure 15.24(c)) is simply an opal hemisphere placed over a horizontal photocell.

The photometer head for measuring scalar illuminance is illustrated in Figure 15.24(d).[20] This consists of a table-tennis ball cemented to a photocell at its base. To compensate for the reduced sensitivity in the direction of the back of the photocell, areas of the ball are masked. These areas are selected in such a way that the sensitivity of the device is independent of direction.

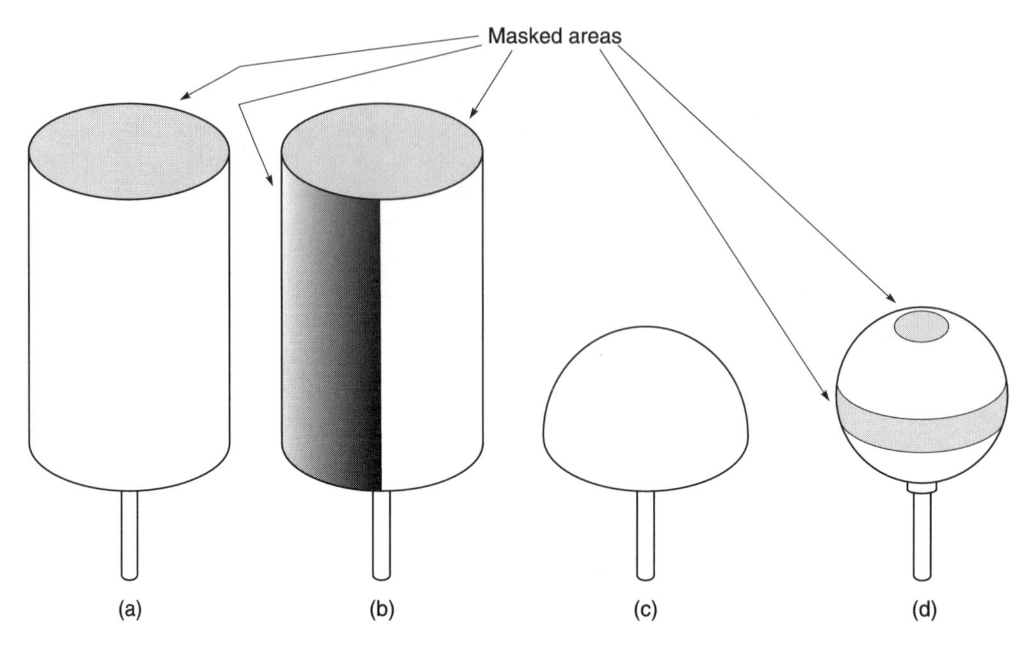

Fig. 15.24 Photometer heads for measuring (a) cylindrical; (b) semicylindrical; (c) hemispherical; and (d) scalar illuminance

15.11.3 VECTOR ILLUMINANCE

The component of the illuminance vector in a particular direction can be measured with a calibrated cosine corrected photometer head. The normal to this is aimed first in the desired direction and the reading recorded. It is then aimed in the opposite direction and the new reading subtracted from the first. When many readings have to be taken, it may be more convenient to use two cells back-to-back and connected electrically in opposition. Care has to be taken that they are matched in sensitivity, by suitable masking, if necessary, or the circuit compensates for any mismatch. The vector direction and magnitude can be calculated by taking six measurements on the faces of a cube[21] and calculating these quantities as explained in Section 8.5.

15.11.4 TESTS ON ILLUMINANCE METERS

The essential tests are as follows.

(1) When the photometer head is covered so that no light reaches it, the reading should be zero.
(2) In a multi-range instrument there should be agreement between the ranges.
(3) The absolute values of the readings should be correct. To test for this, luminous intensity standards will be needed or comparisons can be made with an illuminance meter known to be accurate.
(4) When the illuminated photometer head is covered and uncovered, the reading should recover in a reasonable time.
(5) The spectral response should be as near as possible to that of the V_λ curve or any other spectral response curve that is desired.
(6) The directional response should accord with that for the photometer head being used.

More specific requirements are laid down in national and international[22] documents. These relate to fatigue, temperature dependence, evaluation of modulated light, peak overload capability, and evaluation of modulated light.

15.11.5 ON-SITE ILLUMINANCE MEASUREMENTS

Before measurements are taken on-site, the light output from the luminaires should be checked for stability. This can be done by taking measurements at a few points, at time intervals, say, of ten minutes and checking that reasonable stability has been reached. No hard and fast rules can be laid down for a definition of stability since, overlaid on the settling down of the lamps, there will be fluctuations due to supply voltage variation.

Instruments brought indoors from the cold may suffer from condensation, so care should be taken to keep them warm if this is likely to happen.

The weak point of many illuminance meters is the plug and socket for the photocell, if this is not directly wired to the meter. To avoid embarrassment, and loss of time, on arrival on site the functioning of the meter should be checked. In addition, there should be evidence of recent calibration.

To find the average illuminance, the area is typically divided into a number of rectangles and the illuminance recorded in the centre of each rectangle. The readings are then averaged. To minimize the labour involved in a survey, the object is to take as few as possible readings consistent with achieving a stated accuracy.

For average illuminance in interiors, Bean and Esterson[23] related the number of readings to the room index. This work has been updated by Carter *et al.*[24] who found that the number of measurement points for 10% accuracy should be nine times the room index (Table 15.1), whilst for 5% accuracy the multiplier should be 30, which represents a large increase in the labour required. Table 15.1 has been adopted in the *CIBSE Code for interior lighting* (1994).[25] The rectangles should be as close in shape to squares as possible. Presumably, this work only applies to regular layouts of luminaires, although this is not stated. Large errors result if the measurement points fall directly below the luminaires. To avoid this, extra lines of measurement may have to be taken. Measurements are generally taken at the height of the working plane, 0.8 m.

The recommendations issued by other countries show wide differences. In the IESNA method[14] detailed samples are taken in a number of sub-grids in the area to be surveyed. Averages are found for each of these sub-grids and, from these, an overall average is calculated. Variations of this method have been adopted by other countries, for instance France, Australia and South Africa.

Of interest, and detailed in the *CIBSE Code for Interior Lighting*, is a method devised by Einhorn.[26] It is applicable to rooms with ‘a reasonably regular layout of luminaires, but not necessarily symmetrical or dissymmetrical, for instance the spacing from opposite walls or their

Table 15.1 Relationship between number of points and room index for 10% accuracy

Room index	Number of points
Below 1	9
1 and below 2	16
2 and below 3	24
3 and above	36

colours need not be the same'. Measurements are taken along two perpendicular lines parallel to the walls. In each line, these measurements are evenly spaced with a half spacing for the readings adjacent to the walls. Call the average readings so obtained $\bar{E}_x$ and $\bar{E}_y$. Let the illuminance at the intersection of the two lines be E_{is}, then the average illuminance is given by

$$E_{av} = \frac{\bar{E}_x \times \bar{E}_y}{E_{is}} \tag{15.2}$$

The result is inversely proportional to E_{is} so the point of intersection should therefore be chosen with care. It should not be directly below a luminaire or below a point midway between two luminaires. This dependence on a single reading may be regarded as a weakness of the method in that, if the reading is near luminaires that are not performing correctly, the result may be distorted.

In daylight studies, readings from two illuminance meters taken simultaneously are often required. For instance, to measure the reduction of transmittance of glass due to deposition of dirt, a small area of the glass is cleaned and a photocell placed on this area and on a dirty area. Readings are then taken at the same time to ensure that the sky conditions are the same. The ratio of the two readings gives the reduction of transmittance due to the layer of dirt, provided that the calibration of the meters is the same. In a method devised by Tregenza[27] and to be described here, the effect of any difference in calibration between the two meters is eliminated by interchanging the two meters and taking two more readings.

For the method to work satisfactorily, it has to be assumed that the response of the meters is linear, there is no zero error, and the cosine correction of the two cells is the same.

Let:

A and B denote the illuminance meters;
C_A and C_B be their calibration factors. These are the factors the readings have to be multiplied by to give calibrated readings;
X_A and X_B be the readings taken by the meters;
E_1 and E_2 be illuminances;
T_c and T_d be the overall transmittances of clean and dirty glazing;
τ_g and τ_d be the transmittances of the glazing material and the dirt layer.

If X_{A1} is the reading taken behind the clean glass when the outside illuminance is E_1 then

$$X_{A1} = E_1 T_c C_A$$

Similarly,

$$X_{B1} = E_1 T_d C_B$$
$$X_{A2} = E_2 T_d C_A$$
$$X_{B2} = E_2 T_c C_B$$

from which we obtain

$$\frac{X_{A1} X_{B2}}{X_{A2} X_{B1}} = \frac{E_1 T_c C_A \times E_2 T_c C_B}{E_2 T_d C_A \times E_1 T_d C_B}$$

$$= \left(\frac{T_c}{T_d} \right)^2$$

Now $T_d = \tau_g \tau_d$ and $T_c = \tau_g$ so,

$$\tau_d = \sqrt{\frac{X_{A2} X_{B1}}{X_{A1} X_{B2}}}$$

Where the photocell is held close to or on the glass, there will be interreflection between it and the glass. Tregenza estimates that the error is just less than 10% of the difference in reflectance between clean and dirty glass, for large photocells and photocells held close to the glass.

This two-photocell method is also useful in the measurement of daylight factor, where simultaneous readings have to be taken inside and outside a building.

15.12 Luminance measurements

Luminance can be calculated from luminous intensity measurements or measured directly.

15.12.1 CALCULATION FROM LUMINOUS INTENSITY MEASUREMENTS

$$L = \frac{I}{A}$$

where

L is the luminance in the specified direction (cd/m^2),
I is the luminous intensity in the specified direction (cd),
A is the orthogonally projected area in the specified direction (m^2).

This method is useful when the average luminance of a luminaire is required and the luminous intensity distribution is available. In many cases the orthogonally projected area of the bright or flashed area can be calculated, otherwise it has to be measured by an imaging device such as a camera. The method is most suitable for diffuser luminaires, where the surface is reasonably evenly bright and the lamp is not visible. It is used in the calculation of discomfort glare (see Section 8.1, page 301). For this purpose, the file format (Section 2.7, page 61) contains information that enables the projected area to be calculated. This is not necessarily accurate but is a convention adopted for the purposes of glare calculations.

15.12.2 USE OF A PHOTOCELL DIRECTLY

In this method, a calibrated photocell is applied directly to the surface to be measured. It is, therefore, only of use for measuring the luminance of self-luminance surfaces or surfaces lit from behind. It does not measure the absolute value of luminance correctly since the photocell collects light from all directions, whereas luminance is directional, and light is interreflected between the photocell and the surface to be measured. However, if the photocell and meter combination is calibrated in lux, an approximate value of luminance can be obtained by dividing this by π to give an answer in candelas per square metre. A better method is to calibrate the photocell and meter by using a surface of known luminance. This method is useful for exploring the variation of luminance over a surface.

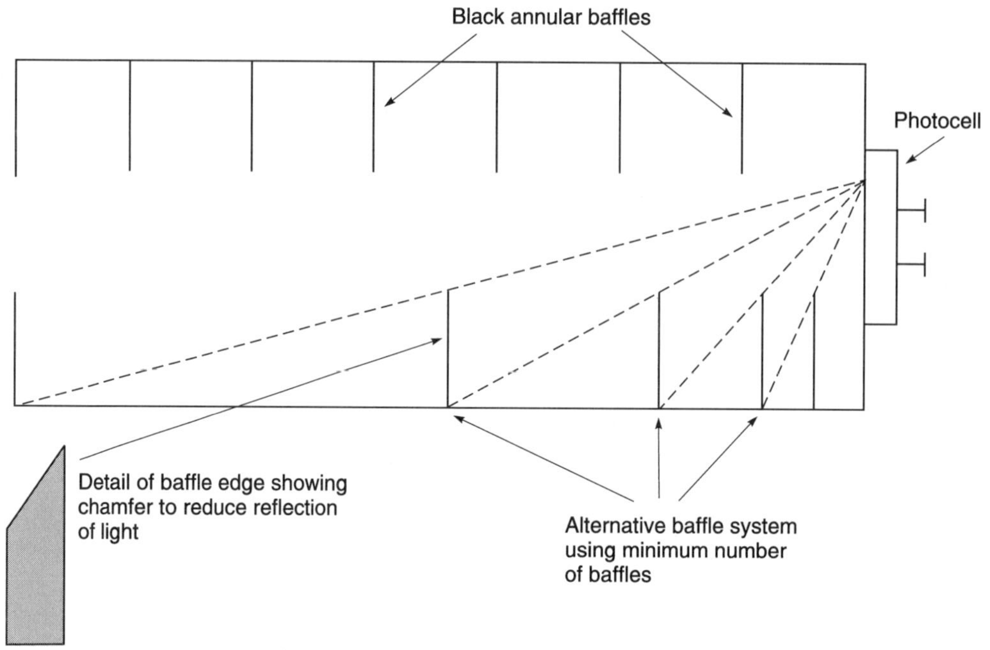

Fig. 15.25 Baffled tube for measuring luminance

15.12.3 USE OF A PHOTOCELL IN A TUBE

The drawbacks of the previous method can be overcome by using a photocell mounted in a baffled tube (Figure 15.25), or one lined with black flock, which is placed on the surface to be measured. The diameter of the baffles should be the same as the diameter of the photocell, and the tube should be long in proportion to its diameter so that the device is selective as regards direction. This method is accurate if the tube is placed on the surface to be measured, which is not always possible because this may obstruct the light. If the tube is not placed on the surface an error may be incurred because different parts of the cell will 'see' different parts of the surface. If the surface is of substantially even luminance, the error incurred will be small. Most of these problems are overcome by an imaging device, which is described next.

15.12.4 IMAGING DEVICES

Basically, these devices produce an image as in a camera, a photocell in the focal plane being used to measure the luminance.

Figure 15.26 shows how a single lens reflex camera might be used as a luminance meter. When the mirror is in the down position, the test surface can be viewed and the camera aimed. The function of the pentaprism is to present to the eye an upright image that is not laterally reversed, as it is on the ground glass screen. The mirror is then lifted to allow the light to fall on the photocell and measure the luminance. Care has to be taken when the camera is modified for its new use that the area seen corresponds to that measured. This can be done by replacing the photocell with a ground glass screen and viewing the image falling on it. It will then be necessary to adjust the mask under the ground glass screen accordingly. Since the illuminance on the

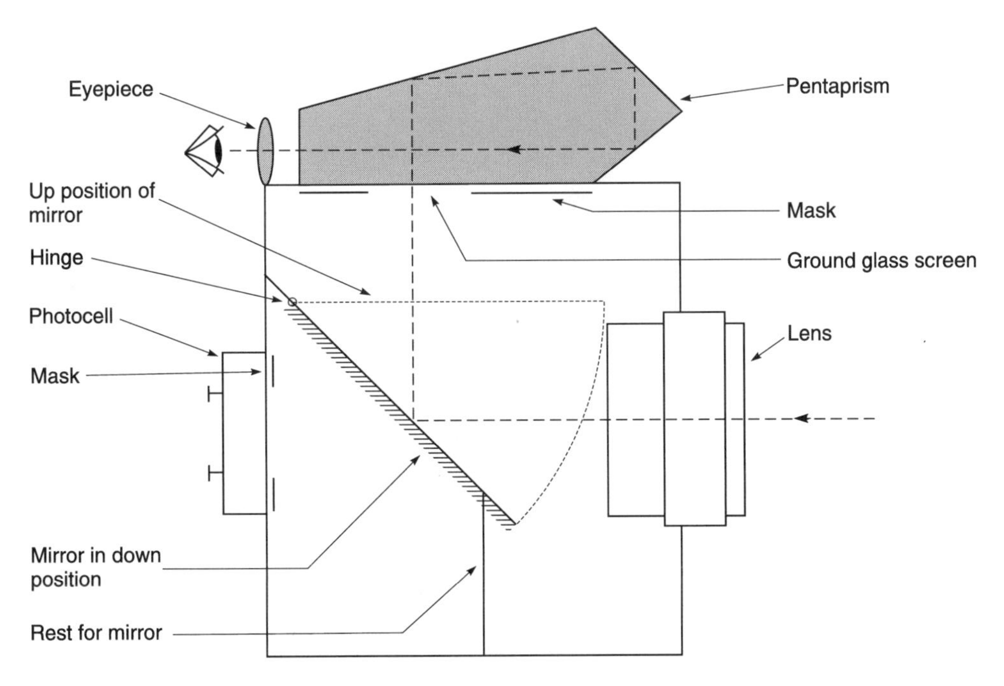

Fig. 15.26 Single lens reflex camera used as a luminance meter

photocell depends on the distance of the lens from it, a correction has to be applied if the lens is moved for focusing. This is a usable instrument but it does suffer from the drawback that, at the moment a measurement is taken, the image disappears because the mirror is lifted.

Some care should be taken in selecting a lens. There is inevitably some scattering of light in the lens, even with modern anti-reflecting coatings. In addition, some lenses produce ghost images. These may become apparent in road lighting photography, where a ghost image of a luminaire may be produced on the image of the road. Similarly, zoom lenses are to be avoided as they may consist of as many as 15 elements and certainly degrade the image.

Commercially available instruments use the general principal of the reflex camera but with arrangements for overcoming the problem of the image disappearing at the moment of measurement. Figure 15.27 shows the arrangement used for the Spectra–Pritchard meter. The objective forms an image of the test surface in the vertical plane passing through the aperture of the mirror and normal to the axis of the optical system. Light that passes through the aperture is directed onto the photomultiplier by the field lens. The remainder of the image is viewed through the eyepiece. The angle of view can be varied by using mirrors with different apertures. In addition, lenses of different focal lengths can be used. The main disadvantage of this type of instrument is that the part of the image that is being measured is not visible. In practice, this causes little inconvenience.

The illuminance on the focal plane is not uniform even if the object portrayed is of uniform luminance. It decreases according to the distance of the point on the image from the centre of the image. It is, therefore, necessary to calibrate the camera so that a correction can be made, which can be carried out by using the luminance box described in Section 15.12.6. It will be necessary to do this for each lens, because the rate of fall-off of illuminance depends on the distance of the lens from the focal plane, and also for a number of focusing distances of the lens. The aperture used will also affect the calibration.

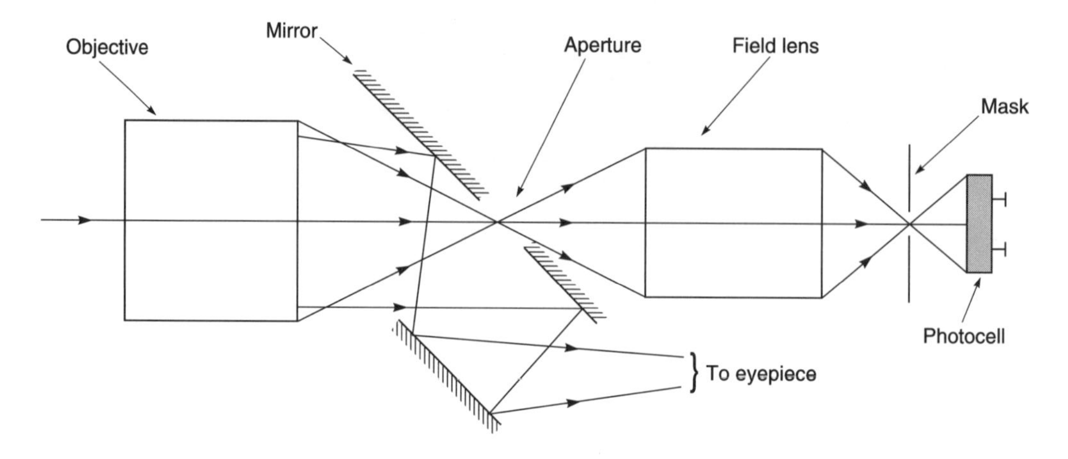

Fig. 15.27 Optical system of Spectra–Pritchard photometer

In some luminance meters the opposite arrangement is used; that is, the light is directed onto the photocell by the mirror and the direct light goes to the eyepiece. This has the disadvantage when measuring polarized light that the amount of light reflected by the mirror varies according to the plane of polarization. In another arrangement an optical fibre is used to collect light for the photocell. The light collected by this may also be susceptible to polarization.

For many purposes, an instrument is required that has a very small field of view, perhaps only 2 minutes of arc, and is capable of measuring low luminances. Such an application would be in road lighting, where luminances as low as 0.1 cd/m^2 have to be measured reliably. An instrument capable of doing this would use a photomultiplier. For these applications a stout tripod is essential. This should have a geared pan and tilt head so that fine angular movements can be made.

15.12.5 INSTRUMENTS DEPENDING ON VISUAL JUDGEMENT

In the past, a number of luminance meters were available that depended on the user making a visual judgement to measure the luminance. This judgement consisted of matching a spot of variable luminance against the luminance of the test surface, both seen side by side in a viewfinder. This matching is difficult to do, especially as the test surface is invariably a different colour from that of the matching surface. There could be a significant variation in the readings obtained by different observers, this in some cases is due to defective colour vision. These instruments have been superseded by the photocell or photomultiplier instruments described previously.

15.12.6 CALIBRATION OF LUMINANCE METERS

There are two methods of calibrating luminance meters:

(1) with an illuminated surface of known reflectance. The surface has to be evenly illuminated by a source of known luminous intensity so that the illuminance can be calculated, or can be measured directly.
(2) with an opal surface illuminated from behind. Here the transmittance of the surface has to be known, or the luminance can be calculated from a measurement of the illuminance it produces. The evenness of the luminance can be measured by the baffled cylinder described

in Section 15.12, page 505. This method has the advantage that a sealed unit can be made, which is convenient to use at any time. By having two compartments, one housing the light source, the other with the opal diffuser, separated by a plate with a variable aperture, it is possible to vary the luminance conveniently and in a reproducible manner.

15.12.7 TESTS ON LUMINANCE METERS

Many of the tests enumerated in Section 15.11.4 for illuminance meters also apply to luminance meters. Some additional ones, however, are worthy of mention.

(1) The effect of the surrounding field. The luminance of the surrounding field should not affect the reading of the luminance of the patch to be measured. Light scattering in the lens is the most likely source of error.
(2) Directional response. A bright patch within the field of measurement of the meter should give the same reading of luminance independently of position.
(3) Polarized light. The light reflected off or transmitted by some surfaces (Section 6.8, page 235) can have a high component of polarized light. In these situations, therefore, it is important that the reading takes into account polarized light.

15.12.8 LUMINANCE PATTERN OF A SCENE

Whilst the luminance of individual points in a scene can be measured by a luminance meter, this process is tedious and lengthy where the whole luminance pattern needs to be recorded. The photocell in a tube described previously (Section 15.12.3) can be used to scan a scene,[28] but for precision an imaging system is needed. Computer processing of the image is an essential adjunct to this and enables correction of most of the lens errors to be made in the final result, with the exception of light scattering or the formation of ghost images.

A photograph on film is able to provide an image where the optical density is a function of the luminance, but the inaccuracies are likely to be large. These arise for a number of reasons:

- the illuminance on the focal plane is not uniform even if the object portrayed is of uniform luminance. It decreases according to the distance of the point on the image from the centre of the image. It is, therefore, necessary to calibrate the camera so that a correction can be made, which can be carried out by using the luminance box described in Section 15.12.6. It will be necessary to do this for each lens because the rate of fall-off of illuminance depends on the distance of the lens from the focal plane, and also for a number of focusing distances of the lens. The aperture used will also affect the calibration. Computer processing of the image eases the application of these corrections in the final result
- the optical density of the image varies according to the evenness of development of the image and the evenness of the photographic emulsion.

However, in spite of these inaccuracies, Hopkinson[29] showed that, provided that the exposure both in taking the photograph and printing it are carefully controlled, good representational photographs can be obtained.

Charge-coupled devices (CCD) have been used successfully for recording the luminance of scenes.[30,31] These enable the scene to be analysed by computer, giving the research worker a very powerful tool, especially for road lighting, since there is no need to close the road. However, up to the present time, human intervention is needed for selection of the part of the scene that is of interest.

References

1. CIE 121–1996. *The Photometry and Goniophotometry of Luminaires.*
2. Dunlop, D. and Finch, D.M. (1962) *Illum Engng (NY)* **57**, 159.
3. Riggs, W.G. and Lampert, A.R. (1963) *Illum Engng (NY)* **58**, 736.
4. Frost, J.W. (1985) *Errors due to length of measuring distance.* PhD thesis, Polytechnic of the South Bank, London.
5. Holmes, J.G. (1990) Photometry from the right distance. *Lighting Research and Technology*, **22**(4), 183.
6. Frederiksen, E. (1981) The collimated light photometer. *Report No 28 of Lysteknisk Laboratorium.*
7. Pritchard, E.H. and Simons, R.H. (1992) High-speed photometer for measuring light intensity distributions. *Lighting Research and Technology*, **24**(2), 107.
8. Walsh, J.W.T. (1958) *Photometry* (Constable).
9. BS 354 (1995) *Recommendations for Photometric Integrators.*
10. CIE 84 (1989) *The Measurement of Luminous Flux.*
11. Keitz, H.A.E. (1955) *Light Calculations and Measurements* (Philips Technical Library).
12. Bean, A.R. (1964) *Light and Lighting,* **57**, 191.
13. BS 5225 (1975) *Photometric Data for Luminaires. Part 1: Photometric Measurements.*
14. Rea, M.S. and Boyce, P.R. (1993) *Lighting Handbook* (IESNA).
15. Baumgartner, G. (1941) *Illum Engng (NY)* **36**, 1340.
16. CIE 30-2 (TC-4.6) (1982) *Calculation and Measurement of Luminance and Illuminance in Road Lighting.*
17. CIE 47 (TC-4.6) (1979) *Road Lighting for Wet Conditions.*
18. Pleijel, G. and Longmore, J. (1952) *J. Sci. Inst.* **29**, 137.
19. Gooding, P.E., Hunt, D.R.G. and Cockram, A.H. (1976) *Lighting Research and Technology*, **4**(4), 225.
20. Lynes, J.A., Burt, W., Jackson, G.K. and Cuttle, C. (1966) *Trans IES (London)* **31**, 65.
21. Cuttle, C. (1997) Cubic illumination. *Lighting Research and Technology*, **29**(1), 1.
22. European Standard EN13032–1 (1999) *Measurement and Presentation of Photometric Data of Lamps and Luminaires. Part 1: Measurement and File Format.*
23. Bean, A.R. and Esterson, D.M. (1966) Average illuminance measurement – a preliminary investigation. *Light and Lighting*, **59**(7), 204–205.
24. Carter, D.J., Sexton, R.C. and Miller, M.S. (1989) Field measurement of illuminance. *Lighting Research and Technology*, **21**(1) 29–37.
25. (1994) *Code for interior lighting*. CIBSE.
26. Einhorn, H.D. (1990) Average illuminance: two line measurement. *Lighting Research and Technology*, **22**(1), 43–47.
27. Tregenza, P.R. (1998) Paired illuminance measurements: Eliminating calibration error. *Lighting Research and Technology*, **30**(3), 133–134.
28. Rowlands, E., Loe, D.L. and Brickman, N.T. (1984) *Instrumentation for measuring the luminance distribution within the visual field.* CIBSE Nation Lighting Conference.
29. Hopkinson, R.G. (1936) The photographic representation of street lighting installations. *The Photographic Journal*, June, p. 323.
30. Green, J. and Hargroves, R.A. (1979) *Lighting Research and Technology*, **11**(4), 197.
31. Glenn, J. (1997) *Electronic evaluation of road lighting performance.* ILE Annual Conference, p. 35.

Appendix

LIGHTING BODIES AND ASSOCIATED STANDARDIZING ORGANIZATIONS

Initials	Name	Comments
AFE	Association Française de l'Eclairage	The French lighting society. Produces recognized recommendations. Linked to the lighting magazine *Lux*.
AFNOR	Association Française de Normalisation	The French standardizing body.
ANSI	American National Standards Institute	The American standardizing body.
BSI	The British Standards Institute	The British standardizing body.
CEN	Comité Européen de Normalisation	The European standardizing body, which deals with all matters not dealt with by CENELEC.
CENELEC	Comité Européen de Normalisation Electrotechnique	European body parallel to IEC.
CIBSE	The Chartered Institute of Building Services Engineers	The UK body concerned with building services. The Society of Light and Lighting formed from the The Lighting Division of CIBSE in June 2000 is responsible for lighting matters, principally as they relate to buildings. Formerly the IES – The Illuminating Society (London). Produces *LR&T – The International Journal of Lighting Research and Technology.* Linked to the lighting magazine, *Light and Lighting.*
CIE	Commission Internationale de l'Eclairage	International body devoted to cooperation on matters related to the art and science of lighting. Produces recommendations, reports, and standards.
DIN	Deutsches Institut für Normung	The German standardizing body.
IALD	The International Association of Lighting Designers	Promotes the lighting design profession and advances lighting design excellence in the built environment internationally.
IEC	The International Electrotechnical Commission	The authority for international standards for electrical and electronic engineering.

Initials	Name	Comments
IESNA	The Illuminating Engineering Society of North America	The lighting society of North America. Produces ANSI recognized standards and practices. Publishes *Journal of the IES* (*JIES*) for scientific papers, and the magazine *Lighting Design + Application* (*LD+A*).
ILE	The Institution of Lighting Engineers	UK lighting body formerly mainly concerned with the lighting of roads but now involved with all branches of lighting. Produces recommendations. Originally APLE – The Association of Public Lighting Engineers. Produces *The Lighting Journal*, a lighting magazine.
ISO	The International Organization for Standardization	The international federation of institutes for national standards.
SANCI	South African Committee on Illumination	The South African illuminating engineering body.

Index